THE
CHRISTIAN PHILOSOPHER

COTTON MATHER

The Christian Philosopher

Edited, with an Introduction and Notes, by
Winton U. Solberg

UNIVERSITY OF ILLINOIS PRESS
Urbana and Chicago

Publication of this work has been supported by a grant from
the Oliver M. Dickerson Fund. The Fund was established
by Mr. Dickerson (Ph.D., Illinois, 1906) to enable
the University of Illinois Press to publish selected works
in American history, designated by the executive committee
of the Department of History

© 1994 by the Board of Trustees of the University of Illinois
Manufactured in the United States of America
C 5 4 3 2 1

This book is printed on acid-free paper.

Library of Congress Cataloging-in-Publication Data

Mather, Cotton, 1663–1728.
   The Christian philosopher / Cotton Mather : edited, with an introduction and notes, by Winton U. Solberg.
     p. cm.
   Includes bibliographical references and index.
   ISBN 0-252-01952-0 (cl)
   1. Natural theology—Early works to 1800. 2. Religion and science—Early works to 1800. 3. Science—Early works to 1800. I. Solberg, Winton U., 1922– . II. Title.
BL180.M4 1993
215—dc20   92-32064
                                                                                   CIP

To
KRISTIN RUTH SOLBERG

O Lord, how manifold are thy works! in wisdom hast thou made them all: the earth is full of thy riches.

Psalm 104:24

There were times when Leibnitzes with their heads buried in monstrous wigs could compose Theodicies, and when stall-fed officials of an established church could prove by the valves in the heart and the round ligament of the hip-joint the existence of a "Moral and Intelligent Contriver of the World." But those times are past; and we of the nineteenth century... know nature too impartially and too well to worship unreservedly any God of whose character she can be an adequate expression. Truly, all we know of good and duty proceeds from nature; but none the less so all we know of evil. Visible nature is all plasticity and indifference—a moral multiverse... and not a moral universe.... If there be a divine Spirit of the universe, nature, such as we know her, cannot possibly be its *ultimate word* to man. Either there is no Spirit revealed in nature, or else it is inadequately revealed there; and... what we call visible nature, or *this* world, must be but a veil and surface-show whose full meaning resides in a supplementary unseen or *other* world.

William James, *The Will to Believe* (1897), pp. 42-43

CONTENTS

Preface	xi
Acknowledgments	xv
Introduction	xix
1. *The Christian Philosopher* in Context	xix
2. Cotton Mather (1663–1728)	xxii
3. The Origins of *The Christian Philosopher*	xxxv
4. The Sources of *The Christian Philosopher*	xlviii
5. The Contents of *The Christian Philosopher*	lxix
6. The Reception of *The Christian Philosopher*	xcii
7. The Significance of *The Christian Philosopher*	cxii
Note on Editorial Method	cxxxv
Abbreviations of Works Cited	cxli

The Christian Philosopher

The Introduction	7
Religio Philosophica; or, The Christian Philosopher	17
1. Of the Light	18
2. Of the Stars	26
3. Of the Fixed Stars	29
4. Of the Sun	35
5. Of Saturn	42
6. Of Jupiter	44
7. Of Mars	46
8. Of Venus	47
9. Of Mercury	48
10. Of Comets	50
Appendix. Of Heat	54
11. Of the Moon	56
12. Of the Rain	60
13. Of the Rainbow	63

14. Of the Snow	67
15. Of the Hail	69
16. Of the Thunder and Lightning	70
17. Of the Air	73
18. Of the Wind	75
19. Of the Cold	80
20. Of the Terraqueous Globe	82
21. Of Gravity	89
22. Of the Water	96
Appendix [Fluids]	101
23. Of the Earth	102
Appendix [A Few Points of the Mahometan Philosophy]	110
24. Of Magnetism	111
25. Of Minerals	122
26. Of the Vegetables	129
27. Of Insects	150
28. Of Reptils	177
29. Of the Fishes	184
30. Of the Feathered	192
31. Of the Four-Footed	212
32. Of Man	236
Endnotes	319
Biographical Register	379
Recapitulation of Mather's Sources	451
List of Biblical References	467
Index	473

PREFACE

Cotton Mather was an extraordinary man who figures prominently in American history. Although most literate Americans recognize his name, they have not thought well of him over the years. Scholars have contributed to the hazing of Mather (Perry Miller's inveterate animus is all too obvious), people who should know better ridicule him, and the public harbors the stereotype of Mather as a persecutor of witches and a typically bigoted Puritan priest. Such an image is not only a caricature but a bad caricature. If the truth sets us free, we need more truth to discern the real Cotton Mather.

Mather's bad press was not in my mind when I began work on this book. I first became vaguely interested in *The Christian Philosopher* when, years ago as a graduate student, I began to marvel at the rich fabric of Cotton Mather's life. Mather's *Christian Philosopher*, I learned, introduced the Enlightenment to America, surely a worthy enterprise. At the time, however, Mather was not high on my list of scholarly interests: he seemed too complex, too forbidding. Years passed, and with the years came several books providing fresh perspectives on the Mather family, Puritan piety, and science in early America, as well as new biographies of Mather and valuable editions of his writings on history, theology, and medicine. While these volumes shed fresh light on the Puritan worthy, none of them emphasized Mather's relation to science.

My earlier interest in *The Christian Philosopher* revived, and I decided that it might be desirable to prepare an edition that would make Mather's book "on the Enlightenment" accessible for the first time to our own generation. Since *The Christian Philosopher* has very thick texture, the preparation of this edition has taken considerably longer than anticipated. Mather's *Christian Philosopher* is the first comprehensive treatise on all the sciences known at the time to be written by an American. The author's purpose was to demonstrate the harmony between religion and the new science associated with the illustrious name of Sir Isaac Newton. The thesis is the ancient design argument. In pursuing his main purpose, Mather transmitted a large part of the humanistic heritage of Western civilization to America. My aim was to prepare an edition that would make his essay valuable to the widest variety of readers.

This book reveals Cotton Mather the scientist, the dimension of his life that has been least well recognized and appreciated. Mather records his own significant contributions to the field in *The Christian Philosopher,* but the book is primarily important in demonstrating Mather's warm receptivity to the new science and in portraying Mather as a popularizer of science. Mather was always a Puritan clergyman as well, and he makes a sustained argument that there can be no conflict between science and religion. I have tried to do justice to Mather's lifelong interest in science and religion, seeking neither to defend nor to defame him.

Mather ransacked the learning of the ages for his book. This edition underscores the need to study early American history in its widest historical, geographical, intellectual, and cultural context. While Mather's treatise is best understood as a product of the Anglo-American culture of the seventeenth and early eighteenth centuries, his account runs back through the Middle Ages to the early Christian church, the Hebrews, and classical antiquity. We should try to appreciate its many dimensions.

The fame of historical figures is often subject to vicissitudes over time. Mather's reputation is now on the ascendant. Perhaps his name will never entirely escape from the unwarranted abuse that has been heaped upon it. But since he occupies a commanding position in the American heritage, we honor ourselves in treating him fairly. Above all, we should try to understand him. This book is intended to help the present generation form a more just opinion of the Puritan priest.

I have taken considerable pains to be accurate in preparing this edition, but errors are likely in a work with such a vast multitude of details. "I know well," as the seventeenth-century worthy John Ray wrote in one of his prefaces, "that the longer a Book lies by me, the perfecter it becomes. Something occurs every day in reading or thinking, either to add or to correct and alter for the better; but should I defer the Edition till the Work were absolutely perfect, I might wait all my life-time, and leave it to be published by my Executors." Since they would not welcome the task, I release my book, echoing Thomas Gataker, another seventeenth-century worthy, who likewise wrote in a preface, "Defects in it (I know) there can not but be many. Of thee (whosoever thou art) I desire but to find an unpartial reader, a judicious discusser, and charitable censurer."

ACKNOWLEDGMENTS

Since Mather's *Christian Philosopher* contains an immense treasury of knowledge, the preparation of an edition that makes his book accessible to the present age put me under a heavy debt to various institutions and individuals. With my task completed, I am happy to acknowledge the assistance given as I proceeded.

My research was conducted mainly in the library of the University of Illinois at Urbana-Champaign, which contains splendid collections in science, religion, and related topics in the early modern period, especially seventeenth- and early eighteenth-century English imprints. N. Frederick Nash, Rare Book Room Librarian, was especially helpful; Mary S. Ceibert, Louise M. Fitton, and Gene K. Rinkel also responded to my requests, as did Nancy L. Romero, Rare Book Room Librarian at a later stage in my work. Many other librarians obtained interlibrary loans and photocopies.

The Research Board of the University of Illinois at Urbana-Champaign provided financial support that enabled me to employ a number of research assistants. Cynthia D. Nichols, Gregory G. Schmidt, Robert H. Lackie, Suzanne E. Esserman, and Steven J. White assisted in such tasks as identifying authors and titles and determining the sources of Mather's material. Randall Stewart offered help in translating Latin and Greek passages. Lisa M. Warne-Magro aided in the final stages of preparing the manuscript for publication.

In 1980, I submitted to the National Science Foundation a proposal that would have funded a leave for me during the summer and fall semester of 1981 to work on "Cotton Mather's *Christian Philosopher* (1720) and Early American Science." This proposal was approved (Grant No. SES-8025384), but in March 1981 NSF officials informed me that budget reductions mandated by the Reagan administration would prevent the NSF from writing a contract for the award. Subsequently, the associate program director of History and Philosophy of Science at the NSF requested that I submit a severely reduced budget (less than 10 percent of the original) along with a statement of what could be done under such a grant. This proposal (bearing the same number) was approved; the grant provided support for a quarter-time research assistant during 1981–82.

In 1983, the National Science Foundation awarded me a Summer Scholar's Award (Grant No. SES-8219062) for a project entitled "American Science at Its Roots: Cotton Mather's Christian Philosopher." This assistance enable me to continue my research during one summer.

The William Andrews Clark Memorial Library of the University of California at Los Angeles awarded a fellowship that enabled me to spend a few weeks in the summer of 1983 in its outstanding collection of books on seventeenth- and early eighteenth-century English civilization. At UCLA I also consulted the rich holdings in the history of the health sciences and the history of the life sciences in the Louise M. Darling Biomedical Library.

I owe a very special debt to the late Dorothy Schullian Adelmann, a classicist trained by Paul Shorey at the University of Chicago who later became a specialist in the history of early modern science. I met her when she replied to a letter of inquiry about Mather that I addressed to her husband, Howard B. Adelmann, a distinguished historian of science, whose health was failing at the time. She immediately took a lively interest in my work, helping me to identify Mather's quotations from the ancient classics, insisting on the importance of my project, and encouraging me when it seemed as if mastery of Mather was hopeless. To our mutual delight, we discovered that her closest friend, dating from their years at Chicago, was my former college Latin teacher and friend, Grace L. Beede. I keenly regret that Dorothy Adelmann did not live to see this volume in print.

Colleagues at the University of Illinois at Urbana-Champaign enlarged my knowledge and saved me from pitfalls by commenting on Mather's essays and my notes on them. They include Robert D. Sard, professor of physics, Essay 21, Of Gravity; Albert V. Carozzi, professor of geology, Essay 23, Of the Earth, and Appendix; Essay 24, Of Magnetism; and Essay 25, Of Minerals; Johannes M. J. de Wet, professor of plant cytogenetics, Essay 26, Of the Vegetables; Robert L. Metcalf and May R. Berenbaum, professors of entomology, Essay 27, Of Insects; Thomas H. Frazzetta, professor of ecology, ethology, and evolution, Essay 28, Of Reptiles; Gregory S. Whitt, professor of ecology, ethology, and evolution, and Lawrence M. Page, an ichthyologist in the Center for Biodiversity, Illinois Natural History Survey, Essay 29, Of the Fishes; David L. Stocum, formerly professor of cell and structural biology at the University of Illinois at Urbana-Champaign, now dean of the School of Science at Indiana University–Purdue University at Indianapolis, and Scott K. Robinson, an ornithologist in the Center for Wildlife Ecology, Illinois Natural History Survey, Essay 30, Of the Feathered [Birds]; Professor Stocum, Essay 31, Of the Four-Footed [Quadrupeds]; and Dr. Julius C. Bonello, associate professor of clinical surgery and head, Department of Surgery, College of Medicine, University of Illinois at Urbana-Champaign, and Dr. Frederick Steigmann, clinical professor emeritus of internal medicine and gastroenterology, University of Illinois Medical School, Chicago, Essay 32, Of Man (the parts dealing with human anatomy and physiology).

Many colleagues who offered advice on particular points are thanked at appropriate places in my book; others include John Buckler (ancient history); Adriaan J. de Witte (Dutch-language materials); Timothy J. Hallett (biblical passages); Valerie J. Hoffman-Ladd (Islam); Howard Jacobson (classics); Miroslav Marcovich (classics); Gary G. Porton (Jewish materials); Kenneth R. Robertson (botany); and Donald P. Rogers (botany).

The two external readers who evaluated my manuscript for the University of Illinois Press made extremely helpful comments. I made revisions in light of their suggestions, and am grateful for their contributions to making a better book. I gladly assume responsibility for the contents of this edition.

The Department of History at the University of Illinois at Urbana-Champaign offered a generous subvention from the Oliver M. Dickerson Fund to make publication of this book possible. This fund was established to support the publication of manuscripts, especially those treating American history prior to 1830, by members of the department or by alumni of the university holding the doctorate in history.

Elizabeth G. Dulany, associate director of the University of Illinois Press, again brightened my way from manuscript to book, while Patricia Hollahan won my deep appreciation for her skillful editing.

Finally, I am acutely aware of a debt that can never be adequately discharged. My long and passionate affair with a jealous mistress, scholarship, took time that might otherwise have been spent with my wife. I can only hope that Connie will understand (and forgive). This book is for our youngest daughter, whose turn it is to have a volume dedicated to her. Fortunately, the timing of both publication and her personal plans is such that I am able to present this book, only slightly late, as a wedding gift.

INTRODUCTION

1. *The Christian Philosopher* in Context

Modern thought is preoccupied with science and mastery of the physical universe, but religion abides. In the contemporary world, many people insist that science and religion are incompatible and strive to maintain them in distinct and watertight compartments. In 1972, for example, the National Academy of Sciences adopted a resolution stating that "religion and science are . . . separate and mutually exclusive realms of human thought whose presentation in the same context leads to misunderstanding of both scientific theory and religious belief."[1]

Such an outlook is a recent development, however, for science and religion have been irretrievably intermixed throughout Western history. Since ancient times, people have believed that God is revealed in both nature and Scripture. Christianity provided the intellectual climate in which empirical natural science developed, as M. B. Foster writes, for the methods of natural science depend upon certain presuppositions about nature, and these in turn depend upon the doctrine of God. Modern natural science could begin only when modern presuppositions displaced Greek ones. This uprooting, a gradual process, was possible only when the Christian conception of God supplanted the pagan conception. The displacement was the work of medieval theology. Foster's analysis makes intelligible the metaphor of the two books—that God reveals himself in both the sacred Scriptures and the physical universe.[2]

The design argument, a fundamental element in Western thought, arose within the framework of the metaphor of the two books. This doctrine holds that one may reasonably infer the existence of a purposeful Creator responsible for the creation from the evidences of order and harmony in the universe. This concept originated in antiquity. Plato first formulated the argument from design, the Stoics emphasized it, and Christians later embraced it. The design argument is one form of a closely related aspect of Western thought—natural theology. This doctrine concerns what can be known of the existence and attributes of God by reason reflecting on experience and the evidences of nature. Natural theology originated in antiquity as the antithesis of civil or political theology (Plato was its leading spokesman), but during the Middle Ages it became the antithesis of revealed theology.[3]

The design argument (along with natural theology, of which the former was one expression), was reinvigorated by the scientific revolution of the early modern period. In seventeenth-century England, these closely related doctrines provided one of the main foundations for popular religion. For revealed religion (or Christianity), the pillar of support was the authority of Holy Scripture. For natural religion (or belief in God), the argument from design was the sufficient foundation, and in the form given it by the science of the day. The evidence of both celestial and terrestrial phenomena implied an intelligent designer, while the biological adaptation of various organisms to their environment along with the established scientific doctrine of the fixity of species testified to purpose in nature. Different species are each designed to fulfill certain functions, and each is fixed in its essential characteristics from the beginning. All must have been created for the purposes they fulfill by a designing mind. Human beings occupy a place of supreme dignity in the scale of nature.

William Paley popularized this cluster of beliefs in his *Natural Theology* (1802), but this famous book, which Darwin once admired and knew almost by heart, merely restated, in a much more mechanistic and superficial way, ancient ideas that had been revitalized by the scientific revolution and given their finest expression by John Ray in the late seventeenth century. The design argument remained a basic belief in Western civilization until Darwin challenged it, and in the late nineteenth century it came under severe attack.[4]

Cotton Mather first transmitted this body of thought to British America. In what way, one may well ask, can Mather have transmitted a set of ideas that had been a central part of the Western intellectual heritage for two thousand years? The answer is readily forthcoming. While the design argument had been around for ages, the rise of the new science reinvigorated it, and the argument found fresh and compelling statement in a book published by John Ray in 1691.

Cotton Mather was the first American to write a book of natural theology with the design argument as its thesis. No doubt more people associate Mather with bigotry than with science, but he was acutely interested in all of the sciences known at the time and in the relation between the new science and religious faith. He devoted one of his major books to this subject, and scholars agree that *The Christian Philosopher* (1721) is an important work by a major figure and of the highest significance for American intellectual history.[5]

Despite this testimony, Mather's book is not well known to the modern age. It has rarely been reprinted; a good, readable text has not hitherto been available. When Mather's work on Newtonian science appeared, a small circle of educated readers could have readily comprehended its argument, but at the time the leading English works on the subject eclipsed Mather's contribution. Hence the New Englander's treatise caused no great stir when it fell from the press. With the passage of years, *The Christian Philosopher* became too formidable for most readers. Its baroque style and learned allusions, its numerous Latin and biblical quotations, its pages of ornate prose, packed with names of authors and titles of books, recondite terms and concepts, seemed impenetrable. In due time the work became a neglected American "classic"—the title familiar to many, the contents known by few.

The Christian Philosopher is the first comprehensive account of the physical and natural sciences written by an American; it reveals how the most prominent intellectual in early eighteenth-century America viewed the relation between science and faith. Mather's book is the point of departure for tracing science and religion in America at their roots; it deserves close attention by everyone interested in American thought and culture.

The modern reader needs both a good text and help in comprehending Mather's treatise. The purpose of this edition is to make *The Christian Philosopher* readily accessible to a wide audience for the first time. The Introduction is designed to place the work in historical and intellectual context. After a biographical sketch that focuses on Mather's interest in science and its relation to religion, the Introduction treats the origins of Mather's book, his sources, the contents of his volume, its reception, and the volume's significance in American intellectual history. In the text itself, footnotes supply information designed to help the reader understand the text and follow the argument of the work. Endnotes describe Mather's sources and provide additional information useful for advanced students.

2. Cotton Mather (1663–1728)

As his names testify, Cotton Mather was a member of two of the most prominent families in colonial New England. His grandfathers, John Cotton and Richard Mather, had been spiritual and intellectual leaders of the first generation of Puritans in New England. His father, Increase, graduated from Harvard College in 1656 and took a master's degree at Trinity College, Dublin, before his ordination in 1664 as minister of the North (Second) Church in Boston. He published numerous works on leading issues of the day, served as president of Harvard from 1685 to 1701 while retaining his church, and in 1687 went to England, where he was successful in obtaining a new charter for Massachusetts. By the time he returned in 1692 he was easily the most powerful divine in the political and ecclesiastical life of the colony. Increase married John Cotton's only daughter, Maria, and Cotton Mather, born in Boston on 12 February 1663, was their first child.[6]

The boy grew up in an atmosphere of piety and intellect. Increase, a wise and strict parent, possessed an extensive library; it was strong in the working tools of the ministerial profession— the ancient classics, the Church Fathers, and the Scholastics. His loving mother was a model of piety. Cotton's spiritual life developed with the precocity exhibited by many Puritan youths. He began to pray as soon as he could speak, he wrote synopses of sermons he heard, and he read copiously in the Bible. From his earliest days he was imbued with the belief that he was a member of a covenanted community on which lay a divine obligation to advance the Kingdom of God in America. In addition, he had an "affectation of pre-eminence" which convinced him that he was destined for greatness.[7]

The youth benefited from the best education available in New England at the time. He learned to read and write before attending school and enjoyed reading in his father's library, especially church history. He began his formal studies at the Boston Latin School,

Cotton Mather in 1727, the year before his death. Mezzotint portrait by Peter Pelham. Courtesy of the American Antiquarian Society.

a free grammar school in the South End, first under Benjamin Thompson, then under Ezekiel Cheever, renowned as a pious, learned, and skillful teacher. The school prepared youngsters for college, emphasizing the classics, the key to all ancient learning apart from the Bible, and there Mather read several Latin authors, went through a great part of the New Testament in Greek, read considerably in Isocrates and Homer, and made "some Entrance" into Hebrew grammar, as well as learning to compose themes and verses in Latin and to take Latin notes on sermons preached in English. Cheever brought the dead languages to life while teaching Mather to love Christ above the classics.[8]

In the summer of 1674 Mather was admitted to Harvard College. President Leonard Hoar himself examined the eleven-year-old boy, who demonstrated that he already knew more Latin and Greek than was required for entrance. Although it was still young and small, Harvard was devoted to both Renaissance and Reformation ideals; its purpose was to give Christian gentlemen a liberal education, not merely to train ministers. The curriculum was designed to provide a mastery of the learned languages as well as some knowledge of classical literature and the original languages of Holy Scripture. Thus, Latin was the official language of college life; scholars were not to use English with each other, unless called upon to do so in public exercises of oratory or the like. Students were also expected to be able to translate the Scriptures as part of daily prayer; at morning prayer every student took a turn in reading some portion of the Old Testament out of Hebrew into Greek (freshmen read it out of English into Greek), and at evening prayer everyone in turn translated a portion of the New Testament from English into Greek.[9]

Putting these ideals of classical and scriptural learning into practice, Mather's class spent most of their first two years studying Greek and Hebrew for their general cultural value as well as their practical use in reading the Scriptures, the latter particularly in the case of Hebrew. Greek literature was available in a fat little anthology published by Jean Crespin, a sixteenth-century Protestant scholar who had a press in Geneva. The emphasis on Hebrew and kindred languages, "the most distinctive feature of the Harvard curriculum," aimed primarily at improving Old Testament scholarship. Although the first two Harvard presidents had been excellent Hebraists, the study of Hebrew had fallen off somewhat

Introduction

before Mather entered. Still, as an undergraduate he composed Hebrew exercises.[10]

In addition to the classics, Mather completed the circle of traditional academic studies required by Harvard—logic, ethics, metaphysics, mathematics and natural philosophy, rhetoric, oratory, and divinity. Disputations and declamations were conducted two afternoons a week after students became proficient in logic. Mather exhibited a keen interest in natural philosophy, that is, science or knowledge of the physical world as distinct from the spiritual universe (metaphysics), and for his declamations he usually chose a topic in this area, "by which Contrivance I did kill Two Birds with One Stone"—that is, he learned both natural history and public speaking at the same time.[11]

The natural philosophy taught at Harvard in the earliest years was Aristotelian, the Peripatetic physics of the Schoolmen. Tutors gave their pupils a manuscript synopsis of physics to begin with; they probably used as reference books the *Systema physica* of Bartholomäus Keckermann and the Encyclopedia of Johann Alsted; and they read Aristotelian textbooks, the most popular being the *Physiologiae Peripateticae libri sex, ex optimis interpretibus* (1597 and many later editions) of Johannes Magirus, professor of physics at the University of Marburg. A student of Harvard's first graduating class (1643) possessed a 1620 edition of Julius Caesar Scaliger's *Exotericarum exercitationum* (1557), the ablest modern exposition of Aristotelian physics in the form of a scathing attack on Jerome Cardan's *De subtilitate* (1550). The Mather library had a copy of a 1615 edition of Scaliger's book. As a student Mather owned a copy of a book with a Greek text of Aristotle's works on astronomy, meteorology, and natural history with a Latin translation and commentary by professors at the University of Coimbra.[12]

A new natural philosophy entered Harvard in the late seventeenth century and completely eclipsed the old before 1700. Bacon, Descartes, and Boyle were among the leading proponents of the new methodologies and theories. The Baconian strand of empiricism worked its way to the fore gradually but steadily, while the Cartesian strand of rationalism enjoyed a tremendous vogue in France and England in the middle of the century before giving way to Boyle's corpuscularian or mechanical philosophy and later to the laws of nature demonstrated by Newton's empiricism. It seems that no copy of Descartes's work on physics, the *Principles of Philosophy* (1644), made its way to Cambridge, but apparently a knowledge of Cartesian physics led to a revulsion against Scholastic physics in 1663, and the class of 1671 struck a blow against the old order when it revolted against a tutor who read lectures to them out of the textbook by Magirus. In that same year, Jacques Rohault published his *Traité de physique,* an exposition of Cartesian natural philosophy which became an immediate success in France and beyond. But there is no evidence that Harvard students had a copy of Rohault at this time.[13]

During Mather's college years, authors attempted to reconcile the new physics with the old. The best known of these at Harvard was Adrian Heereboord of Leiden, a disciple of Descartes, who published *Meletemata philosophica* or "Philosophical Exercises" in 1654. Samuel Sewall read it to his Harvard pupils in 1673–74. Charles Morton introduced a new but not Newtonian phase in natural science at Harvard when his "Compendium Physicae" was introduced as a textbook in physics in 1687, the very year that Newton's *Principia Mathematica* appeared. Morton's manuscript "Compendium" discarded all Aristotelian hypotheses inconsistent with the latest discoveries. The author admired Robert Boyle, who synthesized several theories about physical reality then current to explain the universe in terms of corpuscles, an empirically based particulate theory of matter that described molecules made up of atoms joined together that either whirled about or remained at rest. Mather graduated before Morton's "Compendium" was taught; he was attracted to Boyle's corpuscularian natural philosophy for a time but later moved beyond it. In any event, while the natural philosophy Mather learned at college was quickly outdated by the scientific revolution culminating in Newton, the interest in science he demonstrated as an undergraduate remained with him all his life.[14]

Cotton's freshman year was especially difficult. The youngest and smallest boy in the college and the son of an influential as well as protective father, he was subjected to severe hazing by older students. They and the tutors were engaged in warfare with President Hoar, who enjoyed the support of the senior Mather. His troubles were exacerbated by a speech impediment which he had had from the cradle, and by October of his first year at Harvard he began to stammer. This exposed him to ridicule from classmates and led him to doubt whether he could continue the family tradition by entering the ministry. But he remained in college and in fact his handicap served to strengthen his scientific interest; for some time he laid aside thoughts about preaching and studied medicine. Help came when a venerable schoolmaster named Elijah Corlet taught him to overcome his speech impediment by speaking with deliberation.[15]

Always an omnivorous reader, Mather read hundreds of books and began to acquire his own library while a student at Harvard. He continued to collect, later inheriting half of his father's books. By 1700 Cotton's library contained between two and three thousand volumes and was still growing. By the end of his life it consisted of probably four thousand volumes and was probably the largest private library in America. He kept a commonplace book, his *Quotidiana,* in which he entered remarkable passages from various authors encountered in his reading, numbered every quotation, and prepared an index of names and things. In later years he could quickly retrieve material from this treasury for *The Christian Philosopher.*[16]

By the time Mather graduated from Harvard with a bachelor of arts degree in 1678, he possessed an excellent knowledge of classical languages, of classical and Christian history and literature, and of the Bible. His preparation in natural philosophy was weak, but he was actively interested in the physical universe, and from his wide reading had begun to gather notes on the most important advances in science. In addition, despite a heavily mannered literary style, he was the master of his own language. Perhaps as a result of his Latin training and his college declamations, he wrote clear, fluent, and graceful prose. These skills and talents served him well in writing his treatise on Newtonian science.[17]

Mather's progress on the road to the ministry was slow and anxious; he spent seven years after leaving Harvard before settling into his life's work. He continued his studies, devoting himself to church history, homiletics, and divinity, but the state of his soul remained a pressing concern. In his quest for salvation the young man fought the snares of pride and sexual temptation while giving days to secret prayer and fasting. At the age of sixteen his perception of the world changed; he experienced a religious conversion and was admitted as a member of the North Church. In that same year he preached his first sermon and began his public works by founding a voluntary religious society to promote piety and benevolence. In 1681 he earned a master of arts degree from Harvard, arguing in his thesis that Hebrew vowel points are of divine origin, a position he later abandoned. The following year he issued the first of his many publications, a poem dedicated to the memory of Urian Oakes.[18]

He became his father's unordained assistant at the North Church in September 1680. Members valued him so highly that they soon increased his preaching duties; in December 1681 they expressed an intent to call him as their pastor. But Cotton had received a call from the New Haven church the previous month, an invitation that was renewed the following February. Cotton dreaded being separated from his father, who had always protected and supported him, and when Increase advised Cotton to accept the New Haven offer, Cotton felt rejected. His stammer threatened to return. Finally, on 8 January 1683, the Boston church by unanimous vote elected Cotton to its pastorate. Nevertheless, he remained unordained for another two and a half years. Finally, just before turning twenty-one, he brought his stammer under satisfactory control by cultivating a deliberate and emphatic style of preaching, which in turn carried over into his writing. At last, on 13 May 1685, Mather was ordained before a vast congregation in the prestigious North Church, the largest in America. He served here until his death forty-three years later.[19]

Mather came to his vocation at a troubled time. New England experienced a spiritual crisis during the late seventeenth century, and it thoroughly conditioned the life of the Puritan priest. This time of trial arose out of anxiety about the validity of the vision that had inspired the founding of the New Israel. The first American Puritans had been convinced that they were God's elect nation engaged in a divine mission to build a holy commonwealth. By the time of Cotton's early manhood, however, conditions had profoundly changed. A concatenation of adverse circumstances led his generation to question the vision of their forefathers; this questioning took place at a time when the new science influenced modes of inquiry in the transatlantic community.

As a result, the people of New England had to choose between two alternative ways of viewing their situation. One held that Massachusetts was simply another colony in the British imperial system, in which case the American Puritans never had been the People of God commissioned to build the New Israel—a conclusion that spelled catastrophe for the very idea of a covenanted community. The other alternative held that the Puritans were still God's elect nation engaged in a divine mission to restore humankind from the curse of the Fall and to open the last stage in the divine scheme, a stage which was to bring a blessing upon all the nations of the earth. Since the new scientific temper emphasized observation and experiment, late seventeenth-century Puritans felt a need to collect experimental proofs that they still enjoyed God's favor.

New England suffered a seemingly unprecedented series of natural adversities from 1675 to 1693. Since Puritans read spiritual significance into everything that happened, they interpreted these physical calamities as divine judgments upon them for religious declension. The adversities began in 1675 with King Philip's War. Before the Wampanoag chief was captured and beheaded the following year, the Indians had burned many towns, left two thousand people homeless, and decimated the male population of New England (more fatalities were inflicted in proportion to the population than in any American war in the next three centuries). The war was hardly over before a dreadful fire destroyed forty-five dwellings in the North End of Boston, including both Increase Mather's house and meetinghouse. Another sign of God's wrath was a smallpox epidemic, the worst of the century, which reached its deadliest height in Boston in September 1678, killing as many as thirty people in one day and carrying off an estimated two to eight hundred souls before running its course. Fire ravaged the town again on the night of 7–8 August 1679, destroying eighty houses and seventy warehouses in the center of Boston. The winter of 1680–81 was unusually cold, and the next summer drought blighted the land. On 17 August 1682, and again a week later, a comet frightened the community—Increase Mather called it "Heaven's Alarm to the World"—and less than three months after that, several sudden deaths within three or four days added to the sense that God had a controversy with New England. The final calamity was the witchcraft mania of 1692–93, which Puritans interpreted as the severest of divine chastisements.[20]

Political adversity as well as natural disasters contributed to New England's spiritual crisis. The Puritans had established a self-governing Bible commonwealth on the basis of the Bay Colony charter of 1629, and they maintained their independence on this constitutional foundation as long as the Puritans remained in power in England. Relations with the mother country began to change about the time of Cotton's birth, however, when the Restoration brought Charles II to the throne. The new king was determined to reduce Massachusetts to subjection. Although Charles was preoccupied until 1676, he then sent Edmund Randolph to Boston to spell out royal policy, and insisted that the colony repeal all of its laws which were repugnant to those of England.[21]

Under Governor John Leverett, a resolute Puritan, the General Court delayed and evaded but yielded nothing to the Crown. Randolph sent home hostile reports and returned to England to prosecute the Bay Colony. After Leverett's death in 1679, two factions arose in the political affairs of Massachusetts. Joseph Dudley emerged as the leader of those favoring submission to the Crown. As Randolph reported, Dudley, "if he finds things resolutely managed, will cringe and bow to anything," for "he hath his fortune to make in the world." The other party drew its strength from the elected deputies and was defiant. Increase Mather, a bold spokesman for preserving the charter, insisted that to submit to the Crown was to sin against God.[22]

The Court of Chancery finally vacated the charter on 21 June 1684. In English law, Massachusetts was now simply another colony in the imperial system, not a holy experiment conducted by the People of God carrying out a divine mission. For a time the former General Court acted provisionally, but when the Catholic James II ascended the throne in February 1685, he commissioned new officials. A council under Joseph Dudley as president and Edward Randolph as secretary took formal possession of the government on 25 May 1686. The colony felt the full brunt of the loss of its charter when Sir Edmund Andros arrived in Boston on 20 December 1686 to rule over the Dominion of New England. Blunt and tyrannical, Andros abolished the Assembly, limited town meetings to one a year, enforced the Navigation Acts, revoked land titles and demanded payment of a quitrent as a condition of regrants, and imposed assessments.

Advised by friends, Increase Mather hastened to London to lay New England's grievances before the Lords of Trade. Eluding detection when Andros tried to prevent his departure, he sailed on 3 April 1687. Cotton, now twenty-four, was thrown on his own by his father's flight; he endured this condition for five years. The son not only took charge of the large North Church but also found himself at the center of political events. He had some preparation for his heavy responsibilities. He had worked as his father's ministerial colleague for seven years, and he knew the political situation intimately, having learned from Increase and other ministers to scorn Randolph and to stand tall in defense of the colony's civil and ecclesiastical liberties. Cotton gained valuable experience as a leader of both church and state during this period of enforced maturation. Twice threatened with arrest during the Andros regime, he was politically active when Boston rose against the hated governor on 18 April 1689.

This revolt was one manifestation of a widespread colonial response to the Glorious Revolution in England, a momentous event that brought the Protestant monarchs William and Mary to the throne and provided a constitutional settlement to the struggle over the locus of sovereignty, the form of religion in the realm, and personal rights that had agitated the nation for most of the century.[23] Cotton Mather remained politically conspicuous in the unsettled period after the Glorious Revolution. "The young Pope" preached to the delegates who met on 22 May and voted to return to the old charter. Meanwhile, Increase continued to press for restoration of the colony's rights and privileges, and in 1691 the new king granted a new charter that restored many of the cherished features of the earlier patent. Increase returned to Boston on 14 May 1692 with the new charter and a new governor whose appointment he had been instrumental in securing. The governor was Sir William Phips, a native-born, rough-hewn, self-made man who had been knighted as a reward for discovering and raising sunken treasure in the Caribbean. Phips was in favor with the Mathers; in March 1690 Cotton had baptized the forty-year-old man, who made his profession and became a member of the North Church. These political events gave American Puritans reason to believe that God once again favored New England.

Religious innovations also contributed to New England's sense of spiritual crisis in the late seventeenth century. Convinced that they understood God's plan for redemption of the human race, the founders of the Bay Colony had erected a uniform ecclesiastical system. They required conformity to the Congregational church order and did not hesitate to punish, banish, and even execute dissenters. After the Restoration, however, the English Crown dismantled religious uniformity and compelled the Puritans to tolerate other faiths. In 1661 Charles II put an end to the punishment of Quakers, and the following year he ordered Massachusetts to permit liberty of conscience. Under Randolph and Andros, worship in the Church of England was made lawful in the colony. On 15 June 1686 an Anglican minister named Ratcliffe organized the first Anglican church in New England. Overriding ministerial objections, he held Anglican worship in the Old South Church when the regular congregation did not use the building.[24]

While adjusting to this external challenge, the Puritans also experienced a general sense of religious declension. Michael Wigglesworth captured the prevailing mood with two poems, *The Day of Doom* and *God's Controversy with New England*. Published in 1662, both won a wide following. During Cotton's youth, people praised the faith of the founders while denouncing their own departure from earlier standards. The jeremiad became a familiar means of analyzing the declension. This rhetorical device performed the dual function of lamenting apostasy while at the same time heralding restoration. Sermons and theological works observed that the colony was justly afflicted by natural adversities and political challenges for its religious backsliding; yet Puritans believed that the punishments visited upon them were corrective rather than vindictive. These views are evident in the Synod of 1679–80. Delegates from the churches considered the evils that had provoked God's judgments upon New England and what could be done to remedy them. Increase Mather drew up the result; the General Court ordered it published under the title *The Necessity of Reformation*. Having denounced their apostasy and vowed to reform, Puritans were confident that God would not forsake them. They were still a Chosen People and would ultimately triumph.[25]

The new science was instrumental in creating a new intellectual climate during Cotton Mather's lifetime, and he had to accommodate himself to the changes it wrought. When he was born, the medieval worldview still had a strong hold on Western thought. The product of both the pagan tradition of classical antiquity and Judeo-Christian thought, this inheritance offered a coherent image of physical reality, one that shaped the Western mind over two thousand years. According to this image, both nature and the supernatural were one system derived ultimately from God, who rules the physical universe by both first and secondary causes. The former were divine decrees of an inscrutable character, the latter were the regular sequences of nature which could be understood by the human reason. Hence the medieval worldview encouraged belief in the supernatural and acceptance of the physical creation as a world of wonders, events that demonstrated the power of God or Satan to suspend the laws of nature. The darker side of this mental universe is well known; it meant belief in folklore, in tall tales, in thunder and blazing stars as portents, in witches and other occult forces, and in a variety of other superstitions. Mather grew up in such an intellectual atmosphere. Nor was he alone. The educated and the uneducated alike shared this image of reality.[26]

During Mather's lifetime, these age-old beliefs were sharply challenged by a scientific revolution that began with Copernicus in the mid-fifteenth century and culminated with Newton in the late seventeenth century. The scientific achievement of these years provided a new model of reality, one that captured the Western mind and prevailed until the twentieth century. Formulators of the new view retained a conventional belief in the Christian religion while employing the experimental method with its emphasis on observation and demonstration. Natural laws were part of the law of God and capable of mathematical proof. Isaac Newton, who mathematically demonstrated the revolution of the planets in their orbits, justifiably gave his name to the age.

Cotton Mather was foremost among a small group of New Englanders who enthusiastically adopted the new science in the late seventeenth century. Theologians had long studied nature in order to understand the will of God; now they and their allies welcomed scientific advances that explained how God's Providence advanced divine purposes in the physical universe. Mather always saw harmony rather than conflict between science and religion.[27]

He was preeminent among those in New England who were able and willing to contribute to the development of experimental philosophy. He kept in touch with the progress made in the centers of science by reading the *Philosophical Transactions* of the Royal Society and recent European scientific literature, and he was quick to discern that the new scientific temper required that facts about the natural world be established by empirical proof. Like his father, who as early as 1681 called on ministers to send him evidence of "illustrious Providences" about which they had knowledge so that he could document his assertions about God's superintending care in his *Essay for the Recording of Illustrious Providences* (1684), Cotton devoted immense labor to compiling the facts necessary to support his convictions. His empiricism informs his efforts to show that witches were exercising their malign influence on the community in the early 1690s and that New England was indeed God's New Israel, an argument he elaborated in his compendious history of New England from its founding, the *Magnalia Christi Americana,* a magnificent folio published in London in 1702.

Comets made their appearance over New England in 1680 and 1682, and the ministerial response to them demonstrated Boston's receptivity to the new science. Increase preached a sermon on the first of these which he published as *Heaven's Alarm to the World* (1682); after viewing the second occurrence, which was Halley's comet, he published a longer treatise entitled *Kometographia, or A Discourse Concerning Comets* (1683). In this essay he did not completely abandon the traditional belief that comets were divine portents of extraordinary events in human affairs, and he insisted that their appearance could not be predicted. But he recognized that comets proceed from natural causes. Cotton's second publication, *The Boston Ephemeris: An Almanack for MDCLXXXIII* (1683), commented on the comet of 1682.

Interest in comets helped spur the formation of the Boston Philosophical Society, a scientific enterprise patterned after the Royal Society of London. Little precise information about it is available. Increase Mather was the founder; Cotton was a member and took an active part in the society's affairs. The club held its first meeting in April 1683 and appears to have met once a month for a period of three months. Perhaps the society was formed in an effort to gather examples of divine providences. To the Puritan clergymen who dominated the society, the collection of evidence on remarkable events which displayed God's glory and served as warnings to sinners—e.g., tempests, earthquakes, apparitions, and the like—was a legitimate phase of scientific activity. The society assembled descriptions of natural phenomena from New England correspondents, and it had at least one foreign correspondent. Increase Mather sent at least one of his and Cotton's papers to Wolferdus Senguerdius or Wolfgang Senkward (1646–1724), a professor of philosophy at the University of Leiden. Senkward used the material in the second edition of his *Philosophia Naturalis* (1685). The last recorded meeting of the society was held on 10 December 1683; interested members continued their activities for another two years.[28]

Cotton Mather was well equipped by family background, education, and personal inclination to respond positively to the new experimental philosophy. He did not entirely succeed in shaking off ancient superstitions; as a result he earned a reputation as a bigot and a reactionary which has haunted him down the corridors of time. But in reality, Cotton Mather was among the most progressive New England Puritans in responding to the new science. He embraced scientific advances wholeheartedly without accepting a completely mechanistic universe.

3. The Origins of *The Christian Philosopher*

Cotton Mather left little direct evidence as to the origins of *The Christian Philosopher*, but he sheds important light on the matter in his book. In the introduction he declares that his essays "will demonstrate, that *Philosophy* is no *Enemy*, but a mighty and wondrous *Incentive* to *Religion*," and in the untitled preface that follows, he says that he proposes to exhibit "the Works of the Glorious GOD in the *Creation* of the World" [7:3–4; 17:11].[29] His purpose is to show the harmony between science and religion.

We can discover a good deal about the origins of his treatise of Newtonian science by tracing the author's career and writings. *The Christian Philosopher* was a natural outgrowth of his strong scientific interest and his continuing efforts to do good for his fellow creatures. Mather was a minister of the New Testament who lived at a time when new discoveries of the physical universe were challenging Christian orthodoxy. Like other New England clergymen of his generation, he welcomed advances in natural philosophy, confident that God was the creator and sustainer of the universe and that to explore nature was to understand the mind of God. An incorrigible scribbler, he began writing about medicine when he was a Harvard undergraduate.

Mather demonstrated his continuing interest in science during the 1680s. Athough his first published work was a memorial poem, the second, as we have seen, was an almanac in which he described the comet (Halley's) that he had observed the previous year. Also during these years he pored over Robert Boyle's *Usefulness of Experimental Natural Philosophy,* published in the year of his birth, and read Richard Bentley and William Whiston. He drew on Boyle in two works published during the next decade. In *The Wonderful Works of God Commemorated* (1690), a Thanksgiving sermon published in Boston, he listed aspects of the visible creation—plants, animals, minerals, the earth, the stars, and the sun—as objects for gratitude. In *Winter-Meditations* (1693), he celebrated an even larger variety of objects which display God's handiwork. In these books he expresses a more positive stance toward nature than he had earlier shown.

The origins of Mather's argument for theism from the design of the universe are closely related to another of his major literary efforts, the "Biblia Americana." This gigantic enterprise was an interpretation of the whole of Scripture in light of the best learning of Mather's own day. "*The Christian Philosopher* was culled out of the *Biblia,*" Perry Miller wrote, but in light of the evidence on the book's composition and sources this statement is untenable. Otho T. Beall, Jr., and Richard H. Shryock were closer to the mark when they viewed Mather's two great works as opposite sides of the same coin. In the "Biblia," Mather examined the Scriptures in terms of science, they said, whereas in *The Christian Philosopher* he surveyed science from a religious perspective. But Beall and Shryock overemphasize the extent to which Mather relied on science in the interpretation of Scripture in the former work. Natural philosophy was one of the disciplines called on to serve scriptural religion, but by no means the only one.[30]

The "Biblia Americana" grew out of Mather's intense devotion to the Bible. In August 1685, shortly after he was ordained, Mather began a course of reading the Scriptures "with such a devout Attention, as to fetch at least one *Observation* . . . out of every *Verse* in all the Bible." In August 1693, having concluded that none of the available Bible commentaries afforded as much illustration as might be given, he announced his intention to produce a "Biblia Americana." He resolved to write upon a portion of Scripture every morning—no day without a line. By "a laborious Ingenuitie" he would gather the best illustrations of sacred texts from "the scattered Books of learned Men." He sought treasures dispersed in volumes of his own age, aiming at a work that would be entertaining for both its novelty and its rarity. He would heap thousands of "delicious *Curiosities*" of the giants of knowledge upon one table. He approached Scripture from every conceivable angle—etymological, grammatical, historical, and philological as well as scientific. His wide reading and his ability to extract new ideas from printed material quickly were indispensable to his heroic project. By means of daily entries he hoped to complete in seven years "one of the greatest Works, that ever I undertook in my Life."[31]

Mather labored on his "Biblia Americana" for many more than seven years, eventually producing a manuscript of six volumes of about a thousand pages each. He started by obtaining large bound blank books in which he wrote entries of varying length on biblical texts in double column, leaving blank space so that he could amplify the document at a later date. As he came across new material he inserted additional glosses at the top and bottom of pages, in the margins, and at right angles to earlier illustrations. Some exposition is in essay form, but most of it is questions and answers. He underlined heavily and on occasion crossed out entries. Thus, the manuscript shows several distinct stages of Mather's work over a long period of time. He used different inks over the years, and in some places the handwriting appears to be that of a copyist. Mather inserted into the bound books smaller leaves of irregular size with glosses that he jotted down "in the Intervals of Business, especially as I walk the Streets." He also pasted in printed pages from publications, including his own *Stone Cut Out of a Mountain* (1716), which offered several maxims of ancient church doctrine dug out of the Scriptures, and copied one large extract from *The Christian Philosopher*.[32]

Mather's massive commentary on the Scriptures has been called "a great indigested mass of material . . . with no evidence of design," but it has a more settled plan than may readily seem obvious.[33] The completed manuscript is organized into several distinct sections built around the Old and New Testaments. Mather first treats the New Testament. He begins with an essay on "The Harmony of the Gospels," then illustrates each of the four Gospels in order, chapter by chapter, after which he proceeds with an essay entitled "Historia Apostolica" or introduction to the Acts of the Apostles. He concludes this part of the "Biblia" with a commentary on Acts. If he wrote in the order found in the manuscript volumes, he then goes on to treat the latter part of the Old Testament, beginning with Job and proceeding through the end with Malachi. All of these parts of the "Biblia" may have been completed by 1700 or shortly after.

Mather then turns to another stage in his study of scriptural interpretation. He places the two parts of the Bible in the context of world history by essays on "The Old Testament in the Order of History" and "The New Testament in the True Order of History." Then he offers a general view of "The Old Testament," after which, about 1702, he turns to an exegesis of Genesis "agreeable to the Modern Discoveries," which is the most extensive of all his biblical commentary. From Genesis he works his way, book by book, through the remainder of the Old Testament down through Esther, thus completing his interpretation of the entire Old Testament.[34] Next, Mather directs his attention to the books of the New Testament following Acts, taking them up systematically from Romans through Revelation. At this point he has completed his interpretation of all of Scripture. The bound manuscript volumes also contain a number of essays on miscellaneous topics related to knowledge and understanding of the Bible.

The longest comments are on Genesis, Daniel, the Gospels, and Revelation. Mather's annotations on the first chapter of Genesis, approximately fifty folio pages, deal with the Mosaic account of creation. He parades many ancient and contemporary authors in these pages, but more than 60 percent of his space is devoted to matter derived from six books published in London between 1677 and 1717. One of these works preferred the Mosaic over the scientific account of the creation; another has a somewhat pre-Newtonian flavor. Part of Mather's exposition is a summary of Edmund Dickinson's *Physica vetus et vera* (1702) which gives an account of creation based on the corpuscularian or atomistic conception of matter devised and named by Robert Boyle. This theory, though widespread by the end of the century, was soon outmoded by Newtonian ideas, and Mather drew on books by Richard Bentley, William Whiston, and Thomas Pyle that reflect Newtonian notions about gravity and motion. As this evidence shows, Mather does not try to refute new scientific hypotheses in his interpretation of Genesis. Instead, he attempts to reconcile the scriptural text with what he regards as scientific fact.[35]

While Mather worked on the "Biblia Americana," he was also

writing his single most important book, the *Magnalia Christi Americana* (1702). At the same time he continued to perform such ministerial duties as preaching and publishing various sermons and treatises and to take an active part in colonial political affairs. Finally, in May 1706, he thought that he had "happily finished" his "great Work" and need no longer amass illustrations upon the divine oracles, although he admitted that doubtless he would add to the manuscript until it left his hands for publication. He anticipated that his manuscript would constitute two large folio volumes; in September he sent to England a promotional piece describing his "Biblia Americana" as "An American offer *to serve the Great Interests of Learning and Religion in Europe.*"[36]

Mather could not actually abandon work on his scriptural commentary, however, and in 1711 he added over a thousand new illustrations. While subsequently continuing to enter even more glosses on texts, he campaigned relentlessly to get the "Biblia" before the public. He advertised it in the pages of his *Bonifacius* (1710), and in 1714 printed and spread throughout Old and New England "*A New Offer, to the Lovers of Religion and Learning*" describing the work and calling for subscriptions toward publication. He looked primarily to English Dissenters for assistance in seeing the manuscript into print.[37]

He still envisioned publication as late as 1715, but in that year he began to acknowledge discouragement over the prospect, and by October 1716 resigned himself to the advice that publication "is to be despaired of." Even so, a few months later he still hoped that the Lord would inspire "some capable Persons" to bring the "Biblia Americana" into the world.[38]

As events fell out, the manuscript was never published. Because it was massive and unwieldy, printing costs would have been prohibitive; English readers lacked interest in a commentary by an American author; and the British market for such works had been preempted by Matthew Poole and Matthew Henry.[39] Although this was a keen disappointment to Mather, the failure of the "Biblia Americana" to see print may have been fortunate. Mather's work in ancient texts was solid, as Thomas H. Olbricht observes, but his achievement was essentially an updating of earlier glosses, and most of the material in his manuscript was available elsewhere. He made the biblical text the occasion for organizing contemporary knowledge but manifested little interest in determining how Scripture is to be read in light of its own character and milieu.[40]

In abandoning the "Biblia," Mather gradually shifted his emphasis from theology to science. His "Curiosa Americana," a series of letters sent to the Royal Society describing natural phenomena in America, help explain the origins of *The Christian Philosopher.* As we have seen, Mather participated in the activities of the Boston Philosophical Society in 1683 and perhaps later, and in 1711, after he thought he had completed his "Biblia Americana," he began to assemble accounts of curiosities of nature in America with a view to sending them to London. For this he probably drew material from the repository of the Boston Philosophical Society. His motives were mixed. "There is one good Interest, which I have never yett served, and yett I am capable of doing some small Service for it," he confessed. "The Improvement of Knowledge in the Works of Nature, is a Thing whereby God, and His Christ is glorified. I may make a valuable Collection of many Curiosities, which this Countrey has afforded; and present it unto the Royal Society. May the glorious Lord assist me, in this Performance." In other words, he would combine the doing of good with the quest for worldly fame.

Between 1712 and 1724 he dispatched at least eighty-two letters to the Royal Society describing American scientific curiosities. His material is of the type that was being sent for publication to learned societies all over the world. Mather stated that he modeled the form of his letters after those appearing in the "German Ephemerides," the scientific yearbook of a German philosophical society, the Deutsche Akademie Naturforscher or Collegium Naturae Curiosum, whose *Miscellanea Curiosa: sive ephemeridum medico-physicarum Germanicarum* appeared in thirty volumes from 1670 to 1706. Mather sent his reports in multiples of four to a dozen at a time; the "Curiosa," many of which are written with great care for style, reveal his wide interest in science in New England. The letters discuss astronomy, natural phenomena such as snow and thunder, geology, and zoology, but medicine and related topics in psychology and botany are the theme in many of his communications. Here we are interested in the letters sent down to 1715, the year in which he completed the manuscript of *The Christian Philosopher.*[41]

The first group of "Curiosa" consisted of thirteen letters written in as many days from 17 to 29 November 1712. Seven are addressed to John Woodward, a medical doctor, man of great learning, and a fellow of the Royal Society with whom Mather had corresponded; and the rest to Richard Waller, secretary of the Royal Society. In a covering letter dated 1 December, Mather asked Waller to accept "these mean things," the best that he could master in the brief time available, and expressed a desire to send more "from these parts of the world that a Secretary of the Royal Society may judge it proper to take some notice of." His own merits, he added, fishing for recognition in his characteristically indirect and obsequious way, hardly made him worthy of membership in that illustrious body, but he would subscribe to the promise which candidates made at their admission—"That I will Endeavour to promote the good of the Royal Society of London for the improvement of Natural Knowledge."[42]

The letters in the first series discussed the existence of antediluvian giants based on some bones and teeth found at Claverack, New York; plants in America and Indian cures using them; birds of America; antipathies and the power of the imagination, with case histories (e.g., a woman who swooned upon seeing anyone cut his or her nails with a knife); monstrous human births; cures revealed in dreams; recoveries from wounds which appeared mortal, with case histories; the Indians' division of time by sleeps, moons, and winters; natural phenomena, such as the sun attended with mock suns and uncommon rainbows; a dream of a murder on the day it occurred; the rattlesnake and its habits; thunder, lightning, earthquakes, and hurricanes; and the population of the world before the Flood, human longevity, and human fruitfulness.[43]

These letters throw valuable light on Mather's interest in natural phenomena in America and on the transitional character of his mind. The early eighteenth century was still a world of wonders; most people found it hard to distinguish between nature and the supernatural. Mather, like many learned contemporaries, continued to mix the factual with the fanciful. Some of his accounts contain superstition, and some have little relation to "natural philosophy" or science. But the letters must be judged from the point of view of the time when they were written. The scientific information in these "Curiosa" came largely from seventeenth-century origins, including the collections of the Boston Philosophical Society, his father's *Essay for the Recording of Illustrious Providences* (1684), and his own "Biblia" and *Magnalia*. A comparison of the letters of 1712 with his sources reveals that Mather was moving from a supernatural explanation of various phenomena toward greater objectivity. Yet he had traveled but partway.[44]

Waller laid Mather's communications and his covering letter before the Royal Society on 16 July 1713. The society cursorily examined the letters and ordered Waller to cultivate Mather's correspondence. Waller's response to Mather, read and approved by the society on 23 July, explained the reasons for the delay in presenting Mather's communications to the members. It went on to assure Mather that the Royal Society was well pleased with his letters and found his offer to send more of them extremely acceptable. Waller would be happy to cultivate "a Philosophical Correspondence with so candid and Learned a person." Waller informed Mather that he had proposed him to be a member of the society. The council approved his candidacy on 27 July, but the formalities of the election had to be deferred until after Michaelmas.[45]

Not until 3 and 10 December 1713, however, were the extracts of Mather's letters prepared by Secretary Waller read to the society and discussed by members. On 23 December Waller addressed to Mather a letter including remarks on his first series of American curiosities, which the society approved on 7 January 1714. In this letter Waller assured Mather that he had been elected Fellow of the society, although his name was not on the printed list of members because distance prevented Mather from participating in the ceremony of admission. Nevertheless, Mather immediately began to add the coveted letters F.R.S. after his name.[46]

Waller's extract of the first series of Mather's "Curiosa Americana" was published in the *Philosophical Transactions* in early 1714. In the first of these letters, according to Waller, the writer (Mather) gives an account of the "Biblia Americana" without naming its author. Waller's extract described Mather's manuscript as a large commentary upon some passages in the Bible interspersed with "large Philosophical Remarks, taken out of Natural Historians, and the Observations of himself [Mather] and others, more particularly as to Matters observ'd in *America*, whence he entitles the Work, *Biblia Americana*." As a specimen of it, Mather transcribed a comment on Genesis 6:4: "There were giants in the earth in those days." Mather's gloss confirmed the biblical text by describing the bones and teeth of antediluvian men of prodigious stature recently found at Claverack, about thirteen miles from Albany.[47]

Mather's second series of the "Curiosa Americana" was begun on 1 December 1713 and completed in January 1714. The ten letters in this series treated woolen snow, the influence of the moon, monsters, the moose, pigeons, fasting, sisterhood, idiots, the discharge of rye through the skin, and the ear.[48]

Although disappointed that the Royal Society did not publish more of his communications in its journal, Mather was eager to do good and gain fame by sending the English savants more of his scientific observations. In 1715, in place of the usual series on American curiosities, he sent Secretary Waller a manuscript entitled "The Christian Virtuoso." The following year he sent the third series of "Curiosa Americana," eleven letters in all, and, except for 1718 and 1719, he sent additional communications every year, ending in 1724, when ten letters were dispatched.[49]

"The Christian Virtuoso" was published in London five years later as *The Christian Philosopher*. It seems obvious by now that this book, Mather's major statement on science, was a product of his lifelong preoccupation with knowing and understanding God and His relation to the physical universe.

Mather's treatise is an American version of the physicotheological literature that enjoyed great popularity in England around the turn of the eighteenth century. Physicotheology is a branch of natural theology that seeks to prove the existence and attributes of God from the evidence of purpose and design in the universe. England was the principal home of literature of this type, and Mather was very familiar with it. The temptation to publish his own physicotheological treatise proved irresistible.

Some passages in his book suggest that Mather was more concerned with the conflict than with the harmony between science and religion. "*Atheism* is now for ever chased and hissed out of the World," he writes in *The Christian Philosopher*, "every thing in the World concurs to a Sentence of *Banishment* upon it" [308:32–34]. Admittedly, Mather himself had on occasion experienced religious doubt, but his own faith was never seriously at risk.[50] Moreover, New England was not troubled by a conflict between science and theology during Mather's lifetime.[51] In Old England, however, scientific rationalism had begun to challenge certain aspects of religious orthodoxy by the late seventeenth century, and many people in the circle around Newton associated science with the rise of skepticism. Mather's expressed fear of atheism, a term which, as used at the time, defies rigorous definition, was taken over from English authors along with so much else that he borrowed from that source.[52]

We do not know how long Mather worked on his treatise on the design argument. In all likelihood, he composed it rapidly. The work is heavily derivative. It consists largely of material paraphrased, condensed, or quoted nearly verbatim from other authors, with some of it culled from his own commonplace book or earlier publications. He must have dispatched his manuscript no later than June or July, because in August Cotton renewed his request that his brother Samuel "would not lett my *Christian Virtuoso* be lost, but, if you know no better way to make it public Lett it pass thro' *Dr. Woodward's* hand, into the Repository of the Royal Society."[53]

The history of the manuscript in England is not entirely clear. According to Raymond P. Stearns, Waller died before the document arrived, and the Royal Society therefore asked the Reverend John Theophilus Desaguliers "to look it over and give an Account of it which he promised to do." Mather, writing to Sir William Ashurst in October 1715 with an account of the "Biblia Americana," expressed the hope that "one considerable Article in the work, namely *The Christian Virtuoso,* one would think, might procure some subscribers to it among the Royal Society, which have allowed my relation to them."[54]

Nothing further is known about the manuscript until shortly before it was published. At some point along the line the title was changed. Almost certainly Mather borrowed his original title from Robert Boyle, who had published *The Christian Virtuoso: Shewing, That by Being Addicted to Experimental Philosophy, A Man is Rather Assisted than Indisposed, to Be a Good Christian* in London in 1690. In the early seventeenth century the word *virtuoso* had been used to describe gentlemen interested in antiquities and precious stones, but by the middle of the century it had come to identify persons devoted to the study of natural philosophy. Boyle equated *virtuoso* with *natural philosopher,* or what later became known as *scientist.* Mather's title reflects this meaning and demonstrates that his

THE
Chriſtian Virtuoſo:

SHEWING,

That by being addicted to *Experimental Philoſophy*, a Man is rather Aſſiſted, than Indiſpoſed, to be a *Good Chriſtian*.

The First Part.

By T. H. R. B. *Fellow of the* ROYAL SOCIETY.

To which are Subjoyn'd,

I. A Diſcourſe about the Diſtinction, that repreſents ſome Things as *Above Reaſon*, but not *Contrary to Reaſon*.

II. The firſt Chapters of a Diſcourſe, Entituled, *Greatneſs of Mind promoted by Chriſtianity*.

By the ſame AUTHOR.

In the SAVOY:

Printed by *Edw. Jones*, for *John Taylor* at the *Ship*, and *John Wyat* at the *Golden-Lion*, in St. *Paul's* Church-yard, 1690.

The title page of Robert Boyle's *Christian Virtuoso* (1690). Mather originally used this title for his own manuscript on the harmony between science and religion. Courtesy of the Rare Book and Special Collections Library, University of Illinois at Urbana-Champaign.

intention was to illustrate how a Christian should respond to new scientific discoveries.[55]

The circumstances surrounding the printing and publishing of the book are fairly mysterious, and even after close inspection, questions remain as to the identity of the publisher. Jeremiah Dummer, a family friend and the agent of Massachusetts in London, assisted with publication. Writing on 12 September 1720, he informed Mather that his book, now bearing the title *The Christian Philosopher,* "is compleatly printed; but I don't yett publish it, because in the Recess of Parlaiment, all people of Distinction are out of Town, and if it should come abroad now, it would be an old Book before the parliament meets. This is a peece of prudence that the best Authors are obliged to use. Besides, I have not yett determined upon the Patron. So we must be willing, both of us, to stay a little."[56]

Finding a patron was complicated. In a letter written after his manuscript was published, Mather said that in sending his *Christian Philosopher* over to England he had "prefixed unto it a dedication to My Lord Barrington." John Shute Barrington, first Viscount Barrington, was the brother of the governor of Massachusetts, and Mather's dedication expressed great admiration for both men. Lord Barrington himself was a prominent Presbyterian and a member of the congregation of Thomas Bradbury, a distinguished dissenting minister in London. Lord Barrington gave the manuscript to Bradbury, with whom Mather carried on a friendly correspondence.[57]

Long after Mather sent his manuscript to England, Lord Barrington informed him that Bradbury "had much offended him"— the two men took opposite sides in the Arian controversy at Salter's Hall in 1719—and that he would never call for nor take any further notice of the manuscript in Bradbury's hands. Barrington counseled Mather to deal with Bradbury as he saw fit. So advised, Mather reports in a letter to Colman that he asked Bradbury "to send me my mean composure home again, or do what he pleased with it." But rather than return it, Mather adds, "Mr. Bradbury laid me under ponderous obligations by publishing the book, and enriching me with extensive opportunities to do my part among the priests of the creation."[58]

In publishing the book, Bradbury replaced the dedication to Barrington with one to Thomas Hollis, an eminent London merchant who had already made two donations to Harvard College. In the dedication to Hollis signed on 22 September 1720, Bradbury observed that Mather had opened the two great books of God—nature and Scripture—and that Hollis's generosity to Harvard made the dedication "more proper" to him than to any other person. Starting in 1721, Hollis poured more of his bounty into Harvard, in time establishing two professorships, one in divinity, the other in mathematics and natural philosophy.[59]

The octavo book of 304 pages was eventually issued under the title *The Christian Philosopher: A Collection of the Best Discoveries in Nature, with Religious Improvements*. The title page listed Mather as D.D.—the University of Glasgow had awarded him a Doctor of Divinity degree in 1710—and Fellow of the Royal Society. The book was printed for Emanuel Matthews, a London printer and bookseller located "at the Bible" near Saint Paul's Churchyard on Paternoster Row, the leading site of printing houses during the reign of Queen Anne. His shop was probably responsible for all the stages in the production of the book. Bradbury may have been the publisher (Mather says as much), but it is difficult to distinguish between publishers and printers at the time, and details are lacking. Probably at most a few hundred copies were printed. The book was probably bound in paneled sprinkled calfskin. The text is remarkably free of errors, a marvel, granting the complexity of the treatise and the opportunities for mishap in a printer's shop in the early eighteenth century. Mather's book was available before the end of the year, but the publisher gave it a 1721 imprint.[60]

By 10 December 1720 Mather himself knew that *The Christian Philosopher* had appeared in London. A shipment of books left England in the winter, but the ship was blown off the New England coast, took refuge in Antigua, and did not arrive in Boston until the week of 26–31 March 1721. Mather received one copy and the bookseller Samuel Gerrish a hundred copies. The bookseller advertised the work in the end pages of Mather's anonymous *The World Alarm'd*, which Gerrish published in Boston in 1721.[61]

Mather demonstrated great solicitude for his scientific treatise. He hoped to glorify God by getting "our Colledges" filled with the book, and he began to promote it as he had his previous publications. He wrote his friend John Winthrop of New London, son of Wait Still Winthrop, grandson of Governor John Winthrop of Connecticut, and great-grandson of Governor Winthrop of

Massachusetts, who had sent him observations of fauna and flora, offering him a copy. "To detach one [from the bookseller] at the small Expence of a little, Dirty, Ragged Ten Shilling Bill, is but a very small Acknowledgment unto an Invaluable Friend, unto whose Generosity I have been indebted for more than as many pounds." Reading *The Christian Philosopher,* he even suggested, might help Winthrop recover his health. Writing to a Scottish correspondent early in 1723, Mather hoped that the students at Glasgow would profit from the book. Shortly thereafter he wrote again to Winthrop, asking if he had perused the work and "how far, (or whether at all?) it agrees with you."[62]

4. The Sources of *The Christian Philosopher*

Cotton Mather was an immensely learned man and a prolific author. He published more than 460 works during his lifetime, mainly in theology, historiography, and science, and left many thousands of pages in manuscript when he died. His *Christian Philosopher* strikes the reader as formidably learned. It cites some 415 authors, gives the titles of many weighty works, frequently quotes Latin and the Scriptures, and introduces many obscure words and recondite concepts.

This parade of erudition is less impressive than it appears, however, since the treatise is highly derivative. Mather borrowed the structure, the argument, and most of the substance of his book. The structure was dictated by the past. Like its predecessors, the work discusses the heavenly bodies first, then terrestrial phenomena and the earth itself, and finally living bodies, including humans. This order follows the conventional scheme for the study of natural philosophy under Scholasticism, which was based ultimately upon the order in which traditional Aristotelian texts were taken up. The argument is the ancient and familiar one that the evidences of design and purpose in the physical universe testify to the existence of God. As will be shown, the contents also were largely taken from others. Mather's compositional task was mainly one of arranging items gathered from his wide reading under appropriate headings and giving them vitality. He generally starts each individual essay by citing ancient authorities, then he relates the "best discoveries" encountered in modern writers (to which he occasionally adds his own scientific observations), after which he closes with a rhapsody as to how the new knowledge redounds to the glory of God.[63]

All authors draw on the knowledge and insights of others, but their method of acknowledging debts changes over time. What the twentieth century would call plagiarism was, if not generally accepted and widely practiced, at least viewed tolerantly during Mather's day. To paraphrase and quote extensively from other works without attribution showed both learning and proper appreciation of source material. While borrowing freely, authors of the early modern period who wished to indicate their authorities usually did so by mentioning sources in the body of their text. The footnote was known but had not yet come into general use.[64]

Despite his heavy reliance upon previous investigators, Mather deserves considerable credit for *The Christian Philosopher*. A colonial living on the periphery of the great centers of learning, he depended on metropolitan writers and publishers for his knowledge of the new learning of his age. In addition, he lived during a period of rapid change in the mental outlook of Western civilization. At the time he wrote, many people still believed in magic and the occult, fabulous creatures, and the curative power of strange and horrible potions. Contemporary books were full of curious facts and conflicting theories; it took an informed and judicious mind to borrow from them intelligently. Mather should be judged in terms of how wisely he used the knowledge available to him.[65]

As he admits at the outset, he relied extensively on John Ray and William Derham, "a delightful pair of brothers" [10:61], and did "very little" of the writing except for the *"Devotionary Part"* of the essays and some descriptions of American curiosities earlier sent to the Royal Society [10:82–11:85]. Despite his candor, Mather's admission fails to prepare one for the truly amazing extent of his borrowing. The original edition of *The Christian Philosopher* contains 304 pages with an average of 37 lines per page for a total of some 11,250 lines of text. Mather derived the bulk of this material, about 79 percent, from other authors. He himself contributed some 2,322 lines, or about 21 percent of the entire work. This material consists of an introductory essay, an untitled preface, introductions to individual essays, connective passages that link together material taken from others and often include commentary of his own that is interspersed within or between what he has lifted from elsewhere, composition taken from his own writings, rhapsodies, and a conclusion.[66]

His two major sources, as he informs the reader, are Ray and Derham. Next in importance is John Harris, whom Mather does not credit. George Cheyne and Nehemiah Grew rank next as sources, and both are generously acknowledged. These five English physicotheological writers together account for approximately 54 percent of *The Christian Philosopher*. Actually, Mather drew most heavily on Derham, taking 2,104 lines or about 19 percent of his text from two works by that author and 1,530 lines or nearly 14 percent of his volume from Ray.

John Ray was the primary source for Derham as well as Mather, so it is important to understand Ray's place in early modern science and in the interpretation of its religious significance. Born in 1627 in Black Notley, near Braintree, Essex, the son of a village blacksmith, Ray grew up in daily contact with nature and got his religious faith from his mother. Living in a time of transition from the old world of legend and superstition to the new world of observation and experiment, he grew up in an atmosphere of folklore and white magic.[67]

As a student at Cambridge University, his chief subject was classical and modern languages. In religion he became indebted to the Platonism of Benjamin Whichcote, Henry More, and Ralph Cudworth. The Cambridge Platonists gave Ray a theology in which both reason and science could find full scope.

After graduating B.A., Ray became a fellow of Trinity College in 1649. His twelve years as a fellow, a time presumably devoted to Latin, Greek, and Hebrew, to disputations, and to sermons and morning divinity exercises, were a period of profound intellectual change in the country. Britain had come late to science, but in the half-century starting about 1650 it made up for its backwardness by significant discoveries in nearly every department. Ray had a natural bent toward science; he found a circle of like-minded friends—many people turned from religious and political controversy to the study of nature in these years—and he made his first studies in science at this time. Ray undertook the study of the Lord's works as a religious duty, starting with the anatomy of lower animals and then turning to botany. His pupils included Francis Willughby, a wealthy virtuoso and later his scientific partner.

After the Restoration Ray decided that his life's work was in the university; he was ordained deacon and priest on 23 December 1660. As he said later, "Divinity is my profession."[68] In 1662, however, he resigned his fellowship rather than take the oath required by the Act of Uniformity and thereby subscribe to a lie. A man of conscience, he retained his integrity while freeing himself to become simply John Ray, naturalist, and to live a life of poverty.

Even under these conditions, however, Ray was one of the founders of modern science. He had neither the inclination nor the resources to be a laboratory scientist; his genius was as a field observer with a gift for description and classification. He helped to lay the foundations for the systematic study of botany and zoology and crowned his descriptive work with two books in which he set out the interpretation of nature.

Intending to produce a complete British phytology, Ray first published a *Catalogue of Cambridge Plants* (*Catalogus plantarum circa Cantabrigiam nascentium*, 1660). This work, according to Charles E. Raven, initiated a new era in British botany.[69]

Ray then traveled in England and Wales and on to the Continent with Willughby and others for three years to extend his observations. Upon returning, he and Willughby began arranging the collections they had made, and at the request of John Wilkins they prepared the tables of plants and animals for Wilkins's last work, *An Essay towards a Real Character, and a Philosophical Language* (1668), an attempt to produce a universal and simplified language. Raven calls their product "a classification superior to anything till then published."[70] On 7 November 1667 Ray was admitted to the Royal Society.

Ray's *Catalogue of English Plants* (*Catalogus plantarum Angliae, et insularum adjacentium*) appeared in 1670. At a time when magic, astrology, and alchemy still entered into medical practice, he pleaded for exact observation and experiment and promoted the scientific investigation of the medical use of plants. When the second edition of the *English Catalogue* was no longer available, Ray prepared an appendix to it, his *Fasciculus stirpium Britannicarum* (1688).

In 1673, Ray followed the *English Catalogue* with his *Observations Topographical, Moral, and Physiological, Made in a Journey through Part of the Low Countries, Germany, Italy, and France*, including a *Catalogus stirpium in exteris regionibus*, a catalogue of plants not native of England and found growing in those countries. Continental botanists immediately recognized it as "a very remarkable achievement."[71] Next came an even more significant work, Ray's *New Method of Plants* (*Methodus plantarum nova*), published in 1682, in which he based the classification of plants solely on structure. This was "his first serious essay in classification," and here for the first time Ray lays down the division of plants into dicotyledons and monocotyledons.[72]

Ray's greatest work, the *Historia plantarum,* was a general history of plants printed by order of the Royal Society. The two folio volumes, published in 1686 and 1688, occupy 1,940 pages plus prefatory material and an index. For this work Ray surveys the whole field of current botanical knowledge, selecting from it contributions of permanent value. Ray's own work, writes Raven, "stands out as scientifically on an altogether different level from that of his sources."[73] In 1704 Ray published a third, supplementary volume containing 10,000 new items.

In 1690 Ray's *Synopsis of British Plants* (*Synopsis methodica stirpium Britannicarum*) was published with the imprimatur of the Royal Society. The culmination of his botanical work, it was more a rearrangement of the *Catalogue of English Plants* and a condensation of the *History* than a record of new discoveries. With this book Ray had succeeded in laying British botany upon a secure foundation.

While studying botany, Ray also laid foundations for progress in zoology. His interest in animals came early, he worked closely with Willughby in this area, and Willughby's death in July 1672 gave him the incentive and opportunity to proceed. He made full use of Willughby's researches and gave his friend and benefactor credit in the titles of two works.

He published Willughby's ornithology (*Francisci Willughbeii de Middleton . . . Armigeri, e Regia Societate, Ornithologiae libri tres*), a history of birds, with the approval of the Royal Society in 1676. When he began his investigation, the subject was entangled, as Ray put it, with "hieroglyphics, emblems, morals, fables, presages."[74] He removed the confusion and inaugurated a new era. His classification, a system based upon real structural differences, is one of the outstanding merits of the book.

From birds Ray turned to fishes. He had studied and written on the subject over the years, and even more than the *Ornithologia* his *History of Fishes* (*Francisci Willughbeii Armig. de historia piscium*) was his own work. The two volumes, one of plates, were printed at the expense of the Royal Society in 1686. In 1694 he completed a *Synopsis of Birds and Fishes* (*Synopsis avium et piscium*), which was not published until 1713.

His study of quadrupeds and reptiles, *Synopsis of Quadrupeds and Reptiles* (*Synopsis methodica animalium quadrupedum et serpentini generis*), appeared in a more timely fashion. It was published in 1693 under the aegis of the Royal Society.

The study of botany brought insects to Ray's notice. He read the literature available but left the detailed study of insects to Willughby. In 1690 he began to collect in earnest; despite many handicaps, he continued to do so until he was too infirm. At his death in 1705 he left a manuscript which, if properly edited, would have been, in Raven's opinion, a pioneer work of permanent value.[75] A preliminary *Methodus insectorum* of ten leaves was published in 1705, and the *History of Insects* (*Historia insectorum*) in 1710.

Especially in the preface to his *Synopsis of British Plants,* Ray catches the excitement of the dawn of modern science. He is thankful to live at a time when empty sophistry has been replaced in the schools by a philosophy solidly founded upon experiment. His is "an age of daily progress in all the sciences." Those who scorn scientific knowledge "should remember that it is knowledge that makes us men, . . . that makes us capable of virtue and of happiness such as animals and the irrational cannot attain." Ray also expresses relief that the Stuarts are gone and that "our heritage of freedom has been restored and secured" under William and Mary. Now he could again take up his profession.[76]

Ray began his great work on natural theology at this time. For thirty years he had written on natural history and refrained from writing on religion because of his scruples, but now, conditions having changed, he felt obliged to write something on divinity. *The Wisdom of God Manifested in the Works of the Creation* was the result; it was his most popular and influential work.

The book contained the substance of some "Common Places" or divinity exercises delivered in Trinity College chapel when he was a fellow, probably in 1659–60, reinforced by close study of nature in the field and by considerable thought and reading over the years. He acknowledged a debt to several learned contem-

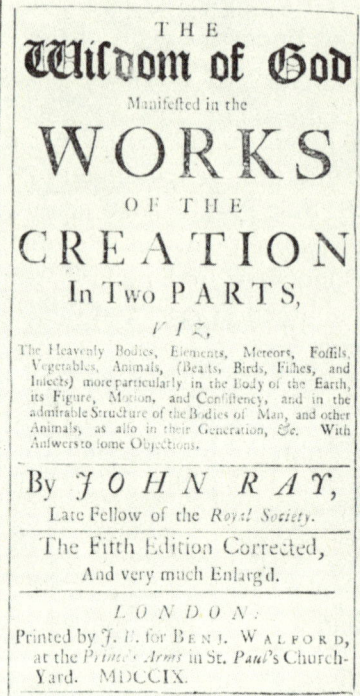

Portrait of John Ray and the title page of *The Wisdom of God,* first published in 1691. Mather drew heavily on Ray's classic statement of the design argument, probably using the fifth edition (London, 1709) shown here. Courtesy of the Rare Book and Special Collections Library, University of Illinois at Urbana-Champaign.

poraries; and in fact his treatise was probably an exposition and commentary on the second book of Henry More's *Antidote against Atheism* (1652), which discusses the phenomena of nature as an argument for the existence of God.[77]

Ray was uniquely qualified to write a book on the interpretation of nature. A disciplined scientist, he had the ability and the determination to separate fact from fiction. A loyal Christian, he worked within the framework of contemporary orthodoxy but was free from the limitations of the old order. Reason was for him a tool in both science and religion, and he refused to accept irrational or supernatural explanations for natural phenomena. For Ray, scientific research had the dignity of a sacred calling; he saw the order of the universe as a manifestation of the mind of God. He was not blind to the weakness of the traditional belief in a direct and omnipotent Creator, and his philosophy did not permit the simplified and crude anthropomorphic teleology of Cicero or of his own contemporary Nehemiah Grew. Ray had faith in the unity and rationality of nature; in Raven's judgment, his book deserves "a primary place in the development of modern science."[78]

The Wisdom of God, first published in 1691, was well received. A second edition, enlarged from 249 to 382 pages, divided into two parts, and provided with a synopsis of the contents, followed the next year. Ray inserted much new material and many personal observations. His references to other scientists, placed in the text, reveal close familiarity with advances in natural history, anatomy, and medicine down to 1692. Two more editions, each containing some new material, appeared in Ray's lifetime. The book remained popular after Ray's death in 1705. Fifteen editions came out between 1709 and 1798; the work was translated into French and German; and two more editions appeared in the nineteenth century, the last in 1845. But by that time William Paley's *Natural Theology* (1802), which imitates and "extensively plagiarise[s]" Ray, had replaced *The Wisdom of God*.[79]

Ray's purpose is to demonstrate the existence of a deity and to illustrate some of his principal attributes as well as to show the adaptation of the parts of animals to various uses. His text is Psalm 104:24: "How manifold are thy Works, O Lord! In Wisdom hast thou made them all."[80] Accordingly, he first describes the number of celestial and terrestrial bodies and then declares that all are wisely contrived and adapted to ends both particular and general.

Before expounding this theme, Ray explains and rejects the theories of atheistic persons who have accounted for the formation of the universe by mechanical hypotheses. He refutes the Aristotelian, Epicurean, and Cartesian hypotheses, arguing not only that God established the laws of motion but that he executes them by a plastic nature. This Platonic idea of an indwelling vital principle that remains active in the physical universe was given full expression by the Cambridge Platonists in the seventeenth century, shortly before Ray published *The Wisdom of God.* Ray also refutes the Cartesian idea that the soul of brutes is material and that animals are mere automata.

Ray then turns to the visible works of God in the creation, which he divides in accordance with commonly received opinion. His arrangement dates from antiquity and is entirely traditional. He begins with inanimate bodies, subdividing them into celestial and terrestrial. He treats heavenly bodies first, accepting the new hypothesis of Copernicus and Galileo. Nothing in the heavens argues chance; on the contrary, everything is the effect of wisdom.

Moving to the terrestrial plane, he first deals with inanimate simple bodies, commonly called the four elements—fire, air, water, and earth. Then he treats inanimate mixed bodies, starting with the imperfectly mixed, that is, meteors, by which he means rain and wind, and proceeding to the perfectly mixed, that is, stones, metals, and minerals.

He proceeds to animate bodies, where he is more at home and speaks with authority. He first treats those of vegetative souls, that is, vegetables or plants. Rays asks such fundamental questions as why plants grow as they do, explaining that there are limits that each plant cannot exceed. "You can by no Culture or Art extend a *Fennel Stalk* to the stature and bigness of an *Oak.*" The reason is "some intelligent *plastick Nature,* which may understand and regulate the whole Oeconomy of the Plant."[81] Then follows an account of plant physiology, a description of the means of adaptation of plants, and a description of nature's care for the propagation of plants.

Ray passes next to animate bodies with sensitive souls, that is, animals. He makes a number of general observations, considering various species only as examples. First, he affirms that nature is not accidental. Providence designs to maintain and continue every species, as the sexual instinct illustrates. Second, he describes strange instincts which demonstrate that animals are directed to ends unknown to them by a wise Superintendent. Third, he notes the care taken for preservation of the weak. Fourth, he treats the relation of form to function in various animals—e.g., the snout of a swine, the tongue of a woodpecker. Since no creature is only volatile, as Aristotle observed, Ray rejects as a fable the story of the Bird of Paradise, which was generally accepted in former times. It is now known that these flying animals have legs and feet as well as wings. Fifth, he discusses the uses of animals. It may be objected, he writes in refuting a crude anthropomorphic teleology, "that these Uses were not designed by Nature in the formation of the things; but that the things were by the Wit of Man accommodated to those Uses."[82]

Ray then interprets the divine intention in an eloquent statement in which the Almighty speaks to man about how his life and character are conditioned by his environment. I have placed you in a spacious world and endued you with the ability to understand what is beautiful and proportional, he has God say, and I have made the earth capable of improvement by your industry. I have made you a sociable creature for improvement of your understanding and have implanted in your nature a desire to learn about other countries and peoples. I am pleased with your industry in turning a barren and desolate wilderness into a polished and civilized country. "If a Country thus planted and adorn'd, . . . be not to be preferred before a barbarous and inhospitable *Scythia*, . . . then surely the brute Beasts Condition, and manner of Living . . . is to be esteem'd better than Man's, and Wit and Reason was in vain bestowed on him."[83]

This passage, Raven writes, "represents an attitude towards Nature which is new and vastly important. This delight in the worth of the world as aesthetically satisfying, intellectually educative and spiritually significant reflects the best Hebrew, Greek and Christian thought, but is in strong and striking contrast to the philosophy and religion both of the Catholic and the Protestant tradition." The antithesis between natural and supernatural had led Christendom to a scorn and depreciation of nature, Raven adds, and as a result a radical difference between nature and grace had prevailed from the third to the sixteenth centuries. According to Raven, "the direct insistence upon the essential unity of natural and revealed, as alike proceeding from and integrated by the divine purpose, had not found clear and well-informed expression until Ray's book was published."[84]

Ray concludes the first part of *The Wisdom of God* with a defense of scientific studies and a plea for their extension, the substance of which was apparently a commonplace delivered to students at Cambridge thirty years earlier. He warns against the study of languages, history, and antiquity to the neglect of natural history and the works of creation. We need to converse with nature as well as books. "Let us not think that the Bounds of Science are fixed like *Hercules*'s Pillars, and inscrib'd with a *Ne plus ultra*. Let us not think we have done, when we have learnt what they have delivered to us. The Treasures of Nature are inexhaustible." Ray then quotes the Roman writer Seneca: *"The People of the next Age shall know many things unknown to us."* Scripture, he adds, calls upon man as a rational being to praise the power, wisdom, and goodness of God manifested in the creation of other beings serviceable to man.[85]

Ray opens the second part of his book with an essay on the terraqueous globe. He defends the "new hypothesis" that the earth moves and discusses both the waters and the dry land, including the uses of mountains. Turning finally to man, he offers eight observations to show the perfect adaptation of the body to its functions, and continues by discussing the various organs separately, devoting special attention to the eye and to spontaneous generation, belief in which he strongly denounces. The book closes with reflections on the religious consequences of the conclusions reached—thankfulness for the perfection of the body, the duty of serving God, and the value of our souls.

"More than any other single book," Raven declares in a fulsome encomium, Ray's *Wisdom of God* "initiated the true adventure of modern science." The novelty of the treatise was twofold. Ray not only presented an interpretation of the significance of organic life, as opposed to merely describing and classifying it, but he also freed himself from the old antithesis between the natural and the supernatural which had led Christendom to depreciate nature.[86]

Soon after the appearance of *Wisdom,* Ray wrote another book on the interpretation of nature. His work in botany and zoology was mainly concerned with description and classification; as such it lay outside speculative science. But his incursions into geology and his study of the nature of fossils plunged him into controversy, which became lively when rejection of the idea of spontaneous generation caused contemporary investigators to distinguish between the living and the nonliving and when the serious study of geology raised problems of cosmogony. Ray had begun his study of geology in 1661; he wrote on the subject both in his *Observations* (1673) and in a long letter to Tancred Robinson in 1684. Breaking with tradition and correcting other scientists who contended that petrified objects found in the earth were "formed stones," Ray argued that fossils were the organic remains of formerly living plants and animals. Their production was due not to a plastic virtue inherent in the earth but to living organisms. In 1681 Thomas Burnet published his *Telluris theoria sacra,* a book that attained great popularity and appeared in an English version, *The Theory of the Earth,* dedicated to the king, in 1684 (and later published as *The Sacred Theory of the Earth*). Burnet's views gave rise to both criticism and rivalry. Interest in the subject was great, and Ray, emboldened by the success of his *Wisdom of God,* decided to produce his own treatise. He published *Miscellaneous Discourses Concerning the Dissolution and Changes of the World* (1692) with the imprimatur of the Royal Society and a dedication to John Tillotson, Archbishop of Canterbury. The book soon sold out, and a second edition, much enlarged, quickly followed. Ray altered the title to *Three Physico-Theological Discourses* (1693). His discussion of the primitive chaos and creation of the world, the general deluge, and the dissolution of the world and future conflagration enabled him to combine cosmogony and geology in treating such topics as the Flood, fossils, the evidence of stratification in the earth, the origin of springs, earthquakes, and the uses of mountains. Ray's *Physico-Theological Discourses* became widely known.[87]

Despite poor health in his later years, Ray remained vigorous until his death in January 1705. His life's work was a reinterpretation of the Christian faith in the light of a sound knowledge of nature. Focusing attention on many fascinating subjects, he convinced the champions of the old outlook that such attention was a legitimate field for Christian inquiry, and he encouraged less adventurous churchmen to support the claims of the new philosophy. As a result, Raven concluded, Britain enjoyed "a century and a half of scientific progress undisturbed by theological controversies and fostered by the spokesmen of religion. . . . During that time there was developed a type of theology, of which *The Wisdom of God* is the first example, capable of giving appropriate expression to the Christian faith in a scientific age. This is John Ray's proper memorial.[88]

Ray's speculative treatises strongly influenced William Derham. Born at Stoulton in Gloucestershire in 1657, Derham entered Trinity College at a time when scientific interest was strong in Oxford. He took his B.A. in 1679, was ordained priest in 1682, took his M.A. in 1683, and became vicar of Upminster in Essex in 1689. During a long tenure there he collected statistics, recorded weather data, conducted experiments, occupied himself with botanical and astronomical observations, and published papers on various natural phenomena. In 1696 he published his first book, *The Artificial Clock-maker*. After Ray's death, he wrote a life of the naturalist and published his *Select Remains*. Ray's *Wisdom of God* forms the basis of Derham's Boyle lectures in 1711–12, which attempted to put "philosophical" or scientific matters to theological uses. They were published as *Physico-Theology: Or, a Demonstration of the Being and Attributes of God, from His Works of Creation* (1713). This book proved very popular; it went through thirteen editions by 1768 and was translated into at least five other languages by the end of the century. "What Christian lives," asked an admirer in 1783, "who can refuse his reverence to Derham's learning?" Derham followed up his success with *Astro-Theology: Or, A Demonstration of the Being and Attributes of God, from a Survey of the Heavens* (1715). A year later he became canon of Windsor.[89]

After Derham and Ray, Mather drew most heavily upon John Harris. Born about 1666, Harris received both bachelor's and master's degrees from Oxford. He taught mathematics while a student there and again in later life. He took holy orders and obtained the degree B.D. at Cambridge in 1699 and the Lambeth degree of D.D. in 1706. A pluralist clergyman who enjoyed the patronage of Sir William Cowper, he held various positions in the Church of England. Like many Anglican clergy of the time, he showed a lifelong interest in natural philosophy and published works which led to his election to the Royal Society in 1696. He used Newtonian natural philosophy and other recent scientific discoveries, including the work of John Ray, to show the harmony between science and religion. Harris defended John Woodward's ideas on geology against Martin Lister and others in *Remarks on Some Late Papers Relating to the Universal Deluge, and to the Natural History of the Earth* (1697). In his Boyle lectures in 1698, he attempted to refute atheistical objections against the being and attributes of God.

Harris's most famous work was *Lexicon Technicum: Or, An Universal English Dictionary of Arts and Sciences* (1704; a second edition in two volumes appeared between 1708 and 1710). This was the first purely English general encyclopedia as well as the first encyclopedia to emphasize science. The *Lexicon* contained alphabetically arranged entries of varying length on a number of diverse fields in what would today be called the humanities, social sciences, and law; its emphasis was on modern scientific advances and technology. Harris drew on earlier reference works but mainly on the published works of leading contemporary scientists, especially fellows of the Royal Society. The science covered most thoroughly is mathematics and its practical applications. In astronomy and some areas of physics, Newton is the authority; in other areas of physics Harris draws chiefly on the work of Boyle, Hooke, and his own research. The authority on botany and the classification of plants and animals is Ray. In geology Harris drew primarily on John Woodward. His discussion of medicine is disappointing, but he is more satisfactory on anatomy, physiology, and surgery.[90]

Mather bases 1,263 lines, over 11 percent of his book, on Harris's encyclopedia. Some of his short essays on the planets are taken completely from the *Lexicon*. Since Mather mentions Harris by name only once, on a substantive point, and since he readily identifies Ray and Derham as major sources, it seems clear that he consciously sought to hide his reliance on Harris.

His silence raises intriguing questions about motives and intended audience. Potential English readers of his book are likely to have known the *Lexicon Technicum,* which was a commercial success; surely they would have regarded Mather as more generous if he had acknowledged "technical Harris" as a source. Americans acquainted with the physicotheological literature of the time often owned the books by Ray and Derham, but few American libraries of the early eighteenth century owned Harris's bulky encyclopedia. Samuel Sewall ordered the work from his London agent in 1711,[91] but the Mather library apparently lacked a copy. It is possible that

Mather viewed his massive borrowing as in no way objectionable. Since Harris himself was a borrower, Mather may have treated the work as a compendium without an author. In any event, he ransacked Harris. One can speculate: perhaps Mather hoped to appear as an expert to fellow Americans by mining the *Lexicon Technicum* without attribution. But one cannot be certain as to why he acted as he did.

After Derham, Ray, and Harris, who together account for about 43 percent of the contents of *The Christian Philosopher,* Mather drew most substantially upon George Cheyne and Nehemiah Grew. Cheyne was born in Aberdeenshire and received a classical education with a view to entering the ministry, but he turned to medicine under the influence of the iatromathematical school, which made medicine a branch of mathematics and drew close analogies between the human body and a machine. In 1702 Cheyne moved to London, was elected a fellow of the Royal Society, and established a medical practice. A prominent spokesman of "Newtonianism," he described the advances in astronomy and physics made by Newton and explored their theological significance in *Philosophical Principles of Natural Religion* (1705). Mather drew on it for 743 lines or nearly 7 percent of his manuscript.[92]

Nehemiah Grew was educated at Pembroke Hall, Cambridge, and earned a medical degree at the University of Leiden. Upon returning to England he relied primarily upon the practice of medicine as a means of livelihood for the rest of his life. He was interested in the structure of both plants and animals; his religious and philosophical beliefs led him to regard both as "the Contrivances of the same Wisdom." An essay he wrote on plant anatomy led to his election as a fellow of the Royal Society in 1671. He moved to London the following year and made the Royal Society the center of his activities. He applied the mechanistic science of his day to the study of botany and attempted to find a mechanistic cause for each phenomenon. Grew's first important book was *The Anatomy of Vegetables Begun* (1672). His outstanding accomplishment was *The Anatomy of Plants* (1682). Grew's last work was *Cosmologia Sacra: Or, A Discourse of the Universe As It is the Creature and Kingdom of God* (1701). It covers the same ground as Ray's *Wisdom of God* and its arguments are often identical, but its general philosophy is different from that of Ray. Grew exhibits a simplified, crude teleology. Mather relied on Grew's treatise for 394 lines or 3.5 percent of his book.[93]

The next five persons in the order of their quantitative importance as sources for Mather were Johann H. Alsted, Joseph Walker, Edmond Halley, Johann Arndt, and Nathaniel Wanley. Little is known about Walker, who translated out of the French a work published in London in 1684 under the title *Astronomy's Advancement*. Mather drew on it for 178 lines, all in the essays dealing with heavenly bodies. Mather never names Walker, but on one occasion he cites the title of Walker's book [32:79].

Edmond Halley demonstrated a wide interest in science, but his most notable achievements were in astronomy. He edited the *Philosophical Transactions* of the Royal Society from 1685 to 1693 and published eighty-four papers in that journal. He also edited *Miscellanea Curiosa,* a collection of articles, including many of his own, that had first appeared in the *Philosophical Transactions.* Mather levied on Halley's *Miscellanea Curiosa,* published in three volumes from 1705 to 1707, with a second edition in 1708, for 159 lines of his book.

Nathaniel Wanley was a Cambridge-educated clergyman who composed various works, the most important being *The Wonders of the Little World: Or, A General History of Man* (1678). His book anecdotally illustrates the prodigies of human nature; Mather pillaged it for 103 lines, all in his essay on man.

Both Alsted and Arndt were dear to Mather, but a purely quantitative assessment of their importance as sources—Mather based 193 lines of his text on Alsted and 114 lines on Arndt—would be misleading. He cites them repeatedly, often by name but more often anonymously, presumably in order to minimize his repeated references to fellow clergymen and to make it appear as if his circle of authorities were larger than it was. They played no part in scientific advancement during the Renaissance and Reformation, but they believed in the harmony between science and religion. Mather used them to buttress his own case.

Johann Heinrich Alsted (1588–1638) is cited eighteen times, with eleven of the references made anonymously (e.g., "An excellent Person," "A famous *German* Doctor of Philosophy"). A native German, Alsted was educated under the influence of Calvin and of Petrus Ramus, a French anti-Aristotelian famous for ordering knowledge in "logical" form according to a dialectical "method."

He held professorships of philosophy and of theology at Herborn Academy before leaving Germany during the Thirty Years' War to head a school in Transylvania. Alsted believed in the fundamental unity of divine and secular knowledge, and he covered nearly every branch of learning. In science he favored a "Christianized" Aristotelianism and declared the Copernican system unacceptable because Scripture and common sense refuted it. Most of his writings were in theology; Mather drew most heavily on his *Theologia naturalis exhibens augustissimam naturae scholam: In quo creaturae Dei communi sermone ad omnes pariter docendos utuntur: Adversus atheos, Epicureos, et Sophistas huis temporis* (1623). His book on millennialism, published in 1627, inspired within English Puritan circles the belief that the reformation of the church would be accompanied by a reformation of learning. Alsted's fame rests ultimately upon his role in perpetuating the idea of an encyclopedia of all human knowledge. This medieval notion was laid on new foundations by Ramus, after whose death Germany became the center of Ramist method, with its diagrammatic approach to knowledge. Ramist influence was represented on a monumental scale by Alsted, whose massive compends of human knowledge mark the beginnings of the first modern encyclopedias. Alsted published an elephantine *Encyclopaedia scientiarum omnium* at Herborn in 1630; Mather praised the four volumes as "a *North-West Passage*" to "all the *Sciences.*"[94]

Johann Arndt (1555–1621), a Lutheran theologian and mystical writer, is drawn on seventeen times, with eleven quotations from this author being made anonymously (e.g., "One," "A devout Writer," "A *German* Writer"). Arndt served in various pastorates and as a superintendent of ecclesiastical affairs at Celle, but his uncompromising Lutheranism aroused hostility. He was steeped in the tradition of the great medieval mystics and sought to awaken German Lutherans from an orthodox formalism to an inward sense of a true, living, and active faith. His most important writing is *Vier bücher vom wahren Christenthum*, the first German Lutheran devotional book for the people and one of the most useful books of devotion produced by Protestants. The first book was published at Jena in 1605, the complete work in four books in 1609 (two more books were later added, but they are in the nature of an appendix). The work was translated into Latin as *De vero Christianismo libri 4* (1625). The first book of *Wahre Christenthum* was translated into English in 1646; a complete translation by Anthony W. Boehm was published as *True Christianity: Or, the Whole Economy of God towards Man and the Whole Duty of Man towards God* (1712). This devotional treatise was also translated into many other languages. Arndt's book aroused opposition among theologians but was eagerly read and performed a valuable service in revitalizing Christianity. German Pietists greatly valued Arndt; Mather deeply sympathized with his emphasis on the work of Christ in the human heart. Mather quotes mainly from book four of *De vero Christianismo*, which treats natural theology.[95]

The thirty-two essays and three appendixes in *The Christian Philosopher* are organized around topics that would now be called astronomy, physics and cognate topics, and the life sciences. Mather's sources in each of these areas deserve comment. He devotes twelve essays, about 15 percent of his book, to the heavenly bodies. His original composition amounts to about 23 percent of the 1,610 lines in this part of his treatise. The sources he relied on most frequently in discussing astronomy are Harris's *Lexicon Technicum*, Cheyne's *Philosophical Principles of Natural Religion*, and Walker's *Astronomy's Advancement*. Not one line of Mather's own composition is found in the essays on Jupiter, Mars, and Venus; only three of his lines appear in the essay on Saturn. The essay on Mars is taken entirely from Harris; those on the other planets named derive from multiple sources.

From astronomy Mather turns to physics, but physics in the broadest sense. Fourteen of the essays, about 23 percent of the volume, are devoted to earthly phenomena. Mather himself composed 25 percent of the 2,574 lines in these essays. About 47 percent of the remainder is drawn from six sources. In quantitative terms, by far the most important of these is Harris's *Lexicon Technicum*. The others in descending order of usefulness to Mather are Ray's *Wisdom of God*, Cheyne's *Philosophical Principles*, Derham's *Physico-Theology*, and Halley's *Miscellanea Curiosa*. Mather drew directly on Robert Boyle for only seventy lines of text, most of which deal with the cold, while the remainder relate to magnetism.

Mather's interest quickens when he turns to what is today called the life sciences, what he calls the vegetable and animal kingdoms and man. He assigns seven chapters to these topics. They account for a total of 6,878 lines or 61 percent of the entire book. The essay on man alone runs to 3,161 lines and takes up over 28 percent of the volume. Mather's hand composed 1,314 lines or 19 percent of this material on the life sciences. In this section he borrows most heavily from that "delightful pair of brothers," drawing upon Derham for 1,974 lines and Ray for over 1,309 lines. To a much lesser extent he also levies upon Harris, Cheyne, and Grew. He uses Wanley only in the essay on man.

A large portion of Mather's own composition consists of introductory passages, rhapsodies, and a conclusion that points the religious significance of the volume. Much of the remaining material Mather culled from his own earlier writings. He recycled some lines from his "Curiosa Americana." His description of the rattlesnake [180:3–12] is a condensed version of the "Curiosa," first series, no. 11, sent to the Royal Society on 27 November 1712. He reused three items from the second series, sent to the Royal Society in 1713 and 1714. His report on a snow of wool [68:26–69:30] was originally no. 1 in this series. The discussion of lunar influence [60:70–82] is a nearly verbatim reprint of no. 2, while the account of vast numbers of pigeons [204:22–207:41] is from the second series, no. 5.[96]

The larger part of the essay on thunder and lightning paraphrases Mather's *Magnalia Christi Americana* (1702); that work in turn reprints one of his earlier publications entitled *Brontologia Sacra: The Voice of the Glorious God in the Thunder* (1695). In the *Magnalia* Mather describes a punishment known as *combustio animae,* or death by lightning, a description repeated in 1712 in the "Curiosa Americana," first series, no. 12, and used again in the paraphrase of the *Magnalia* in *The Christian Philosopher* [72:47–73:58].[97]

Mather's essay on the rainbow reprints, with the addition of some lines from Halley and Harris, his own *Thoughts for the Day of Rain* (1712). In his essay on fishes [191:18–20], Mather borrows a Latin phrase that he had used in a sermon which he published anonymously as *The Nets of Salvation* (1704).

Although Mather often makes his prose do double or even triple duty, two of his observations on American science in *The Christian Philosopher* are fresh and original. His description of cold in New England [82:73–81], and a brief comment on the productivity of corn [134:90–92] do not seem to turn up again elsewhere.

Just as his earlier composition became grist for *The Christian*

Philosopher, so too some of his original material in that work appeared again in later writings. He took four items from his book for letters to the Royal Society. One was his account of the nidification of birds [201:1–15], sent to John Woodward on 4 July 1716 as the third series, no. 3 of his "Curiosa Americana." Another was his description of experiments with corn and squashes [131:17–35], sent to James Petiver, a fellow of the Royal Society and a close associate of John Ray during Ray's last years, on 24 September 1716 as the third series, no. 12 of his reports on American science. This brief account is a pioneering contribution to the study of plant hybridity. Another was his report of a prodigious 150-foot worm thrown up by a man which he claims to have seen with his own eyes [168:64–70], sent to London on 12 December 1717. Finally, in a letter sent to the Royal Society in 1720, Mather recycled at least the lines of verse from his account of whales [189:53–58]. In addition, Mather copied a long extract on the strength of humans, a commentary on Judges 15:20, from *The Christian Philosopher* into the manuscript of the "Biblia Americana" as he continued making additions to that work after he completed his book on Newtonian science. He also incorporated passages on vegetables, worms in the human body, the eye, ear, and tongue nearly verbatim from *The Christian Philosopher* into his medical treatise, *The Angel of Bethesda.* This material is identified in the Endnotes,[98] which also contain further details regarding Mather's dependence upon other authors and his own original composition in *The Christian Philosopher.*

Some additional observations on his use of sources deserve attention here. Consider his reliance on Scripture and the ancient classics. He quotes and paraphrases the Bible throughout his discourse. Sometimes he simply incorporates a scriptural passage used by an intermediate source, but usually he himself supplies the biblical text. Apparently he knew the Bible so well that he could summon up passages pertinent to his literary purposes at will. Mather was also thoroughly familiar with the authors of classical antiquity, as we saw earlier, but he took the vast majority of his Latin quotations from intermediate sources rather than citing them from memory or from the original works.[99]

An ardent Christian who hoped for the conversion of the Jews and shared the negative attitude toward Islam which prevailed in Western culture at the time, Mather nevertheless drew upon the literature of both of these faiths. He cites Jewish authors and legends on a number of occasions. These brief but frequent references are most likely the product of a knowledge of Judaism and rabbinical writings acquired over a long period of time. Interspersed throughout his book, they are probably taken from many authors, and it is difficult if not impossible to trace his sources. Mather's quotation from "a *Mahometan* Writer" [111:7–12:39] is readily traceable to a Western translation. His discussion of "the *Mahometan Philosophy*" concerning the physical universe [110:57–111:89] draws upon the Koran in part, but it is impossible to determine the source of the rest of this passage.

Despite Mather's extensive pillaging of English writers on physicotheology and of many other learned men, his firsthand knowledge of books and authors is impressive, especially when one recalls that he lived on the periphery of the great English and Continental centers of learning. To cite some examples, he knows and quotes from the following: Jacques Gaffarel (1601–81), a French priest and author of a book on curiosities; François de Salignac de La Mothe Fénelon (1651–1715), archbishop of Cambrai, a prolific author famous for *Les Aventures de Télémaque fils d'Ulysse* (1699) who also wrote an influential treatise on physicotheology, demonstrating the existence of God based on the study of nature; Jean Mercier (d. 1570), a French commentator on Genesis; Paul Eggers (1570–1655), a German Protestant who was an ardent Pietist; Johann Bicker (fl. c. 1612), a German who wrote on astrological medicine; Otto de Guericke (1602–86), a German who invented the air pump; Samuel Bochart (1599–1667), a French Protestant theologian and orientalist whom Mather frequently quotes in his "Biblia Americana"; Girolamo Rorario (1485–1556), an Italian diplomat who wrote on reason in brute beasts; Philipp Camerarius (1537–1624), a German jurisconsult whose work touched on the same subject; Giovanni Piero Valeriano Bolzani (1477–1560), an Italian who published a book entitled *Hieroglyphica* (1556); Raymond of Sebonde (d. 1436?), a medieval Spaniard who wrote a book on natural theology; Matthaeus Tympe (fl. 1588–1618), a Roman Catholic priest and professor who wrote on theology; the *Corpus Juris Civilis* of Justinian; Jacques Cujas (1522–91), a French expert on Roman law who commented on the Code of Justinian; and, most curiously, Camillo Baldi (1550–1637), a professor at the University of Bologna who wrote a commentary on the physiognomy attributed to Aristotle.

As the Endnotes show, I have been able to identify, line by line, the sources of most of Mather's *Christian Philosopher*. Unfortunately, a few passages still elude me. Among the most challenging are a description of the Portuguese Inquisition [40:12–18], a Latin definition of the rainbow [63:54–57], a quotation on flying toward heaven [91:40–49], a quotation about the abyss [109:29–110:42], a bishop of Paris on the spider [170:10–11]; lines of verse on the whale [189:53–58], the identification of the horses Bucephalus and Lethargus as well as the source for these lines [228:30–230:33], a Latin quotation on the eye [258:92–94], a quotation from the *Asceticks* [296:65–297:77], and the work in which John Edwards writes on the stench of sick rooms [302:22–35].

While we acknowledge the derivative character of his treatise, Mather nevertheless deserves to be remembered as the author of *The Christian Philosopher*. He was closely in touch with the new science associated with Newton and Ray, and at a time when it was difficult to distinguish between science and superstition, he was fairly discriminating in his choice of authorities. His book is therefore the best record we have of an advanced understanding of the roots of science in America in the early eighteenth century.

5. The Contents of *The Christian Philosopher*

The Christian Philosopher is important mainly for its argument, which is the harmony between science and religion.[100] Mather expounds this thesis within the framework of the ancient metaphor of the two books—that God reveals himself in both nature and the Scriptures. He also employs a core doctrine of Western thought to advance his views—the argument from design, which, as noted earlier, is a form of natural theology.[101] In all this, his greatest intellectual debt is to a tradition that runs from Plato and Augustine through the Cambridge Platonists to Newton and the physico-theological writers who followed in his wake.

This tradition begins with pagan authors in classical antiquity. Plato inferred the existence of God from the harmony and beauty of the heavens and from general consent, and he used the design argument to combat the atheistic notion that the world was mere mechanism or the product of chance. He went far to identify divinity with the processes of nature, insisting that nothing but good may be traced to God. Plato argued for the priority of Spirit. Nature is a system of symbols reflecting the nonsensible realities that lie beyond it. The visible world, patterned on the universal idea of the living thing, is a product of intelligent design and benevolent purpose. The soul is the source of motion in the material world; all corporeal movements are dependent upon its motions. Plotinus elaborated the philosophy of Plato as a basis for Hellenism in its struggle with Christianity, and Neoplatonists devised the concept of the scale of emanations which connected the supreme good with the whole hierarchy of imperfect "images."

The Christian church became the heir of Platonic philosophy as Christian apologists appealed to Plato in their controversy with the gentiles. Origen extravagantly read the eschatology of Plato's myths into Christianity, while Augustine was instrumental in importing Platonism into the main current of Christian orthodoxy. The Augustinian doctrine, derived from Plato and influenced by Neoplatonism, held that God is the author of both nature and Scripture. God reveals himself through the symbolism of nature, which fallen humanity can never adequately decipher, and through Scripture, which is authoritative. The natural and the supernatural must be concordant. Through nature the invisible things of God are made known by things that are visible.[102]

Theologians had long speculated on the attributes as well as the existence of God, and theorists of design devised the convention of describing divinity in terms of three qualities—omnipotence, omniscience, and benevolence. This practice was probably very old; it is already evident in Raymond of Sebonde, a medieval Spanish physician, philosopher, and theologian whose *Theologia naturalis* (1484) was the first book to use the words "natural theology" as a title. It became customary for writers to describe the physical universe as demonstrating the power, wisdom, and goodness of God.[103]

Thus, the argument from design or final causes was deeply embedded in Western thought by the time that modern science began its takeoff, starting with the work of Copernicus in 1543. The late seventeenth century was the most fertile period in the history of English science up to that time. A core of serious investigators along with numerous virtuosi pursued knowledge of the natural world primarily for its own sake and also for religious enlightenment. Robert Boyle and the Cambridge Platonists had a hand in developing the mechanical philosophy, the most influential doctrine of the scientific revolution, but Isaac Newton was most closely identified with it. His *Philosophiae Naturalis Principia Mathematica* (1687) put a capstone on the design argument. Here Newton described the world system in mechanical and mathematical terms: physical nature is a machine, but the mathematical order evident in the world implies a designer. Secondary causes do not explain the universe sufficiently; a first cause or intelligent agent is needed. As Newton said, "this most beautiful system of the sun, planets, and comets, could only proceed from the counsel and dominion of an intelligent and powerful Being."[104]

The idea of design was abstract as applied to the inorganic world, whereas it was linked to purpose and served an explanatory function as applied to the organic world. The idea of a first cause, God, was invoked as the ultimate explanation of the adaptation of organic beings to their environment. Adaptation had to be by design. The Cambridge Platonists saw an element of imperfection in the processes of nature; to combat Cartesian mechanism they drew on Platonic and Christian sources to develop the idea of God's continuous operation within the universe. Their concept of plastic nature described a life force that operates blindly in the organic world to accomplish divine ends. John Ray, the leading spokesman of biological vitalism, viewed all living creatures as to some degree free and autonomous. Newton came under the influence of these ideas. He brought purpose into nature by arguing that since the bodies of animals show an amazing degree of contrivance for certain goals and uses, a contriver must have constructed them in such ways.[105]

Scientists and theologians cooperated in elaborating the design argument in terms of the new mechanical philosophy. Many works of physicotheology were published starting in the late seventeenth century, with England producing far more than the Continent. John Ray's *Wisdom of God* was perhaps the most influential book of this type; many similar works followed in the next quarter-century.

Some of the most important physicotheological treatises originated as Boyle lectures. These lectures, endowed by Robert Boyle through a codicil to his will in 1691, provided a stipend for annual selection of a clergyman to preach eight monthly sermons in a fashionable London church "for proving the Christian religion against notorious Infidels"—that is, atheists and non-Christians. The lectures were entrusted to four trustees whose theological tenets covered a broad spectrum. The early lectures were from all levels of the Anglican clergy. Between 1692 and 1713 the most important and famous lectures were the ones by Richard Bentley, Samuel Clarke, and William Derham, and these men incorporated the new natural philosophy associated with Newton and Ray to defend Christian theology. The published versions of their sermons were extremely popular and widely read in England during the eighteenth century.[106]

These lectures used the design argument triumphantly against atheism, which took many forms at the time. It was a compound of elements drawn from Hobbes, Descartes, Spinoza, and Gassendi. The main offender was Hobbes, who was widely believed to have espoused mechanistic materialism. He was regarded as the chief source of irreligion. Descartes, whose influence was not widespread in England, was suspect for strengthening the materialist, mechanical view. Benedictus Spinoza was seen as a patron of atheism because he taught there is no other God but the universe. The French priest Pierre Gassendi contributed to the climate of atheism by reviving the doctrines of the ancient atomists, Epicurus and Lucretius, who held that the universe had evolved from the random scramble of small, hard particles in a void. Gassendi, however, posited a God who created the atoms and continually regulated them. Many scientists and latitudinarians considered it vital to refute those who viewed the world as a machine governed solely by natural causes. Scientific iconoclasm, the tendency to preoccupy oneself with secondary causes rather than first or final causes in explaining natural phenomena, also was regarded as atheistic. Radical free thought posed another danger of atheism, and opponents combated its various forms—pantheism, enthusiasm, and deism.[107]

In addition, proponents of Newtonian science promoted natural theology as part of their social ideology. The new science aimed at a broad range of reforms, and Anglican clergy were deeply conditioned by Protestant millenarianism, which was largely the influence of Alsted. Latitudinarians believed that they were preparing for new heavens and a new earth, and that Providence requires harmony between the natural and human worlds. Atheism threatened a breakdown of public order; thus influential low church leaders, followers of Newton, promoted natural religion as a basis of national unity. People must cooperate with Providence, they insisted, in effecting social reconstruction.[108]

The Christian Philosopher is avowedly a treatise on the design argument, and Mather introduces the related concept of natural theology at the very beginning of his book by quoting a Mahometan writer. The Mahometan is Abu Bakr Ibn Al-Tufayl, a twelfth-century Arab-Spanish physician who wrote one of the most celebrated books of the Middle Ages. This philosophical romance, translated into English as *The History of Hai Eb'n Yockdan, An Indian Prince: Or, the Self-Taught Philosopher* (1686), describes a religion of nature. The hero of the book, cast on a desert island as a child, grows up to discover the sciences, the existence of the soul, and God by simply contemplating the physical creation. When, in mature life, he comes across the revelation in the Koran, he is willing to accept it as merely confirming what he has already learned by the light of nature.

In turning to the sciences, Mather treats astronomy first. He does so because this was the conventional order in the study of natural philosophy dictated by tradition, not because Newton's great discoveries gave astronomy much prestige in the early eighteenth century. Mather's essays on the heavenly bodies deal with light, the stars, the fixed stars, the sun, the planets (the order of treatment is Saturn, Jupiter, Mars, Venus, and Mercury), comets, heat, and the moon. New England Puritans had long accepted the Copernican theory; Mather took it for granted.[109] No copy of Newton's *Principia* was available in New England when Mather wrote, but in any event astronomy involved complex mathematics that were beyond him, and his knowledge of Newton depended on intermediate sources.

Mather begins by suggesting that it would be proper to lay down the laws of nature by which the material world is governed, laws which "we have in the Rank of *Second Causes*," and go no further. For if we lay down the basic laws of nature, then "all *Mechanical Accounts* are at an end; we step into the Glorious GOD *Immediately*," and the very next thing to do is "to Acknowledge Him, who is the *First Cause of all*" [18:35–41]. These laws, which Mather states without naming their author, are Newton's three laws of motion and a corollary on the principle of universal attraction. For Mather, these are the laws of God, who formed all things.

Mather next discusses the physical properties of light. He mentions Römer's demonstration of the velocity of light; then, to offset surprise at such swift motion, he summons four Renaissance scientists—John Dee, Simon Stevin, Marin Mersenne, and Jerome Cardan—to recount some remarkably slow motion: e.g., a constantly moving wheel that took seven thousand years to complete its revolution. The essay names Isaac Newton in passing [20:85], and ends with a blaze of light imagery drawn from Scripture.

The Copernican revolution led to the concept of an infinite universe, and in treating the fixed stars and the moving stars Mather considers the implications of the concept for religious belief. He may have been familiar with Fontenelle's *Plurality of Worlds* (1686), which appeared in English translation in 1687 and popularized the new cosmology in dialogue form;[110] he certainly knew Huygens's *Celestial Worlds Discover'd* (1698). The fixed stars may be like our sun, Mather writes, and their planets, analogous to our system, may have inhabitants both rational and irrational.

An infinite universe was a novel idea, for the Hebrew Scriptures were based upon the earth as the center of the universe, and the universe as built for man. The problem of size and purpose suggested the unspeakable insignificance of man in a plenitude of space. Mather acknowledges these revolutionary possibilities with equanimity, but immediately quotes George Buchanan, a renowned Scottish Renaissance humanist, in order to preserve perspective. Buchanan drew much of his knowledge in astronomy from classical authors whose texts were newly printed in Greek and Latin editions starting in the last third of the fifteenth century, but he also read modern writers. While praising observational science, he rejected Copernicus and Tycho Brahe. His *De sphaera* (1584), a hexameter poem in five books, defended the older Ptolemaic cosmology and the central position of man in the creation. Mather quotes lines from this work which assert how small the earth would seem looking down from the lofty height of the heavens, if it could be seen at all from that vantage point. Like Buchanan, Mather emphasizes man's supreme dignity and moral responsibility in a finite rather than an infinite universe. Then he quotes Plato, albeit anonymously, to emphasize that the world is the temple of God.

Mather goes on to describe the stars in the firmament. The

Introduction lxxv

ancients reckoned only 1,022 stars in 50 constellations, whereas Kepler raised the number to 1,392, and Bayer to 1,709. Mather comments on new and disappearing stars, including the *nova* of 1572, but fails to note that Tycho Brahe undermined old cosmological theories with a brief tract on *The New Star* published in 1573.

The Reformation and Counter-Reformation had spurred attempts to give the constellations Christian rather than pagan mythological names; Mather urges completion of this program. He commends Ambrosius Novidius Fraccus, an Italian author of neo-Latin verse, who made such an effort in 1547, and Julius Schiller, a German lawyer and monk, whose *Coelum stellatum Christianum* (1627) gave the twelve signs of the Zodiac the names of the Apostles. Mather dismisses astrology with contempt, insisting that what one reads in the stars is the power and grandeur of God.

In relating various scientific views on the nature of the sun, Mather reveals his belief as to the age of the universe. If the sun is a burning body, he asks, why has it been so little altered by intense heat for nearly six thousand years? Mather notes that the quantity of light and heat in the sun is daily decreasing by the emission of rays which never return to their source. This evidence shows that the sun could not have existed from eternity, for eternity would have wasted it. Thus he indirectly refutes Aristotle's belief in the eternity of the world.

In discussing the earth's path around the sun, Mather observes that all astronomers before Kepler supposed this orbit a perfect circle, but Kepler proved it an ellipsis. The discovery of elliptical rather than circular orbits dealt a deathblow to the Ptolemaic-Aristotelian cosmology, giving point to the waggish remark that Newton's greatest discovery was Kepler's laws. At this point Mather does not mention Copernicus or *De revolutionibus orbium coelestium* (1543), his famous book on planetary motion and the heliocentric system.

Mather describes each of the planets—Saturn, Jupiter, Mars, Venus, and Mercury—scientifically and offers a religious improvement on the evidence. That one law of motion should be observed in bodies so vastly distant from each other demonstrates that they were all put into motion by the power and wisdom of God. "*Chance,* or dull *Matter,* could never produce such an harmonious

Regularity in the Motion of Bodies so vastly distant: This shews a Design and Intention in the *First Mover*" [46:10–12].

Comets had long attracted attention from both theologians and natural philosophers. Mather's *Boston Ephemeris* (1683) describes the comet he had seen the previous year. Now he observes that prior to Tycho everyone considered comets as nothing other than vapors below the moon. But Kepler discovered a true system of comets: he found that they moved freely through the planetary orbits with a largely rectilinear motion. Mather's view of comets largely follows the work of Newton and Halley, but while he is advanced in this sense he resists the idea of a wholly mechanistic universe. He cites Newton's speculation that vapors from comets may fall into the earth's atmosphere by gravity and nourish vegetation. If the best part of our air comes chiefly from comets as Newton imagined, "the Appearance of *Comets* is not so dreadful a thing, as the *Cometomania,* generally prevailing, has represented it" [52:1–2]. The element of doubt causes him to cite the opposing opinions that comets are either "Ministers of *Divine Justice*" that may bring about the catastrophe of our system or the habitation of animals in a state of punishment [53:49–54].

The essay on the moon is devoted almost entirely to scientific description. Since we should be in danger of being drowned or stifled by a stagnating ocean if the size or location of the moon differed even slightly, Mather remarks, our satellite is most wisely contrived by God for human purposes. The moon has a considerable apparatus for human habitation, but by what creatures it may be inhabited cannot be solved without divine revelation. Mather believes that the moon's influences on sublunary bodies are wonderful. Lunatics are one instance of that effect, husbandmen can testify to others, and, as Borelli observed, medicine for worms in human bodies works when taken in the wane of the moon.

From the heavenly bodies *The Christian Philosopher* turns to terrestrial phenomena, or what would today be called physics, but physics in the broadest sense. As with astronomy, this field was one in which significant advances were being made in the sixteenth and seventeenth centuries. These essays treat rain, the rainbow, snow, hail, thunder and lightning, air, wind, cold, the terraqueous globe, gravity, water, the earth, magnetism, and minerals.[111] As he proceeds, Mather offers his own interpretation of design to underscore the self-evident character of the facts. With single-mindedness, he selects evidence to advance his thesis while ignoring anything to the contrary. The world he portrays is sunny, one in which everything is designed to promote human welfare. The design argument underlay a philosophical theory of optimism. Some decades after Mather wrote, Voltaire's *Candide* (1759), following the devastating Lisbon earthquake of 1755, mercilessly satirized the notion that everything is for the best in this best of all possible worlds.

In discussing rain, Mather notes that winds distribute clouds so equally that no part of the earth lacks convenient showers except when a special Providence withholds them to punish a sinful people. If a land is without rain, it has a supply of water some other way, as in Egypt with the annual flooding of the Nile. This provision witnesses to God's power and goodness.

Acknowledging that slow progress had been made in knowledge of the rainbow, Mather gives an account of the work of Marco Antonio de Dominis, a somewhat unscrupulous archbishop of Split, who described how the interior and exterior bows are formed by refraction and reflection of the sun's light. Though mistaken in his explanation of the secondary rainbow and incomplete in his explanation of the primary rainbow, Dominis gave a better explanation of the primary rainbow than anything published before 1637, when Descartes gave the correct elementary theory of both. Newton's *Opticks* (1704) incorrectly credits Dominis with explaining how the secondary rainbow is formed; Newton appears to support the charge that Descartes plagiarized Dominis, perhaps to depreciate the importance of Descartes's work. Unaware of these scientific developments, Mather offers a rhapsody on the rainbow as a sign and seal of the covenant that God will not again drown this sinful world.

Mather offers a scientific account of thunder and lightning by describing what both the ancients and the moderns understand about these phenomena. In Mather's view, knowing the natural causes of thunder does not release one from realizing that thunder has in it the voice of God. Death by lightning, a punishment known as "*Combustio Animae*," he adds, has frequently been inflicted upon the "Children of Men" [73:53; 72:47].

As for air, writes Mather, it was Galileo who first determined its weight. He found that water could not be raised any higher than thirty-five feet by pumping, and, concluding that the old notion of an infinite *"Fuga Vacui"* would never do, came up with the idea of the counterbalance of the weight of air [74:15–19]. Actually, Galileo not only failed to suggest the weight of air to explain the siphon phenomenon but also rejected that explanation when Giovanni Battista Baliani offered it to him. Galileo was familiar with the weight of air, however; he himself had devised practicable methods for its determination.[112] Air, Mather adds, is a necessity to the whole animal world, as demonstrated by the work on respiration by Guillaume Rondelet.

Relentless in pursuing the theme of design and in demonstrating divine governance of the physical universe, Mather insists that the wind has many uses. It dissipates contagious vapors, distributes clouds for the more commodious watering of the earth, tempers the heat, carries vessels to remote countries, and drives windmills. Yet it is rarely so violent as to destroy all before it and overwhelm the world. All this, too, proclaims the power, wisdom, and benevolence of God.

Mather's essay on the cold is based on Robert Boyle, whose *New Experiments and Observations Touching Cold, or an Experimental History of Cold* (1655) is a scientific report supplemented by travelers' accounts of adventures in icy regions. Mather is uncertain whether cold is a mere privation or something positive, the result of saline corpuscles floating in the air. But he is impressed with the force of cold. God, "by that one Part of his *Artillery,* the *Cold* alone, can soon destroy his Enemies!" [81:59–60]. Mather is ever ready to include examples of natural phenomena from his own country, and the remainder of this essay, an original contribution, describes conditions in America. The people who settled New England came with the notion that climate was uniform in any given latitude around the world, but Mather realizes that "the Degrees of *Cold* in several Climates are not according to their Degrees of *Latitude*" [81:65–66]. He learned this lesson from Martin Martini, a Jesuit missionary who had described his experience in China, and also from personal experience. The Puritans who settled New England found it cold, and believing that people are responsible for their own environment, they took active measures to improve their living conditions. By opening and clearing the woods, Mather reports, the American Puritans had moderated the force of the cold.

Mather begins his essay on the terraqueous globe by describing its size and physical characteristics. He then affirms the heliocentric theory by quietly adding, "the *Copernican* Hypothesis is now generally preferred" [84:2]. He defends this notion, arguing that it is impossible to account for the whole system of astronomy as we know it without admitting the stability of the sun and the motion of the earth. The motion of the planets, performed so regularly for six thousand years, obliges us to acknowledge that God is the governor as well as the creator of the world. As even pagan authors testify, the order and harmony of the heavens first led men to acknowledge a God. The steady axis of the globe and its inclination to the plane of the ecliptic make the oblate spheroid planet inhabitable in all its parts. The collection of the waters on the globe into vast oceans wherein innumerable fishes are nourished and whereon voyages are performed, and the distinction of the dry land, furnished with so many vegetables and animals, "What can it be any other than the Result of Counsel, of Design; of *Infinite Wisdom!*" [86:67–69].

The new science from Copernicus to Newton dealt with bodies in motion and raised the perplexing question of what held the universe together. Mather discusses these matters in his essay on gravity, which he describes as "a most noble Contrivance" to keep the several globes of the universe from shattering to pieces, as they would do from the centrifugal force of their revolutions if it were not for this power of attraction [89:74]. Philosophers of nature had assigned various causes "for this great and catholick Affection of Matter, the *Vis Centripeta*," but Mather merely notes them in passing in order to endorse Halley's declaration that "*Gravity* is an Effect insolvable by any *philosophical Hypothesis;* it must be religiously resolv'd into the *immediate Will* of our most wise Creator" [89:93–94; 90:99–2].

After describing the operations of gravity, Mather quotes Samuel Clarke to support this viewpoint. Clarke, a prominent Anglican clergyman whose superior mathematical knowledge enabled him to understand Newton better than most of his contemporaries, applied Newtonian principles to metaphysical and theological problems. He established his reputation in his Boyle lectures of 1704–5, which were first published separately and afterwards together. As Mather quotes Clarke, gravitation must of necessity be caused by something which penetrates the very substance of all bodies and puts forth in them a force entirely different from that by which matter acts upon matter; this is an evident demonstration that the world was originally made by a supreme intelligent Cause and depends every moment on some superior Being for the preservation of its frame. All the great motions in it are caused by some immaterial Power which perpetually exerts itself in every part of the world, and this *"preserving and governing Power gives a very noble Idea of* PROVIDENCE" [92:94–95].

Gravity cannot be accounted for mechanically, Mather adds, for matter cannot act at a distance, whereas gravity does so without any medium for the conveyance of it. Mechanical explanations of gravity advance the notion of a subtle fluid whose motions account for gravity, but the cause of its motion must then be immechanical. There must be a First Cause. All systems of natural philosophy require postulates that are not to be accounted for mechanically. The universal force of gravitation is the effect of the Divine Power by which the operations of all material agents are preserved.

Inertia, like gravity, is for Mather something of a mystery that is understood only by reference to God. The force of inertia is not essential to matter, it is rather a positive faculty implanted in matter by the Author of nature. The preservation of a body at rest or in motion, after the first instant, absolutely depends on God as the cause.

As he works his way from the atmosphere down to the surface of the globe, Mather comes next to the water. His essay on this substance reflects the relentless determination of Mather and other physicotheological writers to see nothing but harmony in nature. The sea is the grand fountain of fresh waters; if it were but half its present size, we should soon be miserable for want of vapors. The equal distribution of waters over the face of the globe illustrates divine wisdom and goodness. Few dry places are found in the world; the planet actually has many prodigious rivers. The earth fortunately has but one moon, for if there were two we would alternately suffer from devastating flood tides or no tides at all.

The essay on the earth follows the pattern evident throughout the book of discussing scientific discoveries and then reflecting on their religious significance. Drawing on the experiments of Johann Baptista van Helmont, an unorthodox Belgian physician, scientist, and mystic, Mather observes that vegetables owe their life and

Introduction

lxxxi

growth not to the earth itself as much "as to some agreeable Juices or Salts lodg'd in it" [103:64–65]. Mather, like Ray before him, ridicules the notion advanced by Thomas Burnet in *The Sacred Theory of the Earth*, that the earth is "but an Heap of Rubbish and Ruins" [104:77–78]. Burnet, whom Mather never names, viewed the terraqueous globe as a mighty ruin, a paradise damaged by sin. As a millenarian, he envisioned a last conflagration before the restoration of the original paradisal earth.

Mather, by contrast, finds the earth acceptable. Its uneven surface is beneficial; the mountains and rivers have many uses. Even subterranean caverns and "*Ignivomous Mountains*" [106:60]—that is, volcanoes—have positive value, for without these spiracles to vent vapors, earth tremors would bring greater destruction than they do. The history of earthquakes would make a large and sad volume, Mather admits, but despite his commitment to the design argument, he is hard pressed to find divine benevolence in them. He evades the problem by quoting "a modern Philosopher" to remind us that the earth is actually a thin crust laid over an unfathomable sea of sulphurous and bituminous matter, and that "upon such a *dreadful Abyss* we walk, and ride, and sleep" [110:39–40]. Earthquakes, Mather concludes, are "very *moving Preachers* unto *worldly-minded Men*" [110:43–44].[113]

Mather's essay on magnetism reports discoveries relating to this phenomenon of nature, including the dip in the magnetic needle. But he fails to mention William Gilbert's *De magnete* (1600), the most important Renaissance treatise on the subject, and his discussion of the effect of heat on magnets and iron tools was written before the modern concept of heat was known. Mather finds magnetism, a principle very different from that of gravity, "unaccountable" [111:91]. He takes refuge by urging, nay demanding, that "*Gentlemen Philosophers* . . . glorify the infinite Creator of this, and of all things as *incomprehensible*." They must acknowledge that "*Human Reason* is too feeble, too narrow a thing to comprehend the *infinite* God." He reinforces this assertion by quoting Robert Boyle's *Some Considerations about the Reconcileableness of Reason and Religion* (1675), adding his own statement about the "*natural Imbecillity* of Reason" in such a way as to make it appear to be Boyle's view [117:92–118:14 passim; see also 119:84]. Mather then offers five "insoluble" mathematical examples to demonstrate the limitations of human reason. Mathematicians, it is worth noting, later solved the problems described.[114]

Mather's interest quickens when he turns to natural history, a field much less advanced than physical science at the time. Its method was largely one of observation and classification rather than of experiment and calculation as in natural philosophy. The prevailing concept in the general understanding of organic nature was the ancient idea of a great chain of being. The notion of a chain of created forms reaching from the lowest types up a ladder of perfection through man to God reached its peak in the early eighteenth century.[115] The organic world, one in which form was adapted to function in both the plant and animal kingdoms, provided the best evidence for the argument from design. God was invoked as the ultimate explanation of the adaptation of organic beings to their environment. John Ray's philosophical exposition of organic life was widely influential. Mather was very familiar with it; he devotes much of his book to the life sciences, treating vegetables, insects, reptiles, fishes, birds, quadrupeds, and man.

Mather's view of the contrivance of God in the vegetables which grow on the earth was largely shaped by the biological vitalism of the Cambridge Platonists. Henry More and Ralph Cudworth, as we saw earlier, devised the concept of a spirit at work in the material world as a means of combating materialism and atheism. Ray embraced it, and Mather takes it from the English Pliny. Mather attributes the growth and magnitude of plants to this spirit, an "intelligent *plastick Nature*" [131:45]. The difficulty here, however, is that the notion of a "plastic nature" explains the unknown by asserting another hypothetical unknown; the concept was outmoded by the time Mather wrote.

Nevertheless, Mather drew upon the idea of divine purpose in nature as a selective principle in presenting evidence on the vegetable world. All plants have their uses, he writes, citing interesting examples of how they supply food, clothing, medicines, and aesthetic delights. Even noxious vegetables are useful. If the bramble pricks the owner, for instance, it also tears the thief. Examples of Mather's passion for teleology could readily be multiplied.

According to Mather, design is also demonstrated by the benevolence of the Creator in providing specifics against diseases.

Frontispiece from Ralph Cudworth's *True Intellectual System of the Universe* (London, 1678), showing the outcome of the debate between theists and atheists in ancient times. Mather borrowed his fierce hostility to atheism from English authors such as Cudworth. Courtesy of the Department of Special Collections, University of Chicago Library.

Every disease has a remedy. A compassionate God furnishes every country with the particular remedies for all the distempers with which it may be affected.

When Mather wrote his book, the concept of hybridity in the vegetable and animal kingdoms was poorly understood; even the word *hybrid* had not yet come into common use.[116] Rudolf J. Camerarius, a German physician and botanist, was first to prove the sexuality of plants by experiments which he described in a letter published in 1694;[117] about this time botanists also demonstrated that pollen was necessary for the production of viable seed. These findings stimulated interest in producing hybrid plants by the crossing of varieties and species. Indian corn (*Zea mays*), which had long been domesticated and was widespread in the Americas, has separate male and female flowers, but since both occur on the same plant and many plants are grown together, wind pollination is effective. Nehemiah Grew noted the mingling of different varieties of maize without recording any views as to the method of admixture. Mather knew from his reading that flowering plants reproduce sexually [134:13–15], and he was a close observer. Eager to contribute his own two mites, he includes a passage which describes a couple of experiments carried on in New England [131:17–35]. Mather is the first person to describe correctly the admixture of colors in *Zea mays*. His observations, a significant contribution to the study of plant hybridization, writes Conway Zirkle, "should take their place in the history of botany." Later (24 September 1716), Mather expanded this account and sent it in a letter to James Petiver, a fellow of the Royal Society.[118]

The essay on insects combines scientific description with a hymn to the Creator. According to *The Christian Philosopher,* insects are not imperfect creatures lacking in parts but complete animals of their kind. Their anatomy, with all their parts in such infinitesimal bodies, manifests divine design. Insects posed the problem of spontaneous generation, a lively issue in Mather's day. He insists that no one of sense could any longer believe in it, because spontaneous generation would be nothing less than a creation. (If an insect arose spontaneously, why not an elephant? And if an elephant, why not a human?) All animals are generated of male and female parents. Mather battles irreligion by contending that the production of insects is superior to the laws of mechanism. The ends for which insects are made are worthy of God, for wisdom makes nothing in vain. In Mather's minutely purposive world, Providence ordains creatures of superior order to destroy insects, lest they become too offensive. When army worms threaten New England's corn, he writes in yet another reference to American natural curiosities, wild pigeons clear the country of them. Even noxious vermin have a purpose: they inflict suffering upon us for our misdeeds.

In his essays on vegetables and insects, Mather deals with the problem of the nature of biological development. At the time investigators were pondering the question of how the seemingly amorphous seed of a plant or the egg of an animal could develop into an adult of its kind. Opinion on the subject was divided, with some scientists subscribing to the doctrine of preformation and others to epigenesis. Adherents of epigenesis believed that an organism develops by the new appearance of structures and functions. Mather takes his stand with the preformationists, who held that there was something encapsulated in the seed or egg which was responsible for turning it into an adult. As he writes, the whole parent vegetable was locked up in the little compass of the seed [134:16-135:18; 135:24-42; 152:3-5]. Preformationists had to choose whether the preexistent embryo was located in the egg or the sperm, and almost all of the leading biologists of the period were ovists as opposed to spermists or animalculists. Mather may not have appreciated this distinction.[119]

Mather's essays on reptiles and fishes often include more of the fabulous than the scientific. He cites various Renaissance and Reformation authorities on the magnitude of whales and serpents, and repeats a marvelous story about an army of serpents which appeared in Hungary in 1564. In addition, these essays are filled with evidences of design in nature and with religious reflection. Venomous creatures have been made an objection against divine Providence for being destructive to the rest of the world, Mather says, but they have medicinal uses. Rattlesnake gall worked into lozenges invigorates the blood and cures fever, and viper's flesh cures leprosies. "*Ubi Virus, ibi Virtus*" [180:17] ("Where there is a disease, there is a remedy"). Indeed, there would be no injustice if God made a set of noxious creatures as scourges to execute divine chastisements on sinful people. It is the way of the sea for the greater to devour the lesser, but fish do not devour those of their own particular kind. Therein they condemn the cursed rapacity too often seen among humans.

Mather frequently names scientists whose importance he fails to recognize. A good example is Conrad Gesner, a Swiss Renaissance polymath who achieved renown in botany, zoology, and humane letters. His massive five-volume *Historia animalium* was published from 1551 to 1587. Mather obtained a reference to Gesner from intermediate sources, but he has no inkling of Gesner's scientific stature.

From the fishes Mather turns "to *soar* and *sing*" with the birds in praise of God [192:42]. He cites cases to demonstrate that birds exhibit the adaptation of the organism to its environment. The legs, toes, tail, and tongue of the woodpecker exactly correspond to the creature's way of life. The tongue, for example, has a sharp, sticky point designed to draw maggots out of wood. The bifurcation of the syrinx, which is not found in other animals, fits birds for singing. Mather mentions William Harvey, who published his famous book on the circulation of the blood, *De motu cordis*, in 1628, only on the grinding capacity of the gizzard in fowls and on the chalazae or treddle of the eggs of flamingos. Mather's credulity permits him to repeat the old but mistaken notion that dyeing eggs will determine the color of chicks hatched from them. Borrowing from Henry More by way of Ray, Mather interprets the convenience observed in oviparous animals as a triumph over atheism because such creatures lay many eggs at once, and as a result are able to fly when "pregnant" and are not constantly burdened with bearing and feeding their young. Piling on the evidence, Mather notes that an accurate internal bracing keeps all the parts of a bird's egg in their due place. The parental affection that hens give their young is contrary to the instinct of self-preservation, he observes; it must come from the Creator of all things.

In this essay, Mather celebrates his own country by two long references to the native bird with which he was most familiar—the passenger pigeon, of which there were countless millions in early New England. One describes the curiosities of their nidification [201:1–15]; here he offers an explanation with no thought of experimental verification. The other [204:22–207:43] includes a delightful description of how the cocks feed their young with "a Substance like a tender *Cheese-Curd*" which he calls *"thickned*

Milk." Mather gives a "much more satisfactory account" than earlier authors, and later authors, starting with John Hunter in 1786, determined the validity of every feature mentioned by Mather. Mather's comparison of crop milk with mammary milk is now well established. He errs only in saying that hens do not produce this *"thickned Milk."*[120]

Mather embroiders his account by happily telling a tale, as with an entertaining description of mechanical birds drawn ultimately from three Renaissance worthies—Petrus Ramus; Regiomontanus or Johann Müller, a German-born astronomer and mathematician; and Guillaume de Saluste du Bartas. The latter's *Divine Weekes*, a poetic description of the world which cast the scientific knowledge of the late Middle Ages into the form of a Christian epic, appeared in France in 1578 and in Sylvester's complete English translation in 1605. As the story goes, a wooden eagle flew out of a city, met the emperor a good way off, saluted him, and returned with him, while an iron fly, at a feast to which the emperor invited his friends, flew out of his hand, made a circle, and flew back to him again before the astonished guests.

Mather comes next to quadrupeds. He classifies them into various categories, such as hoofed or clawed, and argues at length that the anatomy and physiology of these creatures demonstrates the adaptation of form to function. A few examples illustrate the point. The prone posture of the body of beasts is beneficial both to them and to humans, since it enables them to perform their own actions "the better," and to serve humans "the better," for both carriage and tillage [216:95]. The location of the ear is different in an owl, a fox, a hare, and a horse so that each animal can easily receive sound. The stomach of the quadruped is adapted to the food intended for it—one kind in carnivorous animals, another in herbaceous animals. The eye of brutes has a nictating membrane which serves to clear the cornea and prevent the eye from drying. Humans and apes are the only animals without this membrane, but they have hands to cleanse their eyes.

The question of reason in brutes had fascinated people since antiquity and was much discussed in Mather's lifetime. Descartes posited a dualism of mind and body and explained it on the basis of mechanistic principles. He considered animals as automata subject to instinctive reflexes.[121] Mather objects to this view because he sees the Creator at work in the creation. The bee and the ant are sagacious in making provision for the winter, he contends, and since these creatures are irrational, their acts must be derived from some wise Being. The same is true of insects in making nests. Their artifice is beyond the reach of animal cunning and must have the concurrence of a superior intelligence. In short, "*The Divine Reason runs like a Golden Vein through the whole Leaden Mine of Brutal Nature*" [172:69–71].

Some writers had made a special case for reason in quadrupeds. As Mather observes, Girolamo Rorario had written two books to prove that beasts often have more reason than humans. Mather quotes from Rorario's *Quod animalia bruta ratione utantur melius homine,* which appeared in 1648, a century after it was written. To Mather, this reasoning is absurd. Animals are not machines, automata without sense, but brutes do have a lower degree of reason than man. Mather illustrates his contention with some "credible Relations" of reasonable dogs, clever foxes, and rational elephants, but he insists that such evidences are mere mechanism, and there must be a power superior to mechanism [227:74–228:20]. To Mather, the reason found in brutes is an immediate effect of Providence, applied for the preservation of God's creatures.

In the Platonic tradition, Mather regards the world as emblematic and exquisitely purposive. He views animal behavior as supplying pictures of virtue with the highest authority of nature. Ants teach forecast, for example, doves, conjugal chastity, and hens, maternal love. Even noxious creatures provide instruction: weasels induce watchfulness; lice, cleanliness; and the fox teaches us to beware of thieves. We should learn the virtues whereof God's creatures are monitors.

Mather's essay on man, the prime letter in the book of nature and "the highest Link in that *golden Chain,* whereby *Heaven* is joined to *Earth*" [237:67–68], is by far the longest in *The Christian Philosopher.* According to Mather, the human body itself is the grand example of design. Man's erect posture is suitable for a rational creature; recent anatomical discoveries by Vesalius, van den Spiegel, and others on the mechanism that makes it possible are enough to counter atheism. There is an end and use for everything in human anatomy. "No sign of *Chance* in the whole Structure of our Body" [240:73–74].

The lodgment of the parts is also admirable. Four of the five senses are located near the brain, and the heart is in the center of the body. That the faces, voices, and handwriting of individuals differ is a providence of God for avoiding confusion and disturbance. Had nature been a blind architect, the faces of humans might have been as much alike as eggs laid by the same hen.

Mather's enduring interest in medicine leads him to give considerable attention to human anatomy and physiology. The prevailing tone in medical thought in the seventeenth and early eighteenth centuries continued to be that of medieval rationalism, as Richard H. Shryock writes, and the success of Newtonian physics encouraged fanciful conjectures in medicine. Since the heart had proved to be a pump, the stomach was no doubt a churn and the body a machine, while everything was subject to measurement. Medical men became iatrophysicists, iatrochemists, iatromechanists, or iatromathematicians; Mather had to steer his way between these contending "schools." Finding the idea of the body as a machine attractive, he tended to the iatromechanical school.[122]

He treats various parts of the body in a somewhat rambling and repetitive way patterned upon leading medical treatises of the day. He starts with the head and the organs located in it, discussing the eyes, ears, nose, tongue, and teeth at length as well as the functions associated with each—seeing, hearing, smelling, tasting, speaking, and chewing. The eye was a favorite subject for proponents of design, since it was useful only when it provided sight and had no survival value in earlier embryonic stages. One could therefore argue that providential care was needed in its development. "No Man who survey'd the *Eye*," Mather declares, "could abandon himself to any *speculative Atheism*" [260:62–63].

From the head, Mather works his way downward on the body, describing the windpipe, lungs, heart, stomach, intestines, liver, bladder, and kidneys. He continues with an analysis of the bones, muscles, blood and lymph, hands, and fingernails. In all this the ancient doctrine of animal spirits figures prominently. He discusses the emunctories, the organs or parts of the body that give off waste products, but does not mention the organs of generation. As he proceeds, he discourses on many related matters, including celebrated feats of human strength.

Mather turns finally from anatomy and physiology to "the stupendous Faculties of the SOUL!" [296:52]. First, he celebrates the human reason, a divinely implanted faculty which enables people to discern the truths which God himself has established. Second, he praises learning, citing as illustrations famous humanists like Vives, Budé, Scaliger, Ussher, and Selden. Third, he exalts the memory, giving examples from antiquity to his own day of persons specially blessed with that faculty. Finally, he admires various practical inclinations that transact the business of the world, especially inventions like printing, the telescope, and clocks as well as advances in anatomy and mathematics. If mathematics improves as much in the next two centuries as it has in the past two centuries, Mather writes, "who can tell what Mankind may come to!" [305:32–33]. As Seneca wrote (here Mather borrows from Ray), the coming age will know many things unknown to us. Mather's discussion of the complete circle of sciences known at the time ends at this point.

The last dozen pages of *The Christian Philosopher* reflect on the religious significance of the scientific material. Here, at last, Mather largely frees himself from his sources. The transition comes in a paragraph based on a French author, Daniel Tauvry. In it Mather declares that the union between the soul and body is "altogether inexplicable, the *Soul* not having any *Surface* to touch the *Body,* and the *Body* not having any *Sentiment* as the *Soul*" [305:35–37]. This statement implicitly refutes Descartes, who made the pineal gland the center of interaction between the two. As Tauvry says, the union consists in the conformity of our thoughts to our corporeal actions; to explain this conformity we must have recourse to a superior power.

Mather goes on to assert that objects make impressions on our senses and that sense impressions terminate in the brain, but there is a law given to the soul by God that "in their doing so there shall be such and such *Thoughts* produced in the *Soul*" [305:48–49]. His point is that humans possess a divinely given intuition.

These considerations lead Mather from the visible to the invisible world. The wonders of the former are as nothing compared to those of the latter. "I do here . . . most religiously affirm," he writes, "that even *my Senses* have been convinced of such a World, by as clear, plain, full *Proofs* as ever any Man's have had of what is most obvious in the *sensible World*" [306:72–75]. This passage, one of the most important in the book, underscores the design argument. The senses together with intuition provide proof for belief in the existence of God. To the extent that Mather makes

Introduction xci

intuition a source of religious faith, he echoes Pascal's statement that nature proves God only to those who believe in him on other grounds.

In the scale of nature, Mather continues, human faculties are more capacious than those of lower creatures; yet humans are but the equator of the universe. The transition from human mind to perfect Mind is gradual, but eventually we arrive at disembodied intellect. The highest perfection that any created mind can arise to is that in the soul of the individual who is personally united to the Son of God. In sum, the chain of being leads from nature to God.

Mather's purpose in writing *The Christian Philosopher* is to enkindle piety, and his conclusion vividly illustrates his intention. "*Atheism* is now for ever chased and hissed out of the World," he triumphantly declares, "every thing in the World concurs to a Sentence of *Banishment* upon it" [308:32–34]. A Being superior to matter is everywhere so conspicuous that there can be nothing more monstrous than to deny God. Atheism is a bundle of contradictions and contrary to common sense, hardly worth a treatise to refute it. Every part of the universe offers evidence to confute a system that deserves "the most *contemptuous Indignation*" [308:48–49]. Moreover, atheism is condemned by the fact that if society embraced it, no integrity would be left in the world.

Having dispatched atheism, Mather proposes two positive steps to promote piety. First, the works of God exhibited in nature demonstrate the power, wisdom, and goodness of God; humans should acknowledge the divine perfections. Creatures endowed with reason are reasonable when they regard everything in the "vast Fabrick of the World" as an incentive to religion [310:3]. The book of nature leads to belief in God.

Second, the Christ of God, who is the Lord of all, must not be forgotten. Having concentrated on the physical universe throughout his treatise, Mather finally declares, almost as an afterthought, that he is not ashamed of the gospel of Christ. A glorious Christ is to be found in the works of nature more fully than philosophers generally realize; the Christian philosopher must consider Christ. Mather draws upon Raymond of Sebonde and Sir Francis Bacon to buttress his point.

Origen urged that the notion of Christ dying for all is to be extended to the very stars and that all things are to be restored to the kingdom of the Father, Mather writes, and the apostle Paul seems to have favored this speculative flight. "If this be so," Mather adds, "we need not break the Glasses of *Galilæo*, the *Spots* may be washed out of the *Sun*, and *total Nature* sanctified to God that made it" [314:26–28].

Origen had advanced the doctrine of the preexistence of souls, rebirth, and universal salvation. His eschatology was based on belief in the justice and goodness of an omnipotent Creator and in the absolute free will of every rational being, including animated stars. His doctrine of universal restoration had been condemned as heretical by Church councils in 400 and 553, but the Renaissance revived interest in Origen and stimulated fresh controversy over the doctrine of the ultimate salvation of all. Mather probably had read a book by the Cambridge Platonist George Rust entitled *A Letter of Resolution Concerning Origen and the Chief of his Opinions* (1661). He goes far in endorsing the speculative flight of Origen, despite the fact that universal salvation had twice been condemned as a heretical doctrine by the collective mind of the Church.

In closing, Mather emphasizes the mediatorial role of Christ in the divine plan of redemption and suggests that all creatures will enjoy felicity in the future state.

6. The Reception of *The Christian Philosopher*

Cotton Mather was instrumental in disseminating the new science associated with Isaac Newton into the American colonies, but an account of how *The Christian Philosopher* worked its way into the culture cannot easily be separated from the reputation of the Puritan priest. Mather made enemies during his lifetime and has not enjoyed a good reputation since.

A horrible contemporary example of the hazing of Mather occurred at the planetarium in Miami, Florida, in early March 1982. The planetarium was showing a film based on a book entitled *The Jupiter Effect* which predicted that the alignment of the planets on 10 March 1982 would, by the special force of gravity, lead to earthquakes and other severe physical disturbances. As background, the film included a history of doomsday predictions in early European and American history, including some comments by Cotton Mather which were introduced with a portrait of the

Puritan accompanied by this advice projected on the dome of the auditorium:

>Cotton
>Feel Free to Boo and Hiss
>Mather

After treating other figures, the film flashed Mather's name and portrait on the dome a second time, now urging the audience to

>Boo and Hiss Again[123]

Whatever his later reputation, as evidenced by this ill-informed, unwarranted, and dastardly assault, Mather was actually the friend rather than the foe of science. Like all authors, he hoped his book would find a large audience. He rejoiced when *The Christian Philosopher* finally arrived in Boston in late March 1721, and immediately began to promote it. The book made no big splash and never became a best-seller. Nor did it sink into oblivion.

The work found some favor in New England shortly after it appeared. By that time works on science made a fairly good showing in the Harvard College Library, as demonstrated by the library catalogue of 1723 and its supplement in 1725. By the former year the library had works by Artistotle, Galen, Hippocrates, Pliny, and Ptolemy among ancient authors, by Gassendi, Scaliger, and Kircher among early modern writers, and by scientific giants such as Tycho Brahe, Kepler, Galileo, Descartes, Harvey, Boyle, and Newton. In addition the library possessed treatises on mathematics and medicine, the early volumes of the *Philosophical Transactions*, several volumes of the *Acta Euriditorum*, and twenty-two volumes of the *Miscellanea Curiosa* or "German Ephemerides." Its holdings also included John Harris's *Lexicon Technicum*, Ray's *Wisdom of God*, and Derham's *Astro-Theology*. As noted earlier, Mather hoped to get "our Colledges" filled with his book. Unaccountably, it was not at Harvard in 1723, but the supplement for 1725 listed it, with the shelf mark "24.4.5."[124]

At the time a bitter struggle between the proponents of Cartesian and Newtonian philosophy was still raging. Harvard students were keenly interested in the new science, and books which were general introductions to the subject, including Ray's *Wisdom of God*, Derham's *Physico-Theology*, and Mather's *Christian Philosopher*, were popular with them. These volumes all spread the Newtonian philosophy. Fourteen young men, including Charles Chauncy, who became a theologian, and Isaac Greenwood, who later taught science at Harvard, organized a student society in October 1722. At their monthly meetings a member gave a discourse of about twenty minutes on any subject he pleased. The society's records show that Ebenezer Turrell read two lectures to the group, one "Upon Light, a Phisico-Theological Discourse," the other upon Providence; Samuel Marshall read a lecture "On God's Wisdom and Power." Turrell could have drawn his first lecture from a number of sources, perhaps most conveniently from Mather's first essay, "Of the Light." Turrell's second lecture and Marshall's remarks could readily have been based on Mather's insistence that nature illustrates the power, wisdom, and benevolence of God. These students also had access to Derham and Ray, so we cannot be sure that they based their knowledge of Newtonian science on *The Christian Philosopher*.[125]

Members of the Boston scientific community familiarized themselves with Mather's volume. Thomas Robie (1689–1729), whom Clifford K. Shipton called "the most famous New Englander in science in his day," may have drawn upon Mather's work. A Harvard graduate (A.B., 1708; A.M., 1711), Robie served as tutor at Harvard from 1713 to 1723, after which he practiced medicine at Salem until his death. He preached for a time but early demonstrated wide scientific interests, turning his attention to mathematics and to astronomical and meteorological observations. Robie published a series of almanacs annually from 1708 to 1720. He often filled the spare end pages with such useful information as the cause of heat in summer and the Copernican theory, and reputable historians have concluded that Robie used Mather's *Christian Philosopher* to explain the cause of thunder, lightning, and hail in his 1716 almanac. Almost certainly they are in error.[126]

Daniel Travis filled up the vacant pages at the end of his *Almanack* for 1723 with a "philosophical account" of thunder, lightning, and hail and on the revolution of the planets as he found them in Mather's *Christian Philosopher*. Travis cited page references for the material he took from the published book.[127]

By the late 1720s Harvard students were reading Mather's work along with other introductions to science. Jonathan Belcher finished "extracting" *The Christian Philosopher* on 4 May 1727, as his commonplace book shows; at about the same time he was reading

Ray's *Wisdom of God*. John Winthrop, the great-great-grandson of the first governor of Massachusetts and the great-grandnephew of the John Winthrop who was the first American colonial fellow of the Royal Society, read and took notes on both Mather and Ray during his freshman year at Harvard. As his commonplace book reveals, his interest in science was kindled by the descriptions of recent European discoveries in Mather's volume. Graduating in 1732, Winthrop became Hollis professor of mathematics and natural philosophy at Harvard in 1738 when Isaac Greenwood, his former teacher, was removed for intemperance; he held the chair until his death in 1779. He became the first important scientific scholar on the Harvard faculty, gaining distinction in mathematics, astronomy, and experimental physics, and publishing many papers in the *Philosophical Transactions* of the Royal Society.[128]

Mather's treatise on Newtonian science also found its way into the hands of other New Englanders. Jonathan Edwards followed scientific advances of the day closely by reading the latest English writers, but no evidence indicates that he knew Mather's book on science and religion. By 1743, however, Yale students had access to this title as well as to related works on natural philosophy and natural religion by the English authors Cheyne, Clarke, Derham, and More (but not Ray), by the French authors Fénelon and Rohault; and to Bernard Nieuwentijdt's *Religious Philosopher*, published in Amsterdam in 1717 and then in an English translation in three volumes in 1718–19.[129]

When Abraham Redwood bestowed a gift of 500 pounds sterling on the Redwood Library, founded in Newport, Rhode Island, in 1747, the funds were spent in London for a collection of useful books for a public library. The volumes assembled included Ray's *Wisdom of God* and *Three Physico-Theological Discourses,* Derham's two tomes, and Nieuwentijdt's treatise, but not the New Englander's *Christian Philosopher*.[130]

Since books on science and its relation to religion attracted increasing attention in these years, libraries added them to their collections. By 1768, the Providence Library in Rhode Island possessed treatises on physicotheology by Derham and Ray as well as by Nieuwentijdt, but not Mather's work on the subject.[131]

Despite the loss of holdings in science in a fire in 1764, the Harvard College Library still held Mather's *Christian Philosopher* in 1773, along with all of Ray's works, Derham's two books, and the volumes of Cheyne and Nieuwentijdt. By 1790 the Harvard College Library possessed two copies of *The Christian Philosopher*.[132]

Mather's treatise was suitable for advanced students, but when Thaddeus M. Harris, the librarian of Harvard, formed a catalogue for a small and cheap library to suit the tastes of common readers in 1793, he intentionally omitted works in the higher departments of science. He included in his model library many books on what would now be called religion and much natural history, but not Mather's book.[133]

Beyond New England, *The Christian Philosopher* was unknown in the eighteenth century. No good bookstore was to be found south of Boston when Mather's volume arrived in America, and public libraries other than parish libraries were few until after the Revolution. The culture of the people was English rather than American. Their connections were with Old, not New, England. The books in colonial libraries were usually brought directly from England. A number of the libraries contained the works of English physicotheological writers.

In Virginia, private libraries of the eighteenth century were small. The average number of titles in about 100 libraries analyzed by George K. Smart was about 106, with nearly half of the libraries containing fewer than 25 titles. Yet the religious interests of Virginians were strong. Smart classified 12 percent of the holdings in these libraries under religion and divinity, with another 11 percent related to science, medicine, and practical arts. Books by Derham and Ray and other works on Newtonian science which Mather himself had used in preparing his manuscript found their way into colonial libraries, but not Mather.[134]

In Virginia, for example, Robert Carter of Nomini Hall in Westmoreland County, one of the wealthiest colonials at the time of his death in 1732, had Ray's *Wisdom of God* in his library. Charles Pasture of Henrico, whose estate was inventoried in 1736, possessed Derham's *Astro-Theology* and a volume of Samuel Clarke's sermons. William Dunlop of Prince William County, the son of a professor of Greek at the University of Glasgow, was a merchant who died in 1739 at the age of thirty-two. He owned both Derham's *Astro-Theology* and Ray's *Wisdom of God*, William Wollaston's *Religion of Nature Delineated* (published in London in 1724), Newton's *Opticks*, Keill's *Astronomy*, and Clarke's sermons. He also had an older Cartesian treatise listed by the copyist of his library as Rohault's

"*Physic.*" Captain Samuel Peachy had a large library by mid-century; it also contained Rohault's book, the title of which was *A System of Natural Philosophy*. When Daniel Parke Custis died in 1757 his library contained about 457 volumes, including both of Derham's treatises. Some of the books in this library went to George Washington in right of his wife, Martha, the widow of Custis; others were set aside for John Parke Custis, who was six when his father died.[135]

In the 1770s, William Fleming, a Scots-born physician who had served in Washington's regiment from 1755 to 1762, had a library strong in medical works, many of which Mather cites, as well as both of Derham's physicotheological treatises, Ray's *Wisdom of God,* and Wollaston's *Religion of Nature Delineated*. Dabney Carr, a lawyer who married the sister of Thomas Jefferson and died in 1773 at the age of thirty, owned a copy of Derham's *Astro-Theology*. Major Charles Dick of Fredericksburg, who served as commissary during the French and Indian War and on the board of a powder factory during the Revolution, owned Derham's *Physico-Theology* and Burnet's *Theory of the Earth*.[136]

It appears that John Parke Custis received the works by Derham which had been in his father's library. When the library of the younger Custis was inventoried in 1782, it contained "Durham's sermons" (presumably the *Physico-Theology*) and "Ditto, Survey of the heavens" (i.e., *Astro-Theology*). Custis also owned Wollaston's *Religion of Nature Delineated,* medical works by Baglivi and Cheyne, and Burnet's *Theory of the Earth*. The library of Thomas Jefferson contained both Derham's *Physico-Theology* and *Astro-Theology* as well as Samuel Clarke's sermons and his Newtonian treatise "On Being and Attributes of God."[137]

While Derham and Ray, the most popular physicotheological writers of the day, eclipsed *The Christian Philosopher* in Virginia, Mather was not totally unknown in that colony: the library of William Byrd III of Westover contained his "History of New England"—that is, the *Magnalia Christi Americana,* published in a large and handsome volume in London in 1702.[138]

The lines of cultural and intellectual influence ran in the same direction in the rest of the South. Joseph T. Wheeler's analysis of a cross section of the libraries of colonial Maryland based on the inventories of estates showed that nearly 60 percent of the free white population possessed books, although three-quarters of the collections contained fewer than ten titles, often only a Bible and a few religious works. Religion was a strong interest; 23 percent of the volumes came under this category. John Jackson, a medical doctor of Queen Anne's County who died in 1768, had in his library a copy of "Demonstrations of the Attributes of God," which was surely Derham's *Physico-Theology*. The Reverend Thomas Bacon, rector of Saint Peter's Church in Talbot County, had what was probably one of the outstanding library collections of a Maryland clergyman. It included Derham's *Physico-Theology*. Wheeler also found Keill's *Introduction to Natural Philosophy* and Cotton Mather's *Life of Governor Phips* in colonial Maryland, but not *The Christian Philosopher*.[139]

The first parish or public library known to exist in North Carolina dates from 1700; other public and private libraries followed. But the historian of this development found "almost no American eighteenth century books in North Carolina libraries outside of law books."[140]

In South Carolina, the Charlestown Library Society, organized in 1748, possessed Ray's *Wisdom of God* and a number of other works on the new science, including Hooke's *Micrographia*, Harris's *Dictionary of the Arts and Sciences*, Keill's *Astronomy*, and Woodward's *Essay toward a Natural History of the Earth*. The society also had Rohault's *System of Natural Philosophy* and Burnet's *Sacred Theory of the Earth*. In 1772 John Mackenzie gave the library Derham's *Physico-Theology* and another copy of Keill's *Astronomy*. No evidence indicates that Mather's *Christian Philosopher* was available in the colony.[141]

In the middle colonies as elsewhere in America, the books found in eighteenth-century libraries reflected the influence of English culture. William Livingston, a Yale graduate from a prominent New York family who published a pastoral poem in which he celebrates the joys of a rural life among friends and books, inferred the existence of God from a view of the beauty and harmony of the creation. He cites Derham as one of the classical and English authors he read for pleasure and profit. The poet Philip Freneau also names Derham but not Cotton Mather on physicotheology.[142]

Students at the College of New Jersey, later Princeton, had access to Mather's *Christian Philosopher* as well as to Ray's *Wisdom of God* by 1760, but the library did not contain the two popular books on natural religion by William Derham.[143]

Introduction   xcix

In 1765 the Association Library Company of Philadelphia, a joint-stock group organized in 1757, published a catalogue of its books. This library was strong in history, the classics, and politics, but weak in the new science. Its holdings included Ray's *Wisdom of God* and various works by Samuel Clarke.[144]

The foundation of the Library Company of Philadelphia was laid in 1731. Benjamin Franklin suggested the plan for this subscription library of Philadelphia tradesmen who wanted a common library of their own, and the company received a charter in 1742. The catalogue of books in the library which Franklin published in 1741 throws light on the intellectual ambitions of the members and manifests a keen interest in the new science. It lists Ray's *Wisdom of God,* Derham's *Astro-Theology* and *Physico-Theology,* along with Mather's *Christian Philosopher,* a gift from David Bush. It was entirely fitting that a library inspired by Franklin, who confessed that he had learned valuable lessons from his fellow Bostonian early in life, should possess a copy of Mather's treatise on science. As subsequent catalogues demonstrate, the Library Company continued to add books in natural philosophy throughout the eighteenth century.[145]

The Loganian Library, a choice collection assembled by James Logan of Philadelphia, was one of the largest private libraries in early America. Its catalogue, published in 1760, included Ray's *Wisdom of God,* Derham's *Physico-Theology,* and Nieuwentijdt's *Religious Philosopher,* but not Mather's *Christian Philosopher.*[146]

The New-York Society Library was organized in 1754 as a subscription library; by 1793 members had secured about five thousand volumes. This collection contained a good representation of works by English physicotheological writers—Ray, Derham, Cheyne, and Clarke's sermons—as well as Nieuwentijdt's *Religious Philosopher.* But it lacked Mather's book and had not acquired it by 1813.[147]

By the end of the eighteenth century a new era was dawning. In 1797 the Library Company of Baltimore owned Ray's pioneering *Wisdom of God,* Derham's two treatises, and Fénelon's *Existence and Attributes of God,* but not Mather's book. It also held a copy of William Paley's *A View of the Evidences of Christianity* (1794), a masterpiece of Christian apologetics that won Paley honors and fame as a theologian. Paley published his more famous book, *Natural Theology: Or, Evidences of the Existence and Attributes of the*

Deity, Collected from the Appearances of Nature in 1802. This book, which drew heavily on Ray's *Wisdom of God*, proved by far the most popular work on natural theology in the nineteenth century.[148]

This survey of the holdings of Mather's introduction to the sciences of the day in colonial libraries reflects the character of intellectual life in eighteenth-century America more than it does the value of Mather's treatise. At that time people imported their books directly from England; buyers undoubtedly preferred a book by a well-known English author to one on the same subject by a colonial author.

Published in London, *The Christian Philosopher* seems to have been intended, at least in part, for an English audience.[149] The book faced stiff competition from local authors, however, and English booksellers are likely to have recommended volumes by the writers they knew best. Works by Ray, Derham, and other early modern English writers on science and religion, as well as Nieuwentijdt, were recommended by the Hull schoolmaster John Clarke in *An Essay upon Study* (1731). But Clarke did not mention Mather.[150]

Presumably Mather's *Christian Philosopher* did find English readers, however; we know that it influenced John Wesley, the founder of Methodism.[151] Wesley wanted a simple, clear compendium of natural philosophy or account of the visible creation to display the power, wisdom, and goodness of God, but he could not find such a treatise in any modern language, not even in English. What came nearest, he thought, were Ray's *Wisdom*, Derham's *Physico-Theology* and *Astro-Theology*, Nieuwentijdt's *Religious Philosopher*, Mather's *Christian Philosopher*, and Wollaston's *Nature Delineated*. Since none of these by itself was adequate to the purpose, Wesley undertook to extract the substance of all of these works. He published a two-volume work entitled *A Survey of the Wisdom of God in the Creation* in Bristol in 1763. The text was in great measure translated from a Latin work of Johann F. Budde, also called Buddeus (1667–1729), professor of philosophy in Halle from 1693 to 1705, then professor of theology in Jena. Wesley found occasion to retrench, enlarge, or alter every chapter and almost every section, to which he added the choicest later discoveries of English and foreign societies. He endeavored to describe the appearances of nature but not to account for them, believing that we may adore but cannot search out God's works to perfection.[152]

On the Continent Mather was read by Christian Wolff (1679–1754), whom Lewis Beck describes as one of two founders of the German Enlightenment (Christian Thomasius was the other). Educated at the universities of Jena and Leipzig, Wolff was appointed professor of mathematics and natural science at the University of Halle in 1706. He won initial fame by lecturing on mathematics and publishing in this field before turning to philosophy. The transition was easy, since mathematics included much natural science (astronomy) and Wolff regarded mathematics as the model for his logic. Moreover, he was dissatisfied with the philosophy taught by Thomasius and Nikolius H. Gundling and began adding lectures in philosophy to undermine their influence on students. His early lectures were regarded as mostly expositions of Leibniz's philosophy, but gradually they took on Wolff's own characteristic scholastic form. His opponents applied the name "Leibniz-Wolffian philosophy" to his system, but both Leibniz and Wolff considered the label erroneous. Wolff claimed that he had taken more from Aquinas than from Leibniz.[153]

While at Halle, Wolff published in German a series of works which made him the leading philosopher in Germany. His study of logic appeared in 1713, of metaphysics in 1719, of ethics in 1720, of politics in 1721, of cosmology in 1723, and his work on natural theology in 1724. These books made him famous. He was honored with membership in European learned academies, including the Royal Society of London, and his philosophy began to be taught in other German universities.

At Halle, nevertheless, he met opposition from several members of the philosophical faculty—Budde, from whom Wesley borrowed for his *Survey*, was one of this group—and from the Pietist August H. Francke of the theological faculty. Wolff served as rector of the university in 1721, and in a formal address defended the view previously taken in his writings that ethics was not dependent upon revelation. His successor as rector called upon the theological faculty to censure Wolff for heresy; Wolff appealed to the court in Berlin for vindication, but King Friedrich Wilhelm I, through the intrigue of a group of military cronies who represented Wolff as teaching a determinism that led to the conclusion that deserters from the army should not be punished, dismissed Wolff on 8

November 1723 and ordered him to leave the realm within forty-eight hours or be hanged.

Wolff went to the University of Marburg; henceforth he wrote for a European audience, and therefore in Latin rather than German, on each of the topics he had earlier published on in German. He belonged to and modified the scholastic tradition going back to Suarez. His works combined with his persecution made him an intellectual hero and brought him European fame and a patent of nobility. In 1740 Frederick the Great offered Wolff a permanent fellowship in the Berlin Academy, but Wolff preferred his old post at Halle. He returned there in 1740 in triumph, but his lectures were no longer successful, and he soon gave up lecturing to write. He died in 1754, full of honors, although his influence had already begun to wane.

Wolff was important not as an original philosopher but as the transmitter and modernizer of a tradition. He changed the old Catholic and Protestant scholasticism and the new mathematical methods and natural science of Leibniz into a conception of philosophy as an instrument of public education. For Wolff, philosophy is world-wisdom. Its scope includes all the human sciences. Its goal is the knowledge of why things must be as they are. Its method is borrowed from mathematics. Both analytic and synthetic, it is concerned with connections between things. Wolff divided philosophy into the theoretical (metaphysical) and the practical. His philosophy was a confused mixture of both rationalistic and empiricist elements, not consistently one or the other. He used the physicoteleological as well as the ontological and cosmological arguments for the existence of God. The world for him was a mechanical whole whose teleology is found in its suitability for human purposes, and nature less a magnificent design than a set of arrangements designed for man's benefit. He viewed this world as the best of all possible worlds. Wolff was a scholastic pedant, a prolix author who lacked humor, but his philosophy was the first comprehensive system to be published in German, and it was the dominant philosophy in Germany from about 1710 to about 1750.

Wolff cites Mather in his *Rational Thoughts on the Purposes of Natural Things* (*Vernünfftige Gedanken von den Absichten der natürlichen Dinge* [1724])—the "German Natural Theology." He opens with a discussion of the purposes and structure of the universe.

In ancient times, he writes, people had childish ideas about the structure because they saw the universe only with bare eyes and imagined it as being very small, a view that prevented them from having a high estimate of God. But after Copernicus, Galileo, Huygens, and Cassini, people better understood the astonishing size of the universe, and their understanding was more appropriate to the infinite God.

In England, Derham, Boyle, and Clarke used this new knowledge for that purpose, Wolff asserted, and "even the theologian Cotton Mather," whose *Christian Philosopher* he cites in a footnote, "who is well known by the English for his piety and his zealous concern with the purity of the Christian religion, has also made use of the truth of the astronomers' discoveries about the structure of the universe when, in his pious zeal, he tried to show how one should use knowledge of nature to praise and glorify God. It would be good if others, too, followed the laudable example of the English theologians."[154]

Wolff was the best German representative of a general movement of Enlightenment thought that swept over Europe in the late seventeenth and eighteenth centuries. New editions of the book in which he discusses Mather were published in 1734 and 1737, and in the latter Wolff repeats the commendation of Mather and adds "contrastive remarks on the intolerance and scientific ignorance of some German theologians." A number of Germans would have learned about the work of the American author from the pages of the leading German philosopher between Leibniz and Kant.[155]

The design argument remained a fundamental component of Western thought in the early nineteenth century. Ray's *Wisdom of God* and Derham's *Physico-Theology* retained their former popularity as expositions of natural theology in 1800, but William Paley, archdeacon of Carlisle, published *Natural Theology: Or, Evidences of the Existence and Attributes of the Deity, Collected from the Appearances of Nature* in London in 1802. It quickly went through several editions and became widely known in both England and America. Now Paley's book rather than Ray's or Derham's eclipsed Mather's work. John Adams, the former president, had Derham's *Physico-Theology* as well as Mather's *Magnalia* in his library. Writing to Thomas Jefferson on 16 July 1814 on what an educated man should read, he said that he would leave theology to "Ray, Derham, Nicuenteyt and Payley, rather than to Luther

Zinzindorph, Sweedenborg[,] Westley, or Whitefield, or Thomas Aquinas or Wollebius." He apparently did not own or know about *The Christian Philosopher,* and no copy of that book was found in the Library of the United States in 1815. In England, shortly thereafter, attempts to reinforce the design argument led to the Bridgewater Treatises, a series of nine works in thirteen volumes by qualified scientists on the power, wisdom, and goodness of God as manifested in various aspects of the creation, published from 1833 to 1837.[156]

The Bridgewater Treatises had wide popular appeal on both sides of the Atlantic, and presumably American interest in the design argument lay behind a new edition of Mather's book which was published in Charlestown, Massachusetts, at the Middlesex Bookstore in August 1815. It bore the same title as the original, followed by a subtitle: *The Style Made Easy and Familiar.* J. M'Kown was the printer. The copyright was in the name of William Collier.

Born in Scituate, Massachusetts, in 1771, Collier went to Boston to learn the carpenter's trade, and at the age of twenty-one he united with the Second Baptist Church. After completing his apprenticeship, he went to Rhode Island College (now Brown University) to prepare for the ministry. He was licensed to preach in 1798 and ordained in 1799. After serving churches in Newport, Rhode Island, and New York City, he went to Charlestown, where he was pastor of the First Baptist Church from 1806 to 1820. Collier was an active and successful minister of the gospel during a time of political conflict; while at Charlestown he shared with the Reverend Jedidiah Morse the chaplaincy of the state prison. In 1820 he resigned on account of health and removed to Boston, where he labored in the City Mission and became a pioneer of temperance reform.

Collier was a man of great humility, teasingly described by a biographer as "eminently conservative" and yet "sufficiently progressive." Although not a bold leader, he was the pioneer of some evangelical enterprises. Starting in 1808 he compiled and published *The Gospel Treasury* in four volumes. The contents were selected chiefly from the *London Evangelical Magazine;* the work was reissued several times under slightly variant titles. In addition, Collier edited for publication eleven sermons by the Reverend James Saurin; he published *A New Selection of Hymns* as well as sermons he himself preached in 1806 and in 1819. After leaving

Charlestown he edited a weekly temperance newspaper, the *National Philanthropist,* the first of the kind ever printed, and *The Baptist Preacher,* consisting of monthly sermons by living ministers.[157]

Surely Collier was the mainspring behind the new edition of Mather's book. In all likelihood he shared the prevailing evangelical conviction that a renewed emphasis on natural theology would strengthen the Christian faith. He probably considered the original edition of Mather's treatise too learned and thus too difficult for most readers. A simpler version would demonstrate anew the ways in which nature testified to the power, wisdom, and goodness of God.

Collier's edition is indeed more "easy and familiar" than the original. In physical appearance it is smaller (eighteen as opposed to twenty-one centimeters high) and longer (324 as opposed to 304 pages) than its predecessor. The "easy" edition has good, readable type; its pages are more "open" than earlier for a number of reasons. Capital letters are used only to begin sentences and for proper names. Quotation marks do not run along the margin of the page throughout a quotation but mark only the beginning and end of quoted material. Mather's use of italics for emphasis is gone. Nothing is italicized. All foreign language material has either been translated into English and run into the text or eliminated entirely, and biblical quotations are placed within quotation marks. All of these revisions are done silently. There is some minor alteration of the text to simplify phrases or make the style conform to contemporary usage. Paragraphs are run on, eliminating many short paragraphs in favor of fewer longer ones. The names of authors of verse are omitted.

The greatest change is in the handling of Latin quotations. Collier either translates them into English or eliminates them. Sometimes he drops only a line or two: e.g., the description of Ray and Derham as *"Fratrum dulce par"* [10:61]; the first line and the last three lines of the essay "Of the Hail" [69:45; 70:74–76]; the references to Vergil and Claudian [116:71]; and the Latin quotation from Raymond of Sebonde [236:53–237:55]. In one case the omission of a Latin quotation unduly simplifies a paragraph. Collier, for example, alters 126:58–62 to read: "He deserves to be herded with the brutes, who is insensible, that the benefits of salt call for very great acknowledgements. My God, save me from what would render me unsavory salt!" In Collier's hands the passage becomes flat and flavorless.

Worse, Collier eliminates entire paragraphs. He drops the first paragraph and the last four lines of the essay on the rainbow [63:54–64:62; 67:75–78], the first paragraph of the essay on minerals [122:65–69], a paragraph containing a long Latin quotation from John Ray [139:49–140:69], and a paragraph with a quotation from Johann Arndt [148:3–13]. Examples could readily be multiplied.

As a result Collier's "easy and familiar" edition differs from Mather's original in two important respects. First, it fails to convey a sense of Mather's contribution in transmitting the legacy of classical antiquity to early modern culture. Mather depended on the learning of the ancients, citing classical authorities in each essay before describing advances made by modern scientists. He also quoted Latin because it was still the universal language of science in the early modern period, and most of the modern natural philosophers whom Mather read wrote in that language. Second and closely related, Collier's edition has a narrower range and less resonance than the original. Both the absence of Latin and the omissions and simplifications yield a book in which Mather's wide learning and supreme confidence in the harmony between science and religion are less evident than in 1715. The book has become a tract for the times.[158]

Scholars began to show keen interest in Mather's book on science and religion in the twentieth century. I. Woodbridge Riley, a pioneer in American intellectual history, discussed Mather's book in *American Philosophy: The Early Schools* (1907) and again in *American Thought from Puritanism to Pragmatism* (1915). Riley interpreted *The Christian Philosopher* as representative of the disintegration of Calvinism and the rise of naturalism in the colonies. The growing success of natural philosophy in the seventeenth century effected a change from a magical to a scientific outlook, and under the impetus of the new science interest in wonder-working providences gave way to interest in the observation of external nature and even to appreciation of the beauty of natural scenery. According to Riley, Mather's *Christian Philosopher* accepts moderate deism in principle if not in name. Its optimism, its emphasis on the purposiveness of nature, its sense of the beautiful, and its cheerful outlook were all at variance with the spirit of Puritanism. For Riley, the book with its scientific argument element lived on, anticipating by a century the Transcendentalists' love of nature for its own sake.[159]

Kenneth B. Murdock, a Harvard professor of American literature, included generous portions of *The Christian Philosopher* in his *Selections from Cotton Mather* (1926, reprinted in 1960, 1965, and 1973). He printed verbatim all of the "preface" and the introduction (reversing the order of their original appearance), all of the four consecutive essays on Earth (including the appendix), magnetism, minerals, and vegetables, and the last portion of the essay on man [from 308:29 to the end], along with footnotes to help the reader.

Murdock emphasized Mather's departure from the theology of the American Puritans. Under the older view, which Mather had earlier embraced, God was a strict ruler who worked through inscrutable decrees. Nature was awful rather than beautiful, a manifestation of God's dread power rather than of his love for mankind. *The Christian Philosopher*, however, argues that the world is well ordered and beautiful and that man can appreciate the goodness of God by the study of nature. According to Murdock, Mather arrived at this advanced intellectual position only late in his career. Mather borrowed his general argument from English writers he mentions as his sources. His content was not original; the intellectual significance of the book lies in the fact that it was written in America, where such a doctrine had not yet been expounded, and by Cotton Mather. *The Christian Philosopher* "expresses as no earlier American book had done the beginning of the more liberal philosophy of the early eighteenth century," according to Murdock, and Mather "was the first man in the Colonies to express in print the dawning of the new ideas."

Murdock also stressed the literary artistry in Mather's treatise. As a piece of writing his book is superior to its sources. His whole attitude toward the physical universe is one of enthusiasm. Again and again he celebrates the beauty of nature, a theme that did not become frequent in American literature until the late eighteenth century. "*The Christian Philosopher* shows constantly not only that Mather saw the wonders of nature with the observant eye of the scientist, but also that his feeling for them was akin to the poet's."[160]

Theodore Hornberger, a professor of American literature at the universities of Michigan and Texas, convincingly demonstrated that Mather's advanced attitude toward science was exhibited at the beginning of his career as an author. In 1940 Hornberger accorded *The Christian Philosopher* high praise. "This book," he writes, "is of the highest importance to the student of American intellectual history. Scholars agree that it is the best summation of Mather's very considerable interest in science, and it is probably, especially when considered together with the *Manuductio ad Ministerium*, the best available index to the state of scientific knowledge in New England at the end of the first quarter of the eighteenth century."[161]

Perry Miller, a Harvard professor of American literature, revitalized the study of American Puritanism in the 1930s and dominated the field for a generation. He harbored a strong animosity which led him to treat Cotton Mather unfairly at times, but in *The Puritan Mind: From Colony to Province* (1953) he acknowledged that *The Christian Philosopher* together with the *Magnalia, Bonifacius,* and the *Manuductio* "justify his [Mather's] place in American literature." Mather had caught the vision of the new science, said Miller, and his book was an effort to capture the Enlightenment, especially Newtonian science, and use it to preserve the New England church order. Mather wished to draw out the meaning of the intellectual revolution already wrought in New England, to demonstrate that in the experimental philosophy which had won the field Puritans had providentially been led to a bulwark of faith. The point of *The Christian Philosopher* is repudiation of the scholastic arts of rhetoric and logic and of Aristotelian physics. Like every religious Newtonian, Mather could explain the operations of the universe by the laws of motion and of gravity, but he could not explain gravity. He religiously resolved it into the immediate Will of the Creator. The natural world is therefore reasonable and yet mysterious, and so infinitely susceptible of spiritualization.[162]

Otho T. Beall, Jr., and Richard H. Shryock stress the growing influence of science on Mather's thought in the early eighteenth century in *Cotton Mather: First Significant Figure in American Medicine* (1954). In his "Biblia Americana," they note, he examines Scripture in terms of science, whereas in *The Christian Philosopher* he surveys science from a religious perspective. In both cases his underlying purpose is that of reconciliation, but there is a significant shift of focus from religion to science. Mather finally brought together his cumulative knowledge of science in *The Christian*

Philosopher. He was the first American to transcend the traditional Calvinistic outlook.[163]

Josephine K. Piercy, professor of American literature at Indiana University, provided an introduction for a facsimile reproduction of the original edition of Mather's *Christian Philosopher* which was published at Gainesville, Florida, by Scholars' Facsimiles and Reprints in 1968. Mather knew that God had created a perfect universe, Piercy wrote, and "his attitude was, in a way, something new, because nature, in the hazardous days of early settlement, had seemed something antagonistic to man." Piercy noted that Mather offered both rational and spiritual explanations for natural phenomena; she concluded that "the lesson of *The Christian Philosopher* is that Man, endowed with soul and intellect, may contemplate, as far as his mind will take him, the infinite wisdom of God." This inexpensive facsimile edition put the book within reach of a new generation, at least in theory, but in practice the book was still unavailable because the contents were too difficult for most readers without editorial help.[164]

Raymond P. Stearns, professor of history at the University of Illinois, offered the best discussion of Mather's place in the development of science in British America in his magisterial *Science in the British Colonies of America* (1970). Stearns views Mather as "a herald of the Enlightenment... in America." *The Christian Philosopher* is a "convincing testimonial" to Mather's "wide-ranging knowledge, if not complete understanding, of the new science," of the world of nature delineated by Newton and scores of lesser figures. "Notwithstanding his fundamental religious motivation," Stearns writes in a judicious appraisal, "Cotton Mather was the first native-born American colonial to advance beyond the status of a mere field agent for European scientists in the New World and to demonstrate a genuine *philosophical* approach to science, with scientific *ideas* and *hypotheses* of his own."[165]

Robert Middlekauff, professor of history at the University of California, Berkeley, touches lightly on *The Christian Philosopher* in *The Mathers: Three Generations of Puritan Intellectuals* (1971). Mather was disenchanted with reason by the time he finished writing this treatise, writes Middlekauff, though he continued to support the uses of experience. Mather's praise of natural theology in his book has often been noted, but his reservations about reason, such as appear in the essay "Of Man," have been ignored. Mather did not expect experimental philosophy or reason to lead men to understand religion. He relied on the Spirit and intuition to persuade them of the truth of Christianity, and his pietistic approach would have amazed the English writers on natural theology from whom he borrowed his general idea.[166]

Sacvan Bercovitch, professor of English at Harvard, evaluates Mather's scientific treatise in a book entitled *Major Writers of Early American Literature* (1972). Bercovitch admits the "transitional importance" of *The Christian Philosopher* "in the transformation of the earlier cosmology. Unquestionably, it is a crucial expression (in the New World) of the configuration of Puritanism, Pietism, and science which has been identified as the mainstream of American thought." For Bercovitch, "what distinguishes *The Christian Philosopher,* and what seems specifically to relate it to later American works, is the theocratic confluence of personal and social eschatology, transferred now to the mind of the awakened observer of Nature." Mather's integration of salvation, serviceableness, and history largely defines the method of his book. "Right perception, then, reconciles man simultaneously with Creation, with history, and with his Redeemer." At the same time, "*The Christian Philosopher* fails in [its] attempt to bridge science and pietism." Its most glaring inadequacies are "in the haphazard proliferation of literary modes." Mather's indiscriminate use of them "tends conspicuously toward chaos," and "it is this collapse of rhetorical distinctions," writes Bercovitch, "that most clearly marks the cultural significance of the book." The work serves to highlight the beginnings of a movement to mid-nineteenth-century American romanticism.[167]

Kenneth Silverman discusses *The Christian Philosopher* in *The Life and Times of Cotton Mather* (1984), a Pulitzer Prize-winning book. According to Silverman, professor of English at New York University, Mather moved effortlessly between theological and scientific modes of explanation. But the piety exemplified in such thinking obscured its dangerous implications and led Mather close to irreligion. Physicotheology tended in practice to displace Christianity. Yet Mather refused to accept a mechanical universe. He emphasized the limitations of human reason and the uncertainties of science, considering mechanical principles inadequate to explain large areas of the physical creation. For Silverman, Mather "figures

Introduction cxi

most importantly as a disseminator and popularizer of new scientific knowledge."[168]

Modern scholars have reasonable access to an original copy of Mather's work. At least sixty copies of the 1721 edition of *The Christian Philosopher* have survived into the late twentieth century. Most of these (fifty-four) are in the United States. A census reveals that thirty-five are on the East Coast. Massachusetts has eleven, located as follows: American Antiquarian Society, Worcester; Boston Public Library (two); Congregational Library, Boston; Essex Institute, Salem; Harvard University (five: one each at Andover-Harvard Theological Library and the Francis A. Countway Library of Medicine—Harvard Medical Library, Boston; and three at Houghton Library); and the Massachusetts Historical Society. Rhode Island has one (at the John Carter Brown Library). Connecticut has three (at Connecticut State Library—Wells Special Collection, Hartford; the Pequot Library, Southport; and Yale University—Beinecke Rare Book and Manuscript Library). New York has nine, located as follows: University of Rochester; Cornell University (two); the New York State Library; Vassar College; Columbia University; Union Theological Seminary; New York Public Library; and New York Academy of Medicine. New Jersey has one (at Princeton University). Pennsylvania has three (the Library Company of Philadelphia has two, with one at the American Philosophical Society). Maryland has three (The Johns Hopkins University has two: one in the William H. Welch Medical Library, for which the Historical Collection of the Institute of the History of Medicine is responsible, and the other in the George Peabody Library; and one is at the U.S. Naval Academy—Nimitz Library, Annapolis). The District of Columbia has one (Library of Congress). Virginia has three at the University of Virginia—Tracy W. McGregor Library. The South and the Midwest have thirteen. Georgia has one at Emory University—Robert W. Woodruff Library, Atlanta. North Carolina has one at Duke University—William R. Perkins Library. Texas has two: one at Rice University, the other at the University of Texas. Ohio has one (University of Cincinnati). Indiana has one (Indiana University—Lilly Library). Michigan has two (Michigan State University, Lansing; University of Michigan—William L. Clements Library). Illinois has two (the Newberry Library and the University of Illinois at Urbana-Champaign). Iowa has one (University of Iowa). Minnesota has one

(University of Minnesota). Kansas has one (University of Kansas—Clendening History of Medicine Library). On the West Coast, California has six: The Huntington Library, San Marino; Stanford University; the University of California (three: the Bancroft Library, Berkeley; UCLA; and Davis); and San Diego State University.

In addition to these fifty-four copies in the United States, six copies of the original edition are in foreign libraries. One is at the University of Western Ontario in Canada and four are in England: one at the British Library, London; one in Dr. Williams's Library, London; another in the Congregational Library, London (both of these libraries are administered by the librarian of Dr. Williams's Library); and one at the Bodleian Library in Oxford. Another copy is at the Royal Library in Copenhagen. Presumably an original edition, the one read by Christian Wolff, is in some German library, but efforts to locate it have been unavailing. Additional copies may exist in the United States, the United Kingdom, and on the Continent.[169] For example, the Wayne State University library reports holding a copy, but it may be among a group of rare books stolen and not yet recovered.

7. The Significance of *The Christian Philosopher*

Mather's treatise on science and religion never became a bestseller, but popular appeal is by no means the chief measure of a book's worth. Several reasons explain the significance of Mather's volume, and a discussion of them demonstrates that *The Christian Philosopher* is deservedly a classic of American intellectual history.

In the first place, Mather's treatise represents a departure from the earlier New England Puritanism. The author moves beyond the Calvinism of his grandfathers' generation with its emphasis on the supernatural, its interpretation of nature as inherently flawed, and its belief that a stern God governs through wonder-working providences. Mather's interest in natural philosophy, which came early in life, helped him transcend this legacy. The Puritan founders focused on Scripture rather than on nature; he reverses the emphasis. His commitment to science carried over to later works. In the *Manuductio ad Ministerium* (1726), for example, he encouraged candidates for the ministry to spend more time studying natural philosophy. "Be sure," he added, "The *Experimental*

Philosophy is that, in which alone your Mind can be at all established." For this purpose, he advised the reading of various authors, especially his own *Christian Philosopher*.[170] His positive attitude toward nature breaks with the past. Taking his cue from John Ray, he praises the beauty of the material world—the order of the heavens, the artifice of insects, the figure of the human body, the stupendous faculties of the soul. Mather's love of nature for its own sake found stronger voices in late eighteenth-century America; indeed, some of his passages anticipate the nature rhapsodies of nineteenth-century romantics, although they did not draw on Mather for their inspiration.

Nevertheless, the Puritan mind was a rope of many strands—Calvinist theology, Scholasticism, Ramist thought, humanism, and folk tradition—and Mather did not break completely with the past. Critics have called him exceptionally credulous. True, he does perpetuate some superstitions. He believed, for example, that the moon exercises "very wonderful" influences on sublunary bodies [60:64–85] and that the edge of a steel axe bitten by a rattlesnake would immediately fall off owing to putrefaction [180:9–12]. And he repeats tales of fabulous creatures, including one about a serpent 120 feet long [179:77–78] and another about a serpent who lifted his head and articulately cried out a warning [181:26–33]. At the same time, Mather abandons other superstitions. He refused to take astrology seriously, he scorned the notion that frogs were generated by rain or in the clouds [155:5–8], and he was quick to insist that the doctrine of spontaneous generation was untenable. The point is, Mather lived in a time of rapid and fundamental intellectual transition. Science had not yet completely emancipated itself from magic, the occult, and folklore in this seminal period of Western intellectual history, and Mather's views are in accord with the best scientific knowledge of the time. He remained a Christian in the Calvinist tradition after writing *The Christian Philosopher*. He believed that advances in understanding the regular operations of nature intensified belief in design, thereby strengthening rather than weakening religious faith. What was weakened was biblical authority in matters of science, a weakening that had been going on for a long time.

In the second place, and closely related, *The Christian Philosopher* is a harbinger of the Enlightenment in America. Nature and reason were the two main commitments of this historical movement; Mather's treatise is a testimonial to both. In modifying his Puritan legacy, he goes far in making nature the ultimate source of authority regarding belief in the existence and attributes of God. Receptive to progress in science since the time of Copernicus, he abandons traditional Scholastic and Ramist authorities who assumed a firm connection between natural and supernatural truth and emphasized the latter. His intention is to popularize the best recent advances in natural philosophy. He conveys a marvelous sense of excitement over discoveries that demonstrate the secondary causes by which God operates in the physical universe. He makes *The Christian Philosopher* a repository of empirical data supporting a scientific outlook. He goes dangerously far at times in characterizing nature in mechanistic terms, but he does not go all the way. There must be, he frequently insists, a power above mechanism.[171]

Reason was also a basic commitment of the Enlightenment. To be sure, Puritan thought had always placed a high valuation on the rational faculty. In the older tradition, the emphasis was on the connection between nature and divine wisdom; a person's epistemological contact with the physical universe was shaped by Scripture. In the eighteenth century, however, the word *reason* took on new meaning. Reason was substituted for Scripture and traditional authorities as the interpreter of the book of nature. Reason, to Mather, is a divinely implanted faculty which enables people to discern God's truth and make true inferences from it. This conviction gave rise to an optimism about the future that became characteristic of the Enlightenment. According to Mather, reason permits discoveries and inventions; thus, the boundaries of science are not fixed, and greater progress is yet to come. Like Ray, he quotes Seneca: "the people of the next age shall know many things unknown to us."[172]

Nevertheless, Mather is ambivalent about the role of reason. Perhaps fearful that the critical use of reason freed from textual authority could be threatening, he describes the "natural Imbecillity" of the human reason and insists that this instrument is too feeble to comprehend the infinite God. Moreover, while celebrating nature and reason, Mather makes his book a hymn of praise to the Creator. He never forgets that the purpose of *The Christian Philosopher* is to enkindle piety.

In the third place, Mather's volume contributed to the under-

Introduction

standing of curiosities of nature in America. Mather's interest in the physical universe was lifelong, but he generally studied books rather than nature. Nevertheless, he was warmly attached to his own country, and in *The Christian Philosopher* he furthered his resolve to do good and glorify God by describing nature as he found it in New England. Some of his most valuable contributions to science appear for the first time in this book.

In the fourth place, *The Christian Philosopher* is a significant document in the transmission of the humanistic legacy to early America. Natural philosophy and theology had traditionally been closely related in Western thought; in the early modern period the study of science was still part of the humanities rather than simply a number of specialized disciplines. Mather was a good citizen of the *respublica litterarum,* the Republic of Letters. His book is a bridge by which the ripe learning of the ages passes from the Old World to the New.

In each essay he first describes what ancient authorities said on the subject, then he recounts what medieval and Renaissance investigators added, and finally he traces the best recent discoveries encountered in his reading. He writes at the end of the period when Latin was the universal language of science, and does not hesitate to quote in both Greek and Latin. His familiarity with classical and Christian antiquity stood him in good stead, and the Renaissance had made ancient texts readily available. Mather names thirteen prominent Greek authors. Those named five times or more are Galen (twenty), Aristotle (sixteen), Plato (seven), Hippocrates (five), and Plutarch (five). Of this group, Plato and Galen are by far the most important to his thought. He names twenty-two prominent Latin writers. Those named five times or more are Pliny (thirty-two), Cicero (sixteen), Vergil (seven), Ovid (six), and Seneca the Younger (five). He names thirteen prominent Christian authors or writers on Scripture. Only Augustine (six) is named five times or more. In addition, he repeatedly quotes and paraphrases the Bible.[173]

The Protestant author skips lightly over the Middle Ages, although he does quote a late medieval author, Raymond of Sebonde. Mather names some 140 persons identified with science during the Renaissance (between 1450 and 1630). He is mainly concerned with recent developments, however, and most of the 415 authors he discusses were active during the late seventeenth and early eighteenth centuries. As a result *The Christian Philosopher* is a valuable piece of humanistic literature. It transmits much of the best of the classical, Christian, and early modern legacy to the American colonies.

In the fifth place, *The Christian Philosopher* merits high consideration for its literary artistry. Mather's "massy" way of writing is the product of his theory of style. A gluttonous reader, he wished to give each sentence a sensible vigor and to embellish every paragraph with profitable references. He loved allusions and copious quotations; his prose is made ornate by them. He makes heavy use of italics, capital letters, and exclamation marks. Like his other compositions, *The Christian Philosopher* is "not only a *Cloth of Gold,* but also stuck with as many *Jewels,* as the Gown of a Russian Embassador."[174] The medium is inseparable from the message. Mather has a gift for pithy statement. A close study of his borrowing shows that he is invariably superior to his sources. He writes succinctly while capturing the essentials. In describing nature, he combines the observant eye of the scientist with the feeling of the poet. He keeps the pace lively and delights in telling a good story. Mather's descriptions of bees at work and at war [164:28–165:60], of the effects of the bite of the tarantula [182:52–183:93], of the nidification of birds [201:1–15], of a kid goat reared by Galen [225:2–16], of a curiosity relating to the elephant [220:49–221:62], and of reason in dogs [226:57–227:73] leave an indelible impression. Mather's *Christian Philosopher* is excellent reading. It is artistically worthy, one of the most impressive productions of early American history.

Finally, and above all, *The Christian Philosopher* is significant because of the author's philosophical approach to his subject. Mather not only describes but also interprets nature, combining both theological and scientific modes of explanation. He interweaves the laws of matter and the decrees of God, confident that the book of nature and the book of Scripture are the two sources of human enlightenment. Nature is both emblematic and purposive. There is a power in the universe superior to mechanism. Mather is not a detached observer; he displays an emotional response to the results of scientific inquiry. He offers a hymn of praise to the Creator. In pursuing his thesis of the harmony between science and religion, he offers the first systematic statement in America of both the design argument and natural theology.

Using nature as a rational foundation for Christianity, Mather's exposition describes a clockwork universe regulated by secondary causes. The human reason is to guide the attempt to unlock nature's secrets. Mather may have taken a step toward deism in embracing natural theology, but he is no Deist. He erects a superstructure of Christian revelation upon a foundation of natural theology. That he did not fully accept the mechanical philosophy is evidenced by his attempt to explain both gravity and magnetism as well as animal instinct. Mather's senses demonstrate the wonders of the invisible as well as of the visible world; he acknowledges intuition as a source of religious faith and urges the Christian philosopher to remember Christ. Mather remains a Puritan even as he embraces the new science.

Mather's thesis is the design argument, an ancient doctrine that met practically universal acceptance in the eighteenth century. It was regarded as the all-sufficient ground for belief and the trump card against atheism. The design argument was so basic during the Enlightenment that its premises were taken for granted. Nevertheless, uncritical acceptance of it was a dangerous foundation for Christian faith. Hume published a penetrating criticism of design in his *Dialogues concerning Natural Religion* (1779). He found the argument untenable as an argument: random variation might have produced order in the world; we have no right to infer design from the facts of the universe.[175] Hume revived the possibility of Epicurean chance, a doctrine Mather loathed. But if the Scot effectively demonstrated the logical deficiencies of the ancient belief, he did not diminish its universal acceptability. Meanwhile, physicotheology bred optimism—all is for the best in this best of all possible worlds—and probably contributed to a type of faintheartedness that prevented Christians from meeting the doubts regarding design which Hume expressed.

In the late eighteenth century, the destructive criticisms of Hume and Kant began to undermine belief in design. By the early nineteenth century many people in the Western world, including Emerson and Theodore Parker in the United States, were disenchanted with the argument from design. Then Darwin dealt his lethal blow. Design was the central point around which the Darwinian controversy later turned. Darwin used natural selection to explain evolution, and natural selection cut the ground from under the age-old design argument.[176]

As the nineteenth century gave way to the twentieth, William James declared the "bankruptcy of natural religion naively and simply taken" and observed that "it is strange . . . to see how little [the argument from design] counts for since the triumph of the darwinian theory. Darwin opened our minds to the power of chance-happenings to bring forth 'fit' results if only they have time to add themselves together." It used to be a question of purpose *against* mechanism, James added, but theologians had stretched their minds so as to accommodate both. Now divine purpose is seen as accomplished through "the sole agency of nature's vast machinery." "This saves the form of the design-argument at the expense of its old easy human content. The designer is no longer the old man-like deity." But since we can hardly comprehend the character of a cosmic mind whose purposes are fully revealed by the strange mixture of goods and evils that we find in this actual world's particulars, the word "design" by itself explains nothing. "The old question of *whether* there is design is idle. The real question is *what* is the *world*, . . . and that can be revealed only by the study of all nature's particulars." No matter what nature may have produced, the means must have been fitted to that production. "We can always say, therefore, in any conceivable world, of any conceivable character, that the whole cosmic machinery *may* have been designed to produce it. Pragmatically, then, the abstract word 'design' is a blank cartridge. It carries no consequences." The only serious questions are what sort of design and designer, and the study of facts is the only way of getting the answers. Meanwhile, pending the slow answer from facts, anyone who insists that there is a designer and a divine one gets a certain pragmatic benefit from the term. " 'Design,' worthless tho it be as a mere rationalistic principle set above or behind things for our admiration, becomes, if our faith concretes it into something theistic, a term of *promise*. . . . This vague confidence in the future is the sole pragmatic meaning at present discernible in the terms design and designer." According to James, "there *is* a Spirit in things to which we owe allegiance," but it is inadequately revealed in visible nature, which is "but a veil and surface-show whose full meaning resides in a supplementary unseen or *other* world."[177]

With James we enter a new era of thought. Even so, the design argument remains deeply embedded in the Western mind. Scientists, theologians, and philosophers continue to debate the cosmological and teleological arguments for the existence of God, while at the popular level reluctance to abandon belief in a divine designer remains strong. "Among biologists and scientists generally," Alvar Ellegård wrote in 1956, "the claims of religious metaphysics [he is referring to the design argument] to determine the map of world experience are no longer upheld."[178] Yet the scientist and theologian Charles E. Raven, writing in 1954, observed a renewed interest in design on the part of many scientists in many different fields. Every mechanism implies purpose in its structure and operations, Raven held, and while refusing to return to the old analogy of the watchmaker and his watch or to the concept of the Creator and his creation which this implies, he observed that an analogy which has served continuously since the time of Cicero (whose *De natura deorum* argues that it is reasonable to infer the existence of God from the order of nature), which is still congruous with our own notion of God as mathematician, architect, and engineer, could not be lightly dismissed.[179] Meanwhile, philosophers were making strong and sophisticated arguments for the existence of God based upon empirical premises about the natural world, and critics were questioning them.[180]

We need not pursue this matter further, for enough has been said to demonstrate the centrality and persistence of the problem. A proper appreciation of the Darwinian controversy and the resulting belief that conflict rather than harmony characterizes the relation between science and religion requires an understanding of the background out of which they arose. In American history, the proper place to begin is with *The Christian Philosopher*.[181]

Mather's book derives much of its importance from the fact that it is the first statement of the design argument written by an American. He popularized a belief that had been an essential element of Western thought for over two thousand years and continues to maintain a powerful hold in modern culture. Mather emphasizes the harmony between science and religion. God is the creator of the world, the universe exhibits order and purpose, and nature demonstrates the power, wisdom, and goodness of God. Human beings possess supreme dignity and responsibility in a finite universe. These views reflect the core values of Western civilization.

Mather is too often remembered as a Puritan bigot who persecuted witches; by now, however, it should be abundantly clear that he deserves better. Cotton Mather was one of the most progressive thinkers in early America. When the name of that remarkable American flashes on the screen, rather than boo and hiss, one should praise and cheer. *The Christian Philosopher* is deservedly an American classic.

NOTES

1. The text of the resolution is in William H. Austin, *The Relevance of Natural Science to Theology* (London: Macmillan Press, 1976), pp. 1–2. See also A. R. Peacocke, *Creation and the World of Science: The Bampton Lectures, 1978* (Oxford: Clarendon Press, 1979), pp. 1–7.

2. M. B. Foster, "The Christian Doctrine of Creation and the Rise of Modern Natural Science," *Mind*, 43 (1934), 446–48, and idem, "Christian Theology and Modern Science of Nature," ibid., 45 (1936), 1–27.

3. Frederick Ferré, "Design Argument," in Philip P. Wiener, ed., *Dictionary of the History of Ideas*, 5 vols. (New York: Charles Scribner's Sons, 1968–74), 1:670–77; Robert H. Hurlbutt III, *Hume, Newton, and the Design Argument*, rev. ed. (Lincoln: University of Nebraska Press, 1985); Thomas McPherson, *The Argument from Design* (London: Macmillan Press, 1972); and Clement C. J. Webb, *Studies in the History of Natural Theology* (Oxford: Clarendon Press, 1915). Natural religion, defined as moral and religious beliefs based upon reason and tied to the notion of nature, may be distinguished from natural theology. But the two terms are often used interchangeably. See Hurlbutt, *Design Argument*, pp. 65–78.

4. Charles Gore, *The Reconstruction of Belief*, new ed. (London: J. Murray, 1926), pp. 5–6; Charles E. Raven, *Organic Design: A Study of Scientific Thought from Ray to Paley* (London: Oxford University Press, 1954), pp. 9–15. See also D. L. LeMahieu, *The Mind of William Paley* (Lincoln: University of Nebraska Press, 1976).

5. For evaluations of Mather's treatise by American scholars in the twentieth century, see section 6.

6. Samuel Mather, *The Life of the Very Reverend and Learned Cotton Mather* (Boston, 1729), pp. 3–5. The most recent biographies are David Levin, *Cotton Mather: The Young Life of the Lord's Remembrancer, 1663–1703* (Cambridge, Mass.: Harvard University Press, 1978), and Kenneth Silverman, *The Life and Times of Cotton Mather* (New York: Harper and Row, 1984). On Increase, see Michael G. Hall, *The Last American Puritan: The Life of Increase Mather, 1639–1723* (Middletown, Conn.: Wesleyan University Press, 1988).

7. Cotton Mather, *Paterna: The Autobiography of Cotton Mather*, ed. Ronald

A. Bosco (Delmar, N.Y.: Scholars' Facsimiles and Reprints, 1976), pp. 6–7; Samuel Mather, *Cotton Mather*, pp. 6, 7.

8. Cotton Mather, *Paterna*, p. 7; idem, *Corderius Americanus: An Essay upon the Good Education of Children . . . in a Funeral Sermon upon Mr. Ezekiel Cheever* (Boston, 1708), pp. 21–33. See also Robert Middlekauff, *Ancients and Axioms: Secondary Education in Eighteenth-Century New England* (New Haven: Yale University Press, 1963), pp. 53, 75–88.

9. Levin, *Cotton Mather*, pp. 23, 25–26; Samuel E. Morison, *Harvard College in the Seventeenth Century*, 2 vols. (Cambridge, Mass.: Harvard University Press, 1936), 1:195–96.

10. "Harvard College Records, Part III," *Collections of the Colonial Society of Massachusetts*, 31 (1925), 325–39; Morison, *Harvard*, 1:200 (quotation); Levin, *Cotton Mather*, pp. 42–48.

11. Morison, *Harvard*, 1:208–84 passim; Levin, *Cotton Mather*, p. 39; Cotton Mather, *Paterna*, p. 8 (quotation).

12. Morison, *Harvard*, 1:224–27, 157–58.

13. Ibid., 1:223–33, 236–51. In 1697, Samuel Clarke, an English Newtonian, published a Latin translation of Rohault's *Traité*, adding notes based on Newton's views in order to counter Rohault's Cartesian tract. He amplified the notes in later editions, producing a book containing both Cartesian and Newtonian physics. A facsimile of Clarke's 1723 English translation was reprinted as *A System of Natural Philosophy*, intro. L. L. Laudan, Sources of Science, no. 50 (New York: Johnson Reprint Corp., 1969).

14. Morison, *Harvard*, 1:250–51. Morton's work has been given a modern edition: Charles Morton, *Compendium Physicae* (1687), Publications of the Colonial Society of Massachusetts, 33 (Boston: Colonial Society of Massachusetts, 1940). Theodore Hornberger collated the variant texts of Morton's manuscript and wrote an introduction; Samuel E. Morison wrote a biographical sketch of the author.

15. Samuel Mather, *Cotton Mather*, pp. 26–27; Levin, *Cotton Mather*, pp. 28–39.

16. Morison, *Harvard*, 1:292; *Diary of Cotton Mather*, ed. Worthington C. Ford, in *Collections of the Massachusetts Historical Society*, 7th ser. vols. 7–8 (Boston: Massachusetts Historical Society, 1911–12), 7:368 (Ford's edition of the *Diary* was published separately as *Diary of Cotton Mather*, 2 vols. [New York: Frederick Unger, 1957]); hereafter Mather, *Diary*, vol. 1 or 2 rather than 7 or 8; Thomas G. Wright, *Literary Culture in Early New England, 1620–1730* (New Haven: Yale University Press, 1920), p. 178; Julius H. Tuttle, "The Libraries of the Mathers," *Proceedings of the American Antiquarian Society*, 20 (1911), 296–356; Henry J. Cadbury, "Harvard College Library and the Libraries of the Mathers," ibid., 50 (1941), 20–48; Samuel Mather, *Cotton Mather*, p. 24.

17. Levin, *Cotton Mather*, pp. 47–48.

18. Cotton Mather, *Paterna*, pp. 9–12; Samuel Mather, *Cotton Mather*, pp. 8–10, 5–6; Levin, *Cotton Mather*, pp. 58–66, 74–77, 83–88.

19. Samuel Mather, *Cotton Mather*, pp. 27–28; Levin, *Cotton Mather*, pp. 88–91.

20. Levin, *Cotton Mather*, pp. 66–70, 80–82; Silverman, *Cotton Mather*, pp. 56–57.

21. The paragraphs on political affairs draw on John Gorham Palfrey, *History of New England*, 5 vols. (Boston, 1865–90), vol. 3, esp. chaps. 8–9, 12–15; Levin, *Cotton Mather*, pp. 66–105 passim; and Silverman, *Cotton Mather*, pp. 59–82.

22. Palfrey, *History*, 3:356 (quotation); Levin, *Cotton Mather*, pp. 102–3.

23. On the American events, see David S. Lovejoy, *The Glorious Revolution in America* (New York: Harper and Row, 1972).

24. Palfrey, *History*, 3:324, 348, 353, 494–95, 499–502, 521.

25. Levin, *Cotton Mather*, pp. 72–74; Silverman, *Cotton Mather*, pp. 57–59. See also Perry Miller, *The New England Mind: From Colony to Province* (Cambridge, Mass.: Harvard University Press, 1953), and Sacvan Bercovitch, *The American Jeremiad* (Madison: University of Wisconsin Press, 1978).

26. C. S. Lewis, *The Discarded Image: An Introduction to Medieval and Renaissance Literature* (Cambridge: University Press, 1964) describes the older view. On the American mind in transition during the period covered here, see Herbert Leventhal, *In the Shadow of the Enlightenment: Occultism and Renaissance Science in Eighteenth Century America* (New York: New York University Press, 1976); David D. Hall, *Worlds of Wonder, Days of Judgment: Popular Religious Belief in Early New England* (New York: Alfred A. Knopf, 1989); and Jon Butler, *Awash in a Sea of Faith: Christianizing the American People* (Cambridge, Mass.: Harvard University Press, 1990), chap. 3.

27. Raymond P. Stearns, *Science in the British Colonies of America* (Urbana: University of Illinois Press, 1970), pp. 150–61.

28. Otho T. Beall, Jr., "Cotton Mather's Early 'Curiosa Americana' and the Boston Philosophical Society of 1683," *William and Mary Quarterly*, 3d ser., 18 (July 1961), 359–63; Mather, *Diary*, 2:85–86; Michael G. Hall, *Last American Puritan*, p. 166.

29. The references in brackets are to page(s) and line(s) in Mather's text.

30. Miller, *New England Mind: Colony to Province*, p. 441; Otho T. Beall, Jr., and Richard H. Shryock, *Cotton Mather: First Significant Figure in American Medicine* (Baltimore: Johns Hopkins University Press, 1954), p. 50; Cotton Mather, *Bonifacius: An Essay upon the Good* (Boston, 1701), pp. 203–6.

31. Mather, *Diary*, 1:103, 169–71, 230–31.

32. The hand of the copyist is evident in commentary on Genesis 1 and on 2 Peter 3. Mather, *Diary*, 2:358 (quotation). In his commentary on Judges 15:20, Mather poses a question about the great strength of Samson and answers it with a long quotation introduced as taken from *The Christian Philosopher* and identified with running quotation marks in the margin. The quoted material runs from 293:67 to 295:22 in the printed version of the book. Mather himself had originally taken the material from an English author. For further details, see the Endnotes pertaining to this portion of *The Christian Philosopher*. Professor Edward H. Davidson called this sentence in the "Biblia Americana" to my attention.

33. The quotation is from Worthington C. Ford, editor of Mather's *Diary*, 1:170n.

34. The bound materials are largely unpaginated, but materials can be located by biblical versification. For the quotation, see the commentary on Genesis 1.

35. See the commentary on Genesis; see also Theodore Hornberger, "Cotton Mather's Annotations on the First Chapter of Genesis," *University of Texas Publications*, no. 3826 (1938), 113-21.

36. Mather, *Diary*, 1:229-30, 545, 563-64 (first quotation at 564), 570 (second quotation).

37. Ibid., 2:40-41, 162, 178, 283 (the quotation), 309-10, 314-15, 330-32; Cotton Mather, *Bonifacius*, pp. 200-206.

38. Mather, *Diary*, 2:311-12, 317-18, 330-32, 376 (first quotation), 377, 413 (second quotation), 416, 436.

39. The first volume of Poole's *Synopsis criticorum aliorumque sacrae Scripturae interpretum* was published in 1669; his completed work, in Latin, appeared in five volumes by 1676. The first volume of Henry's *Exposition of the Old and New Testament* was published in 1708; four other volumes bringing his commentary to the end of the Gospels appeared in a uniform edition in 1710. Friends completed his exposition after his death in 1714.

40. Thomas H. Olbricht, "Biblical Primitivism in American Biblical Scholarship, 1630-1870," in Richard T. Hughes, ed., *The American Quest for the Primitive Church* (Urbana: University of Illinois Press, 1988), p. 88.

41. Stearns, *Science in the British Colonies of America*, pp. 405-6; Beall, "Cotton Mather's 'Curiosa Americana,'" pp. 364-66; George L. Kittredge, "Cotton Mather's Scientific Communications to the Royal Society," *Proceedings of the American Antiquarian Society*, n.s., 26 (1916), 18-57. It appears that Mather rarely draws directly on the "Miscellanea Curiosa" for *The Christian Philosopher;* he takes his information from the "German Ephemerides" from an intermediate source.

42. Stearns, *Science in the British Colonies of America*, pp. 405-6.

43. "An Extract of Several Letters from Cotton Mather, D.D. to John Woodward, M.D. and Richard Waller, Esq; S.R. Secr," *Philosophical Transactions*, 29 (Apr.-June 1714), 62-71; Beall, "Cotton Mather's 'Curiosa Americana,'" pp. 369-71; David Levin, "Giants in the Earth: Science and the Occult in Cotton Mather's Letters to the Royal Society," *William and Mary Quarterly*, 3d ser., 45 (Oct. 1988), 751-70.

44. Kittredge, "Cotton Mather's Scientific Communications to the Royal Society," pp. 18-19; Beall, "Cotton Mather's 'Curiosa Americana,'" pp. 371-72. On the persistence of superstition, see Leventhal, *In the Shadow of the Enlightenment;* David D. Hall, *Worlds of Wonder;* and Butler, *Awash in a Sea of Faith*, chap. 3.

45. Stearns, *Science in the British Colonies of America*, pp. 407-9 (quotation at 407); Mather, *Diary*, 2:245-46. On Mather's election and admission into the Royal Society and the ensuing problems over the matter, see George L. Kittredge, "Cotton Mather's Election into the Royal Society," *Publications of the Colonial Society of Massachusetts*, 14 (1913), 81-114, and idem, "Further Notes on Cotton Mather and the Royal Society," ibid., pp. 281-92.

46. Stearns, *Science in the British Colonies of America*, pp. 408-9.

47. "Extract of Several Letters from Cotton Mather," pp. 62–63.

48. Kittredge, "Cotton Mather's Scientific Communications to the Royal Society," pp. 27–35.

49. Ibid., pp. 36–57; Stearns, *Science in the British Colonies of America*, p. 409.

50. Silverman, *Cotton Mather*, p. 188; *The Diary of Cotton Mather, D. D., F.R.S. for the Year 1712*, ed. William R. Manierre II (Charlottesville: University Press of Virginia, 1964), pp. 58, 60, 61, shows Mather's concern to battle deism.

51. David D. Hall, *Worlds of Wonder*, pp. 17, 19, 162.

52. "Criticism of science and an awareness of areas of conflict between faith and scientific rationalism begin to appear between 1690 and 1740," wrote Theodore Hornberger, "perhaps because of echoes in America of the English controversy over Deism, or as a result of the visits of George Whitefield, the evangelist, and the beginnings of Methodism." Thus, any conflict arose largely *after* Mather's book appeared. Charles Chauncy of the Harvard class of 1721 "went the way of scientific rationalism," added Hornberger, but Chauncy was not the "central figure among the graduates of American colleges in the transition period of scientific instruction." In the colonies, said Hornberger, the outstanding intellectual successors to the Mathers were Samuel Johnson and Jonathan Edwards. Both were graduates of Yale and "both unquestionably understood the New Science as the Mathers never did." But both men lacked the enthusiasm for science displayed by their predecessors. See Hornberger, *Scientific Thought in the American Colleges, 1638–1800* (Austin: University of Texas Press, 1945), pp. 82–83. On the growth of Deism and atheism in England, see Margaret Jacob, *The Newtonians and the English Revolution, 1689–1720* (Ithaca: Cornell University Press, 1976), pp. 201–50.

53. Mather, *Diary*, 2:324. Mather cites books published in England as late as 1715 (the year he sent his manuscript to England). He may have used the English translation of David Gregory's *Elements of Astronomy* (London, 1715), although it is possible that he used the Latin edition (London, 1702). Mather paraphrases a long passage from George Cheyne's *Philosophical Principles of Religion: Natural and Revealed* (London, 1715), which was not in the first edition of Cheyne's work, *Philosophical Principles of Natural Religion* (London, 1705). Mather drew heavily on the 1705 edition. The 1715 edition may have appeared early and been postdated.

54. Stearns, *Science in the British Colonies of America*, p. 409; Mather, *Diary*, 2:332. Note that here Mather views "The Christian Virtuoso" as part of the "Biblia Americana." Perhaps he initially conceived of his manuscript on Newtonian science as part of his biblical commentary and mistakenly referred to it in those terms on this occasion.

55. Walter E. Houghton, "The English Virtuoso in the Seventeenth Century," *Journal of the History of Ideas*, 3 (Jan. 1942), 51–73; (Apr. 1942), 190–219.

56. Mather, quoting Dummer, in a letter to John Winthrop, in *The Mather Papers, Collections of the Massachusetts Historical Society*, 4th ser., 8 (1868), 445.

57. Mather to Benjamin Colman, 14 September 1722, in *Selected Letters of Cotton Mather*, ed. Kenneth Silverman (Baton Rouge: Louisiana State University Press, 1971), pp. 353–54; A. H. Grant, "Barrington, John Shute," *DNB*, 1:1209–11; Alexander Gordon, "Bradbury, Thomas," *DNB*, 2:1058–61.

58. Mather to Colman, 14 September 1722, in *Selected Letters of Cotton Mather*, pp. 353–54; Mather, *Diary*, 2:800.

59. Josiah Quincy, *The History of Harvard University*, 2 vols. (Cambridge, Mass., 1840), 1:231–32.

60. Marjorie Plant, *The English Book Trade: An Economic History of the Making and Sale of Books* (London: George Allen and Unwin, 1939), pp. 82, 94, 223; Henry R. Plomer, *A Dictionary of the Printers and Booksellers Who Were at Work in England, Scotland and Ireland from 1688 to 1725* (Oxford: Oxford University Press, 1922), pp. 165–200.

61. Mather, *Diary*, 2:610; *Mather Papers*, p. 447; Kittredge, "Cotton Mather's Election into the Royal Society," p. 98n.

62. Mather, *Diary*, 2:610; *Mather Papers*, pp. 447–48 (first quotation at 447), 455 (second quotation); *Selected Letters of Cotton Mather*, p. 357.

63. Theodore Hornberger, "Notes on The Christian Philosopher," in Thomas J. Holmes, ed., *Cotton Mather: A Bibliography of His Works*, 3 vols. (Cambridge, Mass.: Harvard University Press, 1940), 1:137, 138.

64. Archibald W. Smith, *A Gardener's Dictionary of Plant Names*, [new ed.] rev. and enlarged by William T. Stearn and Isadore L. L. Smith (New York: St. Martin's Press, 1972), p. 153. Most of Mather's sources incorporated their references in the text, but William Derham, the author he relied on most heavily, used copious footnotes.

65. A considerable literature describes the European mind in transition during the period covered here. The American story is told in a number of works, including Leventhal, *In the Shadow of the Enlightenment*; David D. Hall, *Worlds of Wonder*; and Butler, *Awash in a Sea of Faith*.

66. For details on Mather and his borrowing, see the "Recapitulation of Mather's Sources." To prepare this recapitulation I counted lines in Mather's 1721 edition, a simple if tedious mechanical task, and presumably accurate for full lines. The handling of partial lines poses a more difficult challenge. I have tried to count to the quarter line and add up totals. This procedure does not guarantee exactitude (e.g., it yields 11,062 lines, whereas a count of full lines gives 11,250), but it should be accurate enough for all practical purposes.

67. The following account of Ray is most indebted to Charles E. Raven, *John Ray, Naturalist: His Life and Work*, 2d ed. (Cambridge: University Press, 1950).

68. Ibid., p. 59.

69. Ibid., p. 81. On Ray's publications, see Geoffrey Keynes, *John Ray: A Bibliography* (London: Faber and Faber, 1951).

70. Raven, *John Ray*, p. 183.

71. Ibid., p. 172.

72. Ibid., pp. 186 (quotation), 195; see also p. 200.

73. Ibid., pp. 221–22, 225 (quotation).

74. Ibid., p. 309.
75. Ibid., pp. 403–4; Keynes, *John Ray*, p. 137.
76. As quoted in Raven, *John Ray*, pp. 251, 252.
77. Ibid., pp. 458–61.
78. Ibid., p. 457.
79. Ibid., p. 452 (the quotation; I have changed the tense); Keynes, *John Ray*, pp. 91–106. See also LeMahieu, *Mind of William Paley*.
80. John Ray, *The Wisdom of God Manifested in the Works of the Creation*, 5th ed. (London, 1709), p. 17. Mather probably used this edition. The fourth edition had appeared in 1704.
81. Ibid., pp. 117, 118.
82. Ibid., p. 187.
83. Ibid., p. 193.
84. Raven, *John Ray*, pp. 466–67.
85. Ray, *Wisdom of God*, p. 201.
86. Raven, *John Ray*, pp. 452–53 (quotation at 452), 467.
87. Ibid., pp. 419–51.
88. Ibid., pp. 481, x, 477–78.
89. Basil Willey, *The Eighteenth Century Background: Studies on the Idea of Nature in the Thought of the Period* (New York: Columbia University Press, 1940), pp. 39–42; A. D. Atkinson, "William Derham, F.R.S. (1657–1735)," *Annals of Science*, 8 (1952), 368–92 (quotation at 369); Jacob, *Newtonians and the English Revolution*, pp. 145–46, 162–63, 172, 178–81.
90. Lael Ely Bradshaw, "John Harris's *Lexicon technicum*," in Frank A. Kafker, ed., *Notable Encyclopedias of the Seventeenth and Eighteenth Centuries: Nine Predecessors of the Encyclopedie*, Studies on Voltaire and the Eighteenth Century, 194 (Oxford: Voltaire Foundation, 1981), pp. 107–21; Robert Collison, *Encyclopaedias: Their History throughout the Ages* (New York: Hafner Publishing Co., 1964), p. 99; Jacob, *Newtonians and the English Revolution*, pp. 145–46, 159, 178–79.
91. Wright, *Literary Culture in Early New England*, p. 175.
92. Theodore M. Brown, "Cheyne, George," in Charles C. Gillispie, ed., *Dictionary of Scientific Biography*, 15 vols. (New York: Charles Scribner's Sons, 1970–78), 3:244–45 (hereafter, *DSB*).
93. Agnes Arber, "Tercentenary of Nehemiah Grew (1641–1712)," *Nature*, 147 (24 May 1941), 630–32; Raven, *John Ray*, pp. 188, 200–201, 455; Charles R. Metcalfe, "Grew, Nehemiah," in *DSB*, 5:534–36.
94. Charles Webster, "Alsted, Johann Heinrich," in *DSB*, 1:125–27; idem, *The Great Instauration: Science, Medicine, and Reform, 1616–1660* (London: Duckworth, 1975), pp. 5–13 passim, 32; E. F. Karl Müller, "Alsted, Johann Heinrich," in Shirley M. Jackson, ed., *The New Schaff-Herzog Encyclopedia of Religious Knowledge*, 12 vols. (New York: Funk and Wagnalls Co., 1908–12), 1:138; Walter J. Ong, *Ramus, Method, and the Decay of Dialogue: From the Art of Discourse to the Art of Reason* (Cambridge, Mass.: Harvard University Press, 1958), pp. 160, 163–65, 298–99; Cotton Mather, *Manuductio ad Ministerium: Directions for a Candidate of the Ministry* (Boston, 1726), p. 33 (quotation).
95. H. Hölscher, "Arndt, Johann," in *New Schaff-Herzog Encyclopedia*, 1:299;

Charles F. Schaeffer, "Introduction" to Arndt's *True Christianity*, new American ed. (Philadelphia, 1868), pp. xi–xxxviii; and Heiko A. Oberman, "Preface," and Peter Erb, "Introduction" to Erb's edition of Arndt, *True Christianity* (New York: Paulist Press, 1979).

96. Kittredge, "Cotton Mather's Scientific Communications to the Royal Society," pp. 26, 29–30, 31; for further detail, see the Endnotes for the appropriate pages of Mather's text.

97. Ibid., pp. 26–27; also see the Endnotes.

98. Ibid., pp. 38, 42–43, 44–45. Mather copied the material from 293:67 to 295:22 in *The Christian Philosopher* into the "Biblia Americana."

99. See Winton U. Solberg, "Cotton Mather, *The Christian Philosopher*, and the Classics," *Proceedings of the American Antiquarian Society*, n.s., 96 (Oct. 1986), 323–66.

100. I have treated some of the material in this section in Winton U. Solberg, "Science and Religion in Early America: Cotton Mather's *Christian Philosopher*," *Church History*, 56 (Mar. 1987), 73–92.

101. The literature on the argument from design is extensive. Representative of the older works on the subject are Paul Janet, *Final Causes*, trans. from the French by William Affleck (Edinburgh, 1878), and L. E. Hicks, *A Critique of Design-Arguments: A Historical Review and Free Examination of the Methods of Reasoning in Natural Theology* (New York, 1883). Among recent volumes are the works by Hurlbutt, McPherson, and Raven cited in nn. 3–4 above. On natural theology, see the volume by Webb cited in n. 3.

102. Alfred E. Taylor, *Platonism and Its Influence* (Boston: Marshall Jones Co., 1924; rpt., New York: Cooper Square Publishers, 1963), pp. 3–56 passim; Hurlbutt, *Design Argument*, pp. 95–99. The influence of the Platonist-Augustinian tradition on American Puritanism is emphasized in Perry Miller, *The New England Mind: The Seventeenth Century* (Cambridge, Mass.: Harvard University Press, 1939).

103. Raymond of Sebonde, *Theologia naturalis: sive liber creaturum* (Deventer, 1484). Chapter 71 is entitled "De omnipotentia, sapientia, et bonitate Dei." Later writers customarily attributed power, wisdom, and benevolence to God. The library of the Mathers possessed a copy of *Theologia naturalis* published at Frankfurt in 1635.

104. On this and the following paragraph, see Michael Hunter, *Science and Society in Restoration England* (Cambridge: Cambridge University Press, 1981), pp. 59–86; Hurlbutt, *Design Argument*, pp. 3–42; Raven, *Organic Design*, p. 8; Alvar Ellegård, "The Darwinian Theory and the Argument from Design," *Lychnos*, 16 (1956), 176–77; and *Sir Isaac Newton's Mathematical Principles of Natural Philosophy and His System of the World*, trans. Andrew Motte, rev. Florian Cajori, 2 vols. (Berkeley: University of California Press, 1946), 2:544 (quotation).

105. See, among relevant works, Frederick J. Powicke, *The Cambridge Platonists: A Study* (London: J. M. Dent and Sons, 1926); W. C. de Pauley, *The Candle of the Lord: Studies in the Cambridge Platonists* (London: Society for the Promoting of Christian Knowledge, 1937); Gerald R. Cragg, ed., *The Cambridge Platonists* (New York: Oxford University Press, 1968); C. A. Pa-

trides, ed., *The Cambridge Platonists* (Cambridge, Mass.: Harvard University Press, 1970); Ernst Cassirer, *The Platonic Renaissance in England*, trans. James P. Pettegrove (New York: Gordian Press, 1970); and William B. Hunter, Jr., "The Seventeenth-Century Doctrine of Plastic Nature," *Harvard Theological Review*, 43 (July 1950), 197–213.

106. John J. Dahm, "Science and Apologetics in the Early Boyle Lectures," *Church History*, 39 (June 1970), 172–76; Jacob, *Newtonians and the English Revolution*, pp. 22–71, 100–200, 273–74.

107. Dahm, "Early Boyle Lectures," pp. 176–86; Michael Hunter, *Science and Society*, pp. 162–87; Jacob, *Newtonians and the English Revolution*, pp. 100–142, 201–70.

108. Jacob, *Newtonians and the English Revolution*, pp. 22–71, 271–72.

109. Donald Fleming, "The Judgment upon Copernicus in Puritan New England," in *Mélanges Alexandre Koyré*, vol. 2: *L'Aventure de l'Esprit* (Paris: Hermann, 1964), pp. 160–75.

110. Bernard Le Bovier de Fontenelle's *Entretiens sur la Pluralité des Mondes* (Paris, 1686) was translated into English by Sir W. D. Knight and published as *A Discourse of the Plurality of Worlds* (Dublin, 1687); a translation by Mrs. A. Behn appeared as *A Discourse of New Worlds* (London, 1688).

111. In the early eighteenth century writers generally distinguished between natural philosophy, which embraced astronomy and physics, and natural history, which included geology as well as botany, biology, and zoology. Mather's *Christian Philosopher*, however, follows the conventional order in which Aristotelian texts were discussed. Thus he moves from the heavenly bodies to terrestrial phenomena. I treat the topics listed here as one unit because this approach best accords with his classification of the sciences.

112. Stillman Drake, "Galilei, Galileo," *DSB*, 5:245–46; idem, "Baliani, Giovanni Battista," ibid., 1:424. See the Endnotes for 74:17 and 74:21 for a fuller discussion of this matter.

113. These passages remind one of Paul Tillich's sermon, "The Shaking of the Foundations," in Tillich, *The Shaking of the Foundations* (New York: Charles Scribner's Sons, 1948), pp. 1–11.

114. I am indebted to Professor Richard Jerrard of the University of Illinois for help on the mathematical examples.

115. Arthur O. Lovejoy published the definitive study, *The Great Chain of Being: A Study of the History of an Idea* (Cambridge, Mass.: Harvard University Press, 1936).

116. According to the *OED*, a few examples of this word occur early in the seventeenth century, but *hybrid* was rarely used until the nineteenth century. The word *hybridous* was known in the eighteenth century; Mather uses *hebricious* for *hybrid* in discussing animals [215:51].

117. Camerarius, *Epistola . . . de sexu plantarum* (Tübingen, 1694).

118. Conway Zirkle, *The Beginnings of Plant Hybridization* (Philadelphia: University of Pennsylvania Press, 1935), pp. vii, 91, 97–98, 103–7 (quotation at 104). Zirkle mistakenly concluded that Mather's 1716 letter to Petiver preceded the account in *The Christian Philosopher* (1721). Zirkle did not know that Mather completed the manuscript of his book in 1715.

119. Ernst Mayr, *The Growth of Biological Thought: Diversity, Evolution, and Inheritance* (Cambridge, Mass.: Harvard University Press, 1982), pp. 106, 645.

120. Frederic T. Lewis, "The Passenger Pigeon as Observed by the Rev. Cotton Mather," *The Auk*, 61 (1944), 587–92 (quotation at 590). Hunter's work is *Observations on Certain Parts of the Animal Œconomy* (London, 1786; 2d ed., 1792).

121. There is a large literature on the subject, of which the most valuable item for present purposes is Lenore Cohen Rosenfield, *From Beast-Machine to Man-Machine: The Theme of Animal Soul in French Letters from Descartes to La Mettrie* (New York: Oxford University Press, 1940).

122. Richard H. Shryock, "Empiricism versus Rationalism in American Medicine, 1650–1950," *Proceedings of the American Antiquarian Society*, 79, pt. 1 (1969), 103–4; Cotton Mather, *The Angel of Bethesda*, ed. Gordon W. Jones (Barre, Mass.: American Antiquarian Society and Barre Publishers, 1972), pp. xxii–xxiii.

123. I attended this presentation on 4 March 1982.

124. *Catalogus librorum collegij Harvardini quod est Cantabrigiae in Nova Anglia* (Boston, 1723); *Continuatio supplementi catalogi librorum bibliothecae Collegij Harvardini, quod est Cantabrigiae in Nova Anglia* (Boston, 1725), p. 110. See also Alfred C. Potter, "The Harvard College Library, 1723–1735," *Publications of the Colonial Society of Massachusetts*, 25 (1924), 1–13; and Frederick C. Kilgour, "The First Century of Scientific Books in the Harvard College Library," *Harvard Library Notes*, 3 (Mar. 1939), 217–19.

125. Clarence P. Shedd, *Two Centuries of Student Christian Movements: Their Origin and Intercollegiate Life* (New York: Association Press, 1934), pp. 10–12.

126. Clifford K. Shipton's continuation of John L. Sibley, *Biographical Sketches of Those Who Attended Harvard College* (cited hereafter as *Sibley's Harvard Graduates*), 5 (1937), 452. Shipton writes that Robie discussed scientific phenomena in his spare end pages, citing Robie's interleaved almanac for 1710 (5:450). Frederick G. Kilgour says that in his almanac for 1716, following a discussion of the cause of heat in summer, Robie discussed thunder, lightning, and hail, basing his explanations of these phenomena in part on Mather's *Christian Philosopher*. Kilgour, "Thomas Robie (1689–1729), Colonial Scientist and Physician," *Isis*, 30 (Aug. 1939), 475. Raymond P. Stearns follows both Shipton and Kilgour in writing that Robie, in his almanac for 1716, drew in part on Mather "(which, obviously, he had seen in manuscript form)," for his discussion of thunder, lightning, and hail. Stearns, *Science in the British Colonies of America*, p. 427. But the almanac for 1716 does not discuss thunder, lightning, or hail, and the interleaved almanac for 1710 contains no references to these subjects. I am indebted to Richard Anders and Nancy Burkett of the American Antiquarian Society for verifying my conclusions on this point. Kilgour may have known of some other place in which Robie based his explanations on Mather's *Christian Philosopher*, but otherwise the matter remains a mystery.

127. Daniel Travis, *An Almanack* (Boston, 1723), pp. 13–16.

128. *Sibley's Harvard Graduates*, 9 (1956), 240.

129. *A Catalogue of the Library of Yale-College in New-Haven* (N[ew] London, 1743), pp. 9–10, 21, 22.

130. *A Catalogue of the Books Belonging to the Company of the Redwood-Library, in Newport, on Rhode Island* (Newport, 1764), pp. 13, 19.

131. *Catalogue of All the Books, Belonging to the Providence Library* (Providence, 1768), pp. 6, 15.

132. *Catalogus librorum in bibliotheca Cantabrigiensi selectus, frequentiorem in usum Harvardinatum, qui gradu baccalaurei in artibus nondum sunt donati* (Boston, 1773), pp. 18, 21, 10, 9, 19; *Catalogus bibliothecae Harvardianae Cantabrigiae Nov-Anglorum* (Boston, 1790), p. 139.

133. Thaddeus M. Harris, *A Selected Catalogue of Some of the Most Esteemed Publications in the English Language, Proper to Form a Social Library* (Boston, 1793); Earl L. Brasher, "A Model American Library of 1793," *Sewanee Review*, 24 (1916), 458–75.

134. George K. Smart, "Private Libraries in Colonial Virginia," *American Literature*, 10 (1938), 24–52.

135. John R. Williams, "A Catalogue of Books in the Library of 'Councillor' Robert Carter, at Nomini Hall, Westmoreland County, Va.," *William and Mary Quarterly*, 1st ser., 11 (1903), 237; "Books in Colonial Virginia," *Virginia Magazine of History and Biography*, 10 (Apr. 1903), 404; "William Dunlop's Library," *William and Mary Quarterly*, 1st ser., 15 (Apr. 1907), 275, 276, 277; Edward W. James, "Libraries in Colonial Virginia," *William and Mary Quarterly*, 1st ser., 3 (1895), 133; see also pp. 111–15; "Catalogue of the Library of Daniel Parke Custis," *Virginia Magazine of History and Biography*, 17 (Oct. 1909), 412, 404.

136. "Library of Col. William Fleming," *William and Mary Quarterly*, 1st ser., 6 (July 1897), 159, 160, 161, 162, 163; W. G. Stannard, "Library of Dabney Carr, 1773, with a Notice of the Carr Family," *Virginia Magazine of History and Biography*, 2 (Oct. 1894), 226; "Library of Charles Dick," *William and Mary Quarterly*, 1st ser., 18 (Oct. 1909), 113.

137. "The Library of John Parke Custis, Esq. of Fairfax County, Virginia," *Tyler's Quarterly Historical and Genealogical Magazine*, 9 (July 1927), 102, 99, 98, 100; E. Millicent Sowerby, ed., *Catalogue of the Library of Thomas Jefferson*, 5 vols. (Washington: Library of Congress, 1952–59), 4:31, 66; 2:135, 149.

138. "The Byrd Library at Westover," *Virginia Magazine of History and Biography*, 12 (Oct. 1904), 207.

139. Joseph T. Wheeler, "Books Owned by Marylanders, 1700–1776," *Maryland Historical Magazine*, 35 (1940), 353, 340–41, 348; idem, "Reading Interests of the Professional Classes in Colonial Maryland, 1700–1776," ibid., 36 (Sept. 1941), 301; idem, "Reading Interests of the Professional Classes in Colonial Maryland, 1700–1776: The Clergy," ibid., 36 (June 1941), 189–91.

140. Stephen B. Weeks, "Libraries and Literature in North Carolina in the Eighteenth Century," *Annual Report of the American Historical Association for the Year 1895* (Washington, 1896), p. 199.

141. *A Catalogue of Books, Belonging to the Incorporated Charlestown Library Society* (Charlestown, 1770), unpaginated; the books are listed alphabetically under the headings Folio, Quarto, Octavo et Infra; *A Catalogue of Books,*

Given and Devised by John Mackenzie Esquire, to the Charlestown Library Society, for the Use of the College When Erected (Charlestown, 1772), p. 12.

142. William Reitzel, "The Purchasing of English Books in Philadelphia, 1790–1800," *Modern Philology*, 35 (1937), 159–71; [William Livingston], *Philosophic Solitude: Or, The Choice of a Rural Life: A Poem* (New York, 1747), p. 39; Nelson F. Adkins, *Philip Freneau and the Cosmic Enigma: The Religious and Philosophical Speculations of an American Poet* (New York: New York University Press, 1949), p. 40.

143. *A Catalogue of Books in the Library of the College of New-Jersey, January 29, 1760* (Woodbridge, 1760), pp. 22, 29.

144. *A Catalogue of Books, Belonging to the Association Library Company of Philadelphia* (Philadelphia, 1765).

145. *A Catalogue of Books Belonging to the Library Company of Philadelphia* (Philadelphia, 1789), pp. 79, 125, 140, 81. See also Margaret B. Korty, "Benjamin Franklin and the Eighteenth-Century American Libraries," *Transactions of the American Philosophical Society*, n.s. 55, pt. 9 (1965), 5–75.

146. *Catalogus bibliothecae Loganianae: Being a Choice Collection of Books* (Philadelphia, 1760), pp. 14, 16, 17.

147. *The Charter, Bye-Laws, and Names of the Members of the New-York Society Library: With a Catalogue of the Books Belonging to the Said Library* (New York, 1793), pp. 33, 34, 36, 71; *A Catalogue of the Books Belonging to the New York Society Library* (New York, 1813).

148. *A Catalogue of the Books, &c. Belonging to the Library Company of Baltimore* (Baltimore, 1797), pp. 2, 4, 6.

149. Mather writes on one occasion, "You, *Gentlemen*, who think your own Country of *England* worth visiting with your *Travels*, as methinks you should before you go abroad" [175:14–16]. On other occasions he identifies with an English audience by reference to English authors as "ours": e.g., "Our *Wharton*," "Our *Willis*," and "Our *Derham*" [270:78–79, 80].

150. John Clarke, *An Essay upon Study* (London, 1731), pp. 316–26.

151. In December 1761, Wesley read a work entitled *The Christian Philosopher*, proclaiming it "a very extraordinary book, containing, among many (as some would be apt to term them) wild thoughts, several fine and striking observations, not to be found in any other treatise." In all probability, Wesley's reference was to a volume by Nicholas Robinson, an English physician and author of a number of medical works which appeared in the 1720s and 1730s. Robinson published *The Christian Philosopher: Or, A Divine Essay on the Principles of Man's Universal Redemption* in London in 1741. Wesley's characterization of *The Christian Philosopher* best fits Robinson, of whom Norman Moore wrote that "all his writings are diffuse, and contain scarcely an observation of permanent value." Robinson's treatise is "very extraordinary" in a way that Mather's is not. In 1742 Robinson published *Appendix to the First Book of The Christian Philosopher. Containing a Physico-Theological Discourse on the Nature, Attributes and Properties of the Serpent that Tempted Eve*. See *The Journal of the Rev. John Wesley*, ed. Nehemiah Curnock, 8 vols. (London: Robert Culley, 1909–16), 4:422; Moore's evaluation is in *DNB*, 17:36.

152. John Wesley, *A Survey of the Wisdom of God in the Creation: Or a Compendium of Natural Philosophy*, 2 vols., 3d American ed. rev. and enlarged (New York, 1823), pp. iv–v; John Dillenberger, *Protestant Thought and Natural Science: A Historical Interpretation* (Nashville: Abingdon Press, 1960), pp. 156–58; Lewis W. Beck, *Early German Philosophy: Kant and His Predecessors* (Cambridge, Mass.: Belknap Press of Harvard University Press, 1969), p. 298. Wesley's *Survey* went through several editions in England and America. In a new edition by Robert Mudie, the subtitle and title were reversed and the book appeared as *A Compendium of Natural Philosophy, Being a Survey of the Wisdom of God in the Creation*, 3 vols. (London, 1836).

153. My discussion of Wolff is based primarily on Beck, *Early German Philosophy*, pp. 243–47, 256–75; see also Dillenberger, *Protestant Thought and Natural Science*, pp. 167–71.

154. Christian Wolff, *Vernünfftige Gedanken von den Absichten der Natürlichen Dinge* (Halle, 1724), pp. 59–60.

155. *German Baroque Literature: A Descriptive Catalogue of the Collection of Harold Jantz and a Guide to the Collection on Microfilm*, 2 vols. (New Haven: Research Publications, 1974), 2:385.

156. *Catalogue of the John Adams Library in the Public Library of the City of Boston* (Boston: Public Library of the City of Boston, 1917), pp. 73, 164; *The Adams-Jefferson Letters: The Complete Correspondence between Thomas Jefferson and Abigail and John Adams*, ed. Lester J. Cappon, 2 vols. (Chapel Hill: Institute of Early American History and Culture by the University of North Carolina Press, 1959), 2:439; U.S. Library of Congress, *Catalogue of the Library of the United States* (Washington, 1815).

157. William B. Sprague, *Annals of the American Pulpit: Or Commemorative Notices of Distinguished American Clergymen of Various Denominations*, 9 vols. (New York, 1859–69), 6:376–79; James F. Hunnewell, *Bibliography of Charlestown, Massachusetts, and Bunker Hill* (Boston, 1880), pp. 24n, 286–88, 290, 293; *The National Union Catalog: Pre-1956 Imprints* (London: Mansell, 1970), 115:512–13.

158. The location of twenty-five copies of the 1815 edition of *The Christian Philosopher* is known by combining information given in Ralph R. Shaw and Richard H. Shoemaker, *American Bibliography: A Preliminary Checklist for 1815* (New York: Scarecrow Press, 1963), p. 175, and *The National Union Catalog: Pre-1956 Imprints* (London: Mansell, 1975), 369:59. But this information should be revised. My census of locations of Mather's 1721 edition disclosed that a few copies earlier identified as such are actually copies of the 1815 edition. See n. 169 below.

159. Riley, *American Philosophy: The Early Schools* (New York: Dodd, Mead and Co., 1907), pp. 197–99; idem, *American Thought from Puritanism to Pragmatism* (New York: Henry Holt and Co., 1915), pp. 55, 57–58.

160. Kenneth B. Murdock, ed., *Selections from Cotton Mather* (New York: Harcourt, Brace and Co., 1926), pp. 285–362, xlviii–liv (quotations at li–lii).

161. Theodore Hornberger, "The Date, the Source, and the Significance of Cotton Mather's Interest in Science," *American Literature*, 6 (Jan. 1935),

413–20; idem, "Notes on *The Christian Philosopher*," in Holmes, *Cotton Mather: A Bibliography*, 1:133–37, 138.

162. Miller, *New England Mind: Colony to Province*, pp. 441–43.

163. Beall and Shryock, *Cotton Mather: First Significant Figure in American Medicine*, pp. 50–52.

164. Cotton Mather, *The Christian Philosopher, A Facsimile Reproduction*, introduction by Josephine K. Piercy (Gainesville, Fla.: Scholars' Facsimiles and Reprints, 1968), pp. vi, x, xiii.

165. Stearns, *Science in the British Colonies of America*, pp. 405, 424–26.

166. Robert Middlekauff, *The Mathers: Three Generations of Puritan Intellectuals, 1596–1728* (New York: Oxford University Press, 1971), pp. 302–4.

167. Sacvan Bercovitch, "Cotton Mather," in Everett Emerson, ed. *Major Writers of Early American Literature* (Madison: University of Wisconsin Press, 1972), pp. 130–34.

168. Silverman, *Cotton Mather*, pp. 249–52.

169. This census was compiled by combining the information in Holmes, *Cotton Mather: A Bibliography*, 1:129, and *The National Union Catalog: Pre-1956 Imprints* (1975), 369:59, and supplementing it with the Eighteenth-Century Short Title Catalogue (ESTC). The ESTC, a special database available on the Research Libraries Information Network (RLIN), is machine readable. I verified locations by writing to the libraries listed by these sources as having copies. Holmes and the *NUC* contain errors. They erroneously report the following libraries as owning a 1721 edition of Mather's book: the library of the Boston Athenaeum (it reports owning only a second American edition of the popularized 1815 volume, published at New York in 1827; this is the only evidence of this edition I have encountered); Zion Research Library, Brookline, Massachusetts (this library, renamed the Endowment for Biblical Research Collection, is on permanent deposit at Boston University; Ms. Katherine Kominis, Rare Book Selector, can find no record that this book was given to Boston University); Franklin and Marshall College; Lehigh College; University of North Carolina at Chapel Hill; and University of Chicago (the latter three have microfilms of the original edition). The University of Kentucky, which is identified as owning a copy, has a catalog card for the title, but the Rare Books Cataloger was unable to locate the book. ESTC on RLIN permitted the identification of the location of many copies of the original edition not noted in Holmes or *NUC*, and it is highly accurate. Nevertheless, it erroneously identifies Mansfield College, Oxford University, as owning a copy. Letters of inquiry to a dozen of the most important libraries in Germany disclosed that none possess an original edition of Mather's work.

170. Cotton Mather, *Manuductio ad Ministerium*, pp. 47–52 (quotation at 50).

171. Jeffrey Jeske, "Cotton Mather: Physico-Theologian," *Journal of the History of Ideas*, 47 (Oct.–Dec. 1986), 587–89. Jeske argues that Mather repudiated mechanism only nominally. For an excellent discussion of the rise of mechanism, see E. J. Dijksterhuis, *The Mechanization of the World Picture*, trans. C. Dikshoorn (Oxford: Clarendon Press, 1961).

172. Jeske, "Cotton Mather," pp. 589–91; Pershing Vartanian, "Cotton Mather and the Puritan Transition into the Enlightenment," *Early American Literature*, 7 (Winter 1973), 213–24.

173. The classics also played an indispensable part in the rhetorical strategy of Mather's *Magnalia Christi Americana*. See Gusthaaf vom Cromphout, "Cotton Mather: The Puritan Historian as Renaissance Humanist," *American Literature*, 49 (Nov. 1977), 327–37.

174. Cotton Mather, *Manuductio ad Ministerium*, pp. 44–47 (quotation at 44); see also Kenneth B. Murdock, "Mather, Cotton," *DAB*, 12:388–89.

175. Norman Kemp Smith, ed., *Hume's Dialogues concerning Natural Religion* (New York: Social Science Publishers, 1948), p. 30 n. 2.

176. Ellegård, "Darwinian Theory," pp. 189–90, 173–75.

177. The first and the last two quotations are from the discussion of natural religion in William James, *The Will to Believe* (1897; Cambridge, Mass.: Harvard University Press, 1979), pp. 42–43. James deals with the design argument in *Pragmatism* (1907; Cambridge, Mass.: Harvard University Press, 1975), pp. 56–59; the other quotations are from there.

178. Ellegård, "Darwinian Theory," p. 191.

179. Raven, *Organic Design*, pp. 5–6.

180. See, for example, F. R. Tennant, *Philosophical Theology*, 2 vols. (Cambridge: University Press, 1928–30), vol. 2, chap. 4; Bruce R. Reichenbach, *The Cosmological Argument: A Reassessment* (Springfield, Ill.: Charles C. Thomas, 1972); William L. Rowe, *The Cosmological Argument* (Princeton: Princeton University Press, 1975); Richard Swinburne, *The Existence of God* (Oxford: Clarendon Press, 1979); William L. Craig, *The Cosmological Argument from Plato to Leibniz* (London: Macmillan Press, 1980); George N. Schlesinger, *Religion and Scientific Method* (Dordrecht: D. Reidel Publishing Co., 1977); and Michael Martin, *Atheism: A Philosophical Justification* (Philadelphia: Temple University Press, 1990).

181. For an excellent analysis of important links in American thought that bear on design and related ideas from the time of Mather to the impact of Darwinism, see Jon H. Roberts, *Darwinism and the Divine in America: Protestant Intellectuals and Organic Evolution, 1859–1900* (Madison: University of Wisconsin Press, 1988).

Note on Editorial Method

In the interest of making Mather's text clear and its presentation consistent with Kenneth B. Murdock's edition of Mather's *Magnalia Christi Americana, Books I and II* (Cambridge, Mass.: Harvard University Press, Belknap Press, 1977) I have, with few exceptions, employed the same editorial practices as Murdock, which he describes in his edition of the *Magnalia* (pp. vi–viii).

This edition does not preserve the pagination of the original edition of *The Christian Philosopher,* an octavo volume of 304 pages, with most pages carrying thirty-eight lines, but familiarity with the 1721 edition underscores the desirability of reproducing Mather's text instead of "modernizing" it. Accordingly, this edition follows the first edition as faithfully as possible, although Mather's "Index" (actually, a table of contents) and a list of "BOOKS *lately Publish'd*" are not included. The modern *s* replaces the elongated *s* wherever used in the original text. Mather's italics have been preserved. As Murdock notes, "he employs them for emphasis, for proper names, for foreign words and phrases. He uses them often in lieu of quotation marks and no less often to allude without direct quotation to proverbs or other popular sayings, or to commonplaces familiar to scholars." Hence, Mather's italics are retained, even though they seem excessive to the modern reader, because they are essential for understanding his prose.

Mather's use of capital letters is often erratic but not without purpose. Frequently, as Murdock writes, Mather's capital letters "serve to make plain the structure of a passage by accenting key words or by marking out the essential elements in a pun or some other rhetorical device." Mather also uses capitals to emphasize biblical phrases or words that are especially relevant to his purposes; they are preserved here.

Eighteenth-century printers used quotation marks not only at the beginning and end of quoted passages but also along the left margin of the entire quotation. This practice serves no useful purpose and jars the modern reader; I have altered it. Mather's single mark at the beginning and end of quotations is changed to a double mark, and in long quotations these are repeated at the beginning of each of the quoted paragraphs. His running marks in the margins are eliminated.

The 1721 edition of *The Christian Philosopher* contained a prompt word at the bottom of each page which, according to the custom of the time, was repeated as the first word on the following page. These prompts I eliminate. Either Mather or his printer inserted short lines—nearly forty of them—in his text. They serve no apparent literary purpose, and are omitted. Mather uses brackets on several occasions, and his book contains a few special printer's devices. These remain. He uses a hyphen with compound words that are treated as one word today; I retain his usage.

Spellings that are correct according to eighteenth-century usage, though now obsolete, I allow to stand unless there is danger of the word being misunderstood or the spellings are inconsistent (in which case I make them consistently obsolete). Abbreviated words and words with an apostrophe where none is needed (e.g., *thro, flamingo's*) stand as written. In the past tense of certain verbs where Mather omits the letter *e* before *-ned* or *-red* (as in *shortned* and *numbred*), his meaning is clear and I accept his practice. I correct Mather when he misquotes a source. Proper names and place names, including those that lack proper accent marks, are allowed to stand as originally printed because correct names appear in the footnotes or the Biographical Register. In a few cases where the correct name is not given, I have supplied it. A few spellings which are obviously printers' errors or slips of the author's pen have been corrected. A few corrections are made on the basis of a printed errata slip mounted on the inside back cover of one of the copies of Mather's 1721 edition in the Houghton Library at Harvard University (call number *AC7.M4208.720c2, copy C). This slip reads: "Tho, THE CHRISTIAN PHILOSOPHER, has been so happy as to have his Press-Work favoured with an uncommon Correction, yet these few Words remain still to be corrected." Nine items follow, some of which permitted corrections which could not otherwise have been made. According to the Houghton Library cataloguer, the slip may have been printed in Boston. It appears to be from the early eighteenth century; Cotton Mather was likely the author. I thank Mr. Dennis C. Marnon of the Houghton Library for providing this information. Words ending in *-ck* that now end in *c* only (e.g., *elastick, plastick*) are allowed to stand. Mather's rare use of A.C. for A.D. is retained. In Latin quotations, accent marks are removed, but the ampersand and ligatures (*æ*, *œ*) remain.

Editorial Method cxxxvii

All alterations are noted at the bottom of the page. Both the emended and the original readings are given so that one may easily reconstruct the text as it was first printed.

Mather cites some 415 authors in his book in addition to naming many other individuals. These persons are identified and their lives and achievements are discussed in the Biographical Register. In a few cases, all of the information on a person is in a brief footnote. Readers will easily discover how to locate information on other individuals named by Mather, but a word of explanation is in order. Where Mather alludes to a character without giving a name (e.g., "A German Writer"), that person is identified in a footnote and discussed in the Biographical Register. Mather often uses Latin surnames. Where his usage differs appreciably from the name in the Biographical Register, I place the name he cites in brackets before the proper entry, using cross-references where necessary. Brackets are seldom required for Latin names whose *us* or *ius* ending is dropped in the English version of the name. Mather also misspells names, but when a misspelled name can readily be located in the Biographical Register (e.g., Wanly, Wanley), no brackets are needed. The Index will help in locating names.

The Greek words and phrases in Mather's 1721 edition are printed in an old and now outmoded font, and accent marks are often incorrect. The present edition prints the Greek passages in a modern typeface with correct accent marks silently supplied. Citations from Greek and Latin authors, unless otherwise indicated, are to the numbered divisions in the Loeb Classical Library (London and Cambridge, Mass., 1912–). The translations of Latin quotations are from the Loeb edition where these are available, without credit being given in each instance. For quotations where this is not the case, the source of the translation is given in the Endnotes. "Derham's translation" indicates that the translation is taken from William Derham, *Physico-Theology,* new ed., 2 vols. (London, 1798). The Galen translations are from May, *Galen* (see the Abbreviations for full publication data).

The Recapitulation of Mather's Sources at the end of the volume provides a quantitative summary of the contents of Mather's treatise that shows the extent of his own contribution and that of the other authors on whom he drew in the composition of his book.

The Index of Biblical References at the end of the volume lists Mather's references to both Old and New Testament passages in *The Christian Philosopher*.

Technical Apparatus

Following Murdock, the lines of the text are numbered in flights of ninety-nine, regardless of divisions within the text. Every fifth line carries a line number, as does every line 99.

Emendations, identified by line numbers, are recorded below the text on the pages to which they pertain.

Footnotes, numbered consecutively in each essay, are printed at the bottom of the appropriate page. They are designed to provide information immediately useful in understanding Mather's text. My usual practice is to identify persons and define words only when such knowledge is not readily available in standard reference works and dictionaries. I provide biblical references for Mather's frequent quotations from and allusions to Scripture.

Endnotes, printed at the back of the volume, are identified by reference to the appropriate page(s) and line number(s) in Mather's text. The endnotes serve a variety of functions; they are not necessarily tied to the footnotes and should be consulted independently of the footnotes. They furnish interesting explanatory detail and historical background as well as elucidating theological and scientific matters. In addition, the endnotes identify Mather's sources, the vast majority of which I have been able to locate.

Edition and Copy Used

In preparing this volume, I worked with a photocopy made by University Microfilms International in Ann Arbor from the American Culture Series (ACS) microfilm of *The Christian Philosopher* (London, 1721) in the William L. Clements Library, University of Michigan. I compared that text, line by line, with the original edition of Mather's book in the University of Illinois Library. The two copies are identical. In verifying the location of other copies of the 1721 edition, I sent a photocopy of the title page of the copy I worked with, asking librarians to compare it with their holding(s). Nearly all of the librarians to whom I wrote responded on this point, reporting that the title pages were identical. All of the available evidence provides good grounds for concluding that there are no variations from copy to copy in the 1721 edition of Mather's *Christian Philosopher*.

Abbreviations of Works Cited

Alsted — Johann H. Alsted. *Theologia naturalis exhibens augustissimam naturae scholam, in qua creaturae Dei communi sermone ad omnes pariter docendos utuntur: adversos atheos, Epicureos, et Sophistas huius temporis, duobus libris pertracta.* Hanover, 1623.

Arndt — Johann Arndt. *De vero Christianismo libri quatuor.* 2 vols. London, 1708.

Bethesda — Cotton Mather. *The Angel of Bethesda.* This manuscript, unpublished during Mather's lifetime, was edited with introduction and notes by Gordon W. Jones. Barre, Mass.: American Antiquarian Society and Barre Publishers, 1972.

Cheyne — George Cheyne. *Philosophical Principles of Natural Religion: Containing the Elements of Natural Philosophy, and the Proofs for Natural Religion, Arising from Them.* London, 1705.

Derham 2 — William Derham. *Astro-Theology: Or, A Demonstration of the Being and Attributes of God, from a Survey of the Heavens.* London, 1715.

Derham — William Derham. *Physico-Theology: Or, A Demonstration of the Being and Attributes of God, from His Works of Creation.* London, 1713.

DNB — *Dictionary of National Biography.* Edited by Leslie Stephen and Sidney Lee. 22 vols. London, 1908–9.

DSB — *Dictionary of Scientific Biography.* Edited by Charles C. Gillispie. 16 vols. New York: Charles Scribner's Sons, 1970–80.

Grew — Nehemiah Grew. *Cosmologia Sacra: Or, A Discourse of the Universe As It is the Creature and Kingdom of God.* London, 1701.

Harris, LT — John Harris. *Lexicon Technicum: Or, An Universal*

	English Dictionary of Arts and Sciences. 2d ed. 2 vols. London, 1708–10.
Kittredge	George L. Kittredge. "Cotton Mather's Scientific Communications to the Royal Society." *Proceedings of the American Antiquarian Society,* n.s., 26 (1916), 18–57.
Mather, *Diary*	*Diary of Cotton Mather.* Edited by Worthington C. Ford. Massachusetts History Society Collections, 7th ser. Vols. 7–8 (Boston: Massachusetts Historical Society, 1911–12).
May, *Galen*	Margaret T. May. *Galen: On the Usefulness of the Parts of the Body.* 2 vols. Ithaca: Cornell University Press, 1968.
Misc. Cur.	*Miscellanea Curiosa: Being a Collection of Some of the Principal Phenomena in Nature.* [Edited by Edmond Halley.] 3 vols. London, 1705–7.
OED	*The Oxford English Dictionary.* Edited by James A. H. Murray, Henry Bradley, W. A. Craigie, and C. T. Onions. 12 vols. and supplement. Oxford: Clarendon Press, 1933.
PG	*Patrologiae cursus completus. Series graeca.* Edited by Jacques Paul Migne. 161 vols. in 166. Paris, 1857–66.
PL	*Patrologiae cursus completus. Series latina.* Edited by Jacques Paul Migne. 221 vols. Paris, 1844–64.
Phil. Trans.	*Philosophical Transactions* of the Royal Society.
Ray	John Ray. *The Wisdom of God Manifested in the Works of the Creation.* 5th ed. London, 1709 (first published in 1691).
Walker	Joseph Walker. *Astronomy's Advancement: Or, News for the Curious: Being a Treatise of Telescopes. Done out of French.* London, 1684.

THE
Christian Philosopher:
A
COLLECTION
OF THE
Best Discoveries in Nature,
WITH
Religious Improvements.

By COTTON MATHER *D.D.*
And Fellow of the ROYAL SOCIETY.

LONDON;
Printed for EMAN. MATTHEWS, at the Bible in
Pater-Noster-Row. M.DCC.XXI.

The title page of Mather's *Christian Philosopher*. The publication date given is 1721, but the volume actually appeared late in 1720. Courtesy of the Rare Book and Special Collections Library, University of Illinois at Urbana-Champaign.

TO
Mr. *THOMAS HOLLIS,*
Merchant in *London.*

SIR,

THE Learned Author of the ensuing Treatise, has already diffus'd his Name and Reputation in a great Variety of Useful Works; by which the better Part of Mankind do sufficiently know him to be *in Labours more abundant.* The Reader will find in this Treatise, a Collection from Writers of the first and best Character, both in our own and other Nations; and every Observation improv'd to the Ends of Devotion and Practice. The Remarks that the Author gives, are so mingled with the Discoveries that he has brought together, that as it shows us with what Spirit He has pursued His Enquiries into the Wonders of the Universe, so it is both an Instruction and a Pattern to a serious Mind. He has generally drawn into his Application, all that the Bible saith upon the several Subjects: And thus he lays open the two great Books of God, Nature and Scripture. In this way, our Curiosity is not only entertain'd, but sanctified; *the Invisible Things of God from the Creation of the World are seen,* and improv'd to the Glory of Him whose they are.

Your surprizing Generosity to the Academy in *New-England,* has made this Dedication more proper to you than any other Person. Such a Beneficence is an Argument how thorowly you desire that the Doctrines of the Gospel, and the Purity of Discipline, may be transmitted to future Generations. And certainly, it is the noblest, and the most divine Application of your Charity, when by it you are *a Fellow-helper to the Truth.* This is given to those from whom you can have no Expectation of Recompence; but as it's all done to the Lord, and not unto Men, so by him it will be remember'd at *the Resurrection of the Just.* You know how much it is against my Temper to give *flattering Words,* and I'm convinc'd that it is against yours to receive 'em. But I have reason to think, that the Reverend Author, and the whole Country where God has placed him, will believe this Dedication well directed, to the BEST of all their Benefactors. *This Administration of Service is abundant, by many Thanksgivings to God, (whilst by this Ministration, they glorify God for your profess'd Subjection to the Gospel of Christ, and for your liberal Distribution to them and to all Men) and by their Prayer for you.*

I have no more to add, but the Apostle's Wish, that *your Faith may grow exceedingly,* and *your Charity* daily *abound;* that whatever you do, may be done *faithfully to the Brethren, and to Strangers.*

London,
Sept. 22.
1720.

I am,
SIR,
Your Sincere Friend,
and Obedient Servant,

Tho. Bradbury.

THE INTRODUCTION.

THE Essays now before us will demonstrate, that *Philosophy* is no *Enemy*, but a mighty and wondrous *Incentive* to *Religion*; and they will exhibit that PHILOSOPHICAL RELIGION, which will carry with it a most sensible *Character*, and victorious *Evidence* of a *reasonable Service*. *GLORY TO GOD IN THE HIGHEST*, and *GOOD-WILL TOWARDS MEN*, animated and exercised; and a Spirit of *Devotion* and of *Charity* inflamed, in such Methods as are offered in these *Essays*, cannot but be attended with more Benefits, than any *Pen* of ours can declare, or any *Mind* conceive.

In the *Dispositions* and *Resolutions* of PIETY thus enkindled, a *Man* most effectually *shews himself a* MAN, and with unutterable Satisfaction answers the grand END of his Being, which is, *To glorify* GOD. He discharges also the Office of a *Priest* for the *Creation*, under the Influences of an admirable Saviour, and therein asserts and assures his Title unto that *Priesthood,*

The first page of the Introduction of the 1721 edition of *The Christian Philosopher*. Courtesy of the Rare Book and Special Collections Library, University of Illinois at Urbana-Champaign.

THE
INTRODUCTION

THE Essays now before us will demonstrate, that *Philosophy*[1] is no *Enemy*, but a mighty and wondrous *Incentive* to *Religion;* and they will exhibit that PHILOSOPHICAL RELIGION, which will carry with it a most sensible *Character,* and victorious *Evidence* of a *reasonable Service.*[2] GLORY TO GOD IN THE HIGHEST, and GOOD-WILL TOWARDS MEN,[3] animated and exercised; and a Spirit of *Devotion* and of *Charity* inflamed, in such Methods as are offered in these *Essays,* cannot but be attended with more Benefits, than any *Pen* of ours can declare, or any *Mind* conceive.

In the *Dispositions* and *Resolutions* of PIETY thus enkindled, a *Man* most effectually *shews himself a* MAN, and with unutterable Satisfaction answers the grand END of his Being, which is, *To glorify GOD.*[4] He discharges also the Office of a *Priest* for the *Creation,* under the Influences of an admirable Saviour, and therein asserts and assures his Title unto that *Priesthood,*[5] which the Blessedness of the *future State* will very much consist in being advanced to. The whole *World* is indeed a *Temple* of GOD, *built* and *fill'd* by that Almighty *Architect;*[6] and in this *Temple,* every such one, affecting himself with the Occasions for it, will *speak of His Glory*[7] He will also rise into that *Superiour Way* of *Thinking* and of *Living,* which the *Wisest* of Men will chuse to take; which the more *Polite Part* of Mankind, and the *Honourable of the Earth,*[8] will esteem it no Dishonour for them to be acquainted with. Upon

20 *fill'd / fitted*

1. That is, natural philosophy or science.
2. Adapted from Rom. 12:1.
3. Luke 2:14.
4. According to the Westminster Catechism (1647), "Man's chief and highest end is, to glorify God, and fully to enjoy him for ever."
5. Perhaps an allusion to Exod. 40:15.
6. This concept of the world as a "temple of God" derives from both classical antiquity and Judeo-Christian thought. Mather employs it throughout his book.
7. Ps. 29:9.
8. Isa. 23:8, 9.

PSALTERIVM תחלים

fertum Cades. Vox domini parturire facit
ceruas & detegit fyluas, atque in templo
eius nullus non dicit gloriam. ᶜ Dominus
supra diluuium sedet, manetq́ dominus
rex in æternum. Dominus dabit fortitudi
nem populo suo: dominus benedicet populo suo in pace.

קוֹל יְהוָה יְחוֹלֵל אַיָּלוֹת
וַיֶּחֱשֹׂף יְעָרוֹת וּבְהֵיכָלוֹ כֻּלּוֹ אֹמֵר כָּבוֹד ׃
יְהוָה לַמַּבּוּל יָשָׁב וַיֵּשֶׁב יְהוָה מֶלֶךְ
לְעוֹלָם ׃ יְהוָה עֹז לְעַמּוֹ יִתֵּן
יְהוָה יְבָרֵךְ אֶת־עַמּוֹ בַשָּׁלוֹם ׃

    a ¶ *Filij procerum.*) Kimhi in radicibus post Chald. interpretem exponit בְּנֵי מַלְאָכִים cœtus angelorum:
hebraismus autem habet filij deorum. Sunt etiam inter Hebræos qui exponunt stellas, quod hæ inuitentur ad laudandum deum, sicut & Iob 28. uocantur filij deorum. Nihil tamen est, quod hic uetat filios deorum accipi pro filijs potentum & mag-
natum. Vtcunq̃; accipias, propheta id hoc loco uoluit, quod nihil est tam sublime tamq́; magnificum, quod non teneatur laudare & magnificare deū creatorem suum.   b ¶ *Vox domini super aquas.*) Mira, inquit, sunt opera dei, qui in hu-
midis nubibus tam stupendas tonitruorum excitat uoces, ut putes cœlos & elementa scindi & disrumpi. Tam uehemens, inquit, est uox domini, ut ad eius sonitum & fulminis ictum, altissimæ & fortissimæ cedri Libani deijciantur & confrin-
gantur: non quidem simpliciter, imò totus mons exilire putetur. Porrò mons Sirion est mons Hermon coniunctus Antiliba-
no.   c ¶ *Succidit flammas igneas.*) Sensus est, ex tonitruo & nube rupta nascitur, seu rumpit flamma ignis.
     d ¶ *Tremere facit desertum.*) Hoc est, nullum est animal tam sæuum & crudele in deserto illo uastissimo, quod
non expauescit & contremiscit ad horribilem illam uocem domini, sicut & animalia quædā, quæ cum difficultate pariunt,
tonitrui errore concussa, illico eijciunt fœtum. Deteguntur & syluæ fulmine ictæ: at in omnibus, pij homines potentiā
dei animaduertentes, dicunt deo gloriam & laudem in templo suo, quod est totus mundus, quem deus implet.
     e ¶ *Dominus supra diluuium sedet.*) Sensus est, deus præsidet, & in potestate sua habet inundationes illas, quæ
comitantur tonitrua terribilia, & c:

xxx ¶ Psalmus &ᵃ canticum dedicationis domus, Dauid. Exaltabo te domine quo-
niā subexistime, & non lætificasti inimi-
cos meos de me. Domine deus meus cla-
maui ad te, & sanasti me. Domine ascendere fecisti ab inferis animam meam, uiuifica-
sti me à descendentibus in foueam. Psalli-
te domino pij (cultores) eius, & confitemi-
ni memoriæ sanctitatis eius. Nā ᵇ momen-
tanea est ira eius, & uita in uoluntate eius:
in uespera morabiꝰ fletus, & (rursum ade-
rit) mane exultatio. Et ego ᶜ dixi in felicita-
te mea: non mouebor in seculū. Domine,
in uoluntate tua ᵈ collocasti fortitudinem
super montem meum: abscondisti faciem
tuam, & factus sum conturbatus. Ad te domine clamabam, & ad dominum meum
supplex currebam. Quæ utilitas in sanguine meo cum descendero in foueam: nun-
quid confitebitur tibi puluis, aut annunci-
abit ueritatem tuam? Audi domine & mi-
serere mei: domine sis tu mihi adiutor. Cō-
uertisti planctum meum mihi in chorum,
ᵉ soluisti saccum meum & accinxisti me læ-
titia. Propterea decātabit tibi (bonus quisque) gloriam & non silebit: deus meus in
æternum confitebor tibi.

מִזְמוֹר שִׁיר־חֲנֻכַּת הַבַּיִת
לְדָוִד ׃ אֲרוֹמִמְךָ
יְהוָה כִּי דִלִּיתָנִי וְלֹא־שִׂמַּחְתָּ אֹיְבַי לִי ׃
יְהוָה אֱלֹהָי שִׁוַּעְתִּי אֵלֶיךָ וַתִּרְפָּאֵנִי ׃
יְהוָה הֶעֱלִיתָ מִן־שְׁאוֹל נַפְשִׁי
חִיִּיתַנִי מיורדי־בוֹר ׃ זַמְּרוּ לַיהוָה
חֲסִידָיו וְהוֹדוּ לְזֵכֶר קָדְשׁוֹ ׃
כִּי רֶגַע בְּאַפּוֹ חַיִּים בִּרְצוֹנוֹ בָּעֶרֶב יָלִין
בֶּכִי וְלַבֹּקֶר רִנָּה ׃ וַאֲנִי אָמַרְתִּי
בְשַׁלְוִי בַּל־אֶמּוֹט לְעוֹלָם ׃
יְהוָה בִּרְצוֹנְךָ הֶעֱמַדְתָּה
לְהַרְרִי עֹז הִסְתַּרְתָּ פָנֶיךָ הָיִיתִי נִבְהָל ׃
אֵלֶיךָ יְהוָה אֶקְרָא וְאֶל־אֲדֹנָי
אֶתְחַנָּן ׃ מַה־בֶּצַע בְּדָמִי
בְּרִדְתִּי אֶל שָׁחַת הֲיוֹדְךָ עָפָר הֲיַגִּיד אֲמִתֶּךָ ׃
שְׁמַע יְהוָה וְחָנֵּנִי
יְהוָה הֱיֵה־עֹזֵר לִי ׃
הָפַכְתָּ מִסְפְּדִי לְמָחוֹל לִי פִּתַּחְתָּ שַׂקִּי
וַתְּאַזְּרֵנִי שִׂמְחָה ׃
לְמַעַן יְזַמֶּרְךָ כָבוֹד וְלֹא יִדֹּם יְהוָה
אֱלֹהַי לְעוֹלָם אוֹדֶךָּ ׃

     a ¶ *Canticum dedicationis domus.*) Vtebantur olim Israëlitæ sub lege Mosi ceremonijs quibusdam, cum nouas
domos inhabitare uellent, sicut & solenni dedicatione domus domini, quam Salomon extruxerat, consecrata fuit deo, da-
taq́; fuit ad huiusmodi initiationem benedictio dei, sicut rursus dominus minatur maledictionem transgressoribus legis, ut domos nouas extruant, & illas minime inhabitent. Sic igitur & Dauid, cum in Ierusalem nouam ædificasset domum, sicut se-
cundo Samuelis quinto legitur, putatur illam cum laudibus dei & huius psalmi decantatione primum inhabitasse: tam et-
si principale argumentum huius psalmi esse uideatur gratiarumactio, pro recuperata post infirmitatem sanitate, quemad-
modum pri-

that Passage occurring in the best of Books, *Ye Sons of the Mighty, ascribe unto the Lord Glory and Strength*;⁹ it is a Gloss and an Hint of *Munster*, which carries with it a Cogency: *Nihil est tam sublime, tamque magnificum, quod non teneatur laudare & magnificare Deum Creatorem suum.*¹⁰ Behold, a *Religion*, which will be found *without Controversy*;¹¹ a *Religion*, which will challenge all possible Regards from the *High*, as well as the *Low*, among the People; I will resume the Term, a PHILOSOPHICAL RELIGION: And yet how *Evangelical!*

In prosecuting this *Intention*, and in introducing almost every Article of it, the Reader will continually find some *Author* or other quoted. This constant Method of *Quoting*, 'tis to be hoped, will not be censured, as proceeding from an *Ambition to intimate and boast a Learning*, which the *Messieurs de* Port-Royal¹² have rebuked; and that the Humour for which *Austin*¹³ reproached *Julian*,¹⁴ will not be found in it: *Quis hæc audiat, & non ipso nominum strepitu terreatur, si est ineruditus, qualis est hominum multitudo, & existimet te aliquem magnum qui hæc scire potueris?*¹⁵ Nor will there be discernible any Spice of the impertinent Vanity, which *La Bruyere* hath so well satirized: "*Herillus*¹⁶ will always *cite*, whether he speaks or writes. He makes the *Prince of Philosophers* to say, *That Wine inebriates*; and the *Roman Orator, That Water temperates it.* If he talks of *Morality*, it is not he, but the Divine *Plato*, who affirms, *That Virtue is amiable, and Vice odious.* The most common and trivial things, which he himself is able to think of, are ascribed by him to *Latin* and *Greek* Authors." But in these *Quotations*, there has been proposed, first, a due *Gratitude* unto those, who have been my *Instructors;* and indeed, *something within me* would

38 de/du

9. Adapted from Ps. 29:1.
10. "Nothing is so elevated, so august, that it is not compelled to praise and glorify God its creator."
11. 1 Tim. 3:16.
12. Les Messieurs de Port-Royal-des-Champs was the formal name of the Jansenists, a brotherhood of intellectuals and religious which occupied a former Cistercian convent near Versailles. Port-Royal was their center.
13. That is, Augustine.
14. Flavius Claudius Julianus (332–63), Roman emperor whose stern enmity toward the Christian church earned him the title of "Julian the Apostate."
15. "Who can hear this list and not be frightened by the clamor of names and the banding of schools, if he, as the majority of men, is not a scholar and think that you, who know such things, must be really important?"
16. Herillus of Carthage, pupil of Zeno and founder of a separate Stoic sect which seems not to have survived past 200 B.C.

have led me to it, if *Pliny,* who is one of them, had not given me a Rule; *Ingenuum est profiteri per quos profeceris.*¹⁷ It appears also but a piece of *Justice,* that the *Names* of those whom the Great GOD has distinguished, by employing them to make those *Discoveries,* which are here collected, should live and shine in every such Collection. Among these, let it be known, that there are especially Two, unto whom I have been more indebted, than unto many others; the Industrious Mr. RAY, and the inquisitive Mr. DERHAM; *Fratrum dulce par:*¹⁸ upon whom, in divers Paragraphs of this *Rhapsody,* I have had very much of my Subsistence; (I hope without doing the part of a *Fidentinus*¹⁹ upon them) and I give thanks to Heaven for them.

'Tis true, some Scores of other *Philosophers* have been consulted on this Occasion; but an *Industry* so applied, has in it very little to bespeak any *Praises* for him that has used it: He earnestly renounces them, and sollicits, that not only *he,* but the *Greater Men,* who have been his *Teachers,* may disappear before the Glorious GOD, whom these *Essays* are all written to represent as *worthy to be praised,*²⁰ and by whose *Grace we are what we are*;²¹ nor have we *any thing but what we have received* from Him.²²

A considerable Body of Men (if the *Jansenists* may now be thought so) in *France,* have learnt of Monsieur *Pascal,* to denote themselves by the *French* Impersonal Particle *On;* and it was his opinion, that an honest Man should not be fond of *naming himself,* or using the word I, and ME; that *Christian Piety* will annihilate our I, and ME, and *Human Civility* will suppress it, and conceal it.

Most certainly there can be very little Pretence to an I, or ME, for what is done in these *Essays.* *'Tis done,* and entirely, *by the Help of God.*²³ This is all that can be pretended to.

There is very little, that may be said, really to be performed by the Hand that is now writing; but only the *Devotionary Part* of these *Essays,* tho they are not altogether destitute of *American*

---

17. "It is a noble gesture to name those by whom you have been assisted."
18. "A delightful pair of brothers."
19. Fidentinus was named by Martial (Marcus Valerius Martialis, c. A.D. 40–c. 104) as a plagiarist of his verses.
20. 2 Sam. 22:4 and Ps. 18:3.
21. Adapted from 1 Cor. 15:10.
22. Adapted from John 3:27 and 1 Cor. 4:7.
23. Perhaps an allusion to Ps. 54:4.

Communications:²⁴ And if the *Virtuoso's*,²⁵ and all the *Genuine Philosophers* of our Age, have approved the Design of the devout RAY and DERHAM, and others, in their Treatises; it cannot be distasteful unto them, to see what was more *generally hinted at* by those Excellent Persons, here more *particularly carried on,* and the more *Special Flights* of the true PHILOSOPHICAL RELIGION exemplified. Nor will they that value the Essays of the memorable Antients, *Theodoret,* and *Nazianzen,*²⁶ and *Ambrose,* upon *the Works of the six Days,* count it a Fault, if among lesser Men in our Days, there be found those who say, *Let me run after them.*²⁷ I remember, when we read, *Praise is comely for the Upright,*²⁸ it is urged by *Kimchi,* that the Word which we render *comely,* signifies *desirable,* and *acceptable;* and the Sense of that Sentence is, that *Qui recti sunt, aliud nihil desiderant quam Laudem & Gloriam Dei.*²⁹ Sure I am, such *Essays* as these, to observe, and proclaim, and publish the *Praises* of the Glorious GOD, will be *desirable* and *acceptable* to all that have a *right Spirit* in them; *the rest,* who are *blinded,* are Fools, and unregardable: As little to be regarded as a *Monster* flourishing a *Broomstick! Vix illis optari quidquam pejus potest, quam ut fatuitate sua fruantur.*³⁰ For such *Centaurs* to be found in the Tents of professed *Christianity! Good God, unto what Times hast thou reserved us!* If the *self-taught Philosopher* will not, yet *Abubeker,* a *Mahometan* Writer, by whom such an one was exhibited more than five hundred Years ago, will *rise up in the Judgment with this Generation, and condemn it.*³¹ Reader, even a *Mahometan* will shew thee one, without any *Teacher,* but *Reason* in a serious View of

24. Between 1712 and 1724 Mather sent at least eighty-two letters describing American curiosities to the Royal Society of London. His "Curiosa Americana" treat a wide range of subjects, especially medicine and related topics. Most of the case histories reported were obtained from others, but Mather also culled material from his *Magnalia Christi Americana* and his "Biblia Americana." Mather improves upon some of these letters in *The Christian Philosopher.*

25. In seventeenth-century England the virtuoso was a gentleman whose major interests were painting, antiquities, and science. The virtuoso's interest focused on nonutilitarian things in themselves—coins, fossils, etc. Starting by mid-century, however, science displaced painting and antiquities as the major interest of the virtuosi.

26. That is, Gregory of Nazianzus.

27. Adapted from 2 Kings 5:20.

28. Ps. 33:1.

29. "Those who are upright desire nothing other than the praise and glory of God."

30. "It is scarcely possible for anything worse to be chosen for those who reap the benefit of their own silliness."

31. Adapted from Matt. 12:41, 42.

Nature, led on to the Acknowledgment of a Glorious GOD. Of a Man, supposed as but using his *Rational Faculties* in viewing the Works of GOD, even the *Mahometan* will tell thee; "There appeared unto him those Footsteps of Wisdom and Wonders in the *Works of Creation,* which affected his Mind with an excessive Admiration; and he became hereby assured, that all these things must proceed from such a *Voluntary Agent* as was *infinitely perfect,* yea, above all Perfection: such an one to whom the Weight of the least Atom was not unknown, whether in Heaven or Earth. Upon his viewing of the *Creatures;* whatever *Excellency* he found of any kind, he concluded, it must needs proceed from the Influence of that *Voluntary Agent,* so illustriously glorious, the *Fountain of Being,* and of *Working.* He knew therefore, that whatsoever Excellencies were by Nature in *Him,* were by so much the greater, the more perfect, and the more lasting; and that there was no proportion between those Excellencies which were in *Him,* and those which were found in the *Creatures.* He discerned also, by the virtue of that more Noble Part of his, whereby he knew the *necessarily existent Being,* that there was in him a certain Resemblance thereof: And he saw, that it was his Duty to labour by all manner of Means, how he might obtain the Properties of that *Being,* put on *His Qualities,* and imitate *His Actions;* to be diligent and careful also in promoting *His Will;* to commit all his Affairs unto *Him,* and heartily to acquiesce in all those *Decrees* of *His* which concerned him, either from within, or from without: so that he pleased himself in *Him,* tho he should *afflict* him, and even *destroy* him." I was going to say, O Mentis aureæ Verba bracteata![32] But the Great *Alsted* instructs me, that we *Christians,* in our valuable Citations from them that are Strangers to *Christianity,* should seize upon the Sentences as containing *our Truths,* detained in the hands of *Unjust Possessors;* and he allows me to say, Audite Ciceronem, quem Natura docuit.[33] However, this I may say, *God has thus far taught a* Mahometan! And this I will say, *Christian,* beware lest a *Mahometan* be called in for thy *Condemnation!*

Let us conclude with a Remark of *Minutius Fœlix:* "If so much Wisdom and Penetration be requisite to *observe* the wonderful

---

32. "O gilded words of a golden mind."
33. "Listen to Cicero, whom Nature taught."

50 Order and Design in the Structure of the World, how much more were necessary to *form* it!" If Men so much admire Philosophers, because they *discover* a small Part of the *Wisdom* that made all things; they must be stark blind, who do not admire that *Wisdom* itself!

Religio Philosophica;

or, the

Christian Philosopher:

BEING

A Commentary, of the more Modern and Certain PHILOSOPHY, upon that Instruction,

JOB XXXVI. 24.
Remember that thou magnify His Work which Men behold.

HE Works of the Glorious GOD in the *Creation* of the World, are what I now propose to exhibit; in brief *Essays* to enumerate *some of them*, that He may be glorified in them: And indeed my *Essays* may pretend unto no more than *some of them*; for, *Theophilus* writing, *of the Creation*, to his Friend *Antolycus*, might very justly say, 'That if he should have a *Thousand Tongues*, and live a *Thousand Years*, yet he were not able

Mather begins the substance of his treatise on this page from *The Christian Philosopher*. Courtesy of the Rare Book and Special Collections Library, University of Illinois at Urbana-Champaign.

Religio Philosophica;

or, the

Christian Philosopher:

being

A Commentary, of the more Modern and Certain Philosophy, upon that Instruction,

Job xxxvi. 24.
Remember that thou magnify His Work, which Men behold.

THE Works of the Glorious GOD in the *Creation* of the World, are what I now propose to exhibit; in brief *Essays* to enumerate *some of them*, that He may be glorified in them: And indeed my *Essays* may pretend unto no more than *some of them;* for, *Theophilus* writing, *of the Creation,* to his Friend *Autolycus,* might very justly say, That if he should have a *Thousand Tongues,* and live a *Thousand Years,* yet he were not able to describe the admirable Order of the Creation, διὰ τὸ ὑπερβάλλον μέγεθος καὶ τὸν πλοῦτον σοφίας τοῦ Θεοῦ. Such a *Transcendent Greatness of God,* and the *Riches of his Wisdom* appearing in it![1]

Chrysostom, I remember, mentions a *Twofold Book* of GOD; the Book of the *Creatures,* and the Book of the *Scriptures:* GOD having taught first of all us διὰ πραγμάτων, by his *Works,* did it afterwards

---

9 Work,/Work   15 Autolycus/Antolycus

---

1. Mather provides a translation of the Greek, which is tied to biblical passages. "Transcendent Greatness" is adapted from Eph. 1:19; "the Riches" passage is adapted from Rom. 11:33.

διὰ γραμμάτων, by his *Words*. We will now for a while read the *Former* of these *Books,* 'twill help us in reading the *Latter:* They will admirably assist one another. The Philosopher[2] being asked, What his *Books* were; answered, *Totius Entis Naturalis Universitas.*[3] All Men are accommodated with that *Publick Library. Reader,* walk with me into it, and see what we shall find so legible there, *that he that runs may read it.*[4] Behold, a Book, whereof we may agreeably enough use the words of honest *Ægardus; Lectu hic omnibus facilis, et si nunquam legere didicerint, & communis est omnibus, omniumque oculis expositus.*[5]

ESSAY I. *Of the* LIGHT.

WOULD it not be proper, in the first place, to lay down those *Laws of Nature,* by which the *Material World* is governed, and which, when we come to consider, we have in the Rank of *Second Causes,*[1] no further to go? All *Mechanical Accounts* are at an end; we step into the Glorious GOD *Immediately:* The very *next Thing* we have to do, is to Acknowledge Him, who is the *First Cause* of all: and the CHRISTIAN PHILOSOPHER will on all Invitations make the *Acknowledgments.* The acute Pen of Dr. *Cheyne* has thus delivered them.

I. All *Bodies* persevere in the same State of *Rest,* or of *Moving* forwards in a *strait Line,* unless forced out of that State, by some *Violence* outwardly impressed upon them.

II. The *Changes* made in the *Motions* of *Bodies,* are always proportional to the *Impressed Force* that moves them; and are produced in the same *Direction* with that of the Moving Force.

2. The reference is to Anthony of Egypt (251?–356). He devoted himself to asceticism and the solitary life of the desert. His holiness and discipline attracted disciples, and he briefly came out of solitude to organize them into a community of hermits.

3. Literally, "the whole universe of natural things"; or "the whole world is my book."

4. Adapted from Hab. 2:2.

5. "This is easy for all to read, even if they have not learned to read, and it is common to all and it is open to the eyes of all."

1. The distinction here is between first and second causes, or between the secret will of God and the revealed will of God which man can understand by observation and the use of reason. Second (or secondary) causes were therefore equated with the laws of nature by which the material world is governed. These laws were described in mathematical terms by Newton and others.

50 III. The *same Force* with which one *Body* strikes another, is *returned* upon the first by that other; but these Forces are impressed by *contrary Directions.*

IV. *Every Part* of every Body *attracts* or *gravitates* towards *every Part* of every other Body: But the *Force* by which one Part attracts
55 another, in different Distances from it, is reciprocally as the *Squares* of those Distances; and at the same Distance, the *Force* of the Attraction or Gravitation of one Part towards divers others, is as the Quantity of Matter they contain.²

These are *Laws* of the Great GOD, *who formed all things.*³ GOD
60 is ever to be seen in these *Everlasting Ordinances*. But now, in proceeding to *magnify that Work of God which Men behold,*⁴ it seems proper to begin with *that* by which it is that we *Behold* the rest.

The Light calls first for our Contemplation. A most marvellous Creature, whereof the Great God is the *Father:*⁵

65 *Illic incipit DEUM nosse.*⁶

The *Verus Christianismus*⁷ of the pious *John Arndt* very well does insist upon that Strain of Piety; GOD and His LOVE exhibited in the *Light.*

It was demanded, *In what Place is the Light contained? By what*
70 *Way is the Light divided?*

Aristotle's Definition of *Light;* Φῶς ἐστιν ἡ ἐνέργεια τοῦ διαφανοῦς, *Light is in the Inworking of a Diaphanous Body,* is worth an attentive Consideration.

Light is undoubtedly produced, as Dr. *Hook* judges, by a *Motion,*
75 quick and vibrative.

It is proved by Mr. *Molyneux,* That *Light* is a *Body.* Its *Refraction,* in passing thro a *Diaphanous Body,* shews that it finds a *different Resistance; Resistance* must proceed from a Contact of *two Bodies.* Moreover, it requires *Time* to pass from one place to another, tho

---

2. These four points are Newton's three laws of motion—the law of inertia, a form of what is today known as Newton's second law, and the law that "to every action there is always opposed an equal and opposite reaction"—plus Newton's principle of universal attraction or gravitation.
3. Prov. 26:10.
4. Adapted from Job 36:24.
5. Adapted from James 1:17.
6. "Here it [the mind] begins to know God."
7. Johann Arndt published *Vier Bücher vom wahren Christenthum* (Jena, 1605–9). The work was translated into Latin as *De vero Christianismo libri 4* (Lüneburg, 1625), with another two-volume Latin edition appearing in London in 1708.

it has indeed the quickest of all Motions. Finally, it cannot by any means be *increased* or *diminished*. If you *increase* it, it is by robbing it of some other part of the Medium which it would have occupied, or by bringing the *Light,* that should naturally have been diffused thro some other Place, into that which is now more enlightened.

Sir *Isaac Newton* judges, 'Tis probable, that *Bodies* and *Light* act mutually on one another. *Bodies* upon *Light,* in emitting it, and reflecting it, and refracting it, and inflecting it: *Light* upon *Bodies,* by *heating* them, and putting their Parts into a *Vibrating Motion.*

All *Hypotheses* of *Light* are too *dark,* which try to explain the *Phænomena* by *New Modifications* of *Rays;* they depend not on any such *Modifications,* but on some *Congenite*[8] and Unchangeable Properties, essentially inherent in the Rays.

The *Rays of Light* are certainly little Particles, actually emitted from the *Lucent Body,* and refracted by some *Attraction,* by which *Light,* and the *Bodies* on which it falls, do mutually act upon one another. It is evident, That as Rays pass by the Edges of Bodies, they are *incurvated*[9] by the Action of these *Bodies,* as they pass by them.

And it is now perceived, That *Bodies* draw *Light,* and this *Light* puts Bodies into *Heat:* And that the Motion of *Light* is therefore swifter in *Bodies,* than *in vacuo,* because of this Attraction; and slower after its being *reflected,* than in its Incidence.

Irradiated by the Discoveries of the Great Sir *Isaac Newton,* we now understand, That every *Ray* of *Light* is endowed with its *own Colour,* and its different Degree of *Refrangibility* and Reflexibility. One Ray is *Violet,* another *Indigo,* a third *Blue,* a fourth *Green,* a fifth *Yellow,* a sixth *Orange,* and the last *Red.* All these are *Original Colours,* and from the Mixture of these, all the intermediate ones proceed; and *White* from an equable Mixture of the whole: *Black,* on the contrary, from the small Quantity of any of them reflected, or all of them in a great measure suffocated. It is not *Bodies* that are *coloured,* but the *Light* that falls upon them; and their *Colours* arise from the *Aptitude* in them, to *reflect* Rays of one Colour, and to *transmit* all those of another. 'Tis now decided, *No Colour in the dark!*

Tho *Light* be certainly a *Body,* it is almost impossible to conceive how *small* the Corpuscles of it are. Dr. *Cheyne* illustrates it with

---

8. That is, innate.
9. That is, bent into a curved form.

Essay 1. Of the Light

an Experiment, That it may be propagated from innumerable different Luminous Bodies, without any considerable Opposition to one another. Their several *Streams of Light* will be together transmitted into a dark Place, thro the least Orifice in the World. Suppose a Plate of Metal, having at the top the smallest Hole that can be made, were erected *perpendicularly* upon an *Horizontal Plane,* and about it were set numberless luminous Objects of about the same Height with the Plate, at an ordinary Distance from it; the *Light* proceeding from every one of these Objects, will be propagated thro this Hole, without interfering.

Mr. *Romer,* from his accurate Observations of the *Eclipses* on the *Satellits* of *Jupiter,* their Immersions and Emersions, thinks he has demonstrated, That *Light* requires one Second of Time to move 9000 Miles. He shews, that the Rays of *Light* require ten Minutes of Time to pass from the *Sun* to us. And yet Mr. *Hugens* hath shewn, That a Bullet from a Cannon, without abating its first Velocity, would be 25 Years passing from us to the *Sun*. So that the Motion of *Light* is above a million times swifter than that of a Cannon-Ball; yea, we may carry the Matter further than so.

We suppose the Distance of the *Sun* from the Earth to be 12000 Diameters of the *Earth,* or suppose 10000, the *Light* then runs 1000 Diameters in a Minute; which is at least 130,000 Miles in a Second. Dr. *Cheyne* shews, That *Light* is about six hundred thousand times more swift than *Sound*. Amazing Velocity!

To chequer the Surprize at so *swift* a Motion, I may propound one that shall be as very surprizingly *slow*. *Dee* affirms, that he and *Cardan* together saw an Instrument, in which there was one Wheel constantly moving with the rest, and yet would not finish its Revolution under the space of seven thousand Years. 'Tis easy to conceive with *Stevinus,* an Engine with twelve Wheels, and the Handle of such an Engine to be turned about 4000 times in an Hour, (which is as often as a Man's Pulse does beat) yet in ten Years time the Weight at the Bottom would not move near so much as an Hair's Breadth: And as *Mersennus* notes, it would not pass an Inch in 1,000,000 Years; altho it be all this while in Motion, and have not stood still one Moment: for 'tis a Mistake of *Cardan*, *Motus valde tardi, necessario quietes habent intermedias*.[10]

---

10. "[That] very slow movements necessarily have intermediate stops."

> "Behold the *Light* emitted from the *Sun;*
> What more familiar, and what more unknown?
> While by its spreading Radiance it reveals
> All Nature's Face, it still itself conceals.
> See how each Morn it does its Beams display,
> And on its golden Wings brings back the Day!
> How soon th' effulgent Emanations fly
> 'Thro the blue Gulph of interposing Sky!
> How soon their Lustre all the Region fills,
> Smiles on the Valleys, and adorns the Hills!
> Millions of Miles, so rapid is their Race,
> To chear the Earth, they in few Moments pass.
> Amazing Progress! At its utmost Stretch,
> What human Mind can this swift Motion reach?
> But if, to save so quick a Flight, you say,
> The ever-rolling Orb's impulsive Ray
> On the next Threads and Filaments does bear,
> Which form the springy Texture of the Air,
> That those still strike the next, till to the Sight
> The quick Vibration propagates the Light:
> Still 'tis as hard, if we this Scheme believe,
> The Cause of Light's swift Progress to conceive."
>
> Sir *Richard Blackmore's Creation,* Book 2.

The *Jews* have a good Saying, *Opera Creationis externæ habent in se Imaginem Creationis internæ.*[11] It will well enough become a *Christian Philosopher,* to allow for that *Image* in his Contemplations, and with devout Thoughts now and then reflect upon it.

Before I go any further, I confess myself unable to *resist* the Invitation, which, I think, that I have, to insert an Observation of *Hugo de Sancto-Victore;* That every Creature does address a *Treble Voice* unto us: *ACCIPE, REDDE, FUGE;*[12] indeed, *there is no Speech nor Language where their Voice is not heard.*[13] It is an Exercise highly becoming the *Christian Philosopher,* to fetch *Lessons of Piety* from the whole Creation of GOD, and hear what *Maxims of Piety* all the Creatures would, in the way of *Reflection* and *Similitude,*

---

11. "The external works of the Creation have in themselves the image of the internal Creation."
12. "Accept, Return, Avoid."
13. Ps. 19:3.

mind us of. In the Prosecution of these *Meleteticks*,¹⁴ what better can be considered, than this *Treble Voice*, from all these Thousands of *Powerful Preachers*, whom we have continually surrounding of us? First, *Accipe Beneficium:*¹⁵ Consider, *What is the Benefit which a Good GOD has, in this Creature, bestowed upon me?* Secondly, *Redde Servitium:*¹⁶ Consider, *What is the Service which I owe to a Gracious GOD, in the Enjoyment of such a Creature?* Lastly, *Fuge Supplicium:*¹⁷ Consider, *What is the Sorrow which a Righteous GOD may inflict upon me by such a Creature, if I persist in Disobedience to Him?* Even a Pagan *Plutarch* will put the Christian *Philosopher* in mind of this, That the World is no other than the *Temple* of GOD; and all the *Creatures* are the *Glasses*, in which we may see the *Skill* of Him that is the Maker of all. And his Brother *Cicero* has minded us, *Deum ex Operibus cognoscimus.*¹⁸ 'Tis no wonder then that a *Bernard* should *see* this; *Verus Dei Amator, quocunque se vertit, familiarem Admonitionem sui Creatoris habet.*¹⁹ The famous Hermite's Book,²⁰ of those three Leaves, the *Heaven*, the *Water*, and the *Earth*, well studied, how nobly would it fill the *Chambers* of the Soul with the most *precious and pleasant Riches?*²¹ *Clemens* of *Alexandria*²² calls the World, *A Scripture of those three Leaves;* and the Creatures therein speaking to us, have been justly called *Concionatores Reales,*²³ by those who have best understood them:

> *Obvia dum picti lustro Miracula Mundi,*
>   *Naturæ intueor dum parientis Opus:*
> *Emicat ex ipsis Divina Potentia Rebus;*
>   *Et levis est Cespes qui probat esse Deum.*²⁴

But the *Light* now calls for me.

¶. How *Glorious* a Body! "But how infinitely, and beyond all

---

14. That is, rules or methods of meditation.
15. "Accept the benefit."
16. "Return the service."
17. "Avoid the suffering."
18. "We recognize God from His works."
19. "The true lover of God, wherever he turns himself, has a familiar reminder of his Creator."
20. Anthony of Egypt.
21. Adapted from Prov. 24:4.
22. Clement of Alexandria (c. 150–c. 215), theologian and head of the Catechetical School at Alexandria, supplemented the Christian faith with Greek philosophy.
23. "Natural Preachers."
24. "While I observe the open wonders of the ornate world, / While I gaze upon the work of Nature, the begetter: / Divine power springs forth from the things themselves; / And it is the trifling turf which proves that God exists."

Comprehension *Glorious* then, the Infinite GOD, who has challenged[25] it as His Glory! Isa. xlv. 7. *I form the Light.* The GOD of whom we have that *Sublime Stroke,* in the History of the Creation; he said, *Let there be Light, and there was Light!*[26] The GOD whose Majesty is within that *Holy of Holies,*[27] where He *dwells in the Light, that no Man can approach unto!*[28] Lord, thou hast in a wondrous Display of thy Benignity, afforded the Benefit of the *Light* unto thy Creatures: *Whatsoever does make manifest, is Light.*[29] How miserable should we be, and in what inexpressible Confusion, if the *Light* were withheld from us! What could be *manifest* unto us; what enjoyed or performed by us! O let all that *walk in the Light of the Living,*[30] unite in Praises to the Creator of the *Light! O! give thanks to the Lord, for He is good, and his Mercy endureth for ever.*[31] But, *Lord,* wilt thou leave my *Soul* in *Darkness!*[32] The *Light* granted unto the *Soul,* in the Knowledge of those things, *which to know is Life eternal,*[33] is more precious and needful, than that in which our *Body* finds itself so much befriended. *O Father of Glory, let me have the Eyes of my Understanding enlightened.*[34]

"I have a most Glorious Redeemer, of whom I am assured, That he is *the true* LIGHT, and *the* LIGHT *of the World.*[35] A *Light* which, like other *Light,* carries its own *Evidence* with it: there needs no more to prove, that our Blessed JESUS is the *Son* of GOD, and the *Saviour* of the World, than attentively to *Behold Him.* He can be no other, than what he asserts Himself to be, *The Light of Men.*[36] *Lord, in thy Light I shall see Light.*[37] When I see the *Truth as it is in JESUS,*[38] in such a Revelation and such an Exhibition, as my JESUS gives of it, then I see every thing

---

25. That is, asserted a natural right or title to.
26. Gen. 1:3.
27. The Authorized Version reads "the most holy place." This phrase is frequently used in several books of the Old Testament.
28. Adapted from 1 Tim. 6:16.
29. Eph. 5:13.
30. Ps. 56:13.
31. Ps. 136:1.
32. Adapted from Ps. 16:10 and Acts 2:27.
33. Adapted from John 17:3.
34. Adapted from Eph. 1:17–18.
35. Adapted from John 1:9, 12:46.
36. John 1:4.
37. Ps. 36:9.
38. Adapted from Eph. 4:21.

in a true Light.[39] *My Saviour,* thou art more precious, and more needful, and more useful to me than the *Light.* I will walk in thee, and under thy Conduct; so shall *I walk in the Light continually.*[40]

"But what signifies the *Light,* unto him that has no *Eyes* to perceive it. *O my Redeemer!* Bestow thou an *Eye* upon me: A *Faculty* to discern the Things that are *Spiritually to be discerned.*[41]

"For the *Light of Reason,* which *enlightens every Man that comes into the World;*[42] every Man has all possible Reason to glorify GOD, and never do any thing, whereof any Man may justly say, *It seems to me unreasonable.*[43]

"But, *O my GOD,* thou hast favoured us with a rich Conglobation[44] of *Light,* in the *Book* of thy lively Oracles, wherein we have a *Light shining in a dark Place.*[45] I would consider every thing in the *Light* wherein this lovely *Book* sets it before me: But, let me not *rebel against the Light!*[46]

"*The Light is truly sweet.*[47] But, what shall I find *the Inheritance of the Saints in Light!*[48] They that are shut out of that *Light,* and cast into *outer Darkness,* and where they shall *never see Light;* Oh! the *Weeping,* and *Wailing,* and *Gnashing of Teeth,* which they must be exposed unto![49] *My Saviour,* I am under thy Conduct, passing through a gloomy Valley into thy *Light;* and when *I sit in Darkness, the Lord will be a Light unto me.*[50]

"How *swift* the Motion of the *Light!* But, *O my Saviour,* why no more *swift* in thy coming to visit and relieve a World lying in the perpetual Night of *Wickedness?* Why thy *Chariot so long in coming?*[51]

"And, *O my soul,* why art thou *slow* in thy Contemplations of GOD, and CHRIST, and HEAVEN; fly thou thither, with a

---

39. Perhaps an allusion to Ps. 36:9.
40. Adapted from 1 John 1:7.
41. Adapted from 1 Cor. 2:14.
42. John 1:9.
43. Acts 25:27.
44. That is, rounded formation or conglomeration.
45. 2 Pet. 1:19.
46. Adapted from Job 24:13.
47. Adapted from Eccles. 11:7.
48. Col. 1:12.
49. Adapted from Matt. 8:12 and Ps. 49:19.
50. Mic. 7:8.
51. Judg. 5:28.

Swiftness beyond that of the *Light*, [for so thou canst] upon all Occasions."

ESSAY II. *Of the* STARS.

LET us proceed, and, conforming to the End of our *Erect Stature*, behold the Heavens, and lift up our Eyes unto the Stars.¹ The learned *Hugens* has a Suspicion, that every *Star* may be a *Sun* to other Worlds in their several *Vortices*. Consider then the vast Extent of our *Solar Vortex,* and into what Astonishments must we find the Grandeur and Glory of the Creator to grow upon us! Especially if it should be so, (as he thinks) that all these Worlds have their *Inhabitants*, whose Praises are offer'd up unto our GOD!

> *Quantula de Cœli spectanti Vertice celso*
> *Terra videretur, si Cœli e Vertice Terra*
> *Ulla videretur!*²   So *Buchanan*.

His Improvement of the Thought is, How *little* of this *little* has vain Man to strive for, and to boast of!

> *O Pudor! O stolidi præceps vesania voti!*³

Mr. *Childrey* mentions two Curiosities, which ought to be a little further enquired into. The one is, That between the two Constellations of *Cygnus* and *Cepheus*, there lies cross the *Milky-Way*, a black, long, little *Cloud*, neither increasing, nor abating, nor changing the Place in which it makes its Appearance.

The other is, That in *February*, and a little before and after that Month, in the Evening, when the Twilight has near deserted the Horizon, there is a very distinguishing Way of the *Twilight;* a *Bright Path* striking up towards the *Pleiades*, and almost reaching them, which is not observed any other time of the Year.

The *Jews* have a Fancy among them, That when the Almighty first bespangled the Heavens with *Stars*, he left a Spot near the *North Pole* unfinished and unfurnished, that so if any other should

---

1. Adapted from Isa. 40:26.
2. "How large would the earth appear to you, as looked down upon from the top of the sky?"
3. "O shame! Precipitate madness of stupid hopes!"

set up for a GOD, there might be this trial made of his Pretensions; *Go, fill up, if you can, that part of the Heavens, which is yet left imperfect.* But without any such Suppositions, we may see enough in the Heavens to proclaim this unto us; *Lift up your Eyes on high, and behold: Who has created these things?*[4] None but an Infinitely Glorious GOD could be the Creator of them!

The TELESCOPE, invented the Beginning of the last Century, and improved now to the Dimensions even of *Eighty Feet,* whereby Objects of a mighty Distance are brought much nearer to us; is an Instrument wherewith our Good GOD has in a singular manner favoured and enriched us: A *Messenger* that has brought unto us, from very distant Regions, most wonderful Discoveries.

My GOD, I cannot look upon our Glasses without uttering thy Praises: By them I see thy Goodness to the Children of Men![5]

By this *Enlightener of our World,* it is particularly discovered,

That all the *Planets* at least, excepting the *Sun,* are *dense* and *dark* Bodies; and that what *Light* these *opake* Bodies have, is borrowed from the *Sun.*

That every one of the *Planets,* excepting the *Sun,* do change their Faces like the *Moon. Venus* and *Mercury* appear sometimes like an *Half-Moon,* and sometimes quite *round,* according as they are more or less opposite to the *Sun. Mars* has his Times of appearing in a Curvi-lined Figure. *Jupiter* has four little Stars, that continually move about him, and in doing so, cast a *Shadow* upon him. *Saturn* has a *Ring* encompassing of him.

That each of these *Planets* have *Spots* in their Superficies, like those of the *Moon.*

That not only each of these *Planets,* but the *Sun* also, besides whatever other Motion they may have, do move themselves upon their own Centers; some of them with a Motion of *Revolution,* others by that of *Libration.*

It was a good Remark made by one of the Antients, *Quid est Cælum, & totius Naturæ Decor, aliud, quam quoddam speculum, in quo summi Opificis relucet Magisterium?*[6]

The Pagan *Tully,*[7] contemplating, *Cælestium admirabilem Ordi-*

---

4. Isa. 40:26.
5. Adapted from Ps. 107:8, 15, 21, 31.
6. "What are heaven and the beauty of all of nature other than a kind of mirror in which the magistracy of the highest Artificer shines back?" The "Ancient" is Plato (c. 429–347 B.C.), who held that the cosmos is a divine artifact but also an intelligent being. Plutarch reaffirmed this cosmological doctrine.
7. That is, Marcus Tullius Cicero.

nem, incredibilemque Constantiam, the admirable Order, and the incredible Constancy of the Heavenly Bodies and their Motions, adds upon it, *Qui vacare Mente putat, ne ipse Mentis expers habendus est:* Whosoever thinks this is not governed by *Mind* and *Understanding,* is himself to be accounted void of all *Mind* and *Understanding.*[8]

According to Mr. *Hugens,* the Distance of the *Sun* from us is 12,000 Diameters of the Earth. A Diameter of the Earth is 7,846 Miles. The Distance of the nearest *Fixed Stars* from us, compared with that of the *Sun,* is as 27,664 to 1: So then the Distance of the nearest *Fixed Stars* is at least 2,404,520,928,000 Miles; which is so great, that if a Cannon-Ball (going all the way with the same Velocity it has when it parts from the Mouth of the Gun) would scarce arrive there in 700,000 Years. *Great GOD, what is thy Immensity!*

The Number of the *Stars!* The learned *Arndt* has a good Thought upon it: *Si Deus tantam Stellarum Multitudinem condidit, quis dubitet, illum multo majorem Copiam habere Spirituum Cælestium, sine intermissione illum laudantium?*[9] If the *Morning-Stars* are so many, how many are the *Sons of GOD!*[10]

¶. "Glorious GOD, I give Thanks unto thee, for the Benefits and Improvements of the *Sciences,* granted by thee unto these our latter Ages. The *Glasses,*[11] which our GOD has given us the *Discretion* to invent, and apply for the most noble Purposes, are Favours of Heaven most thankfully to be acknowledged.

"The World has much longer enjoyed the *Scriptures,* which are *Glasses,* that bring the *best of Heavens* much nearer to us. But, tho the *Object-Glasses* are here, the *Eye-Glasses* are wanting. *My GOD,* bestow thou that *Faith* upon me, which, using the *Prospective*[12] of thy Word, may discover the *Heavenly World,* and acquaint me with what is in that World, which, I hope, I am going to.

"I hear a *Great Voice* from the *Starry* Heavens, *Ascribe ye*

---

8. Cicero's complete thought is as follows (Mather omits the second clause): "Anyone who thinks that the marvellous order and incredible regularity of the heavenly bodies, which is the sole source of preservation and safety for all things, is not rational, himself cannot be deemed a rational being."

9. "For if God has created so great a multitude of stars, who can doubt that he has a much greater multitude of celestial spirits, who praise him without ceasing?"

10. Adapted from Job 38:7.

11. That is, the telescope and the microscope.

12. That is, telescope. Mather uses "Prospective" in this sense again in Essay 3.

Greatness to our GOD.[13] *Great GOD,* what a Variety of *Worlds* hast thou created! How astonishing are the Dimensions of them! How stupendous are the Displays of thy *Greatness,* and of thy *Glory,* in the Creatures, with which thou hast replenished those Worlds! Who can tell what *Angelical Inhabitants* may there see and sing the *Praises* of the Lord! Who can tell what *Uses* those *marvellous Globes* may be designed for! Of these *unknown Worlds* I know thus much, *'Tis our Great GOD that has made them all.*"[14]

ESSAY III. *Of the* FIXED STARS.

OUR Great Prospective[1] having made Enquiry, finds a far greater Number of *Stars,* than what we can discern with the naked Eye. The Antients reckon'd only *One Thousand and Twenty Two* Stars in their *Fifty* Constellations. *Kepler* augments the Number to *One Thousand Three Hundred and Ninety Two. Bayer* carries it on to *One Thousand Seven Hundred and Nine.* Travellers to the Southward increased the Number of their Constellations to Sixty Two. The Number of the *Stars,* brought down into our *latest Globes,* is about *Nineteen Hundred;* but those in the Heavens are inconceivably more. Among the *Pleiades,* in a Circle of but one Degree diameter, where our naked Eye sees but *Six,* thus assisted we see *Forty Six.*

The *Milky-Way* is nothing but an infinite Number of *Stars,* which are so small, and lie so thick, as to give but a confused Glare unto us: And so the *Nebulosæ,* in the Head of *Orion.*

The *Præsepe* is a Cluster of more than Forty Stars. Those adjacent unto the Sword and Girdle of *Orion* about Fourscore. Mr. *Derham* suspects, that the *Whiteness* of the *Milky-Way* is not caused by the great Number of the *Fixed Stars* in that Place, but partly by their *Light,* and partly by the Reflections of their *Planets,* which blend their *Light,* and mix it.

It is a little surprizing, that all the *Planets* appear *greater* in the Glass than to the naked Eye; but the *Fixed Stars* appear *smaller* there.

---

13. Deut. 32:3.
14. Adapted from Ps. 104:24.
1. That is, the telescope.

The Words of the ingenious Dr. *Cheyne* are worth considering: "Since our *Fixed Stars* are exactly of the same Nature with our *Sun,* it is very likely that they have their *Planets;* and these *Planets* have *Satellits;* and these *Planets* and *Satellits* have Inhabitants, rational and irrational; Plants and Vegetables, Water and Fire; analogous to those of our System." *Ascribe ye Greatness to our God!*[2]

That which renders it probable, that the *Fixed Stars* are Bodies like our *Sun,* is this: 'Tis plain they shine by their *own Light.* It is impossible they should appear so lucid as we see them, from the Light of our *Sun* transmitted unto them. 'Tis their astonishing Distance from us that causes the best of our *Telescopes* to lessen them. Tho we in this Globe approach nearer to them, some 24,000 Diameters of the Earth, or 188,304,000 Miles, one time of the Year than another; yet their *Parallax* is hardly sensible, or any at all: which could not be, if the Distance were not wonderful.

Hence also, it is impossible they should be all in the Surface of the same Sphere, since our *Sun,* which is one of them, cannot be reduced unto this Rule. They are doubtless at as immense Distances from one another, as the nearest of them is from us. Were we at such a Distance from the *Sun,* we should not have the least Glimpse of the *Planets* that now attend it. Their Light would be too weak to affect us, and all their Orbs would be united in that one lucid Point of the *Sun.*

There are discovered *New Stars* in the Firmament, which having appeared a certain Time, do again disappear.

A *New Star* appeared about 125 Years before the Birth of our Saviour.

Claudian mentions one which appeared, A. C. 388. *Albumazer Haly* mentions one, which appeared in the fifteenth Degree of *Scorpio,* and continued four Months.

In the Year 1571, and the Month of *November,* there appeared in that Constellation, which we call the Chair of *Cassiopeia,* a most notable and wonderful Star of the first Magnitude, which held a Place among the other Stars, not having any Parallax, and kept a Course like theirs: It continued sixteen Months; then decreased;

29 *from* / frome

2. Deut. 32:3.

Essay 3. Of the Fixed Stars 31

anon grew quite invisible.³ A Noble Person⁴ affirms, there was a *black Spot* remaining in the Place where that *Star* appeared.

In the Year 1601, there appeared a *New Star* of the third Magnitude, in the *Swan*'s Breast, which continued visible twenty five Years, and then disappeared. Thirty three Years after, it appeared again in its former Magnitude; but went away again in a Year or two. It re-appeared five Years after, and was extant for several Years, but of no more than the sixth or seventh Magnitude.

In the Year 1671, another *Star,* which arrived unto the third Magnitude, appeared in the *Swan*'s Bill; it increases, and then decreases, and is about a Month making its Revolution.

There is an admirable Star in the *Whale*'s Neck: This first appears as one of the sixth Magnitude, and then increases by little and little, for one hundred and twenty Days together, till it arrives to its full Bigness and Brightness, which is that of the third Magnitude; wherein it continues fifteen Days together: after which, it then decreases until it becomes invisible. It appears every Year in its greatest Lustre, thirty two or thirty three Days earlier than in the foregoing Year; so that its Revolution is compleated in about three hundred and thirty three Days.

In the Years 1612, and 1613, there appeared a *Cloudy Star* in the Girdle of *Andromeda;* which disappeared until the Year 1664, and then appeared again.

There is another Star, between *Eridanus* and the *Hare,* which also shows itself, and then withdraws, like the former.

There is one Star of the fourth Magnitude, with two of the fifth, in *Cassiopeia,* which in all probability are new ones.

Mr. *Cassini* has observed four towards the *Arctick* Pole, which are probably new ones too.

Some Stars formerly appearing, do now disappear. One such there was in *Ursa Minor*. Another or two in *Andromeda*. One which *Tycho Brahe* inserts in his Catalogue, for the twentieth of *Pisces*. For time out of mind, there were *Seven Stars* observed in the

73 Arctick/Artick

3. Mather's reference is to the famous nova of 1572.
4. According to Mather's source, the "Noble Person" was the marquis of Villena in his book on "the Centiloqui of Ptolemy." Juan Lopez Pacheco, marquis of Villena (c. 1660–c. 1730), the founder and original director of the Royal Academy of Spain, does not appear to have written on astronomy. Enrique de Aragon, marquis of Villena (1384–1434), touched on astronomy, but he wrote before 1572.

Pleiades. The Writer of *Astronomy's Advancement*[5] enquires, whether the *Seven Stars* in the First of the *Revelation* have no Allusion to them. However, at present there are but *Six* to be seen, probably one of them is retired.

Mr. *Derham* thinks these *New Stars* may be Planets, belonging to some of the Systems of the Fixed Stars, and those Planets become visible, when they are in that part of their Orbits which is nearest the Earth, and again gradually disappear, as they move in their Orbits farther from us.

It is surprizing Observation of Dr. *Cheyne:* "Supposing that every *Fixed Star* is a *Sun,* and governs in a *Mundane Space,* equal to our System, then there must be only as many *Fixed Stars* of the *First Magnitude,* as there are Systems that can stand round ours. But there are but about twelve or thirteen *Spheres* that can stand round a middle one, equal to them: And so many are the Stars of the first Magnitude. Again, if we examine how many *Spheres* can stand round this first Range of *Spheres,* we shall find their Number between Forty-Eight and Fifty-Two. And so we find the Number of the *Stars* of the *second Magnitude.* As for the several other Magnitudes, it is not altogether possible to determine their Number, because they are not so distinguishable from those of the other Magnitudes, as the first and second are." He adds most reasonably and religiously: *It is impossible for any body seriously to consider in his Mind, what is certain about these Heavenly Bodies, and to hinder himself from being ravished with the Power and Wisdom of the Great GOD of Heaven and Earth!*

Mr. *Derham* supposes the particular Star *Syrius* to be above two Millions of Millions of Miles distant from us.

Dr. *Grew,* from a very probable Computation, makes the Distance of the *Pole-Star* from the Earth to be Four Hundred and Seventy Millions, and Eight Hundred and Forty Thousand Miles.

Considering the mean and vile Fables of the *Pagan Poetry,* yea, and the scandalous Actions of some *Greater Devils* among the *Pagans,* which are commemorated and celebrated in the Names which our *Globes* give unto the *Constellations,* I cannot but move you, O *Christian Astronomers,* to attempt a Reformation of so

---

5. Joseph Walker, *Astronomy's Advancement: Or, News for the Curious; Being a Treatise of Telescopes: and an Account of the Marvelous Astronomical Discoveries of Late Years throughout Europe; with the Figures of the Sun, Moon, and Planets; with Copernicus His System . . . Done out of French* (London, 1684). Walker (fl. 1684–88) was an English translator about whom little is known.

This plate from Julius Schiller's *Coelum stellatum Christianum concavum* (Augsburg, 1627) depicts the archangel Gabriel, alias Pegasus. Mather too urged that Greco-Roman constellations be converted into Judeo-Christian ones. Courtesy of the Rare Book and Special Collections Library, University of Illinois at Urbana-Champaign.

shameful an Abuse. For shame, let those Glorious Bodies no longer suffer the Affronts of our *Base Denominations.* To put *Christian Names* on the *Constellations,* and allowing the present Figures upon our *Globes* to remain still as they are, nevertheless to transfer them into *Scriptural Stories,* was a thing endeavoured by *Schillerus,* and by *Novidius.*

The Caution used in the antient *Hebraick* and *Arabick* Astronomy, about the Names of the Constellations, is well known to all that are versed in *Antiquities.* Dismissing that Reflection, what remains is this: A learned *Frenchman*[6] pretends to tell us, That the *Stars* in the Heavens do stand ranged in the Form of *Hebrew Letters,* and that it is possible to *Read there, whatever is to happen of Importance throughout the Universe.* Amazing! That so much Learning should be *Consistent* with, and much more, that it should be *Subservient* to such *Futilities!* The true *Reading of the Stars* is to look up, and spell out, the glorious Perfections of that GOD, who is the *Father of those Lights,*[7] and who *made* and *moves* them all.

¶. "I would by no means look up unto the *Stars,* with the foolish *Astrology* of the *Star-gazers,* who try to *read,* what the Great GOD that made them has not *written* there. But there is very plainly to be read there, the Power and the Grandeur of the Glorious GOD. This, this I will observe, prostrate in the Dust before Him. *The Heavens declare the Glory of GOD;*[8] and shall not I *observe* it? *When I consider thy Heavens, O Lord, and the Stars which thou hast ordained,* I cannot but cry out, *What is Man, that thou art mindful of him, and the Son of Man, that thou visitest him!*[9]

"Unto the Father of the Faithful,[10] my GOD said, *Look now toward Heaven, and tell the Stars, if thou be able to number them; so shall thy Offspring be.*[11] *Glorious Lord,* make me one of them. A *Worm* of the Dust, filled with the Love of GOD and of his Neighbour, becomes a *Star* in the Eye of the Glorious GOD: And if he be one of much Grace, and one of much *Use,* he is then a *Star* of the *greater Magnitude.*

"GOD, *my Maker* and theirs, gives me that *Song for the Night,*[12]

6. That is, Jacques Gaffarel.
7. Adapted from James 1:17.
8. Ps. 19:1.
9. Ps. 8:3–4.
10. That is, Abram or Abraham.
11. Adapted from Gen. 15:5.
12. Adapted from Job 35:10.

wherein I view them; *He tells the Number of the Stars; He calls them all by their Names.*[13] 'Tis true of the *Just,* who are to *shine as the Stars for ever and ever.*[14] May I be known by the Lord as one of that *Number,* and have a *Name* in *His Book of Life!*[15]

"Are the very *Stars* themselves liable to *Vicissitudes*? And shall not I look for them in this our miserable World?

"How little can I comprehend the Condition and Intention of the *Stars? O Incomprehensible GOD,* I will not cavil, but adore, when I find *Mysteries* in thy *Providence,* altogether beyond my *Penetration!*"

ESSAY IV. *Of the* SUN.

A Most Glorious and most Useful *Creature!* But still a *Creature!* By Old Astronomers call'd, *Cor Planetarum.*[1]

There will be no *Athenians* now to araign me for it, if I call it, *The Carbuncle of the Heavens.*[2] *Kircher* supposes the *Sun* to be a Body of wondrous *Fire,* unequal in Surface, composed of Parts which are of a different Nature, some fluid, some solid: The Disque of it, a *Sea of Fire,* wherein Waves of astonishing Flame have a perpetual Agitation.

Sir *Isaac Newton,* as well as Dr. *Hook,* takes the *Sun* to be a solid and opake Body. Dr. *Hook* thinks this Body to be encompassed with a vast Atmosphere, the Shell whereof is all that shines. The *Light* of the Sun he takes to be from the Burning of the more superficial Parts, which are set on fire, which may be without hazard of being burnt out in a vast Number of Ages. And Sir *Isaac Newton* thinks the *Sun* to be a sort of a mighty Earth, most vehemently hot; the Heat whereof is conserved by the marvellous Bigness of the Body, and the mutual Action and Re-action between *That,* and the *Light* emitted from it. Its Parts are kept from fuming away, not only by its *Fixity,* but also by the *Density* of the

---

13. Ps. 147:4.
14. Adapted from Dan. 12:3.
15. Probably adapted from Dan. 12:1.
1. "Heart of the planets."
2. A reference to the philosopher Anaxagoras (c. 500–c. 428 B.C.), who was tried, fined, and exiled from Athens for saying that the sun was stone and the moon earth at a time when people regarded sun, moon, and stars as divinities.

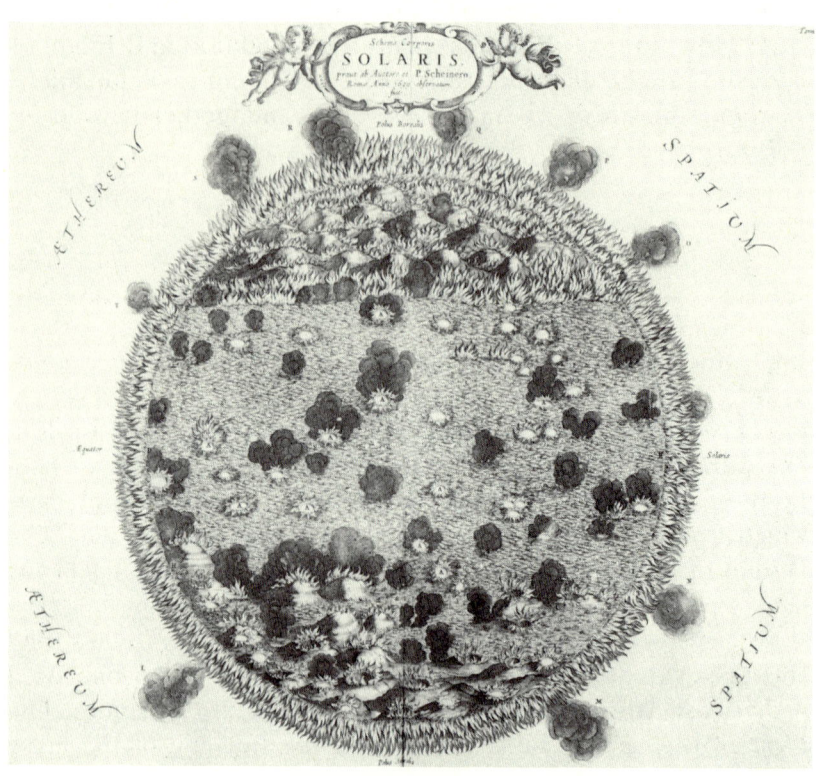

This plate from Athanasius Kircher's *Mundus subterraneus* (Amsterdam, 1664–65, 1678) illustrates Kircher's belief that the sun is a sea of fire whose waves are in perpetual agitation and that solar (black) spots are the froth of the fire which the sun exhales. Mather saw a reproduction of this image in Joseph Walker's *Astronomy's Advancement* (London, 1684). Courtesy of the Rare Book and Special Collections Library, University of Illinois at Urbana-Champaign.

Atmosphere incumbent on it, and the vast Weight thereof. The *Light* seems to be emitted much after the manner as *Iron,* when heated unto such a Degree, as to be just going into Fusion, by the vibrating Motion of its Parts emits with Violence plentiful Streams of liquid Fire. So great a Body will continue its Heat a great while, perhaps in proportion to its Diameter.

Upon the Convexity of the Body of the *Sun,* there are observed *black Spots,* which are moveable, and changeable. These move regularly towards the West, and finish their Revolution in about five and twenty Days; and so testify unto us, that the *Sun* turns upon its own Center: the *Axis* of the Motion inclining to the *Ecliptick.*

These *Maculæ Solares*[3] are probably Evaporations, which arise from the Body of the *Sun,* somewhat as Vapours do from the Earth; and they form themselves into *Clouds.* That which adds to this Probability, is, that the *Spots* are always changeable in their Bulk, and Form, and Configuration. Sometimes their *Number* is greater, and sometimes lesser, and sometimes there are none at all. Some of them shine, and others that shone, become dark. Diligent Astronomers,[4] who have waited on them for nine or ten Years together, have never found them in all this time to return unto the *same Configuration.* In *Charlemain's* time,[5] every one saw a *Spot* in this great Luminary. And there have been divers Days together, [as in the Year 1547,] wherein the *Sun* has appeared little brighter than the *Moon* in her total Eclipse, and the *Stars* have been visible at Noon-Day. *Virgil*[6] and *Ovid*[7] intimate such a Darkness upon the *Sun* once for a whole Year together, that the Fruits of the Earth could not be ripened.

The apparent Diameter of the *Sun* being sensibly greater in *December* than in *June,* it is plain, and Observation confirms it, that the *Sun* is proportionably nearer to the Earth in *Winter* than in *Summer.* It is also confirmed, by the Earth's moving swifter in

10 greater/shorter

3. "Sun spots."
4. They were Charles Malapert (1581–1630), a Belgian Jesuit who was also a poet and mathematician, and Christoph Scheiner (1573–1650), a German Jesuit.
5. Charlemagne (742–814), king of the Franks from 768 on, united almost all the Christian lands of western Europe and assumed the title of emperor in 800.
6. Publius Vergilius Maro (70–19 B.C.), the Roman poet.
7. Publius Ovidius Naso (43 B.C.–A.D. 17), after Vergil one of the greatest Roman poets.

December than in *June;* which it does about five Fifteenths. And for this reason there are about eight Days more from the *Sun's* vernal *Equinox* to the autumnal, than from the autumnal to the vernal.

Mr. *Tompion*'s[8] Observations, from the *Equation of natural Days,* render it evident, That the Motion of the *Sun* (if we must speak in those Terms) must be swifter at some times, than at others. *Great GOD, the Motion is always under thy Glorious Guidance!*

According to *Cassini,* the Sun's mean Distance from the Earth is 22,000 Semidiameters of the Earth. And the Sun's Diameter is equal to 100 Diameters of the Earth: And therefore the Body of the Sun must be 1,000,000 times greater than the Earth.

Cassini more directly expresses himself; That the *Sun's* Distance from the Earth is 172,800,000 *English* Miles.

Take Mr. *Derham's* Computation; *Saturn* is computed at 93,451 Miles in Diameter, and consequently 427,318,300,000,000 Miles in Bulk: *Jupiter* at 120,653 Miles in Diameter, and by consequence 920,011,200,000,000 Miles in Bulk. But yet, as amazing Masses as these all are, they are all far outdone by that Globe of Fire, the *Sun:* which, as it is the *Fountain* of *Light* and *Heat* unto all the Planets about it, by its kind Influences affording them the great Comforts of Life; so does it in *Bulk* surpass them all. Its Diameter is computed at 822,148 Miles; and so there must be 290,971,000,000,000,000 Miles in the solid Content of it.

Dr. *Grew* is of opinion, that for ought we know, the *Sun* may afford us his *Light,* without such an intense *Heat,* as has been imagined. The Beams of the *Sun,* he thinks, may first conceive their *Heat,* when they come to be mixed with our *Atmosphere.* There are things intensely *hot,* which give no *Light* at all; but *Rotten Wood,* or *Fish,* and the *Glowworm,* and some other Bodies, give a brisk *Light,* without any *Heat. Light* and *Heat,* he thinks, have no necessary Conjunction, at least not in any sensible Proportion. It is known also, how necessary the *Air* is to produce *Fire,* and even *Light* itself, in some of those Bodies that shine in the dark. If the Sun were a *burning Body,* and the *Heat* of it so much greater than that we feel of it, as to be in proportion to its

43 *Glowworm / Gloworm*

8. Thomas Tompion (1639–1713), renowned English watchmaker and clockmaker, also a maker of barometers and sundials.

Distance; how comes the Substance of it so little to be altered by so intense an *Heat,* and to hold this *Heat* with so great an *Equality* for near six Thousand Years? One way or t'other; either so *luminous* a Body without *Fire,* or so *burning* a Body, not *consumed* or *altered;* it is wonderful!

But Sir *Isaac Newton* supposes, That a very large, dense, and fixed Body, when *heated* beyond such a degree, may emit *Light* so copiously, that by such Emission, and by the Re-action of its *Light,* and by the Reflection and Refraction of the Rays within its hidden *Meatus,* it may come to grow still hotter and hotter, as deriving more *Degrees of Heat* by those Ways, than it can of *Cold* by any other. Thus, he supposes the *Sun* a vast Globe that is vehemently heated, and the Heat thereof preserved by its great Magnitude, and the mutual Action and Re-action which there is between it, and the *Light* emitted by it. And its Parts are preserved from evaporating in *Flame* and *Fume,* not only by the Great *Fixity* of its Nature, but also by the mighty Weight and Thickness of the Atmosphere, which environs it, and condenses its Vapours, whenever they are emitted.

However, behold the *Sun* seated by the Glorious GOD, like a powerful *Monarch,* on his Throne, (as Dr. *Cheyne* expresses it) from thence distributing Light, and Life, and Warmth, in a plentiful Effusion, to all the Attendants that surround him; and that so equally, that the nearest have not too much, nor the farthest too little: His Bulk and Situation so contrived, in respect of the *Planets,* as to have Quantity of Matter just enough to draw round him these Massy Bodies, and their *Satellits,* who are so various in their Quantities, and their Distances, and that in regular and uniform Orbits. The Doctor says well, *These are things that clearly speak the Omnipotence and Omniscience of their Author.*

What a Fancy is that of Dr. *Wittie!* That the SUN is probably the *Seat of the Blessed;* the *Sun,* which is the Center of the *Heavens,* and the Seat of *inherent Light.* It is true, of the Blessed we read, *They shall shine as the Sun;*[9] and their Blessedness is called, *The Inheritance of the Saints in Light.*[10] But this is very short of Demonstration, that the Saints must be lodged there. Tho the Church Militant were once represented as *clothed with the Sun,*[11]

---

9. Adapted from Matt. 13:43.
10. Col. 1:12.
11. Rev. 12:1.

it follows not, that the Church Triumphant must be *Dwelling in the Sun.*[12]

And Mr. *Arndt* propounds a Thought, which cannot be too much dwelt upon: *Sicut Sol Ornamentum est Cœli, ita CHRISTUS est Ornamentum suæ Ecclesiæ.*[13]

Dr. *Cheyne* with good reason apprehends, That the Quantity of *Light* and *Heat* in the *Sun* is daily decreasing. It is perpetually emitting Millions of *Rays,* which do not return into it. *Bodies* attract them, and suffocate them, and imprison them; and they go no more back into their Fountain.

Mr. *Bernoulli,*[14] from the Flashes of the *Light,* in the Vacuity of a Tube accommodated with *Mercury,* whereby a dark Room is enlightned, renders it likely that our Atmosphere, and all the Bodies on our Globe, are saturated at all times with Rays of *Light,* which never do return unto their Fountain.

'Tis true, this Decrease of the *Sun* is very inconsiderable. It shews that the Particles of *Light* are extremely small, since the *Sun* for so many Ages has been constantly emitting Oceans of *Rays,* without any very sensible Diminution. However, 'tis from hence evident, that the Sun had a Beginning; it could not have been from *Eternity; Eternity* must have wasted it: It had long ere now been reduced unto less than the *Light of a Candle.*

Glorious GOD, thou art the Father of Lights,[15] *the Maker of the Sun!*

In a late *Act of the Faith,* as they call their inhuman Butcheries, performed by that execrable *Hell upon Earth,* the *Inquisition* in *Portugal;* a Confessor being brought forth to die a grievous Death, as soon as he came into the *Light* of the *Sun,* which he had not seen in some Years before, he broke forth into this Expression, *Who that has Reason in him, could worship any but the Maker of that Glorious Creature!* They *gagg'd* him immediately!

My Pen shall not be served so. Enjoying the Benefits of the *Sun,* I will glorify him that made it: *Thou alone art for ever to be adored, O thou Maker of that Glorious Creature!*[16]

8 ere/e'er

12. Cf. 1 Cor. 15:40–41.
13. "As the sun is the ornament of heaven, so Christ is the ornament of the church."
14. That is, Johann Bernoulli.
15. Adapted from James 1:17.
16. Adapted from the Song of the Three Young Men, 40.

An eminent Writer of *Natural Theology*[17] has this Remark, That the *Sun* is *Imago illorum qui aliis præsunt.*[18] And that all *Superiours* in every Station, looking towards the *Sun*, should have shot into their Minds the Rays of such Thoughts as these; *What good Influences ought I to dispense unto those that have Dependance on me!*

The Apocryphal Book of *Wisdom* does wisely, to call the *Light* of the SUN, *An Image of the Divine Goodness.*[19]

The *Diameter* of the *Earth* is near Eight Thousand Miles; and the Diameter of the *Orbis Magnus* Ten Thousand Diameters of the Earth. This *Orbis Magnus,* or the Orbit of the Earth, in its annual Revolution about the *Sun;* Dr. *Gregory* makes the Semidiameter of it 94,696,969 *English* Miles: which is the Distance of the *Earth* from the *Sun*. But the Semidiameter of *Saturn's* Orbit is no less than ten times as great. All Astronomers before *Kepler* supposed this Orbit a *perfect Circle;* but he has proved it an *Ellipsis.* If our *Solar System* have such large Dimensions, and if every *Fixed Star* be a *Sun,* that has a *System,* of the like dimensions perhaps, belonging to it:

Great is our GOD, and greatly to be praised: His Greatness is unsearchable![20]

How is it possible to consider the *Grandeur* of our GOD, without *Annihilating* ourselves before Him, or without Horrour at the View of the *matchless Evil,* in sinning against so Glorious a Majesty!

It is a Passage in a little Treatise, entitled, *The Book of Nature;*[21] not unworthy to be transcribed here: "If thou never observe the Sky with thine Eyes, but to guess at Rain and Fair Weather; or if thy looking up to Heaven be bounded with the *Starry Firmament;* and, if thou removest from thee the Love and Honour of GOD, and the Contemplation of Him who dwelleth in the Heavens, thou hast no cause to raise thyself above the *Brutes,* thy Fellow-Inhabitants of this Lower World."

And now, let *Hugo de S. Victore* conclude for us: *Quis Solem per hyberna descendere Signa præcipit? Quis rursum per æstiva Signa*

---

17. That is, Johann H. Alsted.
18. "The image of those who preside over others."
19. Adapted from Wisd. of Sol. 7:26.
20. Adapted from 1 Chron. 16:25, Ps. 96:4, and Ps. 145:3.
21. *Theologia ruris, sive schola et scala naturae: Or, the Book of Nature* (London, 1686).

55 *ascendere facit? Quis eum ab Oriente in Occidentem ducit? Quis iterum ab Occidente in Orientem revehit? Hæc cuncta sunt mirabilia, sed soli Deo possibilia.*²²

How Glorious will the Righteous be in that World, when they shall *shine as the Sun?*²³

ESSAY V. *Of* SATURN.

ALL the *Master Planets*,¹ as they may be called, move about the *Sun*, as their *Common Center*. They move with different Velocities: but there is this Common Law observed in all of them; *That the Squares of the Times of their Revolutions, are proportional to the Cubes of their Distances.*² And the *Lunar Planets* observe the same Law in their Motions about their *Master Planets*. And another Common Law with them, is, That Lines drawn from the *Foci* of the Curves they move in, to their Bodies, will sweep over equal *Areas* in equal Times on the Planes of other Orbits.³ Who but the Great GOD could make and fix these Laws? *Lord, they continue this day according to thine Ordinances, for all are thy Servants.*⁴

It is now found, that *Saturn*, besides his round *Body*, has also a luminous *Ring*, which encompasses him, as the Horizons of our Artificial Globes do usually encompass them; and is flat upon the Verge, as they use to be. The *Ring* shews itself in an *Oval*, and at certain times it wholly disappears.

It appears not, however, that *Saturn* revolves upon his own *Center*.

When this *Planet* appears at 20 degr. 30 min. of *Pisces*, and of

69 Areas / Area's

22. "Who commandeth the sun to descend through the winter signs? And who again causeth him to ascend through the summer signs? Who leads him from East to West? And who again brings him back from the West to East? All these things are very wonderful, but to God alone possible."
23. Adapted from Matt. 13:43.
1. Mather uses the terminology of Nehemiah Grew, who calls the lunar planets "sub-planets." Many authors contemporary to Mather distinguished between primary and secondary planets, with the latter being satellites of the former. Mather begins his account with those planets farthest from rather than those nearest to the sun.
2. This is Kepler's third law of planetary motion, first published in 1618.
3. This is Kepler's second law of planetary motion, first published in 1609.
4. Ps. 119:91.

Libra, then 'tis that he appears round; or without his *Ansæ*,[5] as they are called, which is once in fifteen Years; or half his Course, which every one knows to be compleated in thirty Years, or 10,950 Days.

The *Ring* seems to be *Opake* and *Solid,* encompassing the *Planet,* but no where touching it. The Diameter of it is two and a quarter of *Saturn*'s Diameters; and the Distance of the *Ring* from the *Planet* is about the Breadth of the *Ring* itself. Mr. *Hugens* takes the Breadth of the Ring to be about Six Hundred *German* Miles.[6]

The Proportion of the Body of *Saturn* to the Earth, is that of 30 to 1.

The Distance of *Saturn* from the *Sun* is about ten times as great as the Distance of our Earth from him; and by consequence, that Planet will not have above an hundredth Part of that Influence from the *Sun,* which this Earth enjoys.

The Ring of *Saturn,* being distant from him no more than two and a quarter of his Semidiameters, it cannot be seen at the Distance of 64 Degrees from *Saturn*'s Equator, in whose Plane the Ring is placed. So that there is a *Zone* of almost 53 Degrees broad, towards either Pole, to which this famous *Ring* does never appear.

Saturn is attended with five *Satellits.*

The First *Satellit* makes a Revolution about *Saturn* in 1 Day, 21 Hours, and 19 Minutes; and makes two Conjunctions with *Saturn* in less than two Days. It is distant from the Center of Saturn 4⅜ of his *Semidiameters.*

The Second makes his Revolution in 2 Days, 17 Hours, and 43 Minutes. It is distant from *Saturn* 5³⁄₇ *Semidiameters* of the Planet.

The Third is distant from *Saturn* eight of his *Semidiameters,* and makes his Revolution in almost 4⅗ Days.

The Fourth revolves in 15 Days, 22 Hours, 41 Minutes. 'Tis distant from the Center of *Saturn* about 18 of his *Semidiameters.*

The Fifth is distant from the Center of *Saturn* 54 of his *Semidiameters,* and revolves about him in 79⅓ Days.

Mr. *Huygens,* who first of all discovered the Fourth, (for which cause 'tis called the *Huygenian Satellit,* tho Dr. *Halley* afterwards corrected the Theory of its Motion) thinks, the mighty Distance

---

5. "A name applied to the apparent ends of Saturn's ring seen projecting like two handles beyond the disk of the planet" (*OED*).

6. A German mile is 4.66 English miles.

between the Fourth and Fifth *Satellits* to be a ground for Suspicion, that there may be a *Sixth* between them, or that the *Fifth* may be attended with some of his own.

On the Revolutions of the *Planets*, the incomparable Sir *Richard Blackmore*, in his Noble Poem of *Creation*, thus drives us to consider the *First Cause* of all:

"*Saturn* in Thirty Years his Ring compleats,
Which swifter *Jupiter* in Twelve repeats.
Mars Three and Twenty Months revolving spends,
The Earth in Twelve her Annual Journey ends.
Venus, thy Race in twice Four Months is run;
For his *Mercurius* Three demands; the *Moon*
Her Revolution finishes in One.
If all at once are mov'd, and by One Spring,
Why so *unequal* is their *Annual Ring?*"

The Motions of the Heavenly Bodies can be produced and governed by none but an Infinite GOD. It is well argued by *Lactantius; There is indeed a Power in the Stars, of performing their Motions; but that is the Power of God who made and governs all things, not of the Stars themselves that are moved.* And by *Plato* before him; *Let us think, how it is possible for so prodigious a Mass to be carried round for so long a time by any natural Cause? For which reason I assert God to be the Cause, and that 'tis impossible it should be otherwise.*

ESSAY VI. *Of* JUPITER.

JUPITER's Globe, according to *Cassini*'s Measures, must be greater than that of the Earth, by 2460 Times. The Periodical Time of his Revolution about the *Sun*, is Twelve Years, or 4380 Days.

In the Body of *Jupiter*, and overthwart his luminous Part, there are observed three darkish *Belts*, like the *Spots* which appear in the *Moon*. These *Belts* or *Girdles* are near strait and parallel, and extending from East to West, after the manner of the Ecliptick. They make a kind of *Equinoctial* with Tropicks. The *Southern* is larger a little than the *Northern,* and a little nearer to the South than the other is to the North.

Dr. *Hook* has observed also a small and a dark Filament, and the *Zones* growing a little darker, as they draw nearer to the Poles. And some have observed in them something of *Curvity,* tho their Borders are perfectly round.

Jupiter has *Four Satellits,* or little Moons, waiting on him.

The nearest is distant from him, according to Mr. *Flamstead*'s most accurate Observations, a little more than Five of his Semidiameters; and finishes his Course in 1 Day, 18 Hours, 28 Minutes, and a few Seconds.

The Second is distant from him about 8 of his Semidiameters, and finishes his Course in 3 Days, 13 Hours, 17 Minutes, and a few Seconds.

The Third is distant from him about 14 of his Semidiameters, and finishes his Course in 7 days, 3 Hours, and 59 Minutes, and some Seconds.

The Fourth is distant from him about 24 of his Semidiameters, and finishes his Course in 16 Days, 18 Hours, 5 Minutes, and some Seconds.

These *Guards* of *Jupiter* cast a Shadow upon him, when they are found interposed between the Sun and him.

The Fourth would appear to an Eye in *Jupiter,* as big as the Moon does to us. A Spectator there would have also four kinds of Months. In one of *Jupiter*'s Years, which is Twelve of ours, there would be 2407 of the least Months; Half that Number for the next *Satellit:* The Months of the Third would be near subduple of the Second, or subquaduple[1] of the First: The Months of the greatest would be about Two Hundred Fifty-four. A Year of *Jupiter* has a great Number of Days; but of the four Sorts of Months, the least contains only *four Days* and a Quarter; the greatest something more than *Forty*.

Mr. *Cassini* has observed a Couple of *Spots* in the Body of *Jupiter,* which make a Revolution on the Center of this Planet, from East to West, in about 9 Hours, 56 Minutes. Others have lately confirmed it by better Observations. This proves, that the Planet moves about upon its own Center. Behold the *shortest Period*

---

1. A subduple is half of a quantity or number. The word "subquaduple" is not in the *OED*, but apparently Mather intends "subquadruple," meaning a fourth of a quantity or number.

that is made in the Firmament! The Days and the Nights, each of them *Five Hours* a-piece.

Campani observed, with a more than ordinary *Telescope,* certain Protuberances and Inequalities in the Surface of this Planet.

We may here insert a Remark upon the Periodical Motions of the *Planets;* both the *Primary* and their *Secondaries.*

One thing very considerable in the Periodical Motion of the Secondary Planets, is, That it is mixed with a kind of *Cochleous*[2] *Direction* towards one or other Pole of its *Primary Planet;* by which means every *Satellit,* by gentle Degrees, changes its Latitude, and makes its Visits towards each Pole of its *Primary.*

We will here break off with the Words of Mr. *Molyneux.* "From hence may we justly fall into the deepest Admiration, that one and the same *Law of Motion* should be observed in Bodies so vastly distant from each other, and which seem to have no Dependance or Correspondence with each other. This doth most evidently demonstrate, that they were all at first put into Motion by one and the same unerring *Hand,* even the infinite Power and Wisdom of GOD, who hath fixed this Order among them all, and hath established a *Law* which they cannot transgress.

"*Chance,* or dull *Matter,* could never produce such an harmonious *Regularity* in the Motion of Bodies so vastly distant: This shews a Design and Intention in the *First Mover!*"

ESSAY VII. *Of* MARS.

MARS borrows his Light from the *Sun,* as well as the rest of his *Planetary Brethren.* He has his Increase and his Decrease of *Light;* like the *Moon;* may be seen almost *bisected,* when in his Quadrature with the *Sun,* or in his *Perigæon;*[1] tho never corniculated or forked, like his Inferiours.

Dr. *Hook* discovered several *Spots* in *Mars,* and particularly a triangular one, which has a Motion. Mr. *Cassini* afterwards discovered *four Spots,* the two first on one Face of *Mars,* afterwards

---

2. That is, shaped like a snail shell; spiral.
1. The Greek word for perigee, the point nearest to the earth in the orbit of a planet.

two more that were larger, on the other Face. Upon further Observation it was found, that the *Spots* of these two Faces turned by little and little from East to West, and returned at the Space of twenty-four Hours and forty Minutes. In such a Term therefore, there is a Revolution of *Mars* upon his own *Axis*.

The Year of *Mars* is near twice as long as ours; his *Natural Day* a little greater than ours: his *Artificial Day* is almost every where equal to his Night, besides what belongs to Twilight. *Mars* as well as *Jupiter* has a perpetual *Æquinox*. Hence there can be but little *Variety of Seasons* in any one particular Place of these Planets.

Whence the *Fasciæ*,[2] or Fillets observed in *Mars*? There appear certain *Swathes*,[3] as we may call them, which are posited paralled to his Æquator. Are they owing to the Heat and Cold there, like our *Clouds* and *Snows*?

It is thought that *Mars* has an *Atmosphere*, because *Fixed Stars* are obscured, and as it were extinct, when they are seen just by his Body.

ESSAY VIII. *Of* Venus.

*V*ENUS has various Appearances; *round* sometimes; anon *half-round;* by and by like a *Crescent*.

Mr. *Cassini* discovered certain *Spots* on this *Planet*, by the motion whereof it appeared that it moved upon its own Center, and upon an *Axis*, which carries it from North to South; a Motion wholly unknown any where else in the Heavens. *Two Spots* it has, which are very thin, long, uncertainly terminated; and a shining Part belongs to one of them.

He discover'd also, as he judges, a *Satellit* attending this Planet; which Dr. *Gregory* assents to, as more than probable. This is not usually seen, perhaps because it may not have a fit Surface to reflect the Light of the *Sun;* which is the Case of the *Spots* in the Moon.

Herigone, and *Keplerus*, and *Rhætensis* conclude, that *Venus* moves about its Axis in about fourteen Hours. Dr. *Cheyne* says in twenty-three.

2. That is, streaks of cloud in the sky.
3. That is, a natural formation constituting a wrapping.

ESSAY IX. *Of* MERCURY.

THE Great *Hevelius* hath observed, That *Mercury* changes his Face, like *Venus,* and like our Moon; appearing sometimes *round,* sometimes *half-round,* sometimes like a *Crescent.*

This Planet has his Abode so near the *Sun,* that as yet there has been little discovered of him.

It appears not yet, whether he revolves upon his own *Axis,* and so what may be the Length of his Days. But it is probable, he may have such a Motion, as well as the other Planets. However, his Year is hardly equal to a Quarter of ours.

Sir *Isaac Newton* has terrible Apprehensions of the Heat in this Planet, as being seven times as much as the Heat of the Summer-Sun in *England;* which according to his Experiments made by the *Thermoscope,*[1] would be enough to make Water boil. If the Bodies in this Planet be not enkindled by this Heat, they must be of a peculiar Density. But Mr. *Azout* pretends to prove, That tho this Planet be so near the Sun, yet the Light there is not capable of burning any Objects.

¶. But let us now entertain ourselves with a *Synopsis,* of certain Matters relating to the Planets, as they are determined by the latest and most accurate Astronomers.

The Distance from the Sun, *in* English *Miles.*

Of *Mercury*	Miles	32,000,000
Venus		59,000,000
The *Earth*		81,000,000
Mars		123,000,000
Jupiter		424,000,000
Saturn		777,000,000

The Diameter in English *Miles.*

Of *Mercury*	Miles	4,240
Venus		7,906
The *Earth*		7,935
Mars		4,444
Jupiter		81,155
Saturn		67,870

---

1. An instrument for indicating without accurately measuring changes in temperature.

Essay 9. Of Mercury

The *Sun* 763,460

The Time of the Periodick Revolution.

	Days	Hours
Of *Mercury*	87	23
Venus	224	17
The *Earth*	365	6
Mars	686	23
Jupiter	4,332	12
Saturn	10,759	7

To this we will add Mr. *Derham*'s Account of their Orbite.

Saturn has an Orb of 1,641,526,386 *English* Miles Diameter.
Jupiter an Orb of 895,134,000 Miles.
Mars an Orb of 262,282,910 Miles.
Venus an Orb of 124,487,114 Miles.
Mercury an Orb of 66,621,000 Miles.

¶. "*Great GOD, thou hast lifted me up to Heaven: Oh! let me not be cast after all down to Hell.*[2]

"The Philosopher,[3] who gazing on the *Stars* with his attentive Observation, tumbled into a Pit that he observed not, was not so unhappy as he that has visited *Heaven* on the noble Intentions of *Astronomy*, but by an ungodly Life, procures to himself a Condemnation to that *Hell*, which is a State and Place of *Utter Darkness*. Wretched Astronomers! *Who are among the wandring Stars, to whom is reserved the Blackness of Darkness for ever.*"[4]

We will conclude what we collect about the *Stars*, with transcribing a Passage out of the *Miscellanea Curiosa*.[5] "The Honour-

2 Orbite/Magnitude

2. Adapted from Matt. 11:23.
3. The reference is to Thales of Miletus (c. 636–c. 546 B.C.), whom Aristotle regarded as the founder of physical science. He looked for a physical rather than mythological understanding of the world. Thales was universally accounted as one of the Seven Sages.
4. Jude 13.
5. *Miscellanea Curiosa: Being a Collection of Some of the Principal Phenomena in Nature, Accounted for by the Greatest Philosophers of This Age. Together with Several Discourses Read before the Royal Society for the Advancement of Physical and Mathematical Knowledge*, 3 vols. (London, 1705–7). This collection of papers was anonymously edited and published by Edmond Halley. A second edition with a slightly variant title was published in three volumes in 1708.

able Mr. *Roberts* computes the Distances of the *Fixed Stars;* which he supposes to be so many *Suns* of a different Magnitude. He thinks, that it seems hardly within the reach of any of our Methods to determine it. The Diameter of the *Earth's Orb,* which is at least One Hundred and Sixty Millions of Miles, is but a Point in comparison of it. At least Nine Parts in Ten, of the Space between us and the *Fixed Stars,* can receive no greater Light from the *Sun,* or any of the *Stars,* than what we have from the *Sun* in a clear Night. *Light* takes up more time in travelling from the *Stars* to us, than we in making a *West-India* Voyage, which is ordinarily performed in six Weeks. A *Sound* would not arrive to us from thence in *Fifty Thousand Years,* nor a Cannon-Bullet in a much longer Time. This is easily computed, by allowing ten Minutes for the Journey of *Light* from the *Sun* hither; and that *Sound* moves about *Thirteen Hundred* Foot in a Second."

ESSAY X. *Of* COMETS.

'TIS an admirable Work of our GOD, that the many *Globes* in the Universe are placed at such Distances, as to avoid all violent Shocks upon one another, and every thing wherein they might prove a prejudice to one another.

Even *Comets* too, move so as to serve the Holy Ends of their Creator! COMETS, which are commonly called *Blazing Stars,* appear unto later Observations to be a sort of *Excentrical Planets,* that move periodically about the *Sun.*

Sir *Isaac Newton,* from whom 'tis a difficult thing to dissent in any thing that belongs to *Philosophy,* concludes, That the Bodies of *Comets* are solid, compact, fixed, and durable, even like those of the other *Planets.*

He has a very critical Thought upon the *Heat,* which these *Bodies* may suffer in their Transits near the *Sun.* A famous one, in the Year 1680, passed so near the *Sun,* that the *Heat* of the *Sun* in it must be twenty-eight thousand times as intense as it is in *England* at Midsummer; whereas the Heat of boiling Water, as he tried, is but little more than the dry Earth of that Island, exposed unto the Midsummer-Sun: and the *Heat* of *red-hot Iron* he takes to be three or four times as great as that of *boiling Water.* Wherefore the *Heat* of that *Comet* in its *Perihelion* was near two thousand times as great as that of *red-hot Iron.* If it had been an Aggregate of nothing but Exhalations, the *Sun* would have render'd it invisible. A Globe of *red-hot Iron,* of the Dimensions of our Earth, would scarce be cool, by his Computation, in 50,000 Years. If then this *Comet* cooled an hundred times as fast as *red-hot Iron,* yet, since his Heat was 2,000 times greater than that of *red-hot Iron,* if you suppose his Body no greater than that of this Earth, he will not be cool in a Million of Years.

The *Tails* of *Comets,* which are longest and largest just after their *Perihelions,* he takes to be a long and very thin Smoke, or a mighty Train of Vapours, which the ignited *Nucleus,* or the Head of the *Comet,* emits from it. And he easily and thoroughly confounds the silly Notion of their being only the *Beams of the Sun,* shining thro the Head of the *Star.*

The Phænomena of the *Tails* of *Comets* depend upon the Motion of their *Heads,* and have their Matter supplied from thence.

There may arise from the Atmosphere of *Comets,* Vapours enough to take up such immense Spaces, as we see they do. Computations made of and from the Rarity of our *Air,* which by and by issue in Astonishments, will render this Matter evident.

That the Tails of *Comets* are extremely rare, is apparent from this; the *Fixed Stars* appearing so plainly thro them.

The Atmosphere of *Comets,* as they descend towards the *Sun,* is very sensibly diminished by their vast running out, that they may afford Matter to produce the *Blaze.* Hevelius has observed, that their Atmosphere is enlarged, when they do not so much run out into *Tail.*

This *Lucid Train* sometimes, as Dr. *Cheyne* observes, extends to four hundred thousand Miles above the Body of the Star.

Sir *Isaac Newton* has an Apprehension, which is a little surprizing, That those Vapours which are dilated, and go off in the *Blazes* of *Comets,* and are diffused thro all the Celestial Regions, may by little and little, by their own proper *Gravity,* be attracted into the *Planets,* and become intermingled with their Atmospheres. As to the Constitution of such an *Earth* as ours, it is necessary there should be *Seas;* thus, for the Conservation of the *Seas,* and Moisture of the Planets, there may be a necessity of *Comets;* from whose condensed Vapours, all that *Moisture,* which is consumed in Vegetations and Putrefactions, and so turned into dry Earth, may by degrees be continually supplied, repaired, and recruited. Yea, he has a suspicion, that the Spirit, which is the finest, the most subtile, and the very best part of our *Air,* and which is necessarily requisite unto the Life and Being of all things, comes chiefly from *Comets.* If this be so, the Appearance of *Comets* is not so dreadful a thing, as the *Cometomania,*[1] generally prevailing, has represented it.

Mr. *Cassini* will thus far allow bad Presages to *Comets,* That if the Tail of a *Comet* should be too much intermingled with our *Atmosphere,* or if the Matter of it should, by its *Gravity,* fall down upon our Earth; it may induce those Changes in our *Air,* whereof we should be very sensible.

Bernoulli,[2] in his *Systema Cometarum,* supposes, That there is a *Primary Planet,* revolving round the *Sun* in the space of four Years and 157 Days; and at the distance of 2,583 Semidiameters of the *Orbis Magnus.* This *Primary Planet,* he supposes, either from his mighty *Distance,* or his minute *Smallness,* to be not visible unto us; but however to have several *Satellits* moving round him, tho none descending so low as the Orbit of *Saturn;* and that these becoming visible to us, when in their *Perigæon,* are what we call *Comets.*

Seneca's[3] Prediction, That a Time should come, when our Mysteries of *comets* should be unfolded, seems almost accomplished. However *Seneca* has not obliged us with the *Phænomena* observed by him, which encouraged this Prediction.

No Histories of *Comets* were of service to the Theory of them, until *Nicephorus Gregoras,* a *Constantinopolitan* Astronomer, described the Path of a Comet in 1337.

All that consider'd *Comets* until *Tycho Brahe,* consider'd them as no other than Vapours below the *Moon.*

Anon, the sagacious *Kepler* improving on *Tycho's* Discoveries, came at a true System of *Comets,* and found, that they moved freely through the Planetary Orbs, with a Motion that is not much different from a *Rectilinear* one.

The incomparable *Hevelius* went on, and though he embraced

---

2 *Cometomania / Cometomantia*

1. That is, preoccupation with comets and their meaning for man and human affairs.
2. That is, Jakob Bernoulli.
3. That is, Seneca the Younger.

Essay 10. Of Comets 53

the *Keplerian* Hypothesis, of the *Rectilinear Motion of Comets,* yet he was aware, *That the Path of a Comet was bent into a curve Line towards the Sun.*

At last the illustrious Sir *Isaac Newton* arrives with Demonstrations, That all the Phænomena of *Comets* would naturally follow from the *Keplerian* Principles. He shewed a Method of delineating the *Orbits* of *Comets* geometrically; which caused Admiration in all that considered it, and comprehended it.

The most ingenious Dr. *Halley* has made Calculations, upon which he ventures to foretell the *Return* of *Comets;* but he observes, that some of them have their *Nodes* pretty near the annual Orb of the Earth. I will transcribe the Words he concludes with: "What may be the *Consequences* of so near an *Appulse,* or of a *Contact,* or lastly, of a *Shock* of the Celestial Bodies, (which is by no means impossible to come to pass) I leave to be discussed by the Studious of Physical Matters."

The Sentiments of so acute a Philosopher as Dr. *Cheyne* upon *Comets,* deserve to be transcribed.

"I think it most probable, that these frightful Bodies are the Ministers of *Divine Justice,* and in their Visits lend us *benign* or *noxious* Vapours, according to the Designs of Providence; That they may have brought, and may still bring about the great Catastrophe of our System; and, That they may be the Habitation of *Animals* in a State of *Punishment,* which if it did not look too notional, there are many Arguments to render not improbable."

And elsewhere: "'Tis most likely, they are the Ministers of Divine Justice, sending baneful Steams, from their long Trains, upon the *Planets* they come nigh. However, from them we may learn, that the Divine Vengeance may find a *Seat* for the *Punishment* of his disobedient Creatures, without being put to the expence of a New Creation."

¶. When I see a vast Comet, blazing and rolling about the unmeasurable Æther, I will think;

"Who can tell, but I now see a wicked World *made a fiery Oven in the Time of the Anger of GOD! The Lord swallowing them up in his Wrath, and the Fire devouring them!*[4]

"What prodigious Mischief and Ruin might such a *Ball of*

4. Adapted from Ps. 21:9.

Confusion bring upon our sinful *Globe*, if the Great GOD order its Approach to us!

"How happy they, that are in the Favour and Friendship of that Glorious Lord, who *knows how to deliver the Pious* out of Distresses, and *reserve the Unjust for a Punishment of a Day of Judgment!*"[5]

Si fractus illabatur Orbis,
Impavidum ferient Ruinæ.[6]

APPENDIX. *Of* HEAT.

WE should be forgetful, if we take our leave of the *Heavenly Bodies*, and say nothing of *Heat*, whereof they have so much among them.

To the *Heat* of Bodies it is requisite, that the small Parts of it be agitated with much Vehemence and Rapidity; and that the Determinations of the insensible Corpuscles thus agitated be also very *various;* and that likewise the variously agitated Particles be so small, as generally speaking to be singly insensible: for unless they be exceeding fine, they cannot penetrate readily into the Pores of contiguous Bodies, and so warm or burn them.

The Operation of *Heat* upon our Senses, the Result of which we commonly call *Heat*, is usually estimated by its Relation to the Organs of our *Feeling*. If the Motion of the small Parts be more languid in the Object than it is in the *Sentient*, we pronounce the Body to be *cold;* but if it be more violent in the *Object* than in the *Sentient*, we say the Body is *hot*.

The *Intenseness* of *Heat* (as of *Light*) always is as the *Density* of the Rays, or Particles of *Fire*, that occasion it; and this *Density* is as the *Distance* from the radiating Point reciprocally.

Dr. *Slare* has published surprising Experiments, of producing

---

5. Adapted from 2 Pet. 2:9.
6. These lines, from an ode by Horace which celebrates justice and steadfastness of purpose, are best understood when the immediately preceding lines are also quoted (the translation of the quoted lines is italicized): "The man tenacious of his purpose in a righteous cause is not shaken from his firm resolve by the frenzy of his fellow-citizens bidding what is wrong, not by the face of threatening tyrant . . . not by the thundering hand of mighty Jove. *Were the vault of heaven to break and fall upon him, its ruins would smite him undismayed.*"

Fire and *Flame,* from the bare Mixture of two Liquors *actually cold;* a vegetable Oil, and a compound Spirit of *Nitre.*¹

The incredible Force of *Burning-Glasses!*

A burning Concave, made at *Lusatia*² in *Germany,* near three *Leipsick* Ells³ in Diameter, made of a Copper-Plate, scarce twice as thick as the Back of a common Knife; makes Wood in the *Focus* (which is two Ells off) to flame in a moment; and Water in an earthen Pot boil immediately: *Tin* three Inches thick, to be melted quite through in three Minutes; a Plate of *Iron* to be presently red-hot, and very quickly perforated: it will run in five or six Minutes; Tiles, and Slates, and earthern Potsherds, melt in a little time, and run into Glass; a Clod of Earth turns into a *greenish Glass.*

Mr. *Tschirnhaus* makes Convex Burning-Glasses of three or four Foot Diameter, the *Focus* at the Distance of twelve Foot Diameter,⁴ which in a moment vitrify Tiles, and Slates, and Pumice-Stones, and earthen Vessels; melt all resinous Things under Water; melt all Metals in a moment, and *Gold* itself is turned into *Glass* of a purple Colour: Of such efficacy are the *Rays,* when strip'd of an *unctious Matter,* which we may suppose them generally clothed with.

¶. "The antient *Persians* were the Worshippers of the *Fire:* But I will abhor their *Fire-Places.* The *Indians* of my Country, while unchristianized, concluded from the strange Effects of the *Fire, It must be a God.* I will adore the Glorious GOD that made the *Fire.* Great GOD, I bless thee for the Benefits, which thy Creatures, and I among them, receive by the *Fire,* which is fetch'd *from Heaven* unto us. May my *Zeal* for thy Service be always kept *boiling* in the *Heat* proper for it.

"Since *Fire* is thus irresistible, and *Heat* so insupportable, surely I should beware of that Impiety, which will expose me to the Revenges of GOD. *Who can dwell with such a devouring Fire, such everlasting Burnings?*⁵ *My GOD,* be not thou unto me

2 Lusatia/Lusace   9 Potsherds/Potsheards

1. That is, nitric acid.
2. A region in eastern Germany and southwestern Poland which includes part of northeastern Saxony and lower Silesia.
3. That is, a unit of measure equal to 22.257 English inches or 565.3 millimeters.
4. That is, the focus is twelve feet distant and of one and one-half-inch diameter.
5. Adapted from Isa. 33:14.

a *Consuming Fire.*[6] *My GOD,* who can abide the *Heat* of *thine Anger!*"[7]

I have seen a Book of Devotion, entitled, *Christianus per Ignem;* or, *A Disciple warming himself, and owning his Lord.*[8] It is there actually evident, and performed, That this one Object, the *Fire on the Hearth,* will afford a whole *Book-full* of profitable Contemplations.

ESSAY XI. *Of the* MOON.

WE are now coming down unto our *Terraqueous Globe.* The MOON, a sort of *Satellit* unto this Globe, salutes us in our Way. Paying an Homage to none but her Glorious Maker, we will now *behold her walking in her Brightness.*[1]

What shall we think of the *Protuberant Parts* observed on that Celestial Body? What of the *Round Hollows,* like Pits or Wells of several Magnitudes, which have been formerly mistaken for *Mountains?*

The Periodical Revolution of the *Moon,* in reference to the *Fixed Stars,* according to Mr. *Flamstead,* is 27 Days, 7 Hours, 43 Minutes, 7 Seconds.

In the same Space, with a strange Correspondence of the two Motions, it revolves the same way about its own *Axis;* by which the *same Side* is always exposed unto our sight. But because in the Space of a Periodical Month,[2] the Earth is also with this her *Satellit,* moved on almost an entire Sign, the *Moon* can't yet come to a new Conjunction with the *Sun,* but wants 2 Days, 5 Hours of it; which must be passed before the entire *Lunation*[3] will be over, and before the Moon has exhibited all her *Phases.* These 2 Days

---

6. Deut. 4:24, 9:3 and Heb. 12:29.
7. Adapted from Nah. 1:6.
8. Here Mather coyly calls attention to his own anonymously published book, *Christianus per Ignem: Or, A Disciple Warming of Himself and Owning of His Lord* (Boston, 1702).
1. Adapted from Job 31:26.
2. "Periodical" pertains to the time taken for the regular revolution of a heavenly body.
3. The time from one new moon to the next, constituting a lunar month.

5 Hours, added unto the Periodical Month, make the *Synodical One*;⁴ which is 29 Days, 12 Hours, and ¾ of an Hour.

Those *Librations*⁵ of the Moon's Body, which occasion that the *Hemisphere* exposed unto our Sight is not always exactly and precisely the same, arise from the Excentricity of the Moon's Orbit, and from the Perturbations it suffers by the Sun's Attraction, and from the Obliquity of the Axis of the Diurnal Rotation of the Moon's own Orbit. Without the Knowledge of these Things, the *Phænomena* of the *Moon* would be inexplicable: but upon the Consideration of these, they are very demonstrable.

'Tis very sure, that although it be almost the same Face which the *Moon* turns to the Earth, yet it is not entirely so. There is a *Libratory Motion,* whence it comes to pass, that sometimes the more Eastern and Western Parts of it, sometimes the more Northern and Southern appear alternately.

According to Sir *Isaac Newton,* the mean Distance of the *Moon* from the Earth, is about 60 Semidiameters of the Earth; or about 24,000 *English* Miles. The mean Diameter of the *Moon* is 32 Minutes, 12 Seconds; as the *Sun's* is 31 Minutes, 27 Seconds. The *Density* of the *Moon,* to that of the Earth, he concludes to be nearly as 9 to 5. And the Mass of Matter in the *Moon,* to that of the *Earth,* to be nearly as 1 to 26.

The *Moon* hath properly no *Atmosphere,* such as belongs to our Earth, of Clouds, Winds, Thunders; her Face is always clear, and by our Telescopes we can see the *Sun's* Light pass regularly and uniformly, from one mountainous Place to another.

The *Light* of the *Moon* reflected on us, is of such a Weakness, that even in the *Full-Moon,* it will be brought by no Burning-Glass to afford the least Degree of *Heat.* The Rays have their Force decreased, at least as the Square of their Distance. The Force of the *Sun's* Rays reflected unto us from the Moon, to those that come to us directly, is decreased, at least in proportion of the Square of the *Moon's* Distance from the Earth, to the Square of the *Moon's* Semidiameter. And by Calculation it will be found, That the *Light* of the *Moon* brought hither, will be in force but

---

4. "Synodical" pertains to the time between two successive conjunctions of a planet with the sun.

5. A libration is an oscillation in the apparent aspect of a secondary body (e.g., the moon) as seen from the primary object around which it revolves (e.g., the earth). The combination of four libration effects causes parts of the side of the moon to be alternately visible and invisible, so that as much as 59 percent of the moon's entire surface can be observed from the earth.

the *Fifty Thousandth* Part of what comes hither directly from the *Sun*.

Dr. *Hook* finds, That the Quantity of *Light* which falls on the Hemisphere of the *Full-Moon,* is rarefied into a Sphere about 288 greater in Diameter than the *Moon,* before it arrive to us. Consequently, the *Moon's Light* is 104,368 times weaker than the *Sun's;* and it would require 104,368 *Full-Moons* to give a *Light* equal unto that of the *Sun* at Noon.

There is a *Secondary Light* of the *Moon;* that is to say, the obscure Part of the Moon appears like to kindled Ashes, just before and after the Change. This is the *Sun's* Rays reflected from the bright Hemisphere of the Earth, to the dark Parts of the *Moon;* and thence again reflected unto the Earth, destitute of the Light of the *Sun.* This is by *Tacquet* and *Zucchius* more largely discoursed on. When the *Moon* is at *Change* to us, the *Earth* is at *Full* to the *Moon;* and the Light of the *Earth* is about fifteen times greater than that of the *Moon.* The *Moon* also being so little, as not to obscure above a twentieth Part of the *Earth,* it may be supposed that the Light from the *Earth* may render her a little visible to us even in *Solar Eclipses.*

The *Moon* is almost one Semidiameter of the Earth nearer to us, when she is in the *Meridian,* than when she is nigh the *Horizon.* But why doth she then appear bigger to our sight when she is nigh the *Horizon,* than when she is in the *Meridian?* Dr. *Wallis* agrees with *Des Cartes* in the Solution: the Horizontal *Moon* is capable of being compared with many intervening Objects, Hills, Trees, and the like; but the Meridian *Moon* hath nothing to be compared with.

Tho the *Moon,* as well as the *Earth,* and probably all the Planets, be of a Figure *oblately spheroidical*, that is to say, having its Diameter at the Æquator, longer than its Axis; yet the Excess of the Æquatorial Diameter in her is so inconsiderable, that she may well enough pass for a Globe. And perhaps this almost spherical Figure of the *Moon* may be the Result of her slow Motion round her Axis; for *Jupiter,* which moves the swiftest of any round its Axis, is of a Figure more *oblate* than any other Planet.

Dr. *Cheyne* observes, If our *Moon* were bigger, or nearer the

24 spheroidical / spheriodical

Earth, or if we had more than one, we should be every now and then in hazard of being drowned. And if our present *Moon* were less, or at a greater distance, or if there were none at all, we should be in hazard of being stifled with the baneful Steams of a stagnating Ocean. It is evident our *Satellit* is most wisely contrived for our Purposes, *by thee, O our Gracious GOD!*

The incomparable Sir *Isaac Newton* has at length obliged the World with a *Theory of the Moon*, which has performed that which all former Astronomers thought almost impossible.[6]

Hugenius had Glasses in perfection, and wrote since the accurate Maps of the *Moon*, taken by *Hevelius* and *Ricciolus;* but he could observe no *Seas* and *Rivers* there. It is also argued, That if any such were there, they could not but raise a mighty *Atmosphere*, and such *Clouds* as must needs darken the Body of the *Moon*, sometimes in one part, sometimes in another. They carry on their Inferences; if no *Waters* in the *Moon*, then there are no *Plants*, nor *Animals*, nor *Men*. About the Constitution of this *Queen of the Night*,[7] there seems a necessity for us to *remain in the dark!*

For Mr. *Derham* has confuted *Hugenius* with his own *Glasses*, and has demonstrated, that there are great Collections of *Waters* in the *Moon*, and by consequence Rivers, and Vapours, and Air; and in a word, a considerable *Apparatus* for *Habitation*.

But by what Creatures inhabited? A Difficulty this, that cannot be solved without *Revelation*.

¶. "My GOD, I bless thee for that *Luminary*, by which we have the uncomfortable Darkness of our *Night* so much abated! That *Luminary*, the Influences whereof have such a part in the *Flux* and *Reflux* of our *Seas;* without which we should be very miserable! That *Luminary*, whose Influences are so sensibly felt in the Growth of our *Vegetables*, and our *Animals!"*

---

6. Newton determined by calculation the moon's place in her quadratures and all other parts of her orbit, besides the syzygies, so accurately that the difference between that and her true place in the heavens was scarcely above three minutes in her quadratures and two minutes in her syzygies, and was usually so small that it could be reckoned only as a defect in the observation. He communicated his theory to David Gregory, who published it in Latin (in his *Astronomiae* [1702], pp. 332–36) and in English (in his *Elements of Astronomy* [1715], 2:563–71). The treatise was separately issued as *A New and Most Accurate Theory of the Moon's Action: Whereby All Her Irregularities May Be Solved* (London, 1702). It was reprinted in *Misc. Cur.*, 1:270–81.

7. The moon was called "the Queen of the Night" as early as 1552 by the Scottish satirical poet David Lindsay. Shakespeare's Silvia says to Porteus, "For me,—by this pale queen of night I swear." *Two Gentlemen of Verona* 4.2.100.

These are some of the *Songs,* which *GOD, the Maker of* us both, has *given me in the Night.*[8]

The Influences of the *Moon* upon *Sublunary* Bodies, are very wonderful. An *History* of them is yet among the *Desiderata* of our Philosophy. With my consent, he shall merit more than the Title of a *Rabbi Solomon Jarchi,* who gives it unto us. Dr. *Grew,* in his *Cosmologia,* has enumerated more than a dozen remarkable *Heads* of *Effects,* and *Motions,* and *Changes* in the World, over which the *Moon* has a sensible Dominion. Our *Lunaticks* are not the only Instances. Our *Husbandmen* will multiply the Instances upon us, till they make a Volume, which neither a *Columella,*[9] nor a *Tom Tusser* have reached unto. The *Georges*[10] of my Neighbourhood just now furnish me with two Instances, which have in them something that is notable. If our *Chesnut-Wood,* whereof we sometimes make our Fuel, be fell'd while the *Moon* is *waxing,* it will so sparkle in the Fire, that there shall be no sitting by it in safety. If it be cut while the *Moon* is *waning,* there will be no such Inconvenience. Moreover, we find, whatever *Timber* we cut, in two *Wanes* of the *Moon* in a Year, the *Wane* in *August,* and the *Wane* in *February,* will be for ever free from *Worms;* no *Worms* will ever breed in it. What Monsieur *Andry* relates, confirming the Observation of *Borellus,* about the Success of Medicines for *Worms* in *Human Bodies,* taken in the *Wane* of the *Moon,* is wonderful.

"I am sure, to be under such Influences of the *Moon,* as to see the Great GOD managing many of his Gracious Intentions by such an *Instrument;* and to be awakened to his Praises in the *Night,* when we see the *Moon walking in her Brightness;*[11] would not be a *Lunacy,* that the most *Rational* of Men could be ashamed of."

ESSAY XII. *Of the* RAIN.

WE are now coming down into our *Atmosphere.* Here we are quickly surrounded with *Clouds.* And here we quickly find

---

8. Adapted from Job 35:10.
9. Lucius Junius Moderatus Columella (fl. first century A.D.), author of *De re rustica.*
10. A pun on the Greek word *georgos,* meaning farmer.
11. Job 31:26.

be turned into *Ice* e'er it came to the Ground. I have sometimes wished, that Wise-Men would make the Reflection of *Petronius* upon this Matter: *Incultis asperisque Regionibus, diutius Nives hærent; ast ubi Aratro domefacta Tellus nitet, dum loqueris levis Pruina dilabitur. Similiter in Pectoribus Ira confidit; Feras quidem Mentes obsidet, Eruditas præterlabitur.*

ESSAY XX. *Of the Terraqueous* GLOBE.

THE Distance at which our *Globe* is placed from the *Sun*, and the Contemperation of our Bodies and other Things to this Distance, are evident Works of our Glorious GOD!

According to the accurate Observations of the *English Norwood*, and the *French Picart*, the Ambit of our Globe will be twenty-four thousand nine hundred and thirty Miles. Wherefore supposing it spherical, the whole Surface will be 197,831,392 Miles; which in the solid Content will be found no less than 261,631,995,920 Miles. The cubick Feet will be 30,000,000,000,000,000,000,000. The *Earth*, with her Satellit the *Moon*, moving about the *Sun*, this *Orbis Magnus*, as 'tis usually called, according to our *Derham*, is a Space of more than 540 Millions of Miles in Circumference, or 172 Millions of Miles in Breadth.

The *Copernican* Hypothesis is now generally preferred, which allows a *Diurnal* and an *Annual* Motion to our *Globe*, rather than to the *Sun*. According to this, the *Diurnal* Motion of our *Globe* is near 1,039 Miles in an Hour.

The Arguments that prove the Stability of the *Sun*, and the Motion of the *Earth*, have now render'd it indisputable. It is impossible to account for the Appearances of the *Planets*, and their *Satellits*, and the *Fixed Stars*, in any tolerable manner, without admitting the Motion of the *Earth*; or to account for *Comets*;

Mather's discussion of lunar influences reveals that he had not entirely freed himself from superstition and prescientific beliefs. Courtesy of the Rare Book and Special Collections Library, University of Illinois at Urbana-Champaign.

ourselves in the midst of that *Rain,* whereof the Great GOD, in his Book, so often claims the *Glory* of being the *Maker* and *Giver.*[1]

The *Rain* is Water by the *Heat* of the *Sun* divided into very small and invisible Parts; which ascending in the *Air,* till it encounters with the *Cold* there, is by degrees condensed into *Clouds,* and thence descends in *Drops.* A *Mist* is a multitude of little, but *solid* Globules; which therefore descend. A *Cloud* is a Congeries of little, but *concave* Globules; which therefore ascend unto that height, wherein they are of equal weight with the *Air,* where they remain suspended, till by a Motion in the Air they are *broken:* and so they come down in *Drops;* either smaller, as in a *Mist;* or bigger, when many of them run together, as in a *Rain.*

Tho the *Rain* be much of it exhaled from the *Salt-Sea,* yet by this *Natural Distillation,* 'tis rendred fresh and drinkable to a degree, which hardly any *Artificial Distillation* of ours has yet effected.

The *Clouds* are so carried about by the Winds, as to be so *equally dispersed,* that no part of the Earth wants convenient Showers, unless when it pleases GOD, for the Punishment of a sinful People, to withhold *Rain,* by a special Interposition of his Providence: Or, if any Land wants *Rain,* they have a Supply some other way; as in the Land of *Egypt,* wherein little *Rain* falls, there is an abundant recompence made for that want, by the annual Overflowing of the River. Mr. *Ray* well observes, That this Distribution proclaims the *Providence* of GOD, and is from a *Divine Disposition.* Without this, there would be either desolating *Floods,* or such *Droughts* as that of *Cyprus,* in which no *Rain* fell for thirty Years together, and the Island was deserted, in the Reign of *Constantine.* The *gradual Falling* of the *Rain* by *Drops,* is an admirable Accommodation of it to the Intention of watering the Earth. 'Tis the best way imaginable. If it should fall in a *continual Stream,* like a River, every thing would be vastly incommoded with it.

¶. When GOD *gives Rain from Heaven,* he will give also *fruitful Seasons* in our Minds,[2] if they be thereby led to due Acknowledgments of him. 'Twill bespeak, 'twill procure, the richest *Showers of Blessings*[3] upon us. How seasonable will it be for us now "humbly

1. See, for example, Lev. 26:4 and Deut. 11:14.
2. Adapted from Acts 14:17.
3. Ezek. 34:26.

to acknowledge the *Witness,* which our GOD gives us of his *Power* and *Goodness!* To see the *Paths of GOD* in the Clouds which *drop Fatness* upon us!⁴ To wish for those Influences of Heaven, which may come upon ourselves *like Rain upon the Grass, as the Showers that water the Earth,*⁵ and *rain down Righteousness* upon the World!⁶ To resolve upon an Imitation of our merciful GOD, who *sends Rain upon the Just, and the Unjust!*⁷ To send up our Desires, that we may not be like the Earth, *which drinks in the Rain that comes often upon it, but bears Thorns and Briars, rejected, and nigh unto cursing!*⁸ In fine, To glorify our GOD with Confessions of this importance; *Can the Heavens give Showers? Art not thou he, O Lord our God? Therefore we will wait upon thee; for thou hast made all of these things!"*⁹

The Archbishop of Cambray¹⁰ shall express our Sentiments. "If I lift up my Eyes, I perceive in the Clouds that fly above us, a sort of hanging Seas, that serve to temper the Air, break the fiery Rays of the Sun, and water the Earth when it is too dry. What Hand was able to hang over our Heads those great Reservatories of Waters! What Hand takes care never to let them fall, but in moderate Showers!"

ESSAY XIII. *Of the* RAINBOW.

AFTER we have given the common Definition of it, *Arcus Cælestis, qui fit ex Solis Luce, in Nubem varie compositam & temperatam, sed ex Diametro Soli ipsi, incurrente ac incidente, pluvioso tempore;*¹ and should add more than there be *Colours in the Rainbow,* and with the modern Corrections of antient Errors, proceed to the Differences between the *Solar Iris* and the *Lunar,* and between

4. Adapted from Ps. 65:11.
5. Adapted from Ps. 72:6.
6. Adapted from Hos. 10:12.
7. Matt. 5:45.
8. Adapted from Heb. 6:7–8. Verse 8 reads (in part): "But that which beareth thorns is rejected, and is nigh unto cursing."
9. Jer. 14:22.
10. That is, François de Salignac de La Mothe Fénelon (1651–1715), archbishop, mystical theologian, and man of letters.
1. "The heavenly bow, which, in a rainy period, comes from the light of the sun as it meets with and intersects a cloud which is variously composed and compounded, but opposite the sun itself."

the *Iris* and the *Halo:* we have yet made so little Progress in real and certain *Knowledge,* that we should be left after all, with the Subject of our Discourse, *still in the Clouds.*

But we are called upon, *To consider the wondrous Works of God;* and particularly that, wherein *he causes the Light of his Cloud to shine,*[2] that is to say, his *Rainbow.*

A famous Clergyman of *Spalato,*[3] in a Book *De Radiis Visus & Lucis,* written before the former Century, began mathematically to describe how the *interiour Bow* of the *Iris* is formed in round Drops of *Rain,* by a Refraction of the Sun's Light, and one Reflection between them; and the *Exteriour* by two Refractions, and two sorts of Reflections between them, in each Drop of Water.

Des Cartes (who don't use to betray his Tutors) took the Hints from *Antonius de Dominis,* and went on *mathematically,* and with much demonstration, to give us a Theory of the *Iris,* from the Laws of *Refraction,* which lucid Rays do suffer in passing through diaphanous Bodies. He clearly demonstrated the *Primary Iris* to be only the *Sun's Image,* reflected from the concave Surfaces of an innumerable Quantity of small spherical Drops of falling Rain; with this necessary Circumstance, That those Rays which fell on the Objects, parallel to each other, should not after one Reflection, and two Refractions, (to wit, at going into the Drop, and coming out again) be dispersed, or made to diverge, but come back again also to the Eye, parallel to each other. The *Secondary Iris,* he supposes produced by those Rays of the Sun, which fall more obliquely, but after the same manner as before: only in these there are two Reflections, before the Sun's Rays, refracted a second time, and tending towards the Eye in a parallel Position, can get out from the aqueous Globules.

The acute and accurate Mr. *Halley* comes after the *French Philosopher,* and shows how the *Cartesian* Problems were more easily solved, than the Author himself imagin'd. He shows how to determine the Angle, by which the *Iris* is distant from the opposite Point of the Sun; and the *Ratio* of the Refraction being given *geometrically,* or *vice versa,* the *Iris* being given, to determine the refractive Power of the Liquor. And he goes on to cultivate the Subject with the Ingenuity proper to so accomplish'd a Gentleman.

---

2. Job 37:14–15.
3. That is, Marco Antonio de Dominis.

Essay 13. Of the Rainbow

But then comes the admirable Sir *Isaac Newton*, whom we now venture to call the *Perpetual Dictator* of the learned World, in the *Principles of Natural Philosophy*;[4] and than whom, there has not yet shone among Mankind a more sagacious Reasoner upon the *Laws of Nature*. This rare Person, in his incomparable Treatise of *Opticks*,[5] has yet further explained the *Phænomena* of the *Rainbow*; and has not only shown how the *Bow* is made, but how the Colours (whereof Antiquity made but *Three*) are formed; how the Rays do strike our Sense with the *Colours*, in the Order which is required by their Degrees of *Refrangibility*, in the Progress from the Inside of the *Bow* to the Outside: the *Violet*, the *Indigo*, the *Blue*, the *Green*, the *Yellow*, the *Orange*, and the *Red*.

In a Book lately published at *Norimberg*, intitled, *Thaumantiadis Thaumasia*,[6] which has not yet reached *America*; the skilful Author lays together whatever is to be found upon this Argument, among the modern, as well as the antient Writers.

It is good Advice given by the Son of *Sirach*;[7] *Look upon the Rainbow, and praise Him that made it.*[8]

The Gospel of the Rainbow, offered by *Frytschius*.
Sic ubi Cœlestem suboriri adspexeris Arcum,
 Quo Cœlum melius non Meteoron habet:
Ille quidem varios ducens e Nube Colores,
 Humano generi conspiciundus adest.
Hunc ita conspicias, seu veri Pignus amoris,
 Ac olim facti fœderis esto Memor.
Quod Deus omnipotens Noah sancto contulit ipsi,
 Se servaturum totius Orbis Opus.
Nec perpessurum submersum Fluminis Unda
 Iri Hominem sicut fecerat ante quidem.

---

4. Newton's *Philosophiae naturalis principia mathematica* was first published in London in 1687. The first English translation, by Andrew Motte, was entitled *The Mathematical Principles of Natural Philosophy* (London, 1729).

5. Newton's *Opticks: Or, a Treatise of the Reflexions, Inflexions and Colours of Light* was first published in London in 1704.

6. Christoph Gottlieb (or Theophil) Volkamer (fl. 1699) was the author of *Thaumantiadis Thaumasia, sive Iridis admiranda* (Nuremberg, 1699), a doctoral dissertation written at Altdorf University under Johann Christophorus Sturm. Mather knew of the book from a reference in a work by Edmond Halley.

7. Jesus the son of Sirach, author of *Ecclesiasticus, or, The Wisdom of Jesus the Son of Sirach*, one of the most highly esteemed of the noncanonical books of the Bible.

8. Ecclus. 43:11.

Englished:

"When you discern the *Bow of Heaven* to rise,
The *brightest Meteor* there salutes your Eyes:
Producing various Colours on the *Cloud,*
Mankind beholds it, and survives the *Flood.*
Behold it, Sirs, a Sign of Heavenly Love,
And of a Covenant made by GOD above:
Almighty GOD did by that *Sign* engage
To keep his *Noah's* World from Age to Age.
'Tis thus engag'd, GOD will no more employ
Deep *Waters,* as of old, Men to destroy."

The *Halo* is of so near kindred unto the *Rainbow,* that it claims a mention with it: A Circle that surrounds the *Sun,* or the *Moon,* (or a *Star;*) sometimes 'tis coloured like a *Rainbow.* According to Sir *Isaac Newton,* it arises from the Sun's or Moon's shining through a thin Cloud, consisting of Globules of Hail or Water, all of the same Size. Mr. *Huygens* conceives it formed by small round Grains of a kind of *Hail,* made up of two Parts; one of which is opake, and inclosed in the other, which is transparent. The same way he accounts for the *Parhelia.* Only there he apprehends, that the icy Grains are of an oblong Figure, and rounding at the Ends like Cylinders, with round convex Tops.

¶. May we *look upon the Rainbow, and praise Him that made it!*[9] My Readers, will you give me leave to *teach you the Use of the Bow?*[10] *Mercer* tells us, the religious *Jews* in many places, upon the appearance of a *Rainbow,* go forth and fall down, and confess their Sins, and own themselves worthy to be drowned with a *Flood* for them.[11] To us *Christians,* our Lord says, *What do you more than they?*[12] "As the sight of the *Rainbow* should bring to remembrance, *What a woful, what a fearful Desolation, once came upon a wicked World, whose Foundation was overflown with a Flood!*[13] So the *Sacramental Importance,* now instamped by the Will of GOD

---

9. Ecclus. 43:11.
10. Adapted from 2 Sam. 1:18.
11. It was reported among Jews that Rabbi Joshua ben Levi (first half of the third century), head of the Academy at Lydda in Palestine, declared that upon seeing the rainbow one should fall on his face, as did Ezekiel (Ezek. 1:28). But in Israel the rabbis disapproved of such action because it appeared as if the person was bowing down to the rainbow.
12. Adapted from Matt. 5:47.
13. Adapted from Job 22:15–16.

upon the *Rainbow*, should be acknowledged with us. It should be considered as a *Sign* and a *Seal* of a *Covenant*, which the Great GOD has made, That He will not have *this World*, though a sinful one, to be *drowned any more;*[14] nor his *Church* in the World. Upon the View of the admirable *Meteor*, how proper this Doxology? *Blessed be our Gracious, and Merciful, and Long-suffering Lord; who hath sworn, that the Waters of* Noah *shall go over the Earth no more!* But then, how can we forget the Glorious CHRIST, who is our *Head* in the *Covenant;* and about whose *Head* there has been the appearance of a *Rainbow*, in the Visions of his Prophets, betokening our Dependance upon Him for all our Preservations! But then we are not excused from, but rather excited to these further Thoughts on this occasion: *That though a watery Flood, which may drown the World, is no more to be feared; yet there is a fiery Flood, for the Depredations whereof, a miserable World is growing horribly combustible.* We are to expect,

Affore Tempus
Quo Mare, quo Tellus, correptaque Regia Cœli
Ardeat, & Mundi Moles operosa laboret."[15]

ESSAY XIV. *Of the* SNOW.

OF the *Snow*, there are many Curiosities observed by the excellent Dr. *Grew.*

It is observed by him, as well as by *Des Cartes*, and Dr. *Hook*, That very many Parts of the *Snow* are of a most regular Figure; they are generally so many Rowels, or Stars of *six Points*, being as real, as perfect, as transparent Ice, as any one may see upon a Vessel of Water: On each of which *six Points*, there are set other *collateral Points*, and those always at the same Angles as are the main Points themselves.

These are of divers Magnitudes; many are large and fair, but some are very minute.

Among these, there are found some irregular ones, which are

14. On the rainbow as the token of the covenant between God and every living creature that no more floods would destroy the earth, see Gen. 9:8–17.
15. "A time would come when sea and land, the unkindled palace of the sky and the beleaguered structure of the universe should be destroyed by fire."

but *Fragments* of the regular. But some seem to have lost their original Regularity, not by being broken, but by various Winds, first gently thaw'd, and then froze into such irregular Clumpers again.

A *snowy Cloud* seems then to be an infinite Mass of *Icicles* regularly figured, not so much as one of the many Millions being irregular. A Cloud of Vapours is gathered into *Drops;* the *Drops* forthwith descend. On the Descent they pass through a *soft Wind* that freezes them, or a cold Region of the Air, by which each Drop is immediately froze into an *Icicle,* that shoots forth into several *Stiriæ*[1] from the Center. But still continuing their Descent, and meeting with some sprinkling little Gales of a warmer Air, or in their continual Motion or Waftage to and fro, touching upon each other; some are a little thaw'd, blunted, frosted, clumper'd; others broken: but the most hank'd and clung in several Parcels together, which we call *Flakes of Snow.*

It should seem, that every *Drop of Rain* contains in it some spirituous Particles. These meeting in the Descent, with others of an acido-salinous Nature, the spirituous Parts are apprehended by them, and with those the watery; and so the whole Drop is fixed, but still according to the Energy of the spirituous, as the *Pencil,* and the determinate Possibility of the saline Parts, as a *Ruler,* into a *little Star.*

Though the *Snow* seem *soft,* yet it is truly *hard;* it is *Ice:* but the *Softness* of it is from this; Upon the first touch of the Finger on the sharp Edges, it thaws immediately; the Points would else pierce the Fingers like so many Lancets.

Again, though the *Snow* be true Ice, and so hard, and so dense a Body, yet it is very *light:* This is because of the extreme *Thinness* of each *Icicle,* in comparison of the *Breadth.* As *Gold,* though the most ponderous of all Bodies, beaten into Leaves, rides on the least Breath of Air.

We read of Heaven *giving Snow like Wool.*[2] I have known it *give a Snow of Wool.* In a Town of *New-England,* called *Fairfield,*[3] in a bitter snowy Night, there fell a Quantity of *Snow,* which covered a large frozen Pond, but of such a *woollen* Consistence, that it can

---

1. A *stiria* is a concretion resembling an icicle (e.g., a stalactite).
2. Ps. 147:16.
3. Fairfield, Conn.

be called nothing but *Wool.* I have a Quantity of it, that has been these many Years lying by me.

Res admiranda Nix, & optimarum Rerum in sacro Sermone Symbolum:[4] 'Tis the Expression of the pious and learned Mr. *Gale.*

¶. "When we see *the Snow, that comes down from Heaven, and returns not thither, but waters the Earth, and makes it bring forth and bud;*[5] we cannot but hope, that the Word of our GOD, which comes like it, will continue with us, and *accomplish* the Intentions of it.

"Whereof one, upon the Soul of thy Servant, *O my GOD!* is, to produce my Desires, That my *Sins,* which have been like *Scarlet,* may become *white like Snow,*[6] in thy free and full Pardon of them. *O wash me in the Blood of my Saviour, and I shall be whiter than the Snow!*[7] But, *Lord,* let a Work of real *Sanctification,* at the same time upon me, render me *purer than the Snow!*"[8]

ESSAY XV. *Of the* HAIL.

'TIS *Gutta Pluviæ acerrimo frigore congelata.*[1]

Hail is very often a Concomitant of *Thunder* and *Lightning.* 'Tis well known, as Dr. *Wallis* observes, That in our *Artificial Congelations,* a Mixture of *Snow* and *Nitre,* or even common *Salt,* will cause a very sudden Congelation of Water. Now the same in the Clouds may cause *Hail-Stones;* and the rather, because not only in some that are prodigiously great, but also in common *Hail-Stones,* there seems to be something like *Snow,* rather than *Ice,* in the midst of them. The large *Hail-Stones,* that weigh half or three quarters of a Pound, by the Violence of their Fall manifest that they have descended from a considerable height. And though perhaps in their first Concretion, their Bulk might not exceed the moderate Size of the common *Hail;* yet in their long descent, if the *Medium* through which they fell, were alike inclined unto Congelation, they might receive a great Accession to their Bulk,

---

4. "Snow is an admirable thing, and a symbol of the best things in the holy sermon."
5. Isa. 55:10.
6. Isa. 1:18.
7. Ps. 51:7.
8. Lam. 4:7.
1. "A drop of rain frozen by severe cold."

by perhaps many of them coalescing and incorporating into one.
¶. "Worse than *Egyptians* they, whom an *Hail-Storm* will not cause to *fear the Word of the Lord*.² The *irresistible* Judgments of GOD are sometimes compar'd unto *Hail-Storms,* and *great Hail-Stones.*³ These things come down upon the World with that Voice, *Tremble to be in ill Terms with a GOD, who with a Tempest of Hail, and a destroying Storm, can immediately crush all that is opposed unto him.*"⁴

Of all the *Meteors,* both the *fiery* and the *watery,* the Poet⁵ has well acknowledged;

Qui Meteora videt liquido radiantia Cælo,
  Hic videt Æterni facta stupenda Dei.

Who sees bright *Meteors* in the liquid Skies,
Has the great Works of GOD before his Eyes.

Christian, take the Advice; ['tis honest *Frytschius's.*]
  Rumpe Moras, Meteoraque suspice Cæli.
  *Illa aliquid semper quo movearis habent.*⁶

ESSAY XVI. *Of the* THUNDER *and* LIGHTNING.

HIS *powerful Thunder, who can understand?*¹ Yet our Philosophy will a little try to see and say something of it.

The Account of *Thunder,* given by Dr. *Hook,* is this. The Atmosphere of the Earth abounds with *nitrous Particles* of a spirituous nature, which are every where carried along with it. Besides which sort of Particles, there are also others raised up into the *Air,* which may be somewhat of the Nature of *sulphureous,* and *unctious,* and other combustible Bodies. We see Spirit of *Wine,* of *Turpentine,* of *Camphire,* and almost all other combustible Bodies,

---

2. The reference is to Exod. 9:18–35 and Rev. 16:21. The passage in Exodus describes a plague of hail upon the Egyptians. It convinced the Egyptians to fear the word of God, but only until the hail ceased.
3. The allusion is to Ezek. 13:13 and perhaps Ps. 18:13.
4. Perhaps adapted from Isa. 28:2.
5. That is, Marcus Frytsche or Fritsche.
6. "End delay, and look at the meteors of the heavens. / They always have something by which you may be moved."
1. Adapted from Job 26:14.

will by *Heat* be rarefied into the Form of *Air,* or *Smoke,* and be raised up into the Air. All these, if they have a sufficient Degree of *Heat,* will catch *Fire,* and be turned into *Flame,* from the *nitrous* Parts of the Air mixing with them; as it has been proved by Thousands of Experiments. There are also other sorts of such Steams, that arise from *subterraneous* and *mineral* Bodies; which only by their coming to mix with the *Nitre*[2] of the Air, though they have no sensible *Heat* in them, will so ferment and act upon one another, as to produce an actual *Flame.* Of this, the *Mines* are too frequent Witnesses and Sufferers. The *Lightning* seems to be very much of such an Original.

Dr. *Wallis* observes, That *Thunder* and *Lightning* have so much resemblance to *fired Gunpowder* in their *Effects,* that we may very well suppose much of the same *Causes.* The principal Ingredients in *Gunpowder,* are *Nitre* and *Sulphur.* Suppose in the Air, a convenient Mixture of *nitrous* and *sulphureous* Vapours, and those to take *fire* by accident, such an *Explosion,* and with such *Noise* and *Light* as that in the firing of *Gunpowder,* may well follow upon it; and being once kindled, it will run from place to place, as the Vapour leads it, like as in a Train of *Gunpowder.* This Explosion, high in the Air, and far from us, will do no considerable mischief. But, if it be very near us, it has terrible Consequences. The Distance of its *Place* may be estimated by the Distance of the *Time,* which there is between seeing the *Flash,* and hearing the *Clap:* For though in their Generation they be simultaneous, yet *Light* moving faster than *Sound,* they come successively to us. That there is a *nitrous* Vapour in it, we may reasonably judge, because we know of no other Body so liable to so sudden and furious Explosion. That there is a *sulphureous* one, is manifest from the Smell that attends it, and the sultry Heat, that is commonly a Forerunner of it.

¶. "The *natural Causes* of the *Thunder* do not at all release me from considering the *Interest* and *Providence* of the Glorious GOD, concerned in it. It is a Note prepared for the Songs of the Faithful, *The GOD of Glory thundereth.*[3] It is He, who

92 *subterraneous/subterraneons*

2. As used at the time, the word referred to a supposed nitrous substance occurring especially diffused through the air. Niter (nitre) is saltpeter, that is, potassium nitrate or sodium nitrate.
3. Ps. 29:3.

> *Fulmina molitur dextra, quo maxima motu*
> *Terra tremit, fugere Feræ, & mortalia Corda*
> *Per Gentes humilis stravit Pavor.*[4]

And indeed, as the *Thunder* has in it *the Voice of God*,[5] [*Paganism* itself owned it, as being Φωνὴ Διὸς] thus there are several Points of *Piety*, wherein I am, as with a *Bath Kol*,[6] instructed from it.

"There is this *Voice* most sensibly to be heard in the *Thunder, Power belongeth unto God.*[7] There is nothing able to stand before those *Lightnings,* which are stiled the *Arrows of God.*[8] We see Castles fall, Metals melt, Bricks themselves vitrify; all flies, when hot *Thunderbolts*[9] are scattered upon them. The very *Mountains* are torn to pieces, when *Feriunt summos sua Fulmina Montes.*[10] It becomes me now to say, *The Thunder of his Power who can understand?*[11] An haughty Emperor[12] shrinks, and shakes, and hides his guilty Head, before the powerful *Thunder* of God.

"How can I hear the *Voice* of the *Almighty Thunderer,* without such Thoughts as these? *Glorious God, let me, through the Blood of a sacrificed Saviour, be in good Terms with One so able to destroy me in a moment!* And, let me be afraid of offending Him, who is possessed of such an *irresistible Artillery!*

"At the same time, do I not see the *Mercy* and *Patience* of a Good God to a sinful World? The Desolations of the World, how wonderfully would they be,

> *Si quoties peccant Homines sua Fulmina mittat!*[13]

It is no rare thing for the Children of Men to die by a *Thunderbolt:* A *King* has been so slain in the midst of his Army. There was a Punishment of old used upon Criminals, by pouring hot Lead into their Mouths, which was called *Combustio Animæ*,[14] and used in imitation of God's destroying Men with *Lightning;*

---

4. "[He] wields his bolts with flashing hand. At that shock the mighty earth shivers; the beasts flee far, and o'er all the world crouching terror lays low men's hearts."
5. Adapted from Job 40:9, 37:5; Pss. 18:13, 77:18, and 104:7.
6. "Heavenly voice"—a term in rabbinic theology.
7. Ps. 62:11.
8. Adapted from Ps. 18:14 and Zech. 9:14.
9. Ps. 78:48.
10. "And 'tis the tops of the mountains that the lightning strikes."
11. Job 26:14.
12. The reference is to Caligula or Gaius Julius Caesar Germanicus (A.D. 12–41), Roman emperor from 37 to 41.
13. "If at every human error [Jupiter] should hurl his thunderbolts."
14. "Burning out of the life."

whereby the *inward* Parts are burnt without any visible Touch upon the *outward*. This *Combustio Animæ*, A Death by *Lightning*, has been frequently inflicted. Their being *asleep* at the time has not preserved them, though there be a Fancy in *Plutarch* that it would; nor would a Tent of *Seal-Skin* have done it, though some great ones have repaired unto such an *Amulet* for their Protection. *My God, I adore thy Sovereign Grace, that such a Sinner as I have not yet been by Lightning turned into Dust and Ashes before thee!*

"I take notice of one thing, That as Guilt lying on the Minds of Men, makes them startle at a *Thunder-Clap;*

*Hi sunt qui trepidant, & ad omnia Fulgura pallent,
Cum tonat, exanimes primo quoque Murmure Cœli:*[15]

So the Miscarriages about which our Hearts do first and most of all misgive us in a *Thunder-Storm,* are those which most of all call for a *thorough Repentance* with us. There are some Writings which I cannot read, except I hold them against the Fire; by having my Heart held up against the *Lightning,* I may quickly read *my own Iniquity.*

"Impious People are *deaf to Thunder!*"

Herlicius, in his *Tractatus de Fulmine,* reckons up a considerable number of those, which might be called *Fælicia Fulmina.*[16] Such will they be that make these Impressions upon us.

ESSAY XVII. *Of the* AIR.

THE *Air* of our Atmosphere, in which we breathe, is a diaphanous, compressible, dilatable *Fluid;* a Body covering the Earth and the Sea, to a great height above the highest Mountains: in this, among other things, differing from the *Æther,* that it refracts the Rays of the Moon, and other Luminaries.

There seem to be three different sorts of *Corpuscles,* whereof the *Air* is composed. There are such as are carried up into the *Air* from other Bodies, as *Vapours* exhaled by the *Sun's Heat,* or

---

15. "These are the men who tremble and grow pale at every lightning-flash; when it thunders, they quail at the first rumbling in the heavens."

16. "Fruitful lightning-bolts."

by subterraneous. There may be also a more subtile kind, mixed with our *Air,* emitted from the *Heavenly Bodies,* and from the *Magnetick Steams* of the Globe on which we sojourn. But there may be a third sort of Particles, which may most properly merit the Name of *Aerial;* as being the distinguishing Parts of the Air, taken in the stricter sense of the Term. These Particles have an *Elasticity* in them; are springy; resemble the *Spring* of a *Watch. Elasticity* is an essential Property of the *Air,* and it is thought no other *Fluid* has any thing of it, but only so far as it participates of *Air,* or has *Air* contain'd in the Pores of it. Our *Air* abounds with Particles of such a nature, that in case they be bent, or press'd by the Weight of the incumbent part of the *Atmosphere,* or of any other Body, they endeavour to free themselves from that Pressure, by bearing against the Bodies that keep them under it; and as soon as the Removal of these Bodies gives them way, they expand the whole parcel of *Air* which they composed.

Dr. *Hook* thinks the Air to be little else than a Tincture or Solution of terrestrial and aqueous Particles, dissolved in, and agitated by the *Æther,* and to have something *saline* in their Nature.

Mr. *Boyle* found, that one and the same Portion of Air may take up 52,000 times the Space it doth at another time. He found, that the same Quantity of Air, by only having the Pressure of the Atmosphere taken off in the *Pneumatick Engine,* and without increasing the Spring with any adventitious Heat, would possess above 13,000 times its natural Dimensions. Dr. *Gregory* proceeds, That accordingly a Globe of *Air,* of one Inch diameter, would at the Distance of the Semidiameter of the Earth from the Earth, fill all the Planetary Regions as far as, and much beyond the Sphere of *Saturn.* Admirable Rarefaction!

The *Weight of Air* was discover'd first by *Galilæus,* who finding that *Water* could not by pumping be raised any higher than 34 or 35 Foot, concluded that the old Notion of an infinite *Fuga Vacui*[1] would never do; and so fell to thinking on the Counterbalance of *the Weight of the Air. Torricellius* afterwards pursued and improved the Thought, and as a further Proof of *the Weight of the Air,* invented that which we call *the Torricellian Experiment.*[2]

---
1. "Avoidance of a vacuum."
2. The Torricellian experiment concerned the concept of the vacuum.

Essay 17. Of the Air

Mr. *Boyle* found by repeated Experiments, that the Weight of *Air* to *Water* is as 1 to 1000.

Dr. *Halley* rather determines the *specific Gravity* of Air to Water, to be about 1 to 800. *Mercury* is to Air as 10,800 to 1. And so, a Cylinder of *Air*, of 900 Feet, is equal to an Inch of *Mercury*.

We will, with Dr. *Wainwright*, suppose a cubical Foot of *Water* to weigh 76 Pounds *Troy* Weight. The Compass of a Foot square upon the Superficies of our Bodies, must sustain a Quantity of *Air*, equal to 2660 Pounds Weight. If the Superficies of a Man's Body contains fifteen square Feet, which is pretty near the Truth, he would sustain a Weight equal to 39,900 Pounds *Troy*, which is above thirteen *Tun*. The difference between the greatest and the least Pressure of the *Air* upon our Bodies, is equal to 3982 Pounds *Troy*. On which the Doctor says, "No wonder then we suffer in our Health by Change of Weather; 'tis surprizing that every such Change does not entirely break the Frame of our Bodies to pieces, and be the constant Harbinger of sudden Death."

My God, it is because I have obtained Help from thee, that I continue to this Day![3]

Sir *Isaac Newton* thinks *true and permanent Air* to be made by Fermentation and Rarefaction of Bodies, that are of a very fixed Nature. And it is plain, those Particles *fly* and *avoid* one another with the greatest Force *at a distance*, which when they are *very near*, do *attract* and *adhere* to one another with the greatest Violence.

The Particles of *true and permanent Air*, being extracted from the densest and most fixed Bodies, will be more dense and crass than those of *Vapour*, and from hence, it's likely, may be heavier than those; and the Parts of an *humid Atmosphere* may be lighter than those of a *dry* one, as in fact they appear to be. He thinks therefore, that the Rarefaction and Condensation of the Air cannot be accounted for from the *Spring*, or Elastick Forms of the Particles, without a Supposition, that they are endued with some *Centrifugal* Force or Power, by which they *fly* and *avoid* one another, and the dense Bodies, from whence they are extracted.

This may be the cause for *Filtration*, and the Ascent of Water in small capillary Tubes, to a much greater height, than the Surface of the Water in the open Vessel, in which they are placed. The Air within the Tubes is much rarer than in more open

---

3. Adapted from Acts 26:22.

Spaces, and by that means not pressing so much on the Surface of the Water within the Tubes, as without.

It is admirable to consider the Necessity of *Air* to the whole *animal* World; how soon the *vital Flame*[4] does languish and expire, if *Air* be withheld from it! Even the Inhabitants of the Water cannot live without the Use of it. It is evident that the *Air,* at the least that part of it which is the Aliment of *Fire,* and the Fuel of the *vital Flame* in Animals, easily penetrates the Body of Water exposed to it, and with a wondrous Insinuation diffuses itself thro every part of it. Put Fishes into a Vessel of a narrow mouth, full of Water, they will continue to live and swim there whole Months and Years. But if with any Covering you stop the Vessel, so as to exclude the *Air,* or interrupt the Communication of it with the Water, they will suddenly be suffocated; which was an Experiment often made by *Rondeletius.* The *Insects* rather need more *Air* than other Creatures, having more *Air-Vessels* for their Bulk, and many Orifices on each side of their Bodies for the Admission of *Air,* which if you stop with Oil or Honey, they presently die, and revive no more. *Pliny* knew not the reason of his own Observation; *Oleo illito Insecta omnia exanimantur.*[5] Yea, *Malpighius* has discovered and demonstrated, that the *Plants* themselves have a kind of Respiration, being furnished with a Plenty of Vessels for the Derivation of *Air* to all their Parts. Dr. *Hulse,*[6] and Mr. *Ray,* and others, have now also render'd it very evident, That the *Fœtus* in the Womb does receive a measure of *Air* from the maternal Blood, by the *Placenta Uterina,* or the *Cotyledons.* When this Communication is broken off, what is it that now, to preserve the Life of the Animal, speedily raises the *Lungs,* and fetches into them an abundance of *Air,* which causes a sudden and mighty Accension in the Blood, for the Maintenance whereof a far greater Quantity of Air is requisite? Certainly some intelligent Being must now interpose, to put the Diaphragm, and all the Muscles that serve to Respiration, into their Motion! *My God, I know thee!*[7] And now, as our ingenious *Waller* sings;

4. The concept of a vital flame held that a fine and kindled substance resides in the heart of animals and that air taken in by respiration is necessary to the preservation of this vital flame.
5. "Olive oil kills all insects when it is smeared upon them."
6. Probably Edward Hulse.
7. Adapted from Mark 1:24.

"Thus wing'd with Praise, we penetrate the Sky,
Teach Clouds and Stars to praise Him as we fly.
For that He reigns, all Creatures should rejoice,
And we with Songs supply their want of Voice.
Angels and we, assisted by this Art,
May sing together, tho we dwell apart."

¶. "The *Syrians* worshipped the *Air* as a *God*. I will worship Him that created it.

"I will give Thanks to the Glorious God, for the Benefits with which the *Air* is replenished by his Bounty. It was long since called the *Paranymph*,[8] by which the Espousal and Communion between *Heaven* and *Earth* is carried on.

"I *breathe* in the *Favours* of God continually. An ungrateful Wretch, if I do not *breathe out* his *Praises!*

"How justly might the Great God fill the Air with invisible *Arrows* of Death, and such deleterious *Miasms,* and pestilential *Poisons,* as might suffer the *Unholy* and *Unthankful* to *breathe* no longer in it!"

ESSAY XVIII. *Of the* WIND.

WHAT better Definition of the *Wind,* than *the Stream of the Air?* *Plato* long since defin'd it, *The Motion of the Air about the Earth.*

Other Hypotheses for this Current of the Air not well answering all *Phænomena,* the learned Mr. *Halley* recommends this to Consideration, as the Cause of it; The Action of the *Sun-beams* on the *Air* and *Water,* as the Sun passes every day over the Oceans, consider'd with the Nature of the Soil, and the Situation of the Continents adjoining.

According to the Laws of *Staticks,* the *Air,* which is less rarefied and expanded by *Heat,* and consequently more ponderous, must have a Motion round those Parts thereof, which are more rarefied and less ponderous, to bring it into an Æquilibrium. The Presence of the *Sun* also continually shifting to the Westward, that Part

---

8. "The friend of the bridegroom or the bridesmaid."

unto which the *Air* tends, by reason of the Rarefaction made by his greatest Meridian Heat, is with him carried Westward, and consequently the Tendency of the whole Body of the *lower Air* is that way. Thus a general *Easterly Wind* is formed. From this Principle, the *Easterly Wind* on the *North* Side of the Æquator, should be to the *Northwards* of the *East;* and in *South* Latitudes, it should be to the *Southwards* thereof: inasmuch as near the Line, the *Air* is much more rarefied than at a greater distance from it. Here all the *Phænomena* of the general *Trade-Winds* are answer'd for; which if the whole Surface of the Globe were Sea, would undoubtedly blow all round the World, as they are found to do in the *Atlantick* and *Ethiopick*[1] Oceans. But since great Continents interpose, and break the Continuity of the *Oceans,* regard must be had to the Nature of the Soil, and the Position of the high Mountains, which cause the Variation of the Winds, from the general Rule that has been proposed. If a Country, which lies near the *Sun,* prove to be low, flat, and sandy, the *Heat* occasion'd by the Reflection and Retention of the Sun-beams there, will so rarefy the Air, that the denser and cooler Air will run thither, to restore the Æquilibrium. Hence may be the *constant Calms* in that part of the Ocean; called *The Rains.* This Tract being placed in the middle, between the Westerly Winds blowing on the hot Coast, and the Easterly Winds that blow to the Westwards, the Tendency of the Air there is indifferent to either, and so stands *in æquilibrio,* between both; and the Weight of the incumbent *Atmosphere* being diminished by the continual contrary Winds blowing from hence, the *Air* here holds not the copious Vapour it receives, but lets it fall into frequent *Rains.*

   It is very hard to conceive, why the Limits of the *Trade-Wind* should be fixed about the thirtieth Degree of Latitude all round the Globe, and that they should so seldom transgress those Bounds, or fall short of them.

   Behold the *Wings of the Wind!*[2]

   The inquisitive and ingenious Mr. *Derham* found by many Trials, That the *Wind* in a great Storm does move about *fifty or sixty Miles* in an Hour; That a common brisk *Wind* moves about *fifteen Miles* an Hour. But so gentle is the Course of many Winds, that they do not exceed *one Mile* an Hour.

   1. That is, the south Atlantic Ocean, between South America and Africa.
   2. 2 Sam. 22:11 and Pss. 18:10, 104:3.

Essay 18. Of the Wind

Dr. *Grew* observes, That there are Winds (besides the *Trade-Winds*) especially from the West, which blow sometimes two or three Days upon one Point, and will in this time drive before them a Ship an hundred and fifty Leagues, or four hundred and fifty *English* Miles.

The *Wind* is of great Use to ventilate the Air, and to dissipate contagious Vapours; which if they should stagnate, would produce grievous Diseases on the animal World. *Si non ventosa, venenosa.*[3] It also transfers the *Clouds* from one place to another, for the more commodious watering of the Earth. It likewise tempers the *Heats* of many Countries, which else would be excessive. It carries *Vessels* on their Voyages to remote Countries. *Windmills* are driven by it, whereof there are many Benefits. But as the excellent Mr. *Ray* observes, That it is rarely so violent, as to destroy all before it, and overwhelm the World; this proclaims a superiour Power moderating of it, the *Wisdom* and *Goodness* of Him, *who brings the Wind out of his Treasures.*[4]

What amazing things the *Winds,* called the *Tuffoon;* (or *Typhons!*) and how irresistibly furious! But our merciful God *stays the rough Winds.*[5]

The *Hurricanes* in the *West-Indies,* and their Brethren the *Monsoons* in the *East;* what shocking Stories do the Travellers give us of them! How direful Effects are sometimes caused by them! They blow down mighty *Trees* by the Roots. They chase mighty *Ships* up into the Woods. They make every thing to tremble, and give way, that is in their way. *Great God, who ridest on the Wind, and makest it move which way thou shalt please; who can stand in thy sight, if thou art angry!*[6]

¶. Whatever Point of the Compass the *Wind* blows upon, it may blow some Good Thoughts into our Minds; and then it will be no *Ill Wind* unto us.

"We ought certainly to consider *the stormy Wind, as fulfilling the Word of God.*[7] And there are *Tempests,* and *Whirlwinds* of the Divine Wrath to be deprecated. But then there are Influ-

---

3. "If not windy, very poisonous."
4. Adapted from Ps. 135:7; Jer. 10:13 and 51:16.
5. Adapted from Isa. 27:8.
6. "Ridest on the wind" is perhaps adapted from the biblical phrase "wings of the wind," for which see n. 2 above. "Makest it move" is perhaps adapted from John 3:8, which is quoted at the conclusion of this essay. "If thou art angry" is adapted from Ps. 76:7.
7. Adapted from Ps. 148:8.

ences of Heaven to be desired, which are, *As the Wind bloweth where it listeth, and we hear the Sound thereof, but cannot tell whence it cometh, nor whither it goeth.*"[8]

ESSAY XIX. *Of the* COLD.

THERE is much Dispute about the *Primum Frigidum*.[1] None, I hope, about the *First Cause* of the *Cold,* which sometimes mortifies us.

It is questioned by some, whether the *Cold* be any thing that is *positive,* and not a mere *Privation*. The *Coldness* of any thing, they say, signifies no more, than its not having its insensible Parts agitated so much as those of our *Sensories,* by which we judge of *Tactile Qualities*. To make a thing become *cold,* there needs no more, than that the *Sun,* or *Fire,* or some other Agent, that more vehemently agitated its Parts before, do now cease to do it.

But then, on the other side, there are Instances of *Cold* produced by vehement Agitations.

To some there seems to be a mighty store of *Corpuscles,* a little a-kin to *Nitre,* exhaled from the terrestrial Globe, (of the Figure which *Philoponus* tells us, *Democritus* assigned to *Frigorifick Atoms*) which may more than a little contribute to our *Cold*.

That *Cold* (and so *Freezing*) may arise from some saline Substance floating in the Air, seems probable from this; That all *Salts,* but some above others, when mixed with *Snow* or *Ice,* do prodigiously increase the Force of *Cold*. And all *saline* Bodies produce a *Stiffness* in the Parts of those Bodies, into which they enter.

The Force of the *Cold* is truly wonderful. *Olearius* tells us, in *Muscovy* their Spittle will freeze e'er it reach the Ground. So violent the *Cold* there, that no *Furs* can hinder it, but sometimes the *Noses,* the *Ears,* the *Hands,* and the *Feet* of Men will be frozen, and all fall off. 'Tis reported by *Fletcher* and *Herberstein,* That not only they who travel abroad, but many in the very Markets of their Towns are so mortally pinched, as to fall down dead with the *Cold*. Captain *James* and *Gerat de Veer* tell us frightful things

---

8. Adapted from John 3:8.

1. "First cold." The term was used to describe a body which is by nature supremely cold and imparts that quality to all other cold bodies.

of the *Cold* they found in their Northern Coasting. *Beauplan* adds, That without good Precautions, the *Cold* produces those *Cancers,* which in a few Hours destroy the Parts they seize upon. What mighty Rands of *Ice* (the *magnum Duramen Aquarum,*² as *Lucretius* calls it) have been encounter'd by such Navigators as *Munchius* and *Baffin,* who found some *Icy Islands* near three hundred Foot high above the Water! In the River of *Canada* sometimes are seen *Icy Islands,* computed four-score Leagues in length.

The irresistible Force of *Congelation!*

Congelation seems to be from the Introduction of the *Frigorifick Particles,* into the Interstices between the Particles of the Water; and thereby getting so near to them, as to be just within the Sphere of one another's attracting Force, on which they cohere into one solid Body.

Was it not then a Mistake in *Pliny,* when *Ice* was defined by him, *Aquæ Copia in Angusto?*³ The *Dimensions* of Water are increased by *Freezing;* and with such a Force in the Expansion, that the *Weights* raised by it, the *Stones* broke in it, the *Metals* obliged to give way to it, were hardly credible, if these Eyes had not seen them!

¶. "When we consider the *Cold,* especially if we have it under our more *sensible* Consideration, we cannot but subscribe to that Word, *Who can stand before his Cold!*⁴ How naturally are we now led to a Dread, and a Deprecation of lying under the *Displeasure* of the *Glorious God,* who by that one Part of his *Artillery,* the *Cold* alone, can soon destroy his Enemies!"

The *Mitigations* of our *Cold,* and our *Comforts* and *Supports* against the Assaults of it, bespeak our thankful Praises to our Glorious *Benefactor:* That we are not, as *Livy* says of the *Alps, Æternis damnati Nivibus!*⁵

It is observable, That the Degrees of *Cold* in several Climates are not according to their Degrees of *Latitude.* Some have met with very tolerable Weather under the *Arctick Pole.* But *Martinius,*⁶ in his *Atlas Chinensis,* reports of *China, Majus in hac Provincia*

---

2. "Great hardener of waters."
3. "A quantity of water in a small space."
4. Ps. 147:17.
5. "Damned by eternal snows."
6. Martin Martini (1614–61), Jesuit missionary to China, published *Novus atlas Sinensis* (Antwerp, 1654).

*Frigus est, quam illius poscat Poli Altitudo.*⁷ The Country lies in little more than *forty Degrees* of Latitude, and yet for four Months together in the Year, the Rivers there are so frozen, that the *Ice* will bear the Passage not only of Men, but of *Horses* and of *Coaches* too upon it. The like Report could I give of my own Country, which lies in the same Latitude. In my warm Study, from the Billets of Wood lying on a great Fire, the *Sap* forced out at the ends of the short Billets by the Fire, has frozen there, and been turned into *Ice,* while the Wood has been consuming. However, our *Cold* is much moderated since the opening and clearing of our *Woods,* and the Winds do not blow such Razours, as in the Days of our Fathers, when *Water,* cast up into the *Air,* would commonly be turned into *Ice* e'er it came to the Ground. I have sometimes wished, that Wise-Men would make the Reflection of *Petronius* upon this Matter: *Incultis asperisque Regionibus, diutius Nives hærent; ast ubi Aratro domefacta Tellus nitet, dum loqueris levis Pruina dilabitur. Similiter in Pectoribus Ira considit; Feras quidem Mentes obsidet, Eruditas præterlabitur.*⁸

ESSAY XX. *Of the Terraqueous* GLOBE.

THE Distance at which our *Globe* is placed from the *Sun,* and the Contemperation¹ of our Bodies and other Things to this Distance, are evident Works of our Glorious GOD!

According to the accurate Observations of the *English Norwood,* and the *French Picart,* the Ambit of our Globe will be twenty-four thousand nine hundred and thirty Miles. Wherefore supposing it spherical, the whole Surface will be 197,831,392 Miles; which in the solid Content will be found no less than 261,631,995,920 Miles. The cubick Feet will be 30,000,000,000,000,000,000,000. The *Earth,* with her Satellit the *Moon,* moving about the *Sun,* this *Orbis Magnus,*² as 'tis usually called, according to our *Derham,* is a Space

---

7. "There is greater cold in this province than its degree of latitude warrants."
8. "On the wild rough uplands the snow lies late, but when the earth is beautiful under the mastery of the plough, the light frost passes while you speak. Thus anger dwells in our hearts; it takes root in the savage, and glides over a man of learning."
1. An obsolete word meaning synchronism.
2. "Great orbit." That is, the distance traveled in the revolution of the earth around the sun.

they could not but raise a mighty *Atmosphere*, and such *Clouds* as must needs darken the Body of the *Moon*, sometimes in one part, sometimes in another. They carry on their Inferences; if no *Waters* in the *Moon*, then there are no *Plants*, nor *Animals*, nor *Men*. About the Constitution of this *Queen of the Night*, there seems a necessity for us to *remain in the dark!*

For Mr. *Derham* has confuted *Hugenius* with his own *Glasses*, and has demonstrated, that there are great Collections of *Waters* in the *Moon*, and by consequence Rivers, and Vapours, and Air; and in a word, a considerable *Apparatus* for *Habitation*.

But by what Creatures inhabited? A Difficulty this, that cannot be solved without *Revelation*.

¶. '*My GOD*, I bless thee for that *Luminary*, by
'which we have the uncomfortable Darkness of our
'*Night* so much abated! That *Luminary*, the Influ-
'ences whereof have such a part in the *Flux* and *Re-*
'*flux* of our *Seas*; without which we should be very
'miserable! That *Luminary*, whose Influences are so
'sensibly felt in the Growth of our *Vegetables*, and our
'*Animals!*'

These are some of the *Songs*, which *GOD*, the *Maker of* us both, has *given me in the Night*.

The Influences of the *Moon* upon *Sublunary* Bodies, are very wonderful. An *History* of them is yet among the *Desiderata* of our Philosophy. With my consent, he shall merit more than the Title of a *Rabbi Solomon Jarchi*, who gives it unto us. Dr. *Grew*, in his *Cosmologia*, has enumerated more than a dozen remarkable *Heads* of *Effects*, and *Motions*, and *Changes* in the World, over which the *Moon* has a sensible Dominion. Our *Lunaticks* are not the only Instances. Our *Husbandmen* will multiply the Instances upon us, till they make a Volume, which neither a *Columella*, nor a *Tom Tusser* have reached unto. The *Georges* of my Neighbourhood just now furnish me with two Instances, which have in them something that is notable. If our *Chesnut-*

In New England, the notion that the Earth rotates daily on its axis and revolves around the sun was accepted almost from the beginning. Mather here quietly reaffirms it in passing. Courtesy of the Rare Book and Special Collections Library, University of Illinois at Urbana-Champaign.

of more than 540 Millions of Miles in Circumference, or 172 Millions of Miles in Breadth.

The *Copernican* Hypothesis is now generally preferred, which allows a *Diurnal* and an *Annual* Motion to our *Globe,* rather than to the *Sun.* According to this, the *Diurnal* Motion of our *Globe* is near 1,039 Miles in an Hour.

The Arguments that prove the Stability of the *Sun,* and the Motion of the *Earth,* have now render'd it indisputable. It is impossible to account for the Appearances of the *Planets,* and their *Satellits,* and the *Fixed Stars,* in any tolerable manner, without admitting the Motion of the *Earth;* or to account for *Comets;* or for that Analogy of the *Periodical Times,* to the *middle Distances,* which is the necessary Consequence of the establish'd Law of *Gravitation.* Unless we would subvert the whole System of *Astronomy,* and (as Dr. *Cheyne* observes) disprove the Causes of all the *Celestial Motions,* we shall never be able to assert, that the *Earth stands unmoved.* Nor is there any Objection against the Motion of the *Earth,* but what has had a full Solution.

These Motions, performed so regularly for near six thousand Years, how much do they oblige us to cry out, *Great GOD, thou that art the Creator, art also the Governour of the World!*

Even a *Pagan Cleanthes,* as his Brother *Cicero* will tell us, would assign this as a sufficient Cause for a Belief of a Deity; *Æquabilitatem Motus, Conversionem Cæli, Solis, Lunæ, Syderumque omnium Distinctionem, Varietatem, Pulchritudinem, Ordinem; quarum rerum Aspectus ipse satis indicaret, non esse fortuita.*[3] And *Plutarch* says, This Observation was the first that led Men to the Acknowledgment of a GOD.

The Prophet *Habakkuk* mentions the Stop to this Course in the Days of *Joshua,* as a real Matter of Fact.[4] The same *Infinite Power* that gave the *Motion,* gave the *Check.*

The Circumvolutions of the *Globe* are of admirable *Conveniency,* yea, of absolute *Necessity,* to the Inhabitants. As *Tully* notes, *Conservat Animantes.*[5]

---

3. "The uniform motion and revolution of the heavens, and the varied groupings and ordered beauty of the sun, moon and stars, the very sight of which was in itself enough to prove that these things are not the mere effect of chance."
4. Hab. 3:11: "The sun and moon stood still in their habitation: at the light of thine arrows they went, and at the shining of thy glittering spear."
5. "[They] preserve the living creatures."

Essay 20. Of the Terraqueous Globe

The *Spherical Figure* of our Globe has numerous and marvellous Conveniencies, whereof no Man that seriously considers it can be insensible. How incommodious must an *Angular Figure* have been; or such an one as many of the Antients, and particularly the *Epicureans*,[6] with Stupidity enough imagin'd?

It is admirably well order'd, (as Dr. *More* observes) That the *Axis* of the Globe should be steddy, and perpetually parallel to itself; not carelesly tumbling this way and that way, as it might happen: and that the Posture of the *Axis* be inclining as it is, and not perpendicular to a Plane going thro the Center of the *Sun*, or coincident. Hence comes the *Globe* to be so habitable in all Parts; and even under the *Line* itself, as 'tis noted by Sir *Walter Raleigh*, the Parts are as pleasant, and as fruitful, and as fit for a *Paradise*, as any in the World. And the *Longevity* of the Natives there does rather exceed the rest of Mankind, as we learn from the Relations of *Piso*, and *Rochefort*, and *Pirard*, and *Le Blanc*, and other Testimonies. Yea, Mr. *Keill*[7] demonstrates that from the present Position of the Globe, and the Inclination of its *Axis* to the Plane of the *Ecliptick*, we reap this Advantage; They who live beyond forty-five Degrees of Latitude, and have most need of it, have more of the Heat of the *Sun* throughout the Year, than if he had shined always in the *Æquator:* Whereas in the *Torrid Zone*, and even in the *Temperate*, almost as far as forty-five, the Sum of the *Sun*'s Heat, in Summer and Winter, is less than it would be, if the Axis of the Globe were perpendicular to the Plane of the *Ecliptick*. He very well adds, This Consideration cannot but lead us into a transcendent Admiration of the *Divine Wisdom!* Yea, were the whole Creation surveyed, it would be every where found, as Mr. *Ray* observes, *That God has chosen better for us, than we could have done for ourselves.*

And then, the Collection of the *Waters* on the *Globe* into such

---

6. The followers of Epicurus (341–270 B.C.), who held that atoms and the void alone are real. Matter is composed of individual atoms which are in continuous motion in space; their motion is downwards, primarily owing to their weight. Atoms collide with others by a "swerve," and as a result of the blow the two set off in new directions. They impinge upon others, which in their turn are deflected at an angle from their straight course. This process starts the infinite series of motions which leads to the creation of "things." An infinite supply of atoms moving in infinite space and time leads (by chance) to every possible combination (or world).

7. That is, John Keill.

vast *Conceptacula*,[8] wherein the innumerable *Fishes* are nourished, and whereon *Voyages* are performed; and the Distinction of the *Dry Land*, furnished with so many *Vegetables* and *Animals:* What can it be any other than the Result of Counsel, of Design; of *Infinite Wisdom!* How blind art thou, O *Man*, and under what a brutal and fatal Darkness, if thou see it not! *The Brutish among the People will not be wise.*[9]

The *Figure* of our Globe is most probably that of an *Oblate Spheroid.* It swells towards the *Æquatorial* Parts, and flats towards the *Polar;* according to Sir *Isaac Newton,* the *Diameter* of the Globe is about thirty-four Miles longer than the *Axis.*

Dr. *Gregory* shews, that this is the reason why the *Axis* of our Globe does twice every Year change its Inclination to the *Ecliptick,* and as often return back again to its former Position.

That most accurate Astronomer, Mr. *Flamstead,* found the Distance of the *Pole-Star* from the Pole, to be greater about the *Summer* Solstice than about the *Winter,* by about forty or forty-five Seconds. He found also, by repeated Observations, a sensible annual Parallax in others of the Fixed Stars. This proves our Globe to move annually about the *Sun.*

Mr. *Halley* shows the annual Motion of the Earth to be so swift, as far to exceed that of a Bullet shot out of a Cannon, and to be after the rate of 210 Miles in a Minute, and 12,600 Miles in an Hour.

Our Globe is nearer to the *Sun* in *December* than in *June.* Its *Perihelion* is in *December.* The *Sun's* apparent Diameter is greater then; and our Globe then has a *swifter Motion* by a twenty-fifth Part. Hence there are about eight Days more in the Summer Half-Year, than in the Winter Half-Year. The colder and more Northern Places of our Globe are indeed brought some hundreds of thousands of Miles nearer the *Sun* in *Winter* than in *Summer.*

Upon the Occurrences of the whole GLOBE.

"O MAN! we are now come down into thy *Territories.* How many SERVANTS may MAN here see himself attended and surrounded with! The most *reasonable Thing* in the World is for MAN hereupon to contrive and resolve in this manner; O

8. That is, a vessel or receptacle.
9. Adapted from Ps. 94:8.

that my *Service* to the Glorious GOD may be as obedient, as willing, as ready, as what his *Creatures* yield to me!

"It has been excellently well proposed; *Cum cæteræ Creaturæ universæ omnibus Viribus, in Hominis Utilitatem connituntur, discat hinc Homo, similiter ex totis Viribus DEO servire, ad illumque se convertere, qui omnes Creaturas usui, servitioque suo destinavit.*[10]

"But then, to this we will annex a further Disposition of *Piety: Can a Man be profitable to GOD?*[11] My *Service* to Him does not advantage Him. When I have done all, I am an *unprofitable Servant*.[12] Wherefore let me study to transfer to my *Neighbour*, the *Service* which by the Creatures of GOD is done to me. Yea, let me so far as my Tenuity[13] can attain to it, labour to do to my *Neighbour* such Things as the Great GOD pleases to do to me. In this *Charity*, there will be that *Image* of the Glorious GOD, which is the *Glory* of the MAN that arrives to it.

"One says well, *Quocunque vertamus Oculos, ecce Testimonia, Oratores, & Laudatores Dei, qui totum Librum Mundi Laudum suarum Historiam, & Panegyricum esse voluit.*[14]

"MAN, let the Glorious GOD have *Praises* from thee, and have thy *Homage* and *Service*. Hereby the Creatures will be returned and united to GOD their Maker, and it will be brought about, that they shall not be made in vain. It was a wise Thought; *"Per Hominem, & illius Religionem, omnes Creaturæ cum Deo connectuntur, ne frustra a Deo sint creatæ.*[15]

"There is another pathetick Remark, made more than an hundred Years ago, but worthy to be for ever thought upon; *Omnes Creaturæ naturaliter Deum plus amant, quam seipsas, dum illius Mandata exequendo, seipsas consumunt; solus autem Peccator seipsum impensius quam Deum amat.*[16] Every *Creature*, but only the wicked Sinner, *loves GOD* more than it *loves itself.*

---

10. "Since all other creatures strive with all their might for the utility of man, let man learn from this to serve God in similar fashion with all his strength and to turn himself to God who directed all creatures to his use and service."
11. Job 22:2.
12. Luke 17:10.
13. That is, weakness.
14. "Wherever we would turn our eyes, there are witnesses, orators, and eulogizers of God, who wanted the entire book of the world to be the history and panegyric of his praises."
15. "Through man and his religion, all creatures are joined to God, in order that their creation by God not be in vain."
16. "All creatures naturally love God more than themselves, seeing that they consume themselves in following his dictates; the sinner alone loves himself more dearly than he loves God."

"Two Instructions of the pious *Ægardus* will be worth remembring here.

"The one; *Dulces tibi sint Creaturæ, propter Deum, a quo sunt; sed dulcior ipse Creator, qui omnibus major & melior.*[17]

"The other; *In quibus plus Dei, in iis plus sanctæ sit Voluptatis, & cum iis te conjungi cupias.*[18]

"GOD must be the *Sweet* of all Creatures to me; and the more of GOD in any Creatures, the more must be my Regard, the more my Relish for them."

¶. As we go along, we cannot well avoid a Touch upon *Cohesion*. We see two very plain, smooth, well-polish'd Bodies, will firmly *cohere*, even in an *exhausted Receiver*. This renders it evident, that *Cohesion* is not owing to the *Gravity*, nor to any other Property of the *Air*. What appears in the Surfaces of cohering Bodies upon their breaking, shows us, That a necessary Condition of *Cohesion* is a Congruity of *Surfaces;* and such as excludes any *Fluid* from lying between them. We may suppose, with Dr. *Cheyne*, that some of the *Primary Atoms,* whereof Bodies are constituted, are terminated with *plain* and *smooth* Surfaces on all sides; which will produce Bodies of the *strongest Cohesion:* Others are partly terminated with *plain* and *smooth,* and partly with *curve* Surfaces, which will produce Bodies of a *meaner Cohesion*. Others are entirely terminated with *curve* Surfaces, which will produce *Fluids;* and between these entirely *plain* and *smooth,* and entirely *curve,* there are infinite *Combinations of Surfaces, plain,* and *smooth,* and *curve,* which will account for all the various Degrees of *Cohesion* in Bodies, in respect of their Figures. But now the *Cement,* which hinders the Separation of Bodies, when the Points of their Surfaces are brought into Contact, [this] can be nothing but the *universal Law of Attraction,* whereby all the Parts of *Matter* endeavour to embrace one another, and cannot be separated but by a *Force,* that shall be superiour to that by which they *attract.*

"Being arrived here, we are gotten within a little of the Glorious GOD. The very *next Step* we take must be into Him,

44 of/of of

17. "The creatures are kind to you on account of God, because of whom they exist, but the Creator himself, who is greater and better than all, is kinder."
18. "In those who have a greater portion of God, there is a greater portion of holy desire, and you desire to join yourself to them."

who is the *immediate Cause* of *Weight* in *Matter*. None but He producing, imprinting, preserving that *Property* in *Matter,* is to be now considered. We will go on to take notice of that Property."

ESSAY XXI. *Of* GRAVITY.

TO our Globe there is one Property so exceedingly and so generally subservient, that a very great Notice is due to it; that is, GRAVITY, or the Tendency of Bodies to the *Center.*[1]

A most noble Contrivance (as Mr. *Derham* observes) to keep the several Globes of the Universe from shattering to pieces, as they would else evidently do in a little Time, thro their swift Rotation round their own *Axes.* Our *Globe* in particular, which revolves at the rate of above a thousand Miles an Hour, would, by the centrifugal Force of that Motion, be soon dissipated, and spirtled[2] into the circumambient Space, were it not kept well together by this wondrous Contrivance of the Creator, *Gravity,* or the *Power of Attraction.* By this Power also all the Parts of the *Globe* are kept in their proper Place and Order; all Bodies gravitating thereto do unite themselves with, and preserve the Bulk of them entire; and the fleeting Waters are kept in their constant Æquipoise, remaining in the *Place which God has founded for them, a Bound which He hath set, that they may not pass, that they turn not again to cover the Earth.*[3] It is by the virtue of this glorious Contrivance of the *great God, who formed all Things,*[4] that the Observation of the Psalmist is perpetually fulfilled: *Thou rulest the raging of the Sea; when the Waves thereof arise, thou stillest them.*[5]

Very various have been the Sentiments of the Curious, what *Cause* there should be assign'd for this great and catholick Affection of Matter, the *Vis Centripeta:*[6] I shall wave them all, and *bury* them

---
1. That is, to the center of the earth.
2. A spirtle is a small spirt or jet, so the meaning is, to be whirled into space.
3. Adapted from Ps. 104:8–9.
4. Prov. 26:10.
5. Ps. 89:9.
6. "Centripetal force."

in the *Place of Silence*, with the *Materia Striata*[7] of *Descartes,* which our *Keil*[8] has very sufficiently brought to *nothing;* and perhaps the *Fluid* of Dr. *Hook* must go the same way. 'Tis enough to me what that incomparable Mathematician, Dr. *Halley,* has declar'd upon it: That, after all, *Gravity* is an Effect insolvable by any *philosophical Hypothesis;* it must be religiously resolv'd into the *immediate Will* of our most wise CREATOR, who, by appointing this *Law,* throughout the material World, keeps all Bodies in their proper Places and Stations, which without it would soon fall to pieces, and be utterly destroy'd.

All Bodies descend still towards a Point, which either is, or lies near to, the *Center* of the *Globe.* Should our Almighty GOD change that *Center* but the two thousandth part of the *Radius* of our Globe, the Tops of our highest Mountains would be soon laid under Water.

In all Places equi-distant from the *Center* of our Globe, the Force of Gravity is nearly equal.

Indeed, as it has been proved by Sir *Isaac Newton,* the *Equatorial* Parts are something higher than the *Polar* Parts; the difference between the Earth's *Diameter* and *Axis* being about thirty-four *English* Miles.

Gravity does equally affect all *Bodies.* The *absolute Gravity* of all is the same. Abstracting from the resistance of the Medium, the most *compact* and the most *diffuse,* the *greatest* and the *smallest,* would descend an equal Space in an equal Time. In an exhausted Receiver a *Feather* will descend as fast as a *Pound of Lead.* But this resistance of the *Medium* has produc'd a *comparative Gravity.* And upon the difference of *specifick Gravity* in many Bodies, the Observations of our Philosophers have been very curious.

According to the exquisite *Halley* and *Huygens,* the *Descent of heavy Bodies* is after the rate of about *sixteen Foot* in *one Second* of Time.

Nevertheless this Power *increases* as you descend to, *decreases* as you ascend from the *Center* of the Globe, and that in proportion to the Squares of the Distances therefrom reciprocally; so as, for instance, at a double distance to have but a quarter of the Force.

---

7. "Grooved material." Mather may use *"Materia Striata"* simply as an example of a discarded theory or substance; at any rate Descartes does not use the term in his *Principia philosophiae* (Amsterdam, 1644).

8. That is, John Keill.

Essay 21. Of Gravity

A *Ton* Weight on the Surface of the Earth, raised Heaven-wards unto the height of one Semidiameter of the Earth from hence, would weigh but one quarter of a *Ton*. At three Semidiameters from the Surface of the Earth, it would be as easy for a Man to carry a *Ton*, as here to carry little more than an hundred Pounds. At the distance of the *Moon*, which suppose to be sixty Semidiameters of the Earth, 3600 Pounds weigh but *one Pound;* and the Fall of Bodies is but sixteen Foot in a whole Minute.

I remember I have somewhere met with such a devout Improvement of this Observation: "The further you fly towards *Heaven*, the more (if I may use the *Falconer's* Word) you must *lessen*. There is great reason why it should be so. *Defamations* particularly will be Things by which you must be *lessen'd:* you must meet with *heavy* Things; *Defamations* are in a singular manner such; they are not easy to *carry;* 'tis not easy to carry it well under them; some of them are a *Ton* Weight. But, *my Friend*, if you were as near *Heaven* as you ought to be, you would make *light* of them; you would bear them wonderfully!"

The *acute Borelli* has demonstrated that there is no such thing as *positive Levity*, and that *Levity* is only a lesser degree of *Gravity*. But how useful is this, not only to divers Tribes of *Animals*, but also to the raising up of the many *Vapours*, which are to be convey'd about the World? The Evaporations, which, according to Mr. *Sedileau*'s Observations, and others, are the fewest in the Winter, and greatest in the Summer, the most of all in windy Weather, and considerably exceed what falls in *Rain*, many being tumbled about and spent by the Winds, and many falling down in Dews.

The ingenious *Halley* has yet a suspicion that there may be some certain Matter, which may have a *Conatus*[9] directly contrary to that of *Gravity;* as in *Vegetation* the Sprouts directly tend against the *Perpendicular*.

Dr. *Gregory* demonstrates, that the antient Astronomers were not ignorant of the heavenly Bodies *gravitating* towards one another, and being preserv'd in their Orbits by the Force of Gravity.

Mr. *Keil* shews, that the Force of *Gravity* to the *centrifugal Force*,

42 Falconer's / Falconers

9. That is, an effort, impulse, or striving.

in a Body placed at the Equator of our Globe, is as 289 to 1; so that by the *centrifugal Force* arising from the Earth's Rotation, any Body placed in the Equator loses a 289th part of the Weight it would have if the Globe were at rest. And since there is no *centrifugal Force* at the *Poles,* a Body there weighs 289 Pounds, which at the Equator would weigh but 288. On our Globe the decrease of *Gravity,* in going from the Poles towards the *Equator,* is always *as the Square of the Cosine of the Latitude. Quod facit Natura* (to use *Tully*'s Words) *per omnem Mundum, omnia Mente & Ratione conficiens.*[10]

Mr. *Samuel Clark* observes, 'Tis now evident that the most universal Principle of *Gravitation,* the Spring of almost all the great and regular inanimate Motions in the World, answering not at all to the *Surfaces* of Bodies, by which alone they can act one upon another, but entirely to their *solid Content;* cannot possibly be the result of any *Motion* originally impressed on *Matter,* but must of necessity be caused by something which penetrates the very Substance of all Bodies, and continually *puts forth in them a Force* or *Power* entirely different from that by which *Matter* acts on *Matter.* This (he adds) is *an evident Demonstration, not only of the World's being made originally by a supreme intelligent Cause, but moreover that it depends every moment on some superior Being, for the Preservation of its Frame, and that all the great Motions in it are caused by some immaterial Power, not having originally impressed a certain Quantity of Motion upon Matter, but perpetually and actually exerting itself every Moment in every Part of the World: which preserving and governing Power gives a very noble Idea of* PROVIDENCE.

Dr. *Cheyne* demonstrates, That *Gravity,* or the *Attraction* of Bodies towards one another, cannot be mechanically accounted for. The *Planets* themselves cannot continue their Motions in their Orbs without it. It is not a Result from the *Nature* of *Matter,* because the Efficacy of *Matter* is communicated by *immediate Contact,* and it can by no means act at a distance. Whereas this Power of *Gravitation* acts at all Distances, without any *Medium* or Instrument for the Conveyance of it, and passes as far as the Limits of the Universe. *Matter* is indeed entirely *passive,* and can't either *tend* or *draw,* with regard unto other Bodies, no more than it can *move itself.* And what is essential to *Matter* cannot be intended

---

10. "Which Nature effects throughout the whole world, accomplishing everything by wisdom and intelligence."

or be remitted; but *Gravity* increases or diminishes reciprocally, as the Squares of the Distances are increased or diminished. 'Tis plain this universal Force of *Gravitation* is the Effect of the *Divine Power* and *Virtue*, by which the Operations of all *material Agents* are preserved. They that press for a *mechanical Account* of *Gravity*, advance a Notion of a *subtile Fluid*, unto the Motion whereof they would ascribe it. But then still those Parts of Matter must be destitute of *Gravity*, which were very unlikely! And this *Hypothesis* would still remove us but one Step further from *immechanical Principles;* for the Cause of the Motion of your *subtile Fluid*, this, *Gentlemen*, you must own to be *immechanical*. Since you must admit a *first Cause*, you had as good be sensible of it in this place. 'Tis *He* who does immediately impress on *Matter* this Property. There never was yet afforded unto the World (as my Doctor observes) a *System of Natural Philosophy* which did not require *Postulates*, that are not *mechanically* to be accounted for. The fewest any one pretends to, are, *the Existence of Matter*, and *the Impression of rectilinear Motions*, and *the Preservation of the Faculties of natural Agents*. No Man has pretended to fetch from the Principles of Mechanism an Account for these. The *Impression of an attractive Faculty upon Matter*, is no harder a *Postulate* than the rest. It is a *Matter of Fact*, that *Matter* is in possession of this Quality. And it can be referred unto nothing, but the Influence of that Glorious ONE, who is the *first Cause* of all Things.

"Behold, a continual Opportunity for a considerate and religious Man, to have a *Sense* of a Glorious GOD awaken'd in him! And what is a *Walk with God*,[11] but that *Sense* kept alive in every Step of our *Walk*? I am continually entertain'd with *weighty Body*, or *Matter* tending to the *Center of Gravity;* I feel it in *my own*. The *Cause* of this *Tendency*, 'tis the Glorious GOD. *Great GOD, Thou givest this Matter such a Tendency, and thou keepest it in its Operation*. There is no other Cause but the *Will* and *Work* of the Glorious GOD. I am now effectually convinc'd of that antient Confession, and must with Affection make it, *He is not far from every one of us*.[12] When I see any thing moving or settling that way that its *heavy Nature* carries it, I may very justly think, and I would often form the Thought, *it is the*

---

11. Perhaps adapted from Gen. 5:22, 24 and Mic. 6:8.
12. Acts 17:27.

Glorious GOD, who now carries this Matter such a way! When *Matter* sinks *downward,* my Spirit shall even *therefore* mount *upward,* in acknowledgment of the God who orders it. I will no longer complain, *Behold, I go forward, but He is not there, and backward, but I cannot perceive Him; on the Left-hand, where He doth work, but I cannot behold Him; He hideth himself on the Right-hand, that I cannot see Him.*[13] No, I am now taught where to meet with Him, even at *every turn. He knows the way that I take.*[14] I cannot stir *forward* or *backward,* but I *perceive* Him in the *Weight* of every *Matter;* on the *Left-hand* and on the *Right* I see Him *at work.* My *way* shall be to improve this as a *weighty* Argument for the Being of a God. I will argue from it, *Behold, there is a God, whom I ought for ever to love, and serve, and glorify.* Yea, and if I am *tempted* to the doing of any wicked thing, I may reflect, that it cannot be done without some Action, wherein the *Weight of Matter* operates. But then I may carry on the Reflection, *How near am I to that Glorious GOD, whose Commands I am going to violate! Matter keeps his Laws; but, O my Soul, wilt thou break 'em! How shall I do this Wickedness,*[15] *and therein deny the God, who not only is above, but also is most sensibly now exerting His Power in the very Matter, upon which I make my criminal Misapplications!"*

¶. Before we go any further, it appears high time to introduce an Assertion or two of that excellent Philosopher Dr. *Cheyne,* in his *Philosophical Principles of natural Religion.* He asserts, and with Demonstration, (for truly without *that* he asserts nothing!) that there is no such thing as an *universal Soul,* animating the vast System of the World, according to *Plato;*[16] nor any *substantial Forms,*[17] according to *Aristotle;* nor any omniscient *radical Heat,*[18] according to *Hippocrates;* nor any *plastick Virtue,*[19] according to

---

13. Job 23:8–9.
14. Job 23:10.
15. Adapted from Gen. 39:9.
16. A leading feature of Plato's thought was the concept of *anima mundi* or "world-soul." Plato held that the soul of the world extends from center to circumference and is coeval with the world's body. This idea is discussed in the *Timaeus.*
17. Substance was for Aristotle the basic ontological unity. He regarded the substantial element in things as form or essence. Substantial form and essence have the same meaning for Aristotle.
18. According to Hippocrates, the essential factor in life is heat. An "innate heat" pervades the entire body and maintains itself in equilibrium.
19. "Plastick virtue" is another name for plastic nature. On this concept see Essay 26, n. 2.

Scaliger;[20] nor any *hylarchick Principle*, according to *More*.[21] These are mere *allegorical* Terms, coined on purpose to conceal the Ignorance of the Authors, and keep up their Credit with the credulous Part of Mankind. These *unintelligible Beings* are derogatory from the Wisdom and Power of the Great GOD, who can easily *govern* the Machine He could *create*, by more direct Methods than employing such subservient *Divinities;* and indeed these Beings will not serve the Design for which we invent them, unless we endow them with Faculties above the Dignity of *secondary Agents*. It is now plain from the most *evident Principles*, that the Great GOD not only has the *Springs* of this immense *Machine*, and all the several Parts of it, in his own Hand, and is the *first Mover;* but that without His *continual Influence* the whole Movement would soon fall to pieces. Yet besides this, He has reserved to Himself the power of *dispensing* with these *Laws*, whenever He pleases.

My Doctor has made it evident, That it is not essential to *Matter* to be either in *Rest* or in *Motion:* But tho there is in *Matter* a *Vis inertiæ*,[22] by which all Bodies resist, to the utmost of their power, any *Change* of their State, whether of *Rest* or *Motion;* yet this *Vis* is not essential to *Matter*, but a *positive Faculty* implanted therein by the Author of Nature. It is therefore evident that the Preservation of a *Body* in *Rest* or in *Motion* (after the first Instant) absolutely depends on the Almighty GOD, as the Cause. No part of *Matter* can move itself, nor when put into *motion*, is this *Motion* absolutely essential to its Being, nor does depend upon itself; and therefore the *Preservation* of this *Motion* must have its Dependance on some other Cause. But there is no other Cause assignable besides the *omnipotent Cause*, who preserves the Being and Faculties of all natural Agents.

Great GOD, on the Behalf of all thy Creatures, I acknowledge in Thee we move and have our Being![23]

---

20. That is, Julius Caesar Scaliger.

21. Hylarchic means "ruling over matter." Henry More introduced the term "hylarchick principle" as early as 1676 to designate a spiritual power which regulates matter. It is approximately equivalent to "plastic nature."

22. "Force of inertia."

23. The first part perhaps alludes to Gen. 1:29-30; the last part is adapted from Acts 17:28.

ESSAY XXII. *Of the* WATER.

PURE *Water* is a Fluid void of all Sapor, and seems to consist of small, smooth, round and porous Particles, that are of equal *Diameters* and equal *Gravities*. There are also between them Spaces, that are so large, and ranged in such a manner, as to be on all sides pervious. Their *Smoothness* accounts for their sliding easily over the Surfaces of one another. Their *Roundness* keeps them from touching one another in more Points than one. So great is their *Porosity*, that there is at least forty times as much *Space* as *Matter* in *Water*. For *Water* is nineteen times specifically lighter than *Gold;* but *Gold* will by Pressure let Water thro its Pores, and has doubtless more *Pores* than *solid Parts*.

Dr. *Wainwright* observes, The compounding Particles of *Water* are less than those of *Air;* the former will pass thro several Bodies that the latter will not; it will force itself thro the *Skins* of Animals, even after they are dried and converted into Leather. Fasten a strong Rope, of what length you please, to an Hook; at the bottom of the *Cord* hang any *Weight* short of what will break it, tho' ever so great; you will find the Weight will *rise* in *moist* Weather, and *sink* in *dry*. You may also raise the *Weight,* by moistening the sides of the *Cord* with a wet Sponge. Thus a few Particles of Water may overcome any *finite Resistance,* if a *Cord* will bear it. Now since there is but a little Quantity of Water in this Experiment, and this is driven into the sides of the *Cord* with a Force no greater than the Weight of a Cylinder of Air incumbent on the *Water,* therefore the *Water* must act by a Property, whereby its Force is greatly augmented; and this can be no other than that of the *Cuneus:*[1] And the Forces of *Wedges* are to one another reciprocally proportional to the *Angles* their Edges do make. But in *Spheres* the greater or lesser Degree of *Curvity* is to be considered as their *Angles*, when *Spheres* are considered as *Wedges,* and the Degrees of *Curvity* in *Spheres* are reciprocally as their *Radii*. Now the Particles of *Water* being so inconceivably small, much less than those of *Air,* they must, when acting as *Wedges,* have their Powers inconceivably increased, so as to overcome any *finite Resistance*.

If such Power is in a Particle of Water, what is Thy Power, O Thou infinite Maker of that, and all things!

---

1. That is, a triangular prism or a wedge.

Dr. *Cheyne* observes, That the Quantity of Water on the outside of our Globe doth daily decrease, part of it being every day turn'd into *Mineral, Vegetable,* and *Animal* Substances, which are not easily dissolved again into their component Parts.

It is a Curiosity demonstrated by *Mariotte,* in his *Du Movement des Eaux,* That a *Jet-d'eau* never will rise as high as its Reservatory, but always fall short of it by a Space, which is the *Subduplicate Ratio* of that *Height.*

In the *Congregations* of *Water,* and the *Distributions* of it over our Globe, we cannot but see the wonderful Wisdom and Goodness of our GOD.[2] *The great and wide Sea, wherein are swimming Things innumerable, 'tis full of Thy Riches, O our GOD!* And the Uses of it are marvellous. *The Waters are in the Place which Thou, O our God, hast prepared for them: Thou hast set a Bound that they may not pass over.*[3]

A fanciful and presumptuous Gentleman having made his Exceptions against the Proportion of *Water* to *dry Land* on our Globe, is well answer'd by Mr. *Keil;*[4] That the Objections proceed from a deep Ignorance of *Natural Philosophy.* For if there were but half the *Sea* that now is, there would be but half the *Vapours;* and we should soon find our miserable want of these.

Mr. *Ray* assures us, That where the bottom of the Sea is not rocky, but Earth, Ouze, or Sand, which is incomparably the greatest part of it, it is by the Motion of the Waters, as far as the Reciprocation of the *Sea* extends to the bottom, every where brought unto a Level; that is to say, it has an *equal and uniform Descent* from the Shores to the Deeps.

That the *Motion of the Water* descends to a good Depth, is proved from the *Plants,* that grow deepest in the Sea; which all generally grow flat, in manner of a *Fan,* and not with Branches on all sides like *Trees:* a thing that is contrived by the Divine Providence, for that the Edges of them do in that posture, with most ease, cut the Water flowing to and fro. Probably in the greater Depths of the Sea there grow no *Plants* at all; the Bottom is probably too remote for the external *Air* to pass in a sufficient

---

2. Adapted from Ps. 104:25,24.
3. Adapted from Ps. 104:8–9.
4. The "presumptuous Gentleman" was Thomas Burnet (c. 1635–1715), author of *Telluris theoria sacra,* 2 vols. (London, 1681). An English translation was entitled *The Theory of the Earth,* 2 vols. (London, 1684); later editions bore the title *The Sacred Theory of the Earth.* Mr. Keil is John Keill.

Quantity thither. Nay, we are told that in those *deep Seas* there are *no Fish* at all; their Spawn would be lost there: being lighter than the Water, it will not sink thither; and the Climate there may be too cold for the quickening of it.

According to Mr. *Halley*'s Experiment, *Water* as warm as the *Air* in the Summer, will in *twelve Hours* exhale the *tenth part* of an *Inch*. This Quantity will be found abundantly sufficient for all the *Rains,* and all the *Dews,* and all the *Springs* in the World; and will account for the *Caspian Sea,* and our vast *Canadian* Lakes, being always at a stand; and for the *Current,* said always to set in at the Straits of *Gibraltar,* tho the *Mediterranean Sea* receive so many Rivers. Every *ten square Inches* of the Surface of the Water, yields in Vapour *per diem* [we allow it only for the time the *Sun* is up] a *Cube Inch* of Water. Every Mile will yield 6914 Tons. A square Degree of sixty-nine *English* Miles will yield thirty-three Millions of Tons. If the *Mediterranean Sea* be estimated at forty Degrees long, and four broad, which is the least, the whole *Mediterranean* must lose in Vapours in a Summer's-day at least 5280 Millions of Tons. And yet sometimes the *Winds* lick up the Surface of Water faster than it exhales by the Heat of the Sun. The *Mediterranean Sea* receives nine considerable Rivers. We will suppose each of them to bring down ten times as much Water as the River *Thames,* which they do not; but this will allow for the small Rivulets. The *Thames,* allowing the Water to run after the rate of two Miles an Hour, may yield 20,300,000 Tons *per diem*. Allow as before, and all the nine Rivers bring down 1827 Millions of Tons in a day. This is but little more than a Third of what is proved to be evaporated out of the *Mediterranean* in twelve Hours time.

The astonishing *Flux* and *Reflux* of the *Sea,* what Benefits it affords unto the World! If the *Ocean* once were stagnated, first all the Places towards the Shore would be turned into a *Mephitis;*[5] and then by degrees it would yet further corrupt, until the whole became as poisonous as the Lake of *Sodom.*[6] The *Fishes* would be

---

90 Straits/Streights

5. That is, a noxious or pestilential emanation, especially from the earth; a noisome or poisonous stench.
6. That is, the Dead Sea, called the Sea of Sodom in the Talmud. Its surface is about 1290 feet below the Mediterranean Sea. The water is a concentrated lye, impregnated with mineral salts.

first hereby destroyed, and by the poisonous *Steams,* anon the *Plants* and *Animals* would share in the Destruction. In the *Tide* of the Sea the Waters are lifted up in an Heap, and then let fall again. So the fear'd Corruption is prevented: And how many Conveniences afforded for our *Navigation!* But what? Oh! what the Original of it? Where's *the Zaphnath Paaneah*[7] who shall enlighten us?

On our Globe all Bodies have a Tendency towards the *Center* of it. And such a *Gravitation* there is towards the Center of the *Sun,* and of the *Moon,* and of all the *Planets.* There is cause to suspect that the Force of *Gravity* is, in the Celestial Globes, proportional to the Quantity of Matter in each of them. The *Sun,* for instance, being more than ten thousand times as big as the *Earth,* its *Gravitation,* and the attracting Force of it, is ten thousand times as much as that of the Earth, acting on Bodies at the same Distances.

If our Globe were alone, or not affected by the Actions of the *Sun* and the *Moon,* the Ocean, equally pressed by the Force of *Gravity* towards the Center, would continue in a perfect *Stagnation,* always at the same height, without ever *ebbing* or *flowing.* But it is demonstrated, that the *Sun* and the *Moon* have a like Principle of *Gravitation* towards their Centers, and our Globe is also within the Activity of their Attractions. Whence it will follow, that the Equality of the Pressure of *Gravity* towards the Center will be thereby disturbed. And tho the Smallness of these Forces, in respect of the *Gravitation* towards the Center of the Earth, render them imperceptible, yet the *Ocean* being fluid, and yielding to the least Force, by its *rising* shews where there is the least Pressure upon it, and where it is most pressed, by *sinking.* Accordingly we shall find, that where the *Moon* is perpendicularly either above or below the Horizon, there the Force of *Gravity* is most of all diminished, and consequently that there the *Ocean* must necessarily swell, by the coming in of the Water from those Parts where the Pressure is greatest, namely, in those where the *Moon* is near the *Horizon.* The *Sea,* which otherwise would be *spherical,* upon the Pressure of the *Moon* must form itself into a *spheroidal* or *oval Figure,* whose longest Diameter is where the *Moon* is vertical, and shortest where she is in the Horizon; and the *Moon* shifting her Position as she turns round our Globe once a day, this *Oval* of

---

7. "Revealer of secrets." See Gen. 41:45.

Water shifts with her, occasioning thereby the two *Floods* and *Ebbs* observable in each five and twenty Hours. The *Spring-Tides* upon the *New* and *Full Moons,* and the *Neap-Tides* upon the Quarters, are occasion'd by the attractive Force of the *Sun* in the *New* and *Full,* conspiring with the Attraction of the *Moon,* and producing a *Tide* by their *united Forces.* Whereas in the Quarters the *Sun* raises the Water where the *Moon* depresses, and on the contrary; so as the *Tides* are made only by the difference of their *Attraction.* The *Sun* and *Moon* being either conjoin'd or opposite in the *Equinoctial,* produce the greatest *Spring-Tides.* The subsequent *Neap-Tides* being produced by the *Tropical Moon* in the Quarters, are always the *least Tides.*

But then from the *Shoalness* of the *Water* in many places, and from the *Narrowness* of the *Straits,* by which the Tides are in many places propagated, there arises a mighty Diversity, which, without the Knowledge of the Places, cannot be accounted for.

Dr. *Cheyne* has taught me to take notice of one thing more. If our *Earth* had any more than *one Moon* attending it, we should receive probably a Detriment from it, rather than an Advantage. For at the *Conjunction* and *Opposition* with one another, and with the *Sun,* we should have *Tides* that would raise the Waters to the Tops of our Mountains, and in their *Quadratures* we should have no *Tides* at all.

O my Soul, beholding the Moon above, look up to God, who hath so wisely proportion'd her, for the Designs on which He placed her there.

The *Sea* is the grand Fountain of those *fresh Waters,* which supply and enrich the *Earth,* and by convenient Channels are carried back to the place from whence they came; the *perpetui Fontes, vitæque perennis Imago:*[8] How equally are these fresh Waters distributed? How few *Antiguas*[9] in the World? How agreeably are they disposed? And what a prodigious Run have many of the Rivers? The *Danube,* in a sober Account, as *Bohun* computes, runs fifteen hundred Miles in a strait Line from its Rise to its Fall. The *Nile,* according to *Varenius,* allowing for Curvatures, runs three thousand Miles; and the *Niger* two thousand four hundred; the

83 *Antiguas / Antigua's*

8. "Perpetual springs, and a likeness of perennial life."
9. The island of Antigua in the West Indies is known for scarcity of water. It has no rivers, few springs, and suffers prolonged droughts.

Ganges twelve hundred; the *Amazonian* above thirteen hundred *Spanish* Leagues.

¶. "But is it not high time for us to hear *the Voice of many Waters!*[10]

"One[11] celebrating the Bounty of our God unto us in the *Water,* so expresses it: *Quo Thesauro vel unicum Elementum Aquæ, si Deus illud in Sanguinem, ut olim in* Egypto, *converteret, possemus redimere?*[12] The Contemplation may be carried unto the Element that is next above it."

An excellent Person, who writes *Augustissimam Naturæ Scholam,*[13] has thus rendred something of it articulate: *O Homo, ne imitare Equos & Mulos, qui me quidem bibunt, sed tantum bibunt. At tu, cui melior est Anima, ita me bibe, ut non tantum bibas, sed benedicentem Deum habeas dum bibis. Habebis autem si agnoscas ipsius Majestatem, eamque colas.*[14]

Long since have we been taught such Notes as these. "*O Lord, how manifold are thy Works! In Wisdom hast thou made them all. The Earth is full of thy Riches. And so is the great and wide Sea, wherein are swimming things innumerable.*[15]

"But can we look on the *Sea,* and not see a Picture of *a troublesome World;* see and be instructed."

APPENDIX.

§ WE can scarce leave the *Water* without some Remarks on our *Fluids;* and we will be more particularly indebted to Dr. Cheyne for hinting them. First, how *frugal* is Nature in *Principles,* and yet how *fruitful* in *Compositions* and in *Consequences!* The *primary*

14 them. First, how/them first. How

10. Rev. 14:2, 19:6; see also Ezek. 43:2 and Rev. 1:15.
11. Mather's reference is to Johann Arndt.
12. "With what treasure would we be able to buy back the unique element of water, if God, as he once did in Egypt, were to turn it into blood?" The reference is to the first of the plagues which God sent upon Egypt to persuade the Pharaoh to let the children of Israel go. See Exod. 7:20.
13. That is, Johann H. Alsted.
14. "O man, do not imitate horses and mules which indeed drink me [water], but which drink so much. But you, whose soul is better, drink me in such a way that you do not drink so much, but so that you may have the benediction of God while you drink. Moreover, you will have it, if you recognize his majesty and worship it."
15. Ps. 104:24–25.

Fluids are but *four*, *Water* and *Air*, and *Mercury* and *Light*. 'Tis but seldom that three of these are much compounded with others. 'Tis *Water* alone, 'tis *Lymph*, that is mostly the *Basis* of all other Mixtures; and it is the Parts of solid Bodies floating in this Fluid that produce all our pleasant and useful Varieties of Liquors.

Again, How vast the difference between the *specifick Gravities* of our *Fluids!* *Mercury* is about eight thousand times heavier than *Air*. *Air* must have choak'd us, if it had been half so heavy as *Mercury*. And yet Mankind, in its present Circumstances of the *Blood-Vessels*, under frequent *Obstructions*, could not well have done without such an *heavy Fluid* as *Mercury*.

Thirdly, All *Fluids* agree in the condition of the direction of their *Pressure* upon the sides of the containing Vessel. This *Pressure* is for ever communicated in Lines *perpendicular* to the sides of the containing Vessel. This beautiful and uniform Property of all *Fluids* necessarily follows from the *Sphericity* of their constituent Particles.

Our Doctor's Conclusion is as I would have it. "Now could any thing but the Almighty *Power* of God have rounded those infinite numbers of small Particles whereof *Fluids* consist? Or could any thing but his *Wisdom* have assigned them their true Dimensions, their exact Weights, and required Solidities?"

I beseech you, *Sirs*, by what Laws of *Mechanism* were all the Particles of the several *Fluids* turned of differing *Diameters*, differing *Solidities*, differing *Weights* from one another; but all of the same *Diameters*, and *Solidities*, and *Weights* among themselves? *This is the Finger of God!*[16] It is a just Assertion of Dr. *Grew*, *The Regularity of Corporeal Principles shews that they come at first from a Divine Regulator.*

ESSAY XXIII. *Of the* EARTH.

THE *Lord by Wisdom has founded the Earth.*[1] A poor Sojourner on the *Earth* now thinks it his Duty to behold and admire the *Wisdom* of his glorious Maker there.

The *Earth*, which is the Basis and Support of so many Vegetables

---

16. Exod. 8:19.
1. Prov. 3:19.

Essay 23. Of the Earth

and Animals, and yields the alimentary Particles, whereof *Water* is the Vehicle, for their Nourishment: *Quorum omnium* (as *Tully* saith well) *incredibilis Multitudo, insatiabili Varietate distinguitur.*[2]

The various Moulds and Soils of the Earth declare the admirable Wisdom of the Creator, in making such a provision for a vast variety of Intentions. *God said, Let the Earth bring forth!*[3]

And yet,

Nec vero Terræ ferre omnes omnia possunt.[4]

It is pretty odd; they who have written *de Arte Combinatoria*,[5] reckon of no fewer than one hundred and seventy-nine Millions, one thousand and sixty different sorts of Earth: But we may content ourselves with Sir *John Evelyn*'s Enumeration, which is very short of *that*.

However, the *Vegetables* owe not so much of their Life and Growth to the *Earth* itself, as to some agreeable Juices or Salts lodg'd in it. Both Mr. *Boyle* and *Van Helmont*,[6] by Experiments, found the Earth scarce at all diminished when *Plants*, even *Trees*, had been for divers Years growing in it.

The *Strata* of the Earth, its *Lays* and *Beds*, afford surprizing Matters of Observation; the *Objects* lodged in them; the *Uses* made of them; and particularly the *Passage* they give to *sweet Waters*, as being the *Colanders* wherein they are sweetned. It is asserted that there are found all to lie very much according to the Laws of *Gravity*. Mr. *Derham* went far to demonstrate this Assertion.

The *vain Colts of Asses*, that *fain would be wise*,[7] have cavill'd at the *unequal Surface of the Earth*, have open'd against the *Mountains*, as if they were *superfluous Excrescences;* but *Warts* deforming the

71 Colanders/Calanders

2. "All of them incredibly numerous and inexhaustibly varied and diverse."
3. Gen. 1:11, 24.
4. "Nor yet can all soils bear all fruits."
5. This art aimed at combining the subjects of knowledge in such a way as to enable humans to acquire a summary, general knowledge of all things. The "combinatorial art" was a stage in the historical development of the encyclopedia. At the time Mather wrote, leading works on the subject were Gottfried Wilhelm von Leibniz's *Dissertatio de arte combinatoria* (Leipzig, 1666) and Athanasius Kircher's *Ars magna sciendi*, 2 vols. in 1 (Amsterdam, 1669).
6. That is, Johannes Baptista van Helmont.
7. Adapted from Job 11:12. The main target of this paragraph is Thomas Burnet (c. 1635-1715), who denied any signs of art and skill in the making of the globe.

Face of the Earth, and Proofs the *Earth* is but an Heap of Rubbish and Ruins. *Pliny* had more of Religion in him.

The sagacious Dr. *Halley* has observed, That the Ridges of *Mountains* being placed thro the midst of their Continents, do serve as *Alembicks*,⁸ to distil fresh Waters in vast Quantities for the Use of the World: And their *Heights* give a Descent unto the *Streams,* to run gently, like so many Veins of the *Macrocosm,* to be the more beneficial to the Creation. The generation of *Clouds,* and the distribution of *Rains,* accommodated and accomplished by the *Mountains,* is indeed so observable, that the learned *Scheuchzer* and *Creitlovius* can't forbear breaking out upon it with a *Mirati summam Creatoris Sapientiam!*⁹

What *Rivers* could there be without those admirable *Tools of Nature!*

Vapours being raised by the *Sun,* acting on the Surface of the *Sea,* as a *Fire* under an Alembick, by rarefying of it, makes the lightest and freshest Portions thereof to rise first; which *Rarefaction* is made (as Dr. *Cheyne* observes) by the insinuation of its active Particles among the porous Parts thereof, whereby they are put into a violent Motion many different ways, and so are expanded into little Bubbles of larger Dimensions than formerly they had; and so they become specifically lighter, and the weightier *Atmosphere* buoys them up. The Streams of these *Vapours* rest in places where the Air is of equal *Gravity* with them, and are carried up and down the *Atmosphere* by the course of that Air, till they hit at last against the sides of the *Mountains,* and by this Concussion are condensed, and thus become heavier than the Air they swum in, and so gleet¹⁰ down the rocky Caverns of these *Mountains,* the inner parts whereof being hollow and stony, afford them a *Bason,* until they are accumulated in sufficient Quantities, to break out at the first *Crany:*¹¹ whence they descend into Plains, and several of them uniting, form Rivulets; and many of those uniting, do grow into *Rivers.* This is the Story of them; this their *Pedigree!*

Minerals are dug out of *Mountains;* which, if they were sought only in level Countries, the Delfs would be so flown with Waters,

---

8. An alembic is anything that distills, refines, or purifies.
9. "We admired the great wisdom of the Creator."
10. That is, ooze or flow slowly.
11. Or cranny. A notch, cleft, or niche; a transverse fissure in strata.

that it would be impossible to make *Addits* or *Soughs*[12] to drein them. Here is, as *Olaus Magnus* expresses it, *Inexhausta pretiosorum Metallorum ubertas.*[13]

A *German* Writer,[14] got upon the *Mountains,* gives this Account of them: *Sunt ceu tot naturales Fornaces Chymicæ, in quibus Deus varia Metalla & Mineralia excoquit & maturat.*[15]

The *Habitations* and *Situations* of Mankind are made vastly the more comfortable for the *Mountains.* There is a vast variety of *Plants* proper to the *Mountains:* and many Animals find the *Mountains* their most proper places to breed and feed in. *The highest Hills a Refuge to the wild Goats!*[16] A Point Mr. *Ray* has well spoken to.

They report that *Hippocrates* did usually repair to the *Mountains* for the *Plants,* by which he wrought the chief of his Cures.

Mountains also are the most convenient Boundaries to Territories, and afford a Defence unto them. One[17] calls them *the Bulwarks of Nature, cast up at the Charges of the Almighty; the Scorns and Curbs of the most victorious Armies.* The *Barbarians* in *Curtius*[18] were confidently sensible of this!

Yea, we may appeal to the Senses of all Men, whether the grateful Variety of *Hills* and *Dales* be not more pleasing than the largest continued *Plains.*

'Tis also a *salutary Conformation* of the Earth; some Constitutions are best suited *above,* and others *below.*

Truly these massy and lofty Piles can by no means be spared.

Galen, thou shalt chastize the *Pseudo-Christians,* who reproach the Works of God. Say! *Accusandi sane mea Sententia hic sunt Sophistæ, qui cum nondum invenire neque exponere Opera Naturæ queant, eam tamen inertia atque inscitia condemnant.*[19]

---

12. A *delf* is a trench or ditch, a hole or cavity dug in the earth for irrigation or drainage. An *adit* is an approach, specifically, a horizontal opening by which a mine is entered or drained. The verb *sough* means to make a drain in land, to drain by constructing proper channels.

13. "An inexhaustible plenty of precious metals."

14. That is, Johann Arndt.

15. "They are like so many natural chemical furnaces in which God tempers and matures various metals and minerals."

16. Adapted from Ps. 104:18.

17. The identity of "One" is almost certainly unknown to Mather.

18. Quintus Curtius Rufus (first century A.D.), a Roman historian, wrote a ten-book history of Alexander the Great.

19. "Justly then, in my opinion, it is something to wonder at that the sophists, who have not as yet been able to discover the works of Nature, still accuse her of a lack of skill."

Say now, *O Man*, say, under the sweet Constraints of Demonstration, *Great GOD, the Earth is full of thy Goodness!*[20]

And Dr. *Grew* shall carry on the more general Observation for us. "How little is the Mischief which the *Air, Fire,* or *Water* sometimes doth, compared with the innumerable *Uses* to which they daily serve? Besides the *Seas* and *Rivers,* how many *wholesome Springs* are there for one that is *poisonous?* Are the Northern Countries subject to *Cold?* They have a greater plenty of *Furs* to keep the People warm. Would those under or near the Line be subject to *Heat?* They have a constant *Easterly Breeze,* which blows strongest in the Heat of the Day, to refresh them: And with this Refreshment *without,* they have a variety of excellent *Fruits* to comfort and cool them *within.* How admirably are the *Clouds* fed with Vapours, and carried about with the *Winds,* for the gradual, equal, and seasonable watering of most Countries? And in those which have less *Rain,* how abundantly is the want of that supplied with noble *Rivers?*"

Even the subterraneous *Caverns* have their Uses. And so have the *Ignivomous*[21] *Mountains:* Those terrible things are *Spiracles,*[22] to vent the *Vapours,* which else might make a dismal Havock. Dr. *Woodward* observes, That tho Places which are very subject unto *Earthquakes* usually have these *Volcano's,* yet without these *fiery Vents* their *Earthquakes* would bring more tremendous Desolations upon them.

Those two flamnivomous[23] Mountains, *Vesuvius* and *Ætna,* have sometimes terrified the whole World with their tremendous Eruptions. *Vesuvius* transmitted its frightful Cinders as far as *Constantinople,* which obliged the Emperor to leave the City; and Historians tell us there was kept an Anniversary Commemoration of it. *Kircher* has given us a Chronicle of what furious things have been done by *Ætna*; the melted Matter which one time it poured forth, spread-

---

66 flamnivomous/flammivomous

20. Adapted from Ps. 33:5 and perhaps 104:24.
21. That is, "fire-vomiting" mountains, or volcanoes.
22. That is, a small opening by which a confined space has communication with the outer air. An opening in the ground affording egress to subterranean vapors or fiery matter.
23. That is, "flame-vomiting."

Frontispiece from *The Vulcano's: Or, Burning and Fire-vomiting Mountains, Famous in the World: With their Remarkables, Collected for the most part out of Kircher's Subterraneous World* (London, 1669). The upper disc pictures subterraneous fire mixed with water, subterraneous aqueducts, and the concoction of subterrestrial waters through fire ducts. The lower disc depicts subterraneous "fire houses" whose "breath holes" are volcanoes. In the corners (clockwise from lower left) are volcanoes of the Phlegraean Plains, Aetna, Vesuvius, and Stromboli. Courtesy of the Rare Book and Special Collections Library, University of Illinois at Urbana-Champaign.

ing in breadth six Miles, ran down as far as *Catania*, and forced a Passage into the Sea.[24]

Asia abounds in these *Volcano's*. *Africa* is known to have eight at least. In *America* 'tis affirmed that there are no less than fifteen, among that vast Chain of Mountains called the *Andes*. One[25] says, "Nature seems here to keep house under ground, and the Hollows of the *Mountains* to be the *Funnels* or *Chimneys*, by which the fuliginous[26] Matter of those everlasting Fires ascends."

The *North* too, that seems doom'd unto *eternal Cold*, has its famous *Hecla*.[27] And *Bartholomew Zenet* found one in *Greenland*, yet nearer to the Pole; the Effects whereof are very surprizing.

A reasonable and religious Mind cannot behold these formidable *Mountains*, without some Reflections of this importance: *Great GOD, who knows the Power of thine Anger? Or what can stand before the powerful Indignation of that God, who can kindle a Fire in his Anger that shall burn to the lowest Hell, and set on fire the Foundations of the Mountains!*[28]

The *Volcano's* would lead us to consider the *Earthquakes*, wherein the *Earth* often suffers violent, and sometimes very destructive Concussions.

The History of Earthquakes would be a large, as well as a sad Volume. Whether a *Colluctation*[29] *of Minerals* in the Bowels of the Earth is the cause of those direful Convulsions, may be considered: As we know a Composition of Gold which *Aqua Regia*[30] has dissolved, *Sal Armoniack*, and *Salt of Tartar*,[31] set on fire, will with an horrible crack break thro all that is in the way. But Mankind ought herein to tremble before the Justice of God. Particular *Cities* and

73 Catania / Catanea

24. [Athanasius Kircher], *The Vulcano's: Or, Burning and Fire-Vomiting Mountains, Famous in the World: with Their Remarkables. Collected for the Most Part out of Kircher's Subterraneous World. . . . Upon the Relation of the Late Wonderful and Prodigious Eruptions of Aetna* (London, 1669).
25. Mather again quotes from a source without identifying the author.
26. That is, resembling soot; sooty.
27. A volcano in southwest Iceland.
28. Adapted from Ps. 90:11, Nah. 1:6, and Deut. 32:22.
29. That is, a wrestling or struggling together; a conflict.
30. "Royal water." A mixture of concentrated nitric acid and concentrated hydrochloric acid in the ratio of one to three respectively, named by the alchemists because of its ability to dissolve platinum as well as the royal metal, gold.
31. Sal armoniack (or sal ammoniac) is ammonium chloride, a white or colorless cubic solid, a nitrogen-containing salt, which is very soluble in water. Salt of tartar or potash is potassium carbonate.

Countries, what fearful Desolations have been by Earthquakes brought upon them!

The old sinking of *Helice* and *Buris,* absorbed by *Earthquakes* into the Sea, mention'd by *Ovid,* or the twelve Cities that were so swallow'd up in the Days of *Tiberius,* are small things to what *Earthquakes* are to do on our Globe; yea, have already done.[32] I know not what we shall think of the huge *Atlantis,* mentioned by *Plato,* now at the bottom of the *Atlantick* Ocean: But I know *Varenius* thinks it probable, that the Northern Part of *America* was joined unto *Ireland,* till Earthquakes made the vast and amazing Separation. Others have thought so of *England* and *France;* of *Spain* and *Africa;* of *Italy* and *Sicily.*[33]

Ah, *Sicily!* Art thou come to be spoken of? No longer ago than t'other day what a rueful Spectacle was there exhibited in the Island of *Sicily* by an *Earthquake,* in which there perished the best part of two hundred thousand Souls!

Yea, *Ammianus Marcellinus* tells us, in the Year 365, *Horrendi Tremores per omnem Orbis Ambitum grassati sunt.*[34]

O Inhabitants of the Earth, how much ought you to fear the things that will bring you into ill Terms with the Glorious GOD! *Fear,* lest the *Pit* and the *Snare* be upon you![35] Against all other Strokes there may some Defence or other be thought on: There is none against an *Earthquake!* It says, *Tho they hide in the top of* Carmel, *I will find them there!*[36]

But surely the *Earthquakes* I have met with will effectually instruct me to avoid the Folly of setting my Heart inordinately on any *Earthly* Possessions or Enjoyments. Methinks I hear Heaven saying, *Surely he will receive this Instruction!*[37]

A modern Philosopher[38] speaks at this rate, "We do not know when and where we stand upon *good Ground:* It would amaze the stoutest Heart, and make him ready to die with Fear, if he could see into the *subterraneous World,* and view the dark Re-

---

32. Helice and Buris were cities in Achaea, which in ancient times was the northeast coast of the Peloponnese and southeast Thessaly. Tiberius Julius Caesar Augustus (42 B.C.–A.D. 37) was Roman emperor from 14 A.D. to his death.
33. The concept of joined continents and their separation by earthquakes in Varenius and others is an interesting anticipation of modern ideas on continental drift.
34. "Horrible phenomena suddenly spread through the entire extent of the world."
35. Adapted from Isa. 24:17.
36. Adapted from Amos 9:3.
37. Perhaps adapted from Zeph. 3:7; see also Prov. 8:10.
38. The identity of this "modern Philosopher" remains elusive.

cesses of Nature under ground; and behold, that even the strongest of our Piles of Building, whose Foundation we think is laid firm and fast, yet are set upon an Arch or Bridge, made by the bending Parts of the Earth one upon another, over a prodigious Vault, at the bottom of which there lies an unfathomable Sea, but its upper Hollows are filled with stagnating Air, and with Expirations of sulphureous and bituminous Matter. Upon such a *dreadful Abyss* we walk, and ride, and sleep; and are sustained only by an *arched Roof,* which also is not in all places of an equal Thickness."

Give me leave to say, I take *Earthquakes* to be very *moving Preachers* unto *worldly-minded Men:* Their Address may be very agreeably put into the Terms of the Prophet; *O Earth, Earth, Earth, hear the Word of the Lord!*[39]

"*Chrysostom* did well, among his other Epithets, to call the Earth *our Table;* but it shall *teach* me as well as *feed* me: May I be a *Deipnosophist*[40] upon it.

"Indeed, what is the Earth but a *Theatre,* as has been long since observed? *In quo Infinita & Illustria, Providentiæ, Bonitatis, Potentiæ ac Sapientiæ Divinæ Spectacula contemplanda!*[41] But I must not forget that this *Earth* is very shortly to be my *sleeping-place;* it has a *Grave* waiting for me: *I will not fear to go down, for thou hast promised, O my Saviour, to bring me up again.*"[42]

APPENDIX.

§. HAVING arrived thus far, I will here make a Pause, and acknowledge the Shine of Heaven on *our Parts of the Earth,* in the Improvements of our *modern Philosophy.*

To render us the more sensible hereof, we will propose a few Points of the *Mahometan Philosophy,* or Secrets reveal'd unto *Mahomet,* which none of his Followers, who cover so much of the Earth at this Day, may dare to question.

39. Jer. 22:29.
40. That is, "master of the art of dining." Note Mather's pun.
41. "In which the infinite and remarkable spectacles of Divine Providence, Goodness, Power, and Wisdom are to be viewed."
42. Adapted from Gen. 46:3-4.

The *Winds;* 'tis an *Angel* moving his *Wings* that raises them.

The *Flux* and *Reflux* of the *Sea,* is caused by an *Angel's* putting his Foot on the middle of the *Ocean,* which compressing the Waves, the Waters run to the Shores; but being removed, they retire into their proper Station.

Falling Stars are the *Firebrands* with which the *good Angels* drive away the *bad,* when they are too saucily inquisitive, and approach too near the Verge of the Heavens, to eaves-drop the Secrets there.

Thunder is nothing else but the cracking of an *Angel's Whip,* while he slashes the dull Clouds into such and such places, when they want *Rains* to fertilize the Earth.

Eclipses are made thus: The *Sun* and *Moon* are shut in a *Pipe,* which is turned up and down; from each Pipe is a Window, by which they enlighten the World; but when God is angry at the Inhabitants of it for their Transgressions, He bids an *Angel* clap to the Window, and so turn the Light towards Heaven from the Earth: for this Occasion *Forms of Prayer* are left, that the Almighty would avert his Judgments, and restore Light unto the World.

The thick-skull'd Prophet sets another *Angel* at work for *Earthquakes;* he is to hold so many *Ropes* tied unto every Quarter of the Globe, and when he is commanded, he is to pull; so he shakes that part of the Globe: and if a City, or Mountain, or Tower, is to be overturned, then he tugs harder at the Pulley, till the Rivers dance, and the Valleys are filled with Rubbish, and the Waters are swallowed up in the Precipices.

May our Devotion exceed the Mahometan *as much as our Philosophy!*

ESSAY XXIV. *Of* MAGNETISM.

SUCH an unaccountable thing there is as *the* MAGNETISM *of the Earth.* A Principle very different from that of *Gravity.*

The Operations of this amazing Principle, are principally discovered in the communion that *Iron* has with the *Loadstone;* a rough, coarse, unsightly Stone, but of more Value than all the *Diamonds* and *Jewels* in the Universe.

It is observed by *Sturmius,* That the *attractive Quality* of the *Magnet* was known to the Antients, even beyond all History. Indeed,

99 besides what *Pliny* says of it, *Aristotle* speaks of *Thales,* as having said, the *Stone* has a *Soul,* ὅτι τὸν σίδηρον κινεῖ *because it moves Iron.*

It was *Roger Bacon* who first of all discovered the *Verticity* of the *Magnet,* or its Property of pointing towards the *Pole,* about four hundred Years ago.

The Communication of its Vertue to *Iron* was first of all discovered by the *Italians.* One *Goia* first lit upon the Use of the *Mariner's Compass,* about A.C. 1300. After this, the various *Declination* of the *Needle* under different Meridians, was discovered by *Cabot* and *Norman.* And then the Variation of the Declination, so as to be not always the same in one and the same place, by *Hevelius, Auzot, Volckamer,* and others.

The inquisitive Mr. *Derham* says, The *Variation of the Variation* was first found out by our *Gellibrand,* A.C. 1634.

And he himself has added a further Discovery; That as the *Common Needle* is continually varying towards the *East* and *West,* so the *Dipping Needle* varies up and down, towards the *Zenith,* or fromwards, with a *magnetick* Tendency, describing a Circle round the Pole of the World, or some other Point; a Circle, whereof the *Radius* is about 13 Degrees.

In every *Magnet* there are *two Poles,* the one pointing to the *North,* and the other to the *South.*

The *Poles,* in divers Parts of the Globe, are diversly inclined towards the *Center* of the Earth.

These *Poles,* tho contrary to one another, do mutually help towards the *Magnet's* Attraction, and Suspension of *Iron.*

If a *Stone*[1] be cut or broke into ever so many pieces, there are these *two Poles* in each of the *pieces.*

If two *Magnets* are spherical, one will conform itself to the other, so as either of them would do to the *Earth;* and after they have so turned themselves, they will endeavour to approach each other: but placed in a contrary Position, they avoid each other.

If a *Magnet* be cut thro the *Axis,* the Segments of the Stone, which before were joined, will now avoid and fly each other.

If the *Magnet* be cut by a Section perpendicular to its *Axis,* the two Points, which before were conjoined, will become contrary Poles; one in one, t'other in t'other Segment.

---

1. From this point on Mather uses stone to mean lodestone (or loadstone). The lodestone is magnetite possessing polarity; its attractive power is not a property of any stone.

Essay 24. Of Magnetism

Iron receives Vertue from the *Magnet*, by application to it, or barely from an approach near it, tho it do not touch it; and the *Iron* receives this Vertue variously, according to the Parts of the Stone it is made to approach to.

The *Magnet* loses none of its own Vertue by communicating any to the *Iron*. This Vertue it also communicates very speedily; tho the longer the *Iron* joins the Stone, the longer its communicated Vertue will hold. And the better the *Magnet*, the sooner and stronger the communicated Vertue.

Steel receives Vertue from the *Magnet* better than *Iron*.

A *Needle* touch'd by a *Magnet*, will turn its Ends the same way towards the Poles of the World as the *Magnet* will do it. But neither of them conform their Poles exactly to those of the World; they have usually some *Variation*, and this *Variation* too in the same place is not always the same.

A *Magnet* will take up much more *Iron* when *arm'd* or *cap'd* than it can alone. And if the *Iron Ring* be suspended by the *Stone*, yet the magnetical Particles do not hinder the Ring from turning round any way, to the Right or Left.

The best *Magnet*, at the least distance from a lesser or a weaker, cannot draw to it a piece of Iron adhering actually to a much weaker or lesser Stone; but if it come to touch it, it can draw it from the other. But a weaker *Magnet*, or even a little piece of *Iron*, can draw away or separate a piece of *Iron* contiguous to a better and greater *Magnet*.

In our Northern Parts of the World, the *South Pole* of a *Loadstone* will raise more *Iron* than the *North Pole*.

A Plate of *Iron* only, but no other Body interposed, can impede the Operation of the *Loadstone*, either as to its attractive or directive Quality.

The Power and Vertue of the *Loadstone* may be impair'd by lying long in a wrong posture, as also by Rust, and Wet, and the like.

A *Magnet* heated *red-hot*, will be speedily deprived of its *attractive* Quality; then cooled, either with the *South Pole* to the *North*, in an horizontal position, or with the *South Pole* to the *Earth* in a perpendicular, it will change its *Polarity;* the *Southern* Pole becoming the *Northern*, and *vice versa*.

By applying the Poles of a very *small Fragment* of a *Magnet* to the opposite vigorous ones of a larger, the Poles of the Fragment have been speedily changed.

Well temper'd and harden'd *Iron* Tools, *heated* by Attrition,[2] will attract Filings of *Iron* and *Steel.*

The *Iron Bars* of *Windows,* which have stood long in an erect position, do grow permanently *magnetical;* the lower ends of such Bars being the *Northern Poles,* and the upper the *Southern.*[3]

Mr. *Boyle* found *English Oker,*[4] heated red-hot, and cooled in a proper posture, plainly to gain a *magnetick* Power.

The illustrious Mr. *Boyle,* and the inquisitive Mr. *Derham,* have carried on their Experiments, till we are overwhelmed with the *Wonders,* as well as with the *Numbers* of them.

That of Mr. *Derham,* and *Grimaldi,* That a piece of well-touch[5] *Iron Wire,* upon being bent round in a Ring, or coiled round upon a Stick, loses its *Verticity;* is very admirable.

The Strength of some *Loadstones* is very surprizing.

Dr. *Lister* saw a Collection of *Loadstones,* one of them weighed naked not above a *Dram,*[6] yet it would raise a *Dram and half* of *Iron;* but being shod,[7] it would raise *one hundred and forty and four Drams.* A smooth *Loadstone,* weighing 65 Grains,[8] drew up 14 Ounces; that is, 144 times its own weight. A *Loadstone* that was no bigger than an Hazel-nut, fetch'd up an huge bunch of Keys.

The *Effluvia* of a *Loadstone* seem to work in a *Circle.* What flows from the *North Pole,* comes round, and enters the *South Pole;* and what flows from the *South Pole,* enters the *North Pole.*

Tho a minute *Loadstone* may have a prodigious force, yet it is very strange to see what a *short Sphere of Activity* it has; it affects not the *Iron* sensibly above an Inch or two, and the biggest little more than a Foot or two. The *magnetick Effluvia* make haste to return to the Stone that emitted them, and seem afraid of leaving it, as a Child the Mother before it can go alone.

2. That is, heated by use. Tools become warm from friction when they are used in work. The modern concept of heat was unknown at the time Mather wrote.

3. The reason for this interesting observation is that lines of magnetism are at a high angle in all high latitudes but not near the equator.

4. English ocher is earthy hematite which is changed to magnetite upon heating.

5. Mather probably means well touched, i.e., well magnetized.

6. A dram is one-sixteenth of an ounce in avoirdupois weight.

7. Applied to things, the word *shod* means tipped or edged with metal.

8. A grain, the smallest United States and English unit of weight, is .036 drams or 1/5760 of an ounce.

On that astonishing Subject, *The Variation of the Compass,* what if we should hear the acute Mr. *Halley*'s Proposals?[9]

He proposes, That our whole Globe should be looked upon as a *great Magnet,* having four *magnetical Poles,* or Points of Attraction, two near each Pole of the Equator. In those Parts of the World which lie near adjacent unto any one of these *magnetical Poles,* the Needle is governed by it; the nearer Pole being always predominant over the remoter. The *Pole* which at present is nearest unto *Britain,* lies in or near the Meridian of the Land's end of *England,* and not above seven Degrees from the *Arctick* Pole. By this *Pole* the Variations in all *Europe,* and in *Tartary,* and in the *North Sea,* are principally governed, tho' with some regard to the other *Northern Pole,* which is in a Meridian passing about the middle of *California,*[10] and about fifteen Degrees from the *North Pole* of the World. To this the Needle pays its chief respect in all the North *America,* and in the two Oceans on either side, even from the *Azores* Westward, unto *Japan,* and further. The two *Southern Poles* are distant rather further from the *South* Pole of the World; the one is about sixteen Degrees therefrom, and is under a *Meridian* about twenty Degrees to the Westward of the *Magellanick* Straits; this commands the Needle in all the South *America,* in the *Pacifick Sea,* and in the greatest part of the *Ethiopick*[11] Ocean. The fourth and last Pole seems to have the greatest Power and the largest Dominions of all, as it is the most remote from the Pole of the World; for 'tis near twenty Degrees from it, in the Meridian which passes thro *Hollandia Nova,*[12] and the Island *Celebes.* This Pole has the mastery in the South part of *Africa,* in *Arabia,* and the *Red Sea,* in *Persia,* in *India,* and its Islands, and all over the *Indian Sea,* from the *Cape of Good Hope* Eastwards, to the middle of the great *South Sea,* which divides *Asia* from *America.*

16 Land's/Lands 17 *Arctick/Artick*
20 *California/Calefornia* 27 Straits/Streights

9. The reference is to Edmond Halley, "An Account of the Cause of the Change of the Variation of the Magnetical Needle, with an Hypothesis of the Structure of the Internal Parts of the Earth," *Phil. Trans.,* 17 (Oct. 19, 1692), 563–78; reprinted in *Misc. Cur.,* 1:43–59.
10. That is, the old Mexican province of California.
11. See Essay 18, n. 1.
12. "New Holland." At the time Mather wrote, the term New Holland was applied to New Guinea and adjacent islands, especially New Britain and Timor.

Behold, the Disposition of the *magnetical Vertue,* as it is throughout the whole Globe of the *Earth* at this day!

But now to solve the *Phænomena!*

We may reckon the external Parts of our Globe as a *Shell,* the internal as a *Nucleus,* or an *inner Globe* included within ours; and between these a *fluid Medium,* which having the same common Center and Axis of diurnal Rotation, may turn about with our Earth every four and twenty Hours: only this outer Sphere having its turbinating Motion some small matter either swifter or slower than the internal Ball, and a very small difference becoming in length of Time sensible by many Repetitions; the internal Parts will by degrees recede from the external, and not keeping pace with one another, will appear gradually to move, either Eastwards or Westwards, by the difference of their Motions. Now if the exterior Shell of our Globe should be a *Magnet,* having its Poles at a distance from the Poles of diurnal Rotation; and if the internal *Nucleus* be likewise a *Magnet,* having its Poles in two other places, distant also from the Axis, and these latter, by a slow and gradual Motion, change their place in respect of the external, we may then give a reasonable account of the *four magnetical Poles,* and of the *Changes of the Needle's Variations.* Who can tell but the *final Cause* of the Admixture of the *magnetical Matter* in the Mass of the terrestrial Parts of our Globe, should be to maintain the concave Arch of this our Shell? Yea, we may suppose the Arch lined with a *magnetical Matter,* or to be rather one great *concave Magnet,* whose *two Poles* are fixed in the Surface of our Globe? Sir *Isaac Newton* has demonstrated the *Moon* to be more solid than our *Earth,* as nine to five; why may we not then suppose four Ninths of our Globe to be Cavity? Mr. *Halley* allows there may be Inhabitants of the lower Story, and many ways of producing *Light* for them. The Medium itself may be always luminous; or the concave Arch may shine with such a Substance as does invest the Surface of the *Sun;* or they may have peculiar *Luminaries,* whereof we can have no Idea: As *Virgil* and *Claudian* enlighten their *Elysian* Fields; the latter,

> *Amissum ne crede Diem; sunt altera nobis*
> *Sydera; sunt Orbes alii; Lumenque videbis*
> *Purius, Elysiumque magis mirabere Solem.*[13]

13. "Think not thou hast lost the light of day; other stars are mine and other courses; a purer light shalt thou see and wonder rather at Elysium's sun and blessed inhabitants."

The Diameter of the Earth being about eight thousand *English* Miles, how easy 'tis to allow five hundred Miles for the Thickness of the Shell! And another five hundred Miles for a Medium capable of a vast Atmosphere, for the Globe contained within it! But it's time to stop, we are got beyond *Human Penetration;* we have *dug* as far as 'tis fit any *Conjecture* should carry us!

It is a little surprizing that the Orb of the Activity of *Magnets,* as Mr. *Derham* observes, is larger or lesser at different times. There is a noble and a mighty *Loadstone* reserved in the Repository at *Gresham*-College,[14] which will keep a Key, or other piece of *Iron,* suspended unto another, sometimes at the distance of eight or ten Foot from it, but at other times not above four.

[A *Digression,* if worthy to be called so!]

§. But is it possible for me to go any further without making an *Observation,* which indeed would ever now and then break in upon us as we go along?

Once for all; *Gentlemen Philosophers,* the MAGNET has quite *puzzled* you. It shall then be no indecent *Anticipation* of what should have been observed at the Conclusion of this Collection, here to demand it of you, that you glorify the infinite Creator of this, and of all things, as *incomprehensible.* You must acknowledge that *Human Reason* is too feeble, too narrow a thing to comprehend the *infinite* God. The Words of our excellent *Boyle* deserve to be recited on this Occasion: "Such is the *natural Imbecillity* of the *Human Intellect,* that the most piercing Wits and excellent Mathematicians are forced to confess, that not only their own *Reason,* but that of Mankind, may be puzzled and nonplus'd about QUANTITY, which is an Object of Contemplation natural, nay, mathematical. Wherefore why should we think it unfit to be believed, and to be acknowledged, that in the *Attributes* of God [it may be added, *and in His Dispensations towards the Children of Men*] there should be some things which our finite Understandings cannot clearly *comprehend?* And we who cannot clearly comprehend how in ourselves two such distant Natures, as that of a *gross Body* and an *immaterial Spirit* should be so united as to make up *one Man,* why should we grudge to have our REASON Pupil to an *omniscient*

---

14. Gresham College, founded in the Elizabethan period to promote education in the City of London, became identified with the Royal Society after the Restoration. It served as headquarters of the Royal Society from 1673 to the early eighteenth century.

Instructor, who can teach us such things, as neither our own mere Reason, nor any others, could ever have discovered to us?"

I will now single out a few plain *Mathematical Instances,* wherein, Sirs, you will find your finest *Reason* so transcended, and so confounded, that it is to be hoped a *profound Humility* in the grand Affairs of our *holy Religion* will from this time for ever *adorn* you.

Mr. *Robert Jenkin* discoursing on *the Reasonableness of the Christian Religion,* gives two Instances *how much we may lose ourselves in the Speculation of material things.*

First, Nothing seems more evident, than that *all Matter is divisible;* yea, the *least Particle of Matter* must be so, because it has the Nature and Essence of *Matter:* it can never be so *divided* that it shall cease to be *Matter.* But then, on the other side, it is plain, *Matter* cannot be *infinitely divisible;* because whatever is *divisible,* is *divisible* into *Parts;* and no *Parts* can be *infinite,* because no *Number* can be so. A *numberless Number* is a Contradiction; all Parts are capable of being *numbred;* they are *more* or *fewer, odd* or *even.* It is not enough to say, that *Matter* is only capable of such a *Division,* but never can be *actually divided into infinite Parts;* for the Parts into which it is *divisible* must be *actually existent,* tho they be not *actually divided.* And last of all to say, these Parts of Matter are *indefinite,* but not *infinite,* is only to confess *we know not what to say.*

Secondly, We all agree that all the *Parts* into which the *Whole* is divided, being taken together are *equal to the Whole.* But it seems any *single Part is equal to the Whole.* It is granted, that in any *Circle* a *Line* may be drawn from *every Point* of the Circumference to the *Center.* Suppose the Circle to be the *Equator,* and a million lesser Circles are drawn within the *Equator,* about the same *Center,* and then a *right Line* drawn from *every Point* of the *Equator* to the Center of the Globe; every such *right Line* drawn from the *Equator* to the *Center,* must of necessity cut thro the million *lesser Circles,* about the same *Center:* consequently there must be the same number of Points in a Circle a million of times less than the *Equator,* as there is in the *Equator* itself. The *lesser Circles* may be multiplied into as many as there are *Points* in the *Diameters;* and so the *least Circle* imaginable may have *as many Points* as the greatest; that is, be as big as the greatest, as big as one that is millions of times as big as itself.

Yet more; What will you say to this? Let a *Radius* be moved as a *Radius* upon a *Circle;* 'tis a Case of Dr. *Grew's* proposing: whether we suppose it *wholly* moved, or but *in part,* the Supposition will bring us to an *Absurdity;* if it be in a part *movent,* and in a part *quiescent,* it will be a *curve Line,* and no *Radius;* if it be wholly *movent,*[15] then it moves either *about* or *upon* the Center; if it moves *about* it, it then comes short of it, and so again is no *Radius:* it cannot move *upon* it, because all motion having parts, there can be no motion upon a *Point.*

More yet; We cannot conceive how the *Perimeter* of a Circle, or other *curve Figure,* can consist without being infinitely *angular;* for the *parts* of a *Line* are *Lines:* But we cannot conceive how those Lines can have, as here they have, a different direction, and therefore an inclination, without making an *Angle.* And yet if you suppose a *Circle* to be *angular,* you destroy the Definition of a *Circle,* and the Theorems depending on it.

Once more; I will offer a Case of my own. The Line on which I am now writing is a *Space* between *two Points;* it will be doubtless allowed me, that my Pen in passing over this Line, from the one point unto the other, must *pass over the half of the Line before it passes over the whole;* and so the *half* of the remaining half, and so the half of the quarter that remains: so still the half of the remaining space, the *half before the whole;* and yet when it comes to execution, you find it is not so. If the Position you allowed me had been true, my Pen would not have reach'd unto the *end* of the *Line* before the *End* of my *Life;* or in a Term wherein it might have written ten Books as big as old *Zoroaster*'s or more Manuscripts than ever were in the *Alexandrian* Library.

It is then evident, that all Mankind is to this day in the dark as to the *ultimate Parts* of *Quantity,* and of *Motion.*

Go on my learned *Grew,* and maintain [who more fit than one of thy *recondite Learning?*] *that there is hardly any one thing in the World, the Essence whereof we can perfectly comprehend.* But then to the *natural Imbecillity* of REASON, add the *moral Depravations* of it, by our Fall from God, and the Ascendant which a corrupt and vicious *Will* has obtain'd over it, how much ought this Consideration to warn us against the Conduct of an *unhumbled Understanding* in things relating to the *Kingdom of God?* I am not out of my

84 add/and

15. That is, moving.

way, I have had a *Magnet* all this while *steering* of this Digression: I am now returning to *that.*

¶. God forbid I should be, *Tam Lapis ut Lapidi Numen inesse putem.*[16] To fall down before a *Stone,* and say, *Thou art a God,* would be an *Idolatry,* that none but a Soul more sensless than a *Stone* could be guilty of. But then it would be a very agreeable and acceptable *Homage* unto the Glorious GOD, for me to see much of Him in such a wonderful *Stone* as the MAGNET. They have done well to call it the *Loadstone,* that is to say, the *Leadstone: May it lead me unto Thee, O my God and my Saviour! Magnetism* is in this like to *Gravity,* that it leads us to GOD, and brings us very near to Him. When we see *Magnetism* in its Operation, we must say, *This is the Work of God!*[17] And of the *Stone,* which has proved of such vast use in the Affairs of the *Waters that cover the Sea,*[18] and will e'er long do its part in bringing it about that the *Glory of the Lord shall cover the Earth,*[19] we must say, *Great God, this is a wonderful Gift of Thine unto the World!*

I do not propose to exemplify the *occasional Reflection* which a devout Mind may make upon all the *Creatures* of God, their *Properties,* and *Actions,* and *Relations;* the *Libri Elephantini*[20] would not be big enough to contain the thousandth part of them. If it were lawful for me here to pause with a particular *Exercise upon the Loadstone,* my first Thoughts would be these of the holy *Scudder,* whose Words have had a great Impression on me ever since my first reading of them in my Childhood: "An upright Man is like a *Needle* touch'd with the *Loadstone;* tho he may thro boisterous *Temptations* and strong *Allurements* oftentimes look towards the Pleasure, Gain and Glory of this *present World,* yet because he is truly touch'd with the sanctifying Spirit of God, he still inclineth *God-ward,* and hath no Quiet till he stand *steady towards Heaven.*"

However, to animate the Devotion of my *Christian* Philosopher, I will here make a Report to him. The ingenious *Ward* wrote a pious Book, as long ago as the Year 1639, entitled, *Magnetis Reductorium Theologicum.* The Design of his Essay, is, to *lead* us from the Consideration of the *Loadstone,* to the Consideration of our

16. "So much a stone that I think there is divinity in a stone."
17. John 6:29; see also Jer. 50:25.
18. Hab. 2:14.
19. Adapted from Hab. 2:14.
20. Mather uses the word *elephantine* to mean big books.

Essay 24. Of Magnetism

SAVIOUR, and of his incomparable *Glories;* whereof the *Magnet* has in it a notable Adumbration. In his Introduction he has a Note, worthy to be transcribed here, as religiously asserting the Design, of which our whole Essay is a Prosecution. *Hic præcipuus & potentissimus Creaturarum omnium Finis est, cum Scalæ nobis & Alæ fiunt, quibus Animæ nostræ supra Dumeta & Sterquilinia Mundi hujus volitantes, facilius ad Cælum ascendunt, & ad Deum Creatorem aspirant.*[21] For what is now before us, if our *Ward* may be our Adviser; *Christian,* in the *Loadstone* drawing and lifting up the *Iron,* behold thy *Saviour* drawing us to himself, and raising us above the secular Cares and Snares that ruin us. In its ready *communication* of its Vertues, behold a shadow of thy *Saviour* communicating his holy Spirit to his chosen People; and his *Ministers* more particularly made Partakers of his *attractive Powers.* When *Silver* and *Gold* are neglected by the *Loadstone,* but coarse *Iron* preferred, behold thy *Saviour* passing over the *Angelical World,* and chusing to take *our Nature* upon him. The *Iron* is also undistinguished, whether it be lodged in a fine Covering, or whether it be lying in the most squalid and wretched Circumstances; which invites us to think how little *respect of Persons*[22] there is with our *Saviour.* However, the *Iron* should be *cleansed,* it should not be *rusty;* nor will our *Saviour* embrace those who are not so far *cleansed,* that they are at least *willing to be made clean,* and have his *Files* pass upon them. The *Iron* is at first *merely passive,* then it *moves* more feebly towards the *Stone;* anon upon Contact it will fly to it, and express a marvellous Affection and Adherence. Is not here a Picture of the Dispositions in our Souls towards our Saviour? It is the Pleasure of our Saviour to work by *Instruments,* as the *Loadstone* will do most when the Mediation of a *Steel Cap* is used about it. After all, whatever is done, the whole *Praise* is due to the *Loadstone* alone. But there would be *no end,* and indeed there should be *none,* of these Meditations! Our *Ward* in his Dedication of his Book to the King, has one very true Compliment. *Hoc ausim Majestati tuæ boni fide spondere; si unicus unicum possideres, Mundi totius te facile Monarcham efficeret.*[23] But what a Great KING is He, who is the Owner, yea,

---

21. "This is the foremost and most important end of all creatures, that they should be as ladders and wings by which our souls, flying above the thickets and dunghills of this world, may ascend to heaven more easily, and aspire toward God their Creator."
22. Rom. 2:11 and Eph. 6:9.
23. "I would dare promise this to your Majesty in good faith: if you alone possessed the only one [magnet], it would easily make you the ruler of the whole world."

and the Maker of all the *Magnets* in the World! *I am a Great KING, saith the Lord of Hosts, and my Name is to be feared among the Nations!*[24] May the *Loadstone* help to carry it to them.

ESSAY XXV. *Of* MINERALS.

65 O*PERUM Dei Cognitionem* (says my dear *Arndt*) *quilibet ex sincero erga Deum amore & gratitudine, sibi acquivere studeat, ut sciat, quæ Deus nostri causa creaverit.*[1] He smiles at the trifling *Logicians*, who, *totam ætatem inter inanes Subtilitates transigentes,*[2] wholly taken up with *Trifles*, overlook the glorious Works of God.

70 Our *Earth* is richly furnished with a Tribe of *Minerals*, called so because dug out of *Mines;* and because *dug*, therefore also called *Fossils.* Many things to be written of these, ought to have a *Nimok*[3] in the Margin!

The *adventitious Fossils*, which are but the *Exuviæ*[4] of *Animals*, 75 have been erroneously thought a sort of *peculiar Stones.* These must be excluded.

But then the *Natives of the Earth* are to be found in a vast variety. The inquisitive Dr. *Woodward* has prepared us a noble *Table* of them.

80 There are near twenty several sorts of *Earth.* Of these, besides the *Potter's Earth,*[5] and the *Fuller's Earth,*[6] how exceedingly useful is the *Chalk* to us! 'Tis a πολύχρηστον.[7]

---

24. Adapted from Mal. 1:14.
1. "Let everyone seek to acquire knowledge of the works of God out of a sincere love for and gratitude towards Him, so that he may know what things God has created for our own sake."
2. "Spend their entire lifetime in empty trifles."
3. Kenneth B. Murdock said that "Nimok" is probably a misprint for Nichols. To read "ch" as "m" and "ls" as "k" is easy in Mather's handwriting. This is plausible. Murdock added that Thomas Nichols, who flourished about 1650, wrote three books on gems and precious stones. But this Nichols left little trace. See Murdock, ed., *Selections from Cotton Mather* (New York: Harcourt Brace and Co., 1926), p. 317.
4. "Exuviae" means "parts sloughed off." By "adventitious Fossils" Mather means accidental fossils, that is, petrified remains of real organisms. In this passage he separates them from minerals. But see n. 17 below.
5. Potter's earth is a plastic clay, kaolinite, suitable for modeling, throwing pottery, and the manufacture of china.
6. Fuller's earth is another clay which resembles potter's earth but lacks plasticity. It is used in fulling cloth and is now important as an absorbent and filler.
7. "Very useful thing."

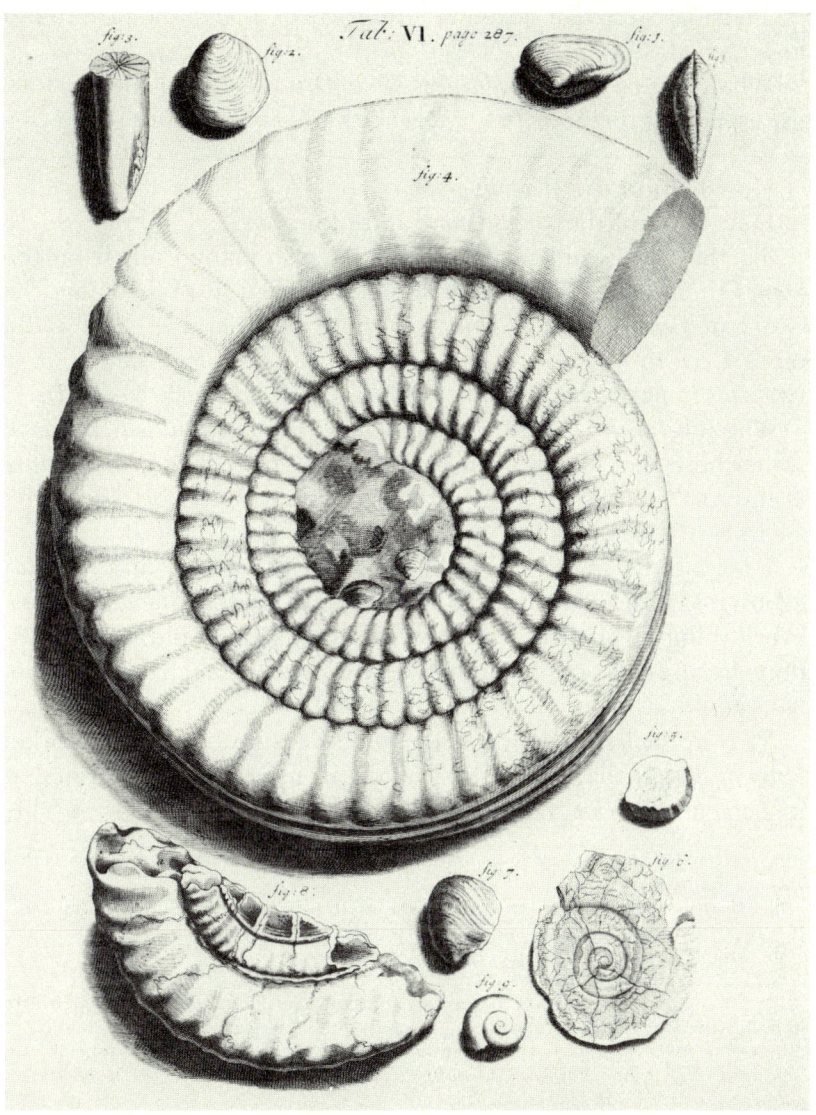

In this plate from Robert Hooke's *Posthumous Works* (London, 1705), figure 4 represents a Cornu-Ammonis, about eighteen inches in diameter, with small petrified shells near the center. Mather's own work reflects the uncertainty common among his authorities as to whether figured stones were the petrified remains of real organisms or minerals. Hooke viewed fossils as the "Monuments" and "Medals" of earlier ages from which the history of the earth can be reconstructed. Courtesy of the Rare Book and Special Collections Library, University of Illinois at Urbana-Champaign.

There are above a dozen several sorts of *Stones*, that are found in *larger* Masses.

What *Vessels*, what *Buildings*, what *Ornaments*, do these afford us; especially the *Slate*, the *Marble*, the *Free-stone*,⁸ and the *Lime-stone*?

How helpful the *Warming-stone*?⁹

How needful the *Grind-stone* and *Mill-stone*?

To the *Service* of our Maker we have so many Calls from the *Stones* themselves, [for if *Men* should be silent at proclaiming the Glory of God, the *very Stones would speak*]¹⁰ that a learned and a pious *German*¹¹ so addresses us: *Audis tibi loquentes Lapides; tu ne sis Lapis in hac parte, sed ipsorum Vocem audi, & in illis Vocem Dei.*¹²

The *Whetstone*¹³ gives me a particular Admonition, which I have somewhere met with: *Multi multa docent alios, quæ ipsi præstare nequeunt.*¹⁴ The worst Motto for a *Divine* that can be! *Lord, save me from it!*¹⁵

How astonishing the *Figures*, which Dr. *Robinson*¹⁶ and Mr. *Ray* report, as naturally delineated upon several kinds of *Stones*;¹⁷ almost every thing in Nature described in them, so as could not be outdone by any Sculptor or Painter! The *Colaptice*,¹⁸ such as no *Human Skill* could arise to!

Yea, in *Stones* there has been sometimes found so much of an *Human Shape*, that every thing really in it has been astonished at it. *Zeiler* and *Kircher* mention some famous *Rocks*, which so resemble

---

8. A stone (such as sandstone or limestone) which can be cut freely in any direction without splitting.
9. Probably coal or any stone that, once heated, would retain heat for a long time. It was put in a warming pan and used to warm a bed.
10. Adapted from Luke 19:40.
11. That is, Johann H. Alsted.
12. This quotation is clearer if seen in context (the lines used by Mather are in italics). Alsted writes (in paraphrase): "There is vitality in everything. One should not despise the rocks. I [God] have created nothing in vain, nothing which is superfluous. The rocks have value. *You hear the rocks speaking to you; do not be a rock in this matter, but hear their voice and in them the voice of God.*"
13. That is, a natural stone such as fine cherty rock used for whetting edge tools.
14. "Many teach much to others which they themselves are unable to perform."
15. Perhaps adapted from Matt. 14:30. Peter, beginning to sink when he walked on the water, called out, "Lord, save me."
16. That is, Tancred Robinson.
17. These are figured or formed stones. They baffled Ray so much that he could never make up his mind whether they were organic remains or not. They are real fossils. Thus Mather contradicts himself, for in describing the *"adventitious Fossils"* a few lines above, he excludes them from the category of minerals.
18. That is, the art of carving or cutting the resemblances and figures of natural things in stone.

Essay 25. Of Minerals

Monks, that all People call them so. *Olaus Wormius* was Possessor of a large *Stone,* which had exactly the Head, Face, Neck, and Shoulders of a *Man. Monconnys* and others relate the several *Parts* of a Man, which many *Stones* have exactly exhibited. *Oh! how happy we, if Men and Stones had less Resemblance!*

There are many sorts of *Stones* found in *lesser Masses.*

Of these there are many who do *not* exceed the hardness of *Marble.*

Seven or eight of these are of an *indeterminate Figure.*

Twice as many have a *determinate Figure.*[19]

Among these the Wonders of the *Osteo-colla,*[20] to join and heal our *broken Bones.*

But then there are others which *do* exceed *Marble* in hardness.

To this Article belong those that are usually called *Gems* or *precious Stones.*

[*Pebbles* and *Flints* are of the *Agate-kind.*]

Some of these are *opake.*

Three of the opake have a Body of *one Colour.*

Here the Wonders of the *Nephritick Stone!*[21]

Three of the *opake* have *different Colours* mixed in the same Body.[22]

Here the Wonders of the *Blood-stone!*[23]

Some are *pellucid.*[24]

Two with *Colours changeable,* according to their different position in the Light.

Nine or ten with *Colours permanent.*

Some are *diaphanous.*[25]

Two *yellow* (or partaking of it.)

Three *red.*

Three *blue.*

Two *green.*

Four *without any Colours.*

19 Bones / Brones

19. By determinate figure Mather means that they are in individual crystals.
20. A deposit of calcium carbonate forming an incrustation on the roots and skins of plants. It was used to cement together the parts of broken bones.
21. As used here, "nephritic" means jade.
22. The different colors are green and red.
23. That is, a stone consisting of green chalcedony sprinkled with red spots resembling blood and resulting from oxidizing of the green. It is also called heliotrope.
24. That is, a mineral admitting maximum passage of light though one cannot see through it; also called translucent.
25. That is, a mineral that one can see through; also called transparent.

"But an excellent Writer[26] observing, *Deus est Figulus Lapidum*,[27] carries on his Observation, That the God who makes *precious* as well as *common Stones*, has made *Men* with as much of a *Difference*, and not altogether without such a *Proportion*.

"Good God, Thy heavenly Graces in the Soul are brighter Jewels than any that are dug out of the Earth! A *poor* Man may be adorn'd with these; those who are so, *they shall be mine, saith the Lord, in the Day when I make up my Jewels*.[28]

"How often have I seen a Jewel in the *Snout of a Swine!*[29]

"And how many *Counterfeits* in the World!"

There are seven sorts of *Salts* to be met withal.

But the *Salt* of our *Table*, of how much consequence this to us! The Uses of it are too many to be by any reckoned: Very many are well known to all. To which add the Experience which *Bickerus* affirms the Army of the Emperor *Charles* V. had, that they must have perish'd on the *African* Shore, if they had not found a Grain of *Salt* in their Mouths; an Antidote not only against *Thirst*, but *Hunger* too.

He deserves to be herded with the Creatures, which *Animam habent pro Sale*,[30] who shall be so *insipid* an Animal, as to be insensible that the Benefits of *Salt* call for very great Acknowledgments. *My God, save me from what would render me unsavory Salt!*[31]

There are three liquid *Bitumens*, six or seven solid.

There are about a dozen *metallick Minerals*. *Mercury* is one of these, but how astonishing an one! The Particles whereof how small, how smooth, how solid! The Corpuscles of it have Diameters much less than those of *Air;* yea, than those of *Water;* and not much greater than those of *Light* itself!

At last we come to *Metals; Iron*, with its Attendants; *Tin, Lead, Copper, Silver* and GOLD.

"I shall not consider the Reasons which moved *Cardan* to assert that *Metals* have a *Soul;* but I am sure that I myself have a *Soul*, and am one that is *reasonable;* if so, what can be more agreeable to me, than a Consideration which I find hinted by

---

26. That is, Johann H. Alsted.
27. "God is the potter who makes the stones."
28. Mal. 3:17.
29. Adapted from Prov. 11:22.
30. "Have a soul instead of salt."
31. Adapted from Matt. 5:13; see also Luke 14:34.

a curious Writer of *natural Theology*:³² We should admire the *Munificence* of one who would bestow a considerable Quantity of enriching *Metals* upon us. But then how much cause have we to adore the *Munificence* of our bountiful GOD, who has enrich'd us with *Metals* in so vast a Quantity, and with so much Profusion from His *hidden Treasures! Quotusquisque est qui non videt, quid Ratio officii sui postulat?*"³³

How amazingly serviceable is our *Iron* to us! In our *mechanical Arts*, in our *Agriculture*, in our *Navigation*, in our *Architecture;* in *all*, I say, *all* our Business! What a *sordid Life* do those *Barbarians* lead, who are kept ignorant of it! Unthankful for this. *O Man*, you deserve *Heaven* should become as *Iron* over you.³⁴

It is from GOD that the *Metals* of most necessary Uses are the most plentiful; others that may be better spared, there is a rarity of them.

That one single *Metal*, Iron, as Dr. *Grew* observes, it sets on foot above an hundred sorts of manual Operations.

Tho the *Love of Money* be the *Root of all Evil*,³⁵ yet the ingenious Dr. *Cockburn* has discoursed very justly on the vast Importance whereof the Use of *Money* is to Mankind. And indeed where the Use of *Money* has not been introduced, Men are brutish and savage, and nothing that is good has been cultivated.

There is a surprizing Providence of GOD in keeping up the Value of *Gold* and *Silver*, notwithstanding the vast Quantities dug out of the Earth in all Ages, ever since the Trade begun of *effodiuntur Opes;*³⁶ and so continuing them fit Materials to make *Money* of.

Among the marvellous Qualities of *Gold*, its *Ductility* deserves to have a particular Notice taken of it.

The *Wire-drawers*, to every 48 Ounces of *Silver*, allow one of *Gold*. Now *two Yards* of the superfine Wire weigh a *Grain*. In the Length of 98 Yards there are 49 Grains of Weight. A single Grain of *Gold* covers the said 98 Yards. The 1000th part of a *Grain* is above one third of an Inch long, which yet may be actually divided into ten; and so the 100000th part of a *Grain of Gold* may be visible without a *Microscope*.

32. That is, Johann H. Alsted.
33. "How few there are who do not see what reason demands."
34. Adapted from Lev. 26:19.
35. 1 Tim. 6:10.
36. "Digging up wealth." That is, mining.

It is a marvellous thing that *Gold,* after it has been divided by corrosive Liquors into *invisible Parts,* yet may presently be so precipitated, as to appear in its own *golden Form* again.

But, as Dr. *Grew* observes, the same *Immutability* which belongs to the Composition of *Gold,* much more belongs to the *Principles of Gold,* and of all other Bodies, when their Composition is destroyed. *Dampier,* an ingenious Traveller all round the Globe, has an Observation; *I know no Place where Gold is found, but what is very unhealthy.*

"Possessor of *Gold!* Beware lest the Observation be verified in the *unhealthy* Influences of thy *Gold* upon thy *Mind;* and lest the *love* of it betray thee into many *foolish and hurtful Lusts,* which will drown thee in *Destruction and Perdition.*"[37]

"The *Auri sacra Fames*[38] is the worst of all Distempers."

My God, I bless Thee; I know something that is better than fine Gold, something that cannot be gotten for Gold, neither shall Silver be weighed for the Price thereof.[39]

If *Gold* could speak, it would rebuke the *Idolatry* wherewith Mankind adores it, in much such Terms as I find a devout Writer[40] assigning to it. *Non Deus sum, sed Dei Creatura; Terra mihi Mater. Ego servio tibi, ut tu servias Creatori.*[41]

¶. "Finally, The antient Pagans not only worshipped the *Host of Heaven,* [justly called *Zabians*][42] but whatsoever they found *comfortable* to Nature, they also *deified,* even, *Quodcunque juvaret.*[43] The River *Nilus*[44] too must at length become a Deity; yea, *Nascuntur in hortis Numina.*[45]

"And according to *Pliny, a Man that helps a Man becomes a God.*

"God save us from the Crime stigmatiz'd by our Apostle, *to*

---

37. The last two italicized phrases are from 1 Tim. 6:9.
38. "Accursed hunger for gold."
39. Adapted from Prov. 8:19 and Job 28:15.
40. That is, Johann H. Alsted.
41. "I am not God, but a creature of God; the earth is my mother. I serve you, so that you may serve the Creator."
42. Zabian, or Sabian, is an erroneous usage for a worshiper of "the host of heaven"; that is, a star-worshiper.
43. "Whatever brought [them] pleasure."
44. The Nile River.
45. Mather's meaning is made clearer by translating the two lines from Juvenal from which his quotation (the italicized phrase) is taken: "But it is an impious outrage to crunch leeks and onions with the teeth. What a holy race to have *such divinities springing up in their gardens!*"

adore the Creatures more than the Creator!⁴⁶ By no means let us be as *Philo* speaks, Κόσμον μᾶλλον ἢ κοσμοποιὸν θαυμάσαντες, *more admiring the World, than the Maker of the World.*

"We will glorify the GOD who has bestowed things upon us; *for the Silver is mine, and the Gold is mine, saith the Lord of Hosts!*"⁴⁷

ESSAY XXVI. *Of the* VEGETABLES.

THE Contrivance of our most Glorious Creator, in the VEGETABLES growing upon this Globe, cannot be wisely observed without Admiration and Astonishment.

We will single out some Remarkables, and glorify our GOD!

First, In *what manner is Vegetation* performed? And how is the Growth of *Plants* and the Increase of their *Parts* carried on? The excellent and ingenious Dr. *John Woodward* has, in the way of nice Experiment, brought this thing under a close Examination. It is evident that *Water* is necessary to *Vegetation;* there is a *Water* which ascends the Vessels of the *Plants,* much after the way of a *Filtration;* and the Plants take up a larger or lesser Quantity of this Fluid, according to their Dimensions. The much greater part of that *fluid Mass* which is conveyed to the Plants, does not abide there, but exhale thro them up into the *Atmosphere.* Hence Countries that abound with *bigger Plants* are obnoxious to greater Damps, and Rains, and inconvenient Humidities. But there is also a *terrestrial Matter* which is mixed with this *Water,* and ascends up into the *Plants* with the *Water.* Something of this Matter will attend *Water* in all its motions, and stick by it after all its Percolations. Indeed the Quantity of this *terrestrial Matter,* which the Vapours carry up into the *Atmosphere,* is very *fine,* and not very *much,* but it is the truest and the best prepared *vegetable Matter;* for which cause it is that *Rain-Water* is of such a singular Fertility. 'Tis true there is in *Water* a *mineral Matter* also, which is usually too scabrous, and ponderous, and inflexible, to enter the Pores of the *Roots.* Be the *Earth* ever so rich, 'tis observed little good will come of it, unless the Parts of it be loosened a little, and separated. And this probably is all the use of *Nitre* and other *Salts* to Plants, to loosen the Earth,

46. Adapted from Rom. 1:25.
47. Hag. 2:8.

and separate the Parts of it. It is this *terrestrial Matter* which fills the *Plants;* they are more or less nourished and augmented in proportion, as their *Water* conveys a greater or lesser quantity of proper *terrestrial Matter* to them. Nevertheless 'tis also probable that in this there is a variety; and all Plants are not formed and filled from the same sort of *Corpuscles.* Every *Vegetable* seems to require a *peculiar and specifick Matter* for its Formation and Nourishment. If the Soil wherein a Seed is planted, have not all or most of the Ingredients necessary for the *Vegetable* to subsist upon, it will suffer accordingly. Thus *Wheat* sown upon a Tract of Land well furnish'd for the Supply of that *Grain,* will succeed very well, perhaps for divers Years, or, as the Husbandman expresses it, *as long as the Ground is in heart;* but anon it will produce no more of that *Corn;* it will of some other, perhaps of *Barley:* and when it will subsist this no more, still *Oats* will thrive there; and perhaps *Pease* after these. When the Ground has lain fallow some time, the *Rain* will pour down a fresh Stock upon it; and the care of the *Tiller* in manuring of it, lays upon it such things as are most impregnated with a Supply for *Vegetation.* It is observ'd that *Spring-water* and *Rain-water* contain pretty near an equal charge of the *vegetable Matter,* but *River-water* much more than either of them; and hence the Inundations of *Rivers* leave upon their Banks the fairest Crops in the World. It is now plain that *Water* is not the *Matter* that composes *Vegetables,* but the *Agent* that conveys that *Matter* to them, and introduces it into the several parts of them. Wherefore the plentiful provision of this Fluid supplied to all Parts of the Earth, is by our *Woodward* justly celebrated with a pious Acknowledgment of that *natural Providence* that superintends over the Globe which we inhabit. The Parts of *Water* being exactly spherical, and subtile beyond all expression, the Surfaces perfectly polite, and the Intervals being therefore the largest, and so the most fitting to receive a *foreign Matter* into them, it is the most proper Instrument imaginable for the Service now assign'd to it. And yet *Water* would not perform this Office and Service to the *Plants,* if it be not assisted with a due quantity of *Heat; Heat* must concur, or *Vegetation* will not succeed. Hence as the *Heat* of several *Seasons* affords a different face of things, the same does the *Heat* of several *Climates.* The *hotter* Countries usually yield the *larger Trees,* and in a greater variety. And in *warmer* Countries, if there be a remission of the *usual Heat,* the Production will in proportion be diminish'd.

That I may a little contribute my *two Mites* to the illustration of the way wherein *Vegetation* is carried on, I will here communicate a couple of Experiments lately made in my Neighbourhood.

My Neighbour planted a Row of Hills in his Field with our *Indian Corn,* but such a Grain as was colour'd *red* and *blue;* the rest of the Field he planted with Corn of the most usual Colour, which is *yellow.* To the most *Windward-side* this Row infected *four* of the next neighbouring Rows, and part of the fifth, and some of the sixth, to render them colour'd like what grew on itself. But on the *Leeward-side* no less than seven or eight Rows were so colour'd, and some smaller impressions were made on those that were yet further distant.

The same Neighbour having his Garden often robb'd of the *Squashes* growing in it, planted some *Gourds* among them, which are to appearance very like them, and which he distinguish'd by certain adjacent marks, that he might not be himself imposed upon; by this means the Thieves 'tis true found a very *bitter Sauce,* but then all the *Squashes* were so infected and embitter'd, that he was not himself able to eat what the Thieves had left of them.

That most accurate and experienc'd Botanist Mr. *Ray* has given us the *Plants* that are more commonly met withal, with certain characteristick Notes, wherein he establishes *twenty-five Genders* of them. These *Plants* are to be rather stiled *Herbs.*

But then of the *Trees* and *Shrubs,* he distinguishes *five Classes* that have their *Flower* disjoined and remote from the *Fruit,*[1] and as many that have their *Fruit* and *Flower* contiguous.

How unaccountably is the *Figure of Plants* preserved? And how unaccountably their *Growth* determined? Our excellent *Ray* flies to an intelligent *plastick Nature,*[2] which must understand and regulate the whole Oeconomy.

Every particular *part* of the *Plant* has its astonishing Uses. The *Roots* give it a Stability, and fetch the Nourishment into it, which lies in the Earth ready for it. The *Fibres* contain and convey the Sap which carries up that Nourishment. The *Plant* has also larger

---

1. These are dioecious plants. They have the male and female reproductive organs on different individuals.
2. The concept of plastic nature arose in mid-seventeenth-century England to explain how spirit and matter interact. The Cambridge Platonists used it to combat the materialism and atheism associated with Descartes and Hobbes.

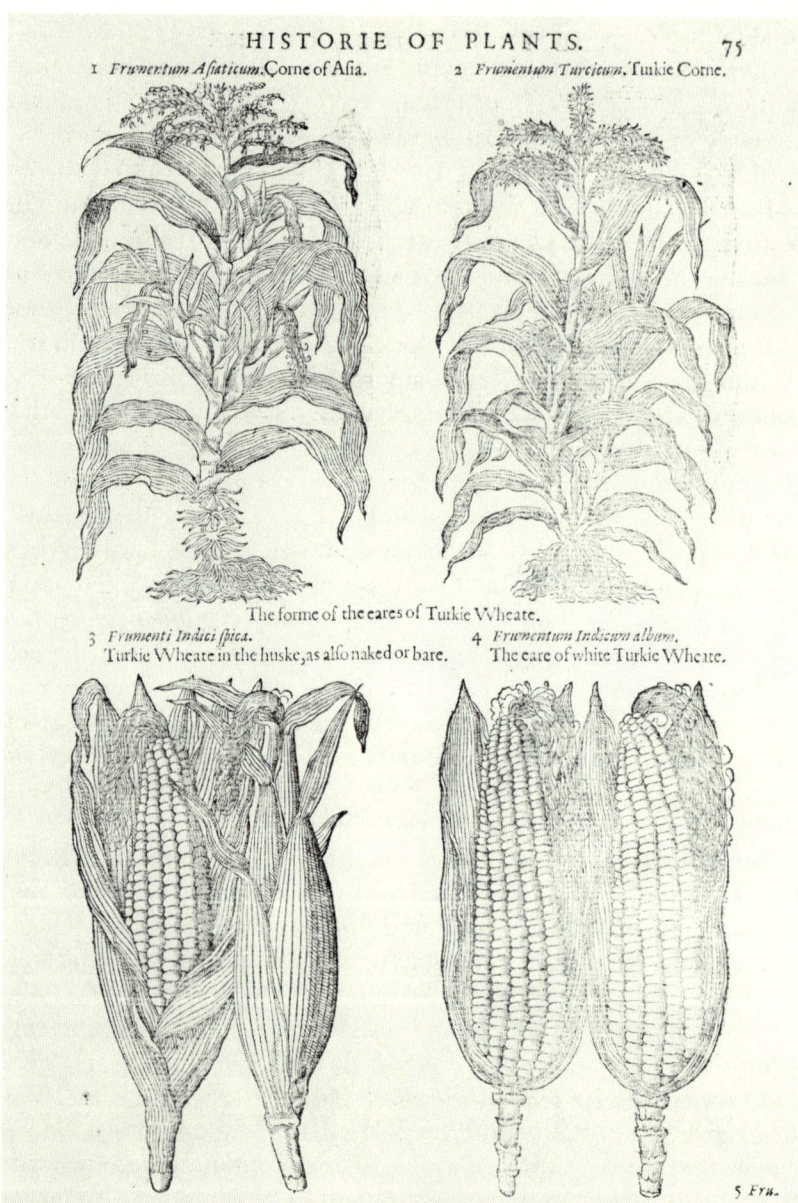

John Gerard treated Indian corn (*Zea mays*) in *The Herball: Or Generall Historie of Plantes* (London, 1597). Mather describes experiments carried on with Indian corn, and he was the first person to describe correctly the admixture of colors in *Zea mays*. Courtesy of the Rare Book and Special Collections Library, University of Illinois at Urbana-Champaign.

Essay 26. Of the Vegetables 133

Vessels, which entertain the proper and specifick Juice of it; and others to carry the Air for its necessary respiration. The outer and inner *Bark* defend it from Annoyances, and contribute to its Augmentation. The *Leaves* embrace and preserve the *Flower* and *Fruit* as they come to their explication. But the principal use of them, as *Malpighi*, and *Perault*, and *Mariotte*, have observed, is, to concoct and prepare the *Sap* for the Nourishment of the *Fruit*, and of the whole *Plant*; not only that which ascends from the Root, but also what they take in from without, from the Dew, and from the Rain. For there is a *regress* of the *Sap* in Plants from above downwards; and this descendent Juice is that which principally nourishes both Fruit and Plant, as has been clearly proved by the Experiments of Signior *Malpighi* and Mr. *Brotherton*.

How agreeable the *Shade* of *Plants*, let every Man say that *sits under his own Vine, and under his own Fig-tree!*[3]

How charming the Proportion and Pulchritude of the *Leaves*, the *Flowers*, the *Fruits*, he who confesses not, must be, as Dr. *More* says, *one sunk into a forlorn pitch of Degeneracy, and stupid as a Beast*.

Our Saviour says of the *Lillies* (which some, not without reason, suppose to be *Tulips*) *that* Solomon *in all his Glory was not arrayed like one of these*.[4] And it is observed by *Spigelius*, that the Art of the most skilful Painter cannot so mingle and temper his *Colours*, as exactly to imitate or counterfeit the *native* ones of the *Flowers* of *Vegetables*.

Mr. *Ray* thinks it worthy a very particular Observation, that *Wheat*,[5] which is the best sort of Grain, and affords the wholesomest Bread, is in a singular manner patient of both Extremes, both Heat and Cold, and will grow to maturity as well in *Scotland*, and in *Denmark*, as in *Egypt*, and *Guiney*, and *Madagascar*. It scarce refuses any Climate. And the exceeding *Fertility* of it is by a Pagan *Pliny* acknowledged as an Instance of the Divine Bounty to Man, *Quod eo maxime Hominem alat;*[6] one Bushel in a fit Soil, he says, yielding one hundred and fifty. A *German* Divine[7] so far plays the Philosopher on this Occasion, as to propose it for a Singularity in *Bread*, that *totum Corpus sustentat, adeo, ut in unica Bucella, omnium Mem-*

3. Adapted from 1 Kings 4:25, Mic. 4:4, and Zech. 3:10.
4. Matt. 6:29.
5. *Triticum aestivum*.
6. "Because it [nature] feeds man chiefly with it."
7. That is, Johann Arndt.

brorum totius externi Corporis, nutrimentum contineatur, illiusque Vis per totum Corpus sese diffundat.[8] A Friend of mine had *thirty-six Ears* of Rye growing from *one* Grain and on *one* Stalk.

But of our *Indian Corn,*[9] one Grain of *Corn* will produce above a *thousand.* And of *Guiney Corn,*[10] one Grain has been known to produce *ten thousand.*

The *Anatomy of Plants,* as it has been exhibited by the incomparable Curiosity of Dr. *Grew,*[11] what a vast *Field of Wonders* does it lead us into!

The most inimitable *Structure* of the Parts!

The particular *Canals,* and most adapted ones, for the conveyance of the lymphatick and essential Juices!

The *Air-Vessels* in all their curious Coylings!

The *Coverings* which befriend them, a Work unspeakably more curious in reality than in appearance!

The strange Texture of the *Leaves,* the angular or circular, but always most orderly Position of their *Fibres;* the various *Foldings,* with a *Duplicature,* a *Multiplicature,* the *Fore-rowl,* the *Back-rowl,* the *Tre-rowl;*[12] the noble Guard of the *Films* interposed!

The *Flowers,* their Gaiety and Fragrancy; the *Perianthium* or *Empalement* of them; their curious Foldings in the *Calyx* before their Expansion, with a *close Couch* or a *concave Couch,* a *single Plait* or a *double Plait,* or a *Plait* and *Couch* together, or a *Rowl,* or a *Spire,* or *Plait* and *Spire* together; and their luxuriant Colours after their *Foliation,* and the expanding of their *Petala!*

The *Stamina,* with their *Apices;* and the *Stylus* (called the *Attire* by Dr. *Grew*) which is found a sort of *Male Sperm,* to impregnate and fructify the Seed!

At last the whole Rudiments and Lineaments of the *Parent-*

---

8. "It sustains the whole body, so that in one little mouthful is contained the nourishment of all the members of the body, and its power spreads through the whole body."

9. *Zea mays.* This plant is known as corn in the United States and as maize in many other countries.

10. *Sorghum bicolor,* a native of Africa, which was introduced into the New World along with slaves. Around 1700, Indians on Bonaire Island planted Guinea corn (*Sorghum bicolor*) and maize (*Zea mays*). In Natal bread made of Guinea corn was a staple of the people's diet, and they made beer from Guinea corn. See William Dampier, *A New Voyage Round the World* (London, 1697), p. 48, and idem, *Voyages and Descriptions* (London, 1699), pt. 3, p. 111.

11. Nehemiah Grew, *The Anatomy of Plants, with an Idea of a Philosophical History of Plants* (London, 1682).

12. The word *rowl* refers to the ability of leaves to roll up. The terms used in the text describe different ways leaves rolled up.

Essay 26. Of the Vegetables

Vegetable, surprizingly lock'd up in the little compass of the *Fruit* or *Seed!*

Gentlemen of Leisure, consult my illustrious Doctor, peruse his *Anatomy of Plants,* ponder his numberless Discoveries; but all the while consider that rare Person as inviting you to join with him in adoring the *God of his Father,* and the God who has *done these excellent things,* which ought to be *known in all the Earth.*[13]

Signior *Malpighi* has maintain'd it with cogent Arguments, that the whole *Plant* is actually in the *Seed;* and he answers the grand Objection against it, which is drawn from a degeneracy of one Plant sometimes into another. One of his Answers is, *Ex morboso & monstroso affectu, non licet inferre permanentem statum a Natura intentum.*[14]

But there is no Objection to be made against *Ocular Observation.* Shew us, *Lewenhoeck,* how it is? He will give us to see, a small Particle no bigger than a Sand, contain the *Plant,* and all belonging to it, all actually in that *little Seed;* yea, in the *Nux vomica*[15] it appears even to the naked Eye, and in an astonishing Elegancy! Dr. *Cheyne* expresses himself with good assurance upon it: "*We are certain* that the *Seeds* of *Plants* are nothing but *little Plants* perfectly formed, with Branches and Leaves duly folded up, and involved *in Membranes,* or surrounded with *Walls* proper to defend them in this tender state from external Injuries; and *Vegetation* is only the unfolding and extending of these Branches and Leaves, by the force of Juices raised by *Heat* in the slender Tubes of the Plant."

Those *capillary Plants,* which all the Antients, and some of the Moderns, have taken to be destitute of *Seeds,* are by *Bauhinus*[16] and others now pronounced *Spermatophorous.* Mr. *Ray* says, *Hanc Sententiam verissimam esse Autopsia convincit.*[17]

Fr. Cæsius claims to be the first who discovered the *Seeds* of these *Plants,* with the help of a *Microscope.* One Mr. *Cole* has prosecuted the Observation, and is astonished at the small Di-

---

13. "God of his father" occurs in Gen. 46:1 and 2 Chron. 17:4; the phrase that follows is adapted from Isa. 12:5.
14. "We ought not, from a morbid and preternatural affection of the plant, to infer a permanent estate intended by Nature."
15. That is, the seeds of a tree native to South Asia, the *Strychnos nux-vomica,* which was first introduced into Germany in the sixteenth century. *Strychnos nux-vomica* is best known for its principal alkaloid, strychnine.
16. That is, Gaspard Bauhin.
17. "My own observation convinces me that this opinion is true."

mensions of the *Seeds*. The *Boxes* or Vessels that hold the *Seeds* are not half, perhaps not a quarter, so big as a Grain of Sand; and yet an *hundred Seeds* are found in one of these. *Tantam Plantam e tantillo Semine produci attentum Observatorem merito in Admirationem rapiat!*[18]

Sir *Thomas Brown* observes, That of the Seeds of *Tobacco*[19] a thousand make not one Grain; (tho *Otto de Gueric*, as I remember, says, fifty-two Cyphers with one Figure will give the Number of those, which would fill the Space between us and the Stars!) A Plant which has extended its Empire over the whole World, and has a larger Dominion than any of all the *Vegetable* Kingdom.

Ten thousand Seeds of *Harts-tongue*[20] hardly make the Bulk of a *Pepper-corn*. But now, as Dr. *Grew* notes, the Body, with the Covers of every Seed, the ligneous and parenchymous Parts of both, the Fibres of those Parts, the Principles of those Fibres, and the homogeneous Particles of those Principles, being but moderately multiplied one by another, afford an hundred thousand millions of Atoms formed in the Space of a *Pepper-corn*. But who can define how many more!

The Uses of *Trees* in various Works were elegantly celebrated, as long ago as when *Theophrastus* wrote his fifth Book of the *History of Plants*.

And what *stately Trees* do sometimes by their glorious *Height* and *Breadth* recommend themselves to a more singular Observation with us! The *Cabbage*-tree[21] an hundred and forty or fifty Foot high, as if it were aspiring to afford a Diet to the Regions above us; how noble a Spectacle!

The *Trees* which are found sometimes near twenty Foot, or perhaps more, in circumference, what capacious *Canoes* do they

---

18. "That such a large plant is produced from such a small seed should strike the attentive observer with astonishment."

19. *Nicotiana tabacum.*

20. Hart's-tongue (*Phyllitis scolopendrium*) is a fern of common growth in England, also naturalized in the northeastern United States. Its name refers to the shape of its fronds. The ferns have spores which germinate to produce gametophytes which are bisexual, zygote, and sporophyte. The plant was formerly considered one of the five great capillary herbs, and older physicians esteemed it a valuable medicine.

21. The Cabbage-tree from New Zealand (*Cordyline australis* also known as *Dracaena australis*) has a single trunk and a terminal crown of leaves suggesting a palm. The cabbage, which is the young leaves before they open, is sweet, tender, and wholesome whether eaten raw or boiled. *Cordyline* is separated from the genus *Dracaena* only by technical characters, and the names are loosely and confusingly applied to both genera. Some of about twenty species of the *Cordyline* genus are popular greenhouse subjects.

afford, when the Traveller makes them change their Element? Near *Scio*²² there is an Island called *Long-Island,* and on this Island (as *Jo. Pitts* tells us) there is a Tree of a prodigious bigness;²³ under it are *Coffee-houses,* and many Shops of several Intentions, and several Fountains of Water; and it has near forty Pillars of Marble and of Timber to support the Branches of it. It is a Tree famous to a Proverb all over *Turkey.*

Even the most *noxious* and the most *abject* of the *Vegetables,* how useful are they! As of the *Bramble*²⁴ Dr. *Grew* notes, *If it chance to prick the Owner, it will also tear the Thief.* Olaus Magnus admires the Benefits which the *rotten Barks* of *Oaks* give to the Northern People, by the *Shine,* with which they do in their long Nights direct the Traveller.²⁵ And Dr. *Merret* celebrates the *Thistles,* and the *Hop-strings,*²⁶ for the *Glass* afforded by their Ashes!

The *frugal Bit* of the old *Britons,*²⁷ which in the bigness of a *Bean* satisfied the most hungry and thirsty Appetite, is now thrown into the Catalogue of the *Res deperditæ.*²⁸

The peculiar Care which the great God of Nature has taken for the Safety of the *Seed* and *Fruit,* and so for the Conservation of the *Plant,* is by my ingenious *Derham* considered as a loud Invitation to His Praises.

They which dare shew their Heads all the Year, how securely is their *Seed* or *Fruit* lock'd up in the Winter in their *Gems,* and well cover'd with neat and close *Tunicks* there!

Such as dare not expose themselves, how are they preserved

---

22. Scio is the Italian name of Chios, a Greek Aegean island off the Karaburun Peninsula of Asiatic Turkey.

23. Perhaps a *Ficus.* This genus includes the banyan tree (*Ficus benghalensis*) of India. Starting from a single main trunk, it sends down aerial roots which themselves become trunks, thus extending the original tree over areas which may be very large.

24. That is, *Rubus,* a genus of perennial herbs, shrubs, or trailing vines, often prickly, with about 250 species. Blackberries and raspberries are important examples of *Rubus.*

25. The reference is probably to the mycelium of *Armilleria mellia,* which grows in "shoestrings" or thin black strips between the bark and wood of oak trees. These young growing fungi are luminescent. I am indebted to Professor Dean A. Glawe of the University of Illinois at Urbana-Champaign for help on this botanical point.

26. That is, *Cirsium,* plumed thistle. The genus includes a great many pernicious weeds. All hop plants belong to the genus Humulus of the family Cannabinaceae. The hop is a quick and tall-growing vine; its flowers are used for brewing, while the ashes of hop-strings were used for making glass.

27. It was said that in necessity the ancient Britons could live upon barks and roots of trees and "with a kind of meat no bigger than a beane," after which for a good time they did not hunger or thirst.

28. That is, ruined or abandoned things.

under the Coverture of the *Earth,* till invited out by the kindly Warmth of the Spring!

When the *Vegetable Race* comes abroad, what strange Methods of Nature are there to *guard* them from Inconveniences, by making some to lie down prostrate, by making others, which were by the Antients called *Æschynomenæ,*[29] to close themselves up at the Touch of Animals, and by making the most of them to shut up under their guard in the cool of the Evening, especially if there be foul Weather approaching; which is by *Gerhard* therefore called, *The Countryman's Weather-wiser!*

What various ways has Nature for the *scattering* and the *sowing* of the *Seed!* Some are for this end winged with a light sort of a *Down,* to be carried about with the *Seed* by the Wind. Some are laid in springy cases, which when they burst and crack, dart their Seed to a distance, performing therein the part of an Husbandman. Others by their good Qualities invite themselves to be swallowed by the Birds, and being fertiliz'd by passing thro their Bodies, they are by them transferred to places where they fructify. *Theophrastus* affirms this of the *Misletoe;*[30] and *Tavernier* of the *Nutmeg.*[31] Others not thus taken care for, do, by their Usefulness to *us,* oblige us to look after them.

It is a little surprizing, that *Seeds* found in the *Gizzards* of *Wildfowl,* have afterwards sprouted in the Earth; and *Seeds* left in the *Dung* of the *Cattel.* The Seeds of *Marjoram*[32] and *Strammonium,*[33] carelesly kept, have grown after seven Years.

How nice the provision of Nature for their Support in *standing* and *growing,* that they may keep their Heads above ground, and

---

29. That is, the "shy" plant, or sensitive plant, *Mimosa pudica.*

30. *Viscum album* is the mistletoe of literature. The mistletoe of North American holiday markets is *Phoradendron serotinum.* Mistletoe seeds are also spread by adhering to the feet, bills, and other parts of birds.

31. *Myristica fragrans.*

32. Sweet marjoram (*Origanum marjorana*), a perennial native of North Africa and southwest Asia, naturalized in southern Europe, which has to be sown annually in colder climates, and wild marjoram (*Origanum vulgare*), a perennial which is distributed across Europe to central Asia and grows freely in England. The marjorams are familiar kitchen herbs.

33. The common thorn apple (*Datura stramonium*), a member of the family Solanaceae, is a large and coarse plant which flowers all summer and gives off a disagreeable odor. Early botanical writers differed as to whether it is native to North America or the Old World. It is a native of North America, but the plant is now spread throughout all but the colder regions of the world. The whole plant is poisonous, the seeds especially so. In North America the plant is a familiar weed, and since the early days of the Virginia Colony has been known as the Jamestown-weed or Jimsonweed. It is sometimes grown as a source of the alkaloid hyoscyamine.

administer to our Intentions! There are some who stand by their own Strength; and the ligneous parts of these, tho' like our Bones, yet are not, like them, inflexible, but of an elastick nature, that they may dodge the Violence of the Winds: and their Branches at the top very commodiously have a tendency to an hemispherical Dilatation, but within such an Angle as makes an Æquilibration there. An ingenious Observer[34] upon this one Circumstance, cannot forbear this just Reflection: *A visible Argument that the plastick Capacities of Matter are govern'd by an all-wise and infinite Agent, the native Strictnesses and Regularities of them plainly shewing from whose Hand they come.* And then such as are too weak to stand of *themselves*, 'tis wonderful to see how they use the Help of their *Neighbours*, address them, embrace them, climb up about them, some twisting themselves with a strange *convolving* Faculty, some catching hold with *Claspers* and *Tendrels*, which are like Hands to them; some striking in rooty *Feet*, and some emitting a natural *Glue*, by which they adhere to their Supporters.

But, Oh! the glorious *Goodness* of our GOD in all these things! Lend us thy Pen, O industrious *Ray*, to declare a little of it. *Plantarum usus latissime patet, & in omni Vitæ parte occurrit. Sine illis caute, sine illis commode, non vivitur; at nec vivitur omnino: quæcunque ad victum necessaria sunt, quæcunque ad Delicias faciunt, e locupletissimo suo Penu abunde subministrant. Quanto ex iis Mensa innocentior, mundior, salubrior, quam ex Animalium Cæde & Laniena! Homo certe Natura Animal carnivorum non est; nullis ad Prædam & Rapinam armis instructum; non Dentibus exertis & serratis, non Unguibus aduncis. Manus ad Fructus colligendos, Dentes ad mandendos comparati. Non legimus ei ante Diluvium Carnes ad esum concessas. At non victum tantum nobis suppeditant, sed & Vestitum, & Medicinam, & Domicilia, aliaque Ædificia, & Navigia, & Supellectilem, & Focum, & Oblectamenta Sensuum Animique. Ex his Naribus Odoramenta & Suffumigia parantur: Horum Flores inenarrabili Colorum & Schematum Varietate & Elegantia Oculos exhilarant, & suavissima Odorum quos expirant Fragrantia, Spiritus recreant. Horum Fructus, Gulæ illecebræ Mensas secundas instruunt, & languentem Appetitum excitant. Taceo Virorem Oculis Amicum, quem per Prata, Pascua, Agros,*

---

34. Charles King (fl. 1705), the supposed author of *An Account of the Origin and Formation of Fossil Shells* (London, 1705). The book is also attributed to H. Rowland.

*Sylvas spatiantibus objiciunt; & Umbras quas contra Æstum & Solis Ardores præbent.*³⁵

70 Indeed *all* the *Plants* in the whole *Vegetable Kingdom* are every one of them so *useful,* as to *rise up* for thy Condemnation, *O Man, who dost little Good in the World.* But sometimes the Uses of one *single Plant* are so many, so various, that a wise Man can scarce behold it without some *Emulation* as well as *Admiration,* or without
75 some wishing, that if a *Metamorphosis* were to befal him, it might be into one of these. *Plutarch* reports, that the *Babylonians* out of the *Palm-tree*³⁶ fetch'd more than three hundred several sorts of Commodities.

The *Coco-tree* supplies the *Indians* with Bread, and Water, and
80 Wine, and Vinegar, and Brandy, and Milk, and Oil, and Honey, and Sugar, and Needles, and Thread, and Linnen, and Clothes, and Cups, and Spoons, and Besoms,³⁷ and Baskets, and Paper, and Nails; Timber, Coverings for their Houses; Masts, Sails, Cordage, for their Vessels; add, Medicines for their Diseases; and what can
85 be desired more? This is more expressively related in the *Hortus Malabaricus,* published by the illustrious *Van Draakenstein.*

The *Aloe Muricata*³⁸ yields the *Americans* all that their Necessities can call for. *De la Vega* and *Margrave* will inform us how this alone furnishes them with Houses and Fences, and Weapons of many
90 sorts, and Shoes, and Clothes, and Thread, and Needles, and Wine,

35. "The uses of plants are most various and extensive in every department of human life. Without them we can neither live comfortably nor conveniently; nay we could not live at all. They furnish us in abundance whatever is necessary for our food, whatever supplies us with delicacies. How much more innocent our repast, how much cleaner, how much more wholesome than what is furnished by the slaughter and butchery of animals? Man certainly is not by nature a carnivorous animal; he is furnished with no arms for prey or rapine; he has neither prominent and serrated tusks, nor crooked claws. He has only hands for gathering the fruits of the earth, and teeth for chewing them. We read not that he ever ate flesh before the deluge. But plants furnish us not only with food; they supply us with clothes, with medicines, with habitations and other structures, with ships, with furniture, with fire, and with numberless gratifications to the senses and to the mind. The sense of smelling is refreshed with their odors; their flowers with the variety and beauty of their forms and colors delight the sight. Their fruits furnish our desserts, and stimulate our languid appetite. I need not mention that most delightful verdure which is so friendly to the eye–sight, and which meets us in our walks through the meads, the pastures, the fields, and woods; nor the grateful shade they furnish against the scorching rays of the sun."
36. The palm tree (*Borassus flabellifer*) is a fan-bearing palm of many uses in India, Burma, and Ceylon.
37. That is, a bundle of twigs used for sweeping.
38. The reference is to the agave, not an aloe. The *Agave americana*, century plant, is used for rope and fiber and also the source of tequila and pulque, the distilled and fermented liquors popular in Mexico.

Essay 26. Of the Vegetables

and Honey, and Utensils that cannot be numbred. *Hernandes* will assure us, *Planta hæc unica, quicquid Vitæ esse potest necessarium facile præstare potest, si esset rebus humanis modus.*[39]

What a surprizing Diversity from the *Cinnamon-tree!*[40]

Some will have the *Plantane*[41] to be the *King of all Fruit,* tho the Tree be little more than ten Foot high, and raised not from *Seed,* but from the *Roots* of the old ones. The *Fruit* a delicate Butter, and often the whole Food that a whole Family will subsist upon.

Among the *Uses* of *Plants,* how surprizing an one is that, wherein we find them used for *Cisterns,* to preserve Water for the needy Children of Men!

The *Dropping-tree* in *Guiney,*[42] and on some Islands, is instead of *Rains* and *Springs* to the Inhabitants.

The *Bandura Cingatensium,*[43] at the end of its Leaves has long Sacks or Bags, containing a fine limpid Water, of great use to the People when they want Rains for eight or ten Months together.

The *wild Pine*[44] describ'd by Dr. *Sloane,* has the Leaves, which are each of them two Foot and an half long, and three Inches broad, so inclosed one within another, that there is formed a large Bason, fit to contain a considerable quantity of Water (*Dampier* says, the best part of a Quart) which in the rainy Season falling upon the utmost parts of the spreading Leaves, runs down by Channels into the Bottle, where the Leaves bending inwards again; come so close to the Stalk, as to hinder the Evaporations of the Water. In the mountainous, as well as in the dry and low Woods,

6 Bandura / Banduca

39. "This single plant could easily supply whatever is necessary for life, if there were moderation in human affairs."
40. *Cinnamomum zeylanicum* is true cinnamon, but the reference may be to *Cinnamomum,* a genus which includes the camphor tree and also cassia, which are widely grown throughout Asia as an adulterant of true cinnamon.
41. That is, perhaps musa species, a plantain or banana tree.
42. This tree was probably a til tree (*Ocotea foetens*). Ray describes the species as found in Guinea and nearby islands. On Hierro (Ferro), the westernmost of the Canary Islands, fresh water is scarce, and in former times there was a celebrated and almost sacred tree which dripped abundant water from its leaves. It was blown down in 1610.
43. Mather and the authors on whom he drew wrote before Linnaean binomial names were established, and botanical names have changed since that time. *Bandura cingalensium* is a case in point.
44. Probably a large *Bromeliad,* a perennial herb with stiff, pineapplelike leaves. Hans Sloane observed the plant in Jamaica and described it by the name of "*Viscum Caryophilloides Maximum flore tripetalo pallide luteo, semine filamentoso.*" See Ray, p. 242.

when there is a scarcity of Water, this *Reservatory* is not only necessary and sufficient for the nourishment of the Plant itself, but it is likewise of marvellous advantage unto Men and Birds, and all sorts of Insects, who then come hither in Troops, and seldom go away without Refreshment.

What tho there are *venomous Plants?* An excellent *Fellow of the College of Physicians*[45] makes a just Remark: "*Aloes* has the Property of promoting *Hæmorrhages;* but this Property is good or bad, as it is used; a *Medicine* or a *Poison:* And it is very probable that the most dangerous *Poisons,* skilfully managed, may be made not only *innocuous,* but of all other Medicines the most *effectual!*"

What admirable Effects of *Opium* well *smegmatized!*[46] Even *poisonous Plants,* one says of them, It may be reasonably supposed that they draw into their visible Bodies that malignant *Juice,* which, if diffused thro the other *Plants,* would make them less wholesome and fit for Nourishment.

In the *Delights* of the *Garden* 'tis not easy to hold a Mediocrity. They afford a Shadow for our *celestial Paradise.* The King of *Persia* has a *Garden* called *Paradise upon Earth.*[47] The antient *Romans* cultivated them to a degree of *Epicurism.* Some confined their *Delights* to a single *Vegetable,* as *Cato,* doting on his *Cabbage.*[48] The *Tulipists* are so set upon their gaudy Flower, that the hard Name and Crime of a *Tulipomania,*[49] is by their own Professors charged upon them; a little odd the Humour of those Gentlemen, who affected Plantations of none but *venomous Vegetables.*

But finally, the vast Uses of *Plants* in *Medicine,* are those which fallen and feeble Mankind has cause to consider, with singular Praises to the merciful God, who so pities us under the sad Effects of our Offences.

Among the eighteen or twenty thousand *Vegetables,* we have ever now and then a single one, which is a *Polychrest,*[50] and almost

45. That is, Nehemiah Grew. Note the homeopathic implications of the point made in the quotation.
46. That is, cleansed, scoured.
47. In Islam the part played by gardens in the life of the people appears to stem from the conception of Paradise, the ideal garden, as portrayed in the Koran.
48. *Brassica oleracea* var. *capitata.*
49. Tulipomania is the name given to the years from 1634 to 1638 in Holland when the price of scarce and choice tulip bulbs often exceeded the price of precious metals. A milder tulipomania occurred in France, and a similar craze occurred in England more than a century later.
50. That is, something adapted to several different uses; especially a drug or medicine serving to cure various diseases.

a *Panacæa;* or at least such an one as obliges us to say of it, as Dr. Morton speaks of the *Cortex Peruvianus;*[51] 'tis *Antidotus in Levamen Ærumnarum Vitæ humanæ plurimarum divinitus concessa.* And, *In Sanitatem Gentium proculdubio a Deo optimo maximo condita.*[52]

Among the Antients there were several Plants that bore the Name of *Hercules,* called *Heracleum,* or *Heraclea;* probably, as *Le Clerc* thinks, to denote the *extraordinary* Force of the Plants, which they compared to the Strength of *Hercules.*

Cabbage was to the *Romans* their grand *Physick,* as well as *Food,* for six hundred Years together.

Mallows[53] has been esteemed such an *universal Medicine,* as to be called *Malva Omnimorbia.*[54]

Every body has heard,

Cur moriatur homo cui Salvia crescit in hortis?[55]

The *six favourite Herbs* distinguish'd by Sir *William Temple* for the many Uses of them, namely, *Sage,*[56] and *Rue,*[57] and *Saffron,*[58]

---

51. Peruvian bark (*Cinchona officinalis*), one of about fifty species of a genus of evergreen trees of the family Rubiaceae which grow abundantly in Andean South America (Colombia to northern Peru). The bark, an important source of quinine, was first introduced into Europe by Jesuits in 1632, and in 1640 by Juan del Vega, physician of the viceroy of Peru. It was made official in the London pharmacopoeia of 1677.

52. "An antidote divinely granted for the alleviation of the many troubles of human life.".... "Undoubtedly established by the most exalted God for the health of the people."

53. Mallows consists of numerous genera (about ninety five) of widely distributed herbs of the mallow (Malvaceae) family, which furnishes many ornamental, fiber (including cotton), medicinal, and some food plants, and are scattered in tropical and temperate regions over the world.

54. "Mallow of all diseases." That is, a plant regarded as a panacea.

55. "Why should a man die who has sage in his garden?"

56. Sage (*Salvia*) is an annual, biennial, or perennial plant comprising about seven hundred species belonging to the mint family and distributed throughout the tropical and temperate world. *Salvia* is cultivated for ornamental, medicinal, and culinary purposes. Common or garden sage (*Salvia officinalis*) is used for seasoning.

57. Rue is a hardy, evergreen, somewhat shrubby plant, native of southern Europe, whose various species comprise the genus *Ruta* of the family Rutaceae. Common rue (*Ruta graveolens*), usually the only one in cultivation, has a disagreeable odor but has been valued for centuries for its medicinal properties. The Romans introduced rue into England, where it is first mentioned in a book in 1562. Rue later became one of the best-known simples for medicinal and homely uses.

58. Saffron (*Crocus sativus*) is a bulbous ornamental known only in cultivation. It is grown for the yellow stigmas, the part used in medicine, cookery, and the arts. Saffron was imported into England from the East, and was once grown extensively around Saffron Walden in Essex.

and *Alehoof*,⁵⁹ and *Garlick*,⁶⁰ and *Elder*,⁶¹ if they were more frequently used, would no doubt be found vastly beneficial to such as place upon *health* the Value due to such a *Jewel*.

The *French* do well to be such great Lovers of *Sorrel*,⁶² and plant so many Acres of it; it is good against the *Scurvy*, and all ill Habits of Body.

The Persuasion which Mankind has imbib'd of *Tobacco* being good for us, has in a surprizing manner prevail'd! What incredible Millions have *suck'd in* an Opinion, that it is an *useful* as well as a *pleasant* thing, for them to spend much of their Time in drawing thro a Pipe the *Smoke* of that lighted Weed! It was in the Year 1585, that one Mr. *Lane* carried over from *Virginia* some *Tobacco*, which was the first that had ever been seen in *Europe;* and within an hundred Years the *smoking* of it grew so much into fashion, that the very Customs of it brought *four hundred thousand Pounds a Year* into the *English* Treasury.

It is doubtless a *Plant* of many Virtues. The *Ointment* made of it is one of the best in the Dispensatory. The Practice of *smoking* it, tho a great part of them that use it might very truly say, *they find neither Good nor Hurt by it;* yet it may be fear'd it rather does more *Hurt* than *Good*.

"May God preserve me from the indecent, ignoble, criminal *Slavery*, to the mean Delight of *smoking a Weed*, which I see so many carried away with. And if ever I should *smoke* it, let me be so wise as to do it, not only with *Moderation*, but also with

---

59. Alehoof is ground ivy (*Glechoma hederacea*, also known as *Nepeta glechoma*), a perennial root of the mint family which throws out long trailing stems that make a dense mat. One of the most common plants, it is suitable for ground cover or hanging over rocks. The leaves of the plant were used in England to clarify and improve the flavor of beer until the reign of Henry VIII; hence it bore the name Alehoof. Ground ivy also has medicinal uses.

60. Garlic (*Allium sativum*) is a rather small, onionlike plant whose bulbs are used in cooking. Common garlic is of great antiquity as a cultivated plant. According to Pliny, garlic and onion (*Allium cepa*) were invocated as deities by the Egyptians at the taking of oaths.

61. Elder (*Sambucus*) is an attractive shrub of the honeysuckle family. The European elder (*Sambucus nigra*) is a tree familiar in the English countryside and gardens. Its bark, leaves, flowers, and berries are put to a variety of medicinal, culinary, and other uses.

62. Sorrel, dock, or dock sorrel (*Rumex*) is a perennial herb of more than a hundred species widely distributed in temperate regions. The plant is not much cultivated, since many species are garden weeds, but some are grown as greens. French sorrel (*Rumex scutatus*) has large succulent leaves and is said to have been introduced into England in 1596. Sorrel has a high vitamin C content and was taken to prevent scurvy.

such Employments of my Mind, as I may make that Action afford me a Leisure for!"

Methinks *Tobacco* is but a poor *Nepenthe*,[63] tho the Takers thereof take it for such an one. It is to be feared the *caustick Salt*[64] in the *Smoke* of this Plant, convey'd by the *Salival Juice* into the Blood, and also the *Vellication*[65] which the continual use of it in *Snuff* gives to the *Nerves*, may lay Foundations for Diseases in Millions of unadvised People, which may be commonly and erroneously ascribed to some other Original.

It is very remarkable, that our compassionate God has furnish'd all Regions with *Plants* peculiarly adapted for the relief of the *Diseases* that are most common in those Regions. 'Tis Mr. *Ray*'s Remark, *Tales Plantarum Species in quacunque Regione a Deo creantur, quales Hominibus & Animalibus ibidem natis maxime conveniunt.*[66]

Yea, *Solenander* affirms, that from the Quantity of the *Plants* most plentifully growing in any place, he could give a probable Guess what were the *Distempers* which the People there were most of all subject to.

Benerovinus[67] has written a Book, on purpose to shew that every Country has every thing serving to its Occasions, and particularly *Remedies* for all the Distempers which it may be afflicted with.

Can we be any other than charmed with the Goodness appearing in it, when we see the *Plants* every where starting out of the *Earth*, and hear their courteous Invitation, *Feeble Man, I am a Remedy, which our gracious Maker has provided for thy Feebleness; take me, know me, use me, thou art welcome to all the Good that is to be found in me!*

Yea, such are the Virtues of the *Vegetable World*, that it is no rare thing to see a whole Book written on the Virtues of one single *Vegetable*.

How long is *Rosenbergius* on the *Rose*, in his *Rhodologia!*[68] *Whitaker* will have the *Vine* to be the *Tree of Life*, in his Treatise on

---

63. That is, a drink or drug supposed to bring forgetfulness of trouble or grief.
64. The word *salt* possessed a multitude of meanings in the early eighteenth century, and "caustic salt" is a term unknown to modern science.
65. That is, a twitching; a convulsive movement of a muscle or other part of the body.
66. "Such species of plants are created by God in every country as are most adapted to the constitutions of its inhabitants."
67. The reference is to Johan van Beverwyck (Johannes Beverovicius).
68. Johann Carl Rosenberg (fl. 1622–28), a Strassburg physician, published *Rhodologia, seu philosophica-medica generosae rosae descriptio*, new ed. (Frankfurt, 1631).

the Blood of it. *Alsted* has entertained us with a yet greater variety on that *Plant of Renown*.⁶⁹

I was going to mention the *Anatomia Sambuci,* written by a *German* Philosopher.⁷⁰

But I presently call to mind such a vast Number of Treatises published, each of them on one *single Vegetable,* by the *Naturæ Curiosi*⁷¹ of *Germany,* that a *Catalogue* would be truly too tedious to be introduced.

If the *Coral* may pass for a *Vegetable, Garencieres* has obliged us with a whole Treatise upon it.

But then we have one *far-fetch'd* and *dear-bought* Plant, on which we have so many Volumes written, that they alone almost threaten to become a *Library.* TEA⁷² is that charming Plant. Read *Pecklinus*'s Book *de Potu Theæ,* and believe the medicinal and balsamick Virtues of it; it strengthens the *Stomach,* it sweetens the *Blood,* it revives the *Heart,* and it refreshes the *Spirits,* and is a Remedy against a World of Distempers. Then go to *Waldschmidt,* and you'll find it also to brighten the *Intellectuals.* When *Prose* has done its part, our *Tate* will bring in *Verse* to celebrate the sovereign Virtues of it.

> *Innocuos Calices, & Amicam Vatibus Herbam*
> *Vimque datam Folio.*⁷³

At last it shall be the very Θεὰ⁷⁴ of the Poet.

> *Whilst TEA, our Sorrows safely to beguile,*
> *Sobriety and Mirth does reconcile:*
> *For to this Nectar we the Blessing owe,*
> *To grow more wise as we more chearful grow.*

There is a Curiosity observed by Mr. *Robinson*⁷⁵ of *Ousby,* that

---

69. The reference is probably to a passage in Alsted (pp. 444–50), which discusses grapevines, the pressing of grapes, and abuses of the grapevines.

70. Martin Blochwitz, a German physician and botanist, published *Anatomia sambuci* (Leipzig, 1631). It was translated as *Anatomia Sambuci: or The Anatomy of the Elder* (London, 1655).

71. That is, naturalists or scientists. Mather is probably thinking of the authors who published in the volumes popularly known as "the German Ephemerides." See Essay 27, n. 83.

72. *Camellia sinensis.*

73. "Harmless cups and the herb friendly to poets and the power given by the leaf."

74. *Thea,* or "goddess." Note the play on words. The Greek word means goddess and suggests tea.

75. That is, Thomas Robinson.

should not be left unmentioned; it is, that *Birds* are the *natural Planters* of all sorts of *Trees;* they disseminate the *Kernels* on the Earth, which brings them forth to perfection. Yea, he affirms, that he hath actually seen a great Number of *Crows* together planting a Grove of *Oaks;* they first made little Holes in the Earth with their Bills, going about and about, till the Hole was deep enough, and then they dropt in the *Acorn,* and cover'd it with Earth and Moss. At the time of his writing, this young Plantation was growing up towards a *Grove of Oaks,* and of an height for the *Crows* to build their Nests in.

In *Virginia* there is a Plant called *The James-Town-Weed,*[76] whereof some having eaten plentifully, turn'd *Fools* upon it for several Days; one would blow up a Feather in the Air, another dart Straws at it; a third sit stark naked, like a Monkey, grinning at the rest; a fourth fondly kiss and paw his Companions, and snear in their Faces. In this frantick State they were confined, lest they should kill themselves, tho there appear'd nothing but Innocence in all their Actions. After eleven Days they return'd to themselves, not remembring any thing that had pass'd.

My Friend, a *Madness* more senseless than that with which this *Vegetable* envenoms the Eaters of it, holds thee in the stupefying Chains thereof, if thou dost not behold in the whole *Vegetable Kingdom* such Works of the glorious Creator, as call for a continual Admiration.

¶. It is a notable Stroke of Divinity methinks which *Pliny* falls upon, *Flores Odoresque indiem gignit Natura, magna (ut palam est) Admonitione hominum.*[77]

"The Man began to be cured of his *Blindness,* who could say, *I see Men, like Trees, walking.*[78] That Man is yet perfectly *blind* who does not *see Men, like Trees,* first *growing* and *flourishing,* then *withering, decaying, dying.*

"The *Rapæ Anthropomorphæ,*[79] and some other *Plants,* that have grown with much of an *Human Figure,* to be fancied on

---

76. The common thorn apple (*Datura stramonium*), also known as Jimsonweed. See n. 33 above.
77. "Nature brings forth blossoms and their perfumes only for a day—a great warning, which is obvious, to men."
78. Mark 8:24.
79. The reference suggests the mandrake (*Mandragora officinarum*) from southern Europe. The root of this plant, like ginseng, resembles the human body, and with such a signature the root was utilized as a cure for practically every malady known to man. Its major importance was as a painkiller and an aphrodisiac.

them, have been *odd things*. But there are Points wherein all *Plants* will exhibit something of the *Human Figure*.

"The *Parts of Plants* analogous to those in an *Human Body*, are notably enumerated by *Alsted* in his *Theologia Naturalis*. The Analogy between their States and ours would be also as *profitable* as *reasonable* a Subject of Contemplation.

"And I hope the *Revival* of the *Plants* in the *Spring* will carry us to the Faith of our own *Resurrection from the Dead*.[80]

"And of the *Recovery* which the *Church* will one day see from a *Winter* of *Adversity;* the *World* from a *Winter of Impiety*: The *Earth* shall one day be filled with the *Fruits of Righteousness*,[81] however barren and horrid may be the present Aspect of it.

"A Man famous in his day (and in ours too)[82] thought himself well accommodated for devotionary Studies, tho he says, *Nullos se aliquando Magistros habuisse nisi Quercus & Fagos*.[83]

"I will hear these *Field-Preachers*, their loud Voice to me from the *Earth*, is the same with what would be uttered by *Angels flying thro the midst of Heaven; Fear God, and glorify him!*[84]

"One[85] thus articulates the Vegetable Sermons: *Ecce nos, O increduli filii hominum, nuper mortui eramus, at nunc revivimus. Vetus nostrum Corpus ac Vestimentum deposuimus, & novæ Creaturæ factæ sumus. Facite vos nunc aliquid simile*. And again, *Dum in hac miserrima Vita estis, nolite de Corpore esse solliciti; nostri memores estote, quas Creator honestissime coloratis Vestibus induit, quotannis per tot Millenarios, jam inde ab exordio Mundi*. And once more, *Ecce vires nostræ, non nobis ipsis, sed vobis deserviunt. Non nostro Bono floremus, sed vestro. Imo Divina Bonitas vobis floret per nos, ut dicere possitis, Dei Benignitatem in nobis florere, suoque Odore suavissimo vos recreare*.[86]

---

80. Luke 20:35, Acts 4:2, and Rom. 1:4.
81. Phil. 1:11.
82. That is, Johann H. Alsted.
83. "That he never had any teachers, except oak trees and birch trees."
84. Adapted from Rev. 14:6–7.
85. That is, Johann Arndt.
86. Mather's meaning is made clearer by quoting (in translation) Arndt's immediately preceding lines first: "So, then, from the earth proceed all the varieties of plants and vegetables, having exchanged their old attire for a new and delicate dress. The tattered garments of the preceding year being decayed and dead, they come forth with exquisite beauty, odor, and color, and, as it were, preach to mankind in words such as these." Then follows the passage which Mather quotes with some omissions: "Look upon us, ye unbelieving sons of men; we were dead, and are now alive again. We have laid aside our old garments and bodies, and are now renewed.

"A famous *German* Doctor of Philosophy[87] declares that he found it impossible for him to look upon the *Vegetable World* without those Acclamations, *Psalm* cxxxix. 6. *The Knowledge of these things is too wonderful for me, it is high, I cannot attain to it.*

"The pious *Arndt* observes, that every Creature is enstamp'd with Characters of the Divine Goodness, and brings Testimonies of a good Creator. Our *Vine* so calls upon us, *Scias, O homo, hanc Liquoris mei Suavitatem, qua Cor tuum recreo, a Creatore meo esse.*[88] Our *Bread* so calls upon us, *Vis ista, qua famem sublevo, a Creatore meo, & vestro mihi obtigit.*[89] It is a Saying of *Austin's*[90] *Deum Creaturas singulas guttula Divinæ suæ Bonitatis aspersisse, ut per illas homini bene sit.*[91]

"A devout Writer[92] treats us with such a Thought as this: Our God is like a tender Father, who, when the Infant complies not presently with his Calls, allures him with the Offer of pleasant *Fruits* to him. Not that the Child should stop in the Love of the *Apple,* the *Plumb,* the *Pear,* but be by the *Fruits* drawn to the Love and Obedience of the *Father* that gives them. Our heavenly Father calling on us in his *Word,* gives us also *Rain from Heaven, and fruitful Seasons,*[93] to engage our Love and Obedience. *Quæ sane Beneficia aliud nihil sunt, quam tot manus & Nuncii Dei, parati ad ipsum Deum nos deducere, illiusque amorem altius animis nostris insinuare, ut ipsum tandem Datorem in Creaturis & Donis suscipere discamus.*[94]

"Among other Thoughts of Piety upon the *Vegetable World,* some have allow'd a room for this; the strong Passion in almost

---

Do ye also imitate us. . . . Consider us, whom the God of nature has annually, for so many thousands of years since the first creation to this time, provided with beautiful clothing, as an argument of his bounty and goodness. . . . Consider our virtues and qualities, which are given not for our, but for your benefit. We bloom and blossom, not for our good, but yours: yea, the blessing of God blossoms through us."

87. That is, Johann H. Alsted.
88. "Consider, O man, that the sweetness of my juice, with which I cheer your heart, is the gift of my Creator."
89. "That power, by which I satisfy your hunger, is bestowed on me by my Creator and yours."
90. That is, Augustine.
91. "That God has, as it were, sprinkled some drops of his divine goodness upon all the creatures, that they might contribute to the happiness of men."
92. That is, Johann Arndt.
93. Acts 14:17.
94. "All these blessings are so many messengers sent from God to draw us to himself, and to instruct us how to taste the goodness of the Giver and Creator in that of the creature."

40 all Children for *Fruit;* by tendring *Fruits* to them, you may draw them to any thing in the World. May not this be a lasting *Signature* of the *first Sin,* left upon the Minds of our Children! An Appetite for the *forbidden Fruit.* When we see our Children greedy after *Fruits,* a remembrance and repentance of *that Sin*
45 may be excited in us."

Add this: *Quid prodest ope Creaturarum vivere, si Deo non vivitur?*[95]

A good Thought of a *German* Writer:[96]

Sol & Luna, totusque Mundus Sydereus, luce sua Deum collaudunt. Terra Deum laudat, cum viret & floret. Sic Herbæ & Flosculi Opificis
50 *sui Omnipotentiam & Sapientiam commendant Odore, Pulchritudine, & Colorum varia Pictura: Aves Cantu & Medulatione; Arbores Fructibus; Mare Piscibus; omnes Creaturæ laudant Deum, dum illius mandata exequuntur. Colloquuntur nobiscum per divinitus ipsis insitas Proprietates, manifestantes opificem suum, & exhortantes nos ad ipsum*
55 *laudandum.*[97]

ESSAY XXVII. *Of* INSECTS.

WE are hastening into the *Animal World.* Here we soon find a Tribe vastly numerous, called by *Aristotle* ἔντομα,[1] and by *Pliny* therefore *Insecta,* because of their having certain Incisures
60 and Indentings about their Bodies.

The *French* Philosopher[2] does well to rebuke us for calling these *imperfect Animals,* for they want no Parts, either *necessary* or *con-*

49 *cum / dum*

---

95. "What does it profit us to live with the wealth of the world unless we live unto God?"

96. That is, Johann Arndt.

97. "The sun, the moon, and all the host of heaven, when they give their light, bear witness at the same time to the majesty and goodness of Him that made them. The earth praises God when it is fruitful and flourishing. The herbs and flowers, by their fragrance, beauty, and variety of colors, show forth the might and wisdom of their Maker. The birds with their songs; the trees with their fruits; the sea with its inhabitants; in short, all the creatures in their several places, praise the God that made them, whilst they fulfill his will, and answer the end for which they were created. And not only so, but they call upon mankind, by the virtues and powers which God has implanted in them, as witnesses of his wisdom and goodness, to praise and glorify God."

1. *Entoma,* or "insects." The classification of living organisms has greatly changed since Mather wrote. In the five-kingdom system of plant and animal classification

venient for them; they are *complete* in their Kind, and the Divine Workmanship is astonishing! *Pliny* shall here correct us, *In his tam parvis atque nullis, quæ Ratio, quanta Vis, quam inextricabilis perfectio!*[3]

Even the poor *Ephemeron*,[4] whose whole Period of Life is but six or seven Hours, who is bred and born, and lives, and goes thro all his Operations, and expires, and goes into his Grave, all within this little Period, must not be thrown into a Class of *imperfect Animals;* nor may it be said of it, that it is *made in vain.*[5]

We enjoy an excellent *Ray*, who in his *Methodus Insectorum*[6] has distinguish'd to us the several Kinds of *Insects*.

Of *Insects*, there are some which do *not change* their Form.

Some of these ἀμεταμόρφωτα[7] are *without Feet;* these are either *terrestrial*, produced *on* the Earth or *in* the Earth, (whereto *Snails* may be referred) or within the *Bowels* of Animals; or else *aquatick*, whereof some are *greater*, which have a peculiar way of moving, by first fixing their *Head* on the Ground, and then drawing up their *Tail* towards it; some are *lesser*, having a different way of crawling; and among these there are both *round* ones and *flat* ones.

But then there are some *having Feet.*

There are *Hexapoda*, or six-footed ones; of these there are some *terrestrial* ones, both of a *larger* sort, and of a *smaller:* of the *smaller*, there are about *five* which molest the Bodies of other living Creatures; and as many that give not that Molestation. There are other *aquatick* ones.

There are also *Octapoda*, or eight-footed ones; of these there are some that have a *Tail*, as the *Scorpion;* and some that have

---

developed by Robert Whittaker in 1969 and since modified, Animalia is one kingdom; its largest phylum is Arthropoda, which includes the class Insecta. Mather discusses many insect species, but some of the creatures he treats are now arranged in other classes of Arthropoda: e.g., crabs (Crustacea); spiders, mites and ticks, and scorpions (Arachnida). Some of the species described here are now classified within other phyla: e.g., flatworms (Platyhelminthes); nematodes and roundworms (Nematoda); snails (Mollusca); earthworms, aquatic worms, and leeches (Annelida), and toads and frogs (Chordata).

2. That is, Nicolas Andry.
3. "Whereas in these minute nothings what method, what power, what labyrinthine perfection is displayed."
4. That is, the mayfly, any member of the insect order Ephemeroptera. The nymphs are aquatic and usually live in clear water for one to four years before emerging in great numbers. The adults usually live only a few hours.
5. Adapted from Ps. 89:47 and Jer. 8:8; cf. Isa. 45:18.
6. John Ray, *Methodus insectorum: seu insecta in methodum aliqualem digesta* (London, 1705).
7. "Things whose form does not change."

none, as the *Spider;* whereof one sort spins no Web; three sorts are *Spinsters.* To these add the *Ticks,* and the *Mites.*

Yea, there are *Tessareskaidecapoda,* or fourteen-footed ones; particularly the three sorts of *Aselli.*[8] More than so, there are *twenty-four-footed* ones, whose eight Fore-Feet are lesser ones, and sixteen Hinder-Feet are larger ones.

More than this, there is a sort of *thirty-footed* ones: but as being tired with specified Numbers for the Feet of these curious things, the rest we call *Polypoda,* or many-footed ones; of these there are some on the *Land,* and others in the *Water.*

Of *Insects,* there are others who do *undergo a Change.* Tho *Squammerdam* (who has given the best Account of these) observes, that this is improperly affirmed of these μεταμορφούμενα,[9] since there is no real *Transformation* of these, but only an Explication of the Parts of the Animal, which were before latent in *Miniature,* and like the Plant in the Seed.

Of these there are some, in whose Transmutation there is no *Rest* or *Stop* between the old and the new Form, and who don't lose their Motion at the time of their shifting the *Pellicula.*[10] And there are some, in whom the *Vermiculus*[11] leaving the former Shape of the *Nympha,* with which it appeared in the Egg, and subsisted without Food, now beginning to feed, hath its Parts visibly increased and stretched out, and takes the Form of a new *Nympha,* which is not without motion, and from thence becomes a *Flyer.*

To the former *Species* of Transmutation there belong many sorts, thirteen at least; to the second a vast multitude more. And among the rest, the multitudinous Armies of *Butterflies,* which being divided into *diurnal* and *nocturnal;* of the former sort alone there is about *fifty* several Kinds observed in *England.*

There is a third *Species* of Transmutation, which is a sensible Change from a *Vermiculus* to a *flying Insect,* but yet with a sensible *Rest* or *Stop* between one Form and the other. The *Flesh-Flies*[12] belong to this, and so do some other Kinds.

Before we go any further, we will make a pause upon an Ob-

---

8. The Aselli (sing. Asellus) are isopods in the class Crustacea. The Asellota, a suborder of the Isopoda, are usually divided into three major groups: Paraselloidea, Aselloidea (which comprises most freshwater asellotes), and Stenetrioidea.
9. "Things whose form is changed."
10. That is, skin or hide.
11. That is, a small worm or grub. The two forms of change described are now known respectively as hemimetabolic and holometabolic.
12. Probably insects in the family Sarcophagidae, order Diptera.

servation, thus expressed by Mr. *Barker* in his *Natural Theology;* for it is upon a Matter which occurs in the View of all Creatures, that now remain for our Contemplation; yea, the *Vegetables* too have themselves exemplified it. "Whence is it that those two natural Principles of *Self-Preservation* and *Self-Propagation,* are so inviolably founded in the Nature of all living Creatures, even those that have *no Reason,* as well as those that have; both which are necessary to the Subsistence of the Universe? May not we hence easily argue, that surely this was done *intentionally* for such an *End?* And if *intentionally,* then it is done by *Reason;* and if by *Reason,* it must be by His *Reason* that first made this Universe."

Dr. *Gorden* adds to the Assurances which all the Inquisitive before him have given us, that no *Insects* are bred of *Corruption,* but all *ex Ovo.*[13]

He also observes, that the Females of all *Flies* put their Spawn in or near those places where the *Eruca's,*[14] which are hatch'd out of them, are to have their *Food.*

He observes likewise, that there is a kind of *Gluten,* by which the Females fasten their *Eggs* to the bearing Buds of Trees, at such a rate, that the *Rain* cannot wash them off.

And he observes that these *Eggs* will not be hurt by the greatest *Frost* that can happen.

Mr. *Andry* in his Book, *De la Generation de Vers dans le corps de l'Homme,* takes notice of a Mistake in the Antients, who denied *Breath* to the *Insects* on the score of their wanting *Lungs;* for *Insects* have a greater number of *Lungs* than other Animals. The Antients also thought that the *Insects* had no *Blood,* because many of them had not a *red Liquor* like ours; but this too was a Mistake, 'tis not the *Colour,* but the *Intent* of the *Liquor* that is to be considered in this Case. It was likewise the Belief of the Antients, that the *Insects* had no *Hearts;* whereas our Microscopes now convince us of the contrary. And the *Silk-worms* particularly have a continued Chain of *Hearts,* from the Head almost to the extremity of the Tail. And it is the number of *Lungs* and *Hearts* that occasions those *Insects*

---

13. That is, from an egg.
14. An eruca is a caterpillar. The body of the larva is cylindrical and with legs. An immature fly is not a caterpillar. Fly larvae are vermiform or maggotlike; the bodies are elongate and legless.

to give signs of Life a long while after they are divided into several parts.

Mr. *Poupart* affirms, that the *Earth-worms* and the *Round-tail'd Worms,* which are found in the *Intestines* of Animals, as also *Snails* and *Leeches,* are *Hermaphrodites;* but such *Worms* as become *Flies* are not so, rather they are of no Sex, but are *Nests* full of Animals.

The *spontaneous Generation* of *Insects* has at last been so confuted by *Redi,* and *Malpighi,* and *Squammerdam,* and our excellent *Ray,* and others, that no Man of Sense can any longer believe it. Indeed such a *spontaneous Generation* would be nothing less than a *Creation.* That all Animals are generated of *Parent Animals,* is a thing so cleared up from Observation and Experiment, that we must speak of it in the Language of those who have lately writ of it, *Nous croyons absolument.*[15] And of their *Generation* any other way, we cannot but use the Language of Dr. *Lyster, Non inducor ut credam.*[16] If an *Insect* may be *equivocally*[17] *generated,* then, as Dr. *Robinson*[18] justly enquires, *why not sometimes a Bird, yea, a Man? Or why no new Species of Animals now and then? For there is as much Art shewn in the Formation of those, as of these.* Dr. *Cheyne* assures us, nobody now-a-days, that understands any thing of Nature, can so much as imagine, that any Animal, how abject soever, can be produced by an *equivocal Generation,* or without the Conjunction of Male and Female *Parents,* in the same or in two different Individuals. And there are very few who have considered the Matter, but what own that every *Animal* proceeds from a præ-existent *Animalcule,* and that the *Parents* conduce nothing but a convenient *Habitation* to it, and suitable *Nourishments,* till it be fit to be trusted with Light, and capable of enjoying the Benefits of the Air. There is nothing in the *Animal Machine,* but an inconceivable number of branching and winding *Canals,* filled with Liquors of different natures, going a perpetual round, and no more capable of producing the wonderful Fabrick of another Animal, than a thing is of making itself. There is besides in the *Generation* of an Animal, a necessity that the *Head, Heart, Nerves, Veins* and *Arteries,* be formed at the same time, which never can be done by the motion of any Fluid, which way soever moved.

63 *Hermaphrodites / Hermophrodites*

15. "We believe it absolutely."
16. "I cannot be persuaded to believe."
17. That is, spontaneously generated.
18. That is, Tancred Robinson.

Essay 27. Of Insects

Great GOD, Thou art the Father of all things; even the Father of Insects, as well as the Father of Spirits:[19] *And Thy Greatness appears with a singular Brightness in the least of Thy Creatures!*

Concerning *Frogs* generated in the *Clouds*, there has been a mighty Noise; the *Thunder* scarce makes a greater! But Mr. *Ray* says well, it seems no more likely than *Spanish* Gennets[20] begotten by the *Wind*, for that has good Authors too. He adds, *He that can swallow the raining of Frogs, hath made a fair Step towards believing that it may rain Calves also; for we read that one fell out of the Clouds in* Avicen's[21] *Time*. *Fromondus*'s Opinion, that the *Frogs* which appear in great multitudes after a Shower, are not indeed generated in the Clouds, but are coagulated of *Dust*, commix'd and fermented with *Rain-water*, is all over as impertinent. It is very certain that *Frogs* are of two different Sexes, and have their spermatick Vessels; and their Copulation is notorious (*per integrum aliquando Mensem continuata*)[22] and after the Spawn must be cast into the Water, where the Eggs lie in the midst of a copious Gelly; then must appear a Feetless *Tadpole*, in which Form it must continue a long while, till the Limbs be grown out, and it arrives to the perfect Form of a *Frog*. To what purpose all this, if your way, Gentlemen, [*Fromondus*, and the rest] may suffice?

Frogs appearing in such multitudes upon *Rains*, do but come forth upon the Invitation which the agreeable Vapor of *Rain-water* gives to them. And for some such reason we are commonly entertain'd with such Armies of them in the cool *Summer-Evenings*, that we wonder where they have been lurking all the Day. Monsieur *Perrault*, upon the Dissection of the *Falling-Frogs*, which the *equivocal Gentlemen* so teaze us with, found their *Stomachs* full of Meat, and their *Intestines* of Excrement. The inquisitive Mr. *Derham*, on his meeting with *Frogs* in a prodigious Number, crossing a sandy Way just after a Shower, pursued the Matter with his usual Exactness, and he soon found the Colony issue from an adjacent Pond, who having pass'd thro their *Tadpole-State*, and finding the Earth moistned for their March, took the opportunity to leave their old *Latibula*,[23] where they had now devour'd their proper

---

19. Adapted from Eph. 4:6 and Heb. 12:9.
20. Gennet is an obsolete form of jennet, meaning a small Spanish horse.
21. Avicenna or Ibn Sina (980–1037), Arabian writer on medicine and philosophy, was noted for his commentaries on Aristotle.
22. "Continuing for a whole month sometimes."
23. That is, hiding places or dens.

Food, and seek a more convenient Habitation. Or what if we suppose them, at least in their Spawn, fetch'd up into the *Clouds* by the *Sun,* and kept there till grown into the State wherein they fall down from thence, as it has been affirmed they have on Vessels at Sea?

As to the *Worms* and other Animals bred in the Intestines of Man and Beast, it is Dr. *Robinson's* Remark, *I think it may be proved, that the vast variety of Worms found in almost all the Parts of different Animals, are taken into the respective Bodies by Meats and Drinks.*

Even the *Maggots* which grow in the Back of the common *Caterpillar,* are by their Parents lodg'd there, as a proper Apartment for them.

The *Toads* found in the midst of *Trees,* nay, and of *Stones,* when they have been sawn asunder, no doubt they grew of a *Toad-Spawn,* which fell into that Matter before the Concretion thereof.

The vulgar Opinion, that the *Heads* or *Clothes* of uncleanly People do breed *Lice,* or that *Mites* are bred in *Cheese,* Mr. *Ray* notes, is a *vulgar Error:* he affirms, that all such Creatures are produced of *Eggs* laid in such places by their *Parents;* Nature has endued them with a wondrous Acuteness of Scent and Sagacity, whereby they can, tho far distant, find out such places, and make towards them; and tho they seem so slow, yet it has been found that in a little time they will march a considerable way to find out a convenient Harbour. Here Mr. *Ray* makes a *Pause of Religion;* says he, "I cannot but look upon the strange Instinct of this noisome and troublesome Creature the *Louse,*[24] of seeking out foul and nasty Clothes to harbour in, as an Effect of Divine Providence, design'd to deter Men and Women from Sordidness and Sluttishness, and provoke them to *Cleanliness.* God himself hates *Uncleanness,* and turns away from it, [*Deut.* xxiii. 12, 13, 14.] But if God requires and is pleased with *Bodily Cleanliness,* much more is He so with the *Pureness* of the *Mind. Blessed are the pure in Heart, for they shall see God!*"[25]

The *Eyes* of *Insects* have in them what is very admirable! Their great necessity for accurate Vision is, in the reticulated *Cornea* of their Eyes, admirably provided for; it is a most curious piece of Lattice-work in which every *Foramen*[26] is of a lenticular nature,

---

24. That is, *Pediculus humanus.*
25. Matt. 5:8.
26. That is, an opening or orifice for the protrusion of an organ.

Essay 27. Of Insects 157

and enables the Creature to see every way without any Time or Trouble; probably every *Lens* of the *Cornea* has a distinct Branch of the *Optick* Nerve ministring to it.

Spiders are mostly *octonocular*;[27] some, as Mr. *Willoughby* thought, *senocular*.[28] *Flies* are *multocular*,[29] having as many Eyes as there are Perforations in their *Corneæ*. The greatest part of the Head of that prædatious Insect, the *Dragon-Fly*,[30] is possessed by its *Eyes*.

Tho we say, As *blind as a Beetle*,[31] Mr. *Lewenhoeck* has discover'd at least *three thousand Eyes* in the *Beetle*.

Insects have their *Antennæ*, by which they not only *cleanse* their *Eyes*, but also *guard* them; their Eyes being fitted mostly to see *distantial Objects*, these *Feelers* obviate the Inconvenience of their too rashly running their Heads against Objects that may be very near to them.[32]

And many of them are, as Mr. *Derham* observes, most surprizingly beautiful.

The Mechanism in those that *creep* is most exquisitely curious. What can exceed the *Oars* of the *Amphibious Insects*, that *swim* and *walk*?[33] Their hindmost Legs are made most nicely, with commodious flat Joints, and Bristles on each side thereof towards the ends, serving for *Oars* to swim; and nearer the Body are two stiff *Spikes*, to enable them to walk, as they have occasion.

An incomparable provision is made in the *Feet* of such as walk or hang on smooth Surfaces; divers of these, besides their acute and hooked *Nails*, have also skinny *Palms* on their Feet, which enable them to stick on Glass, and other smooth Bodies, thro the Pressure of the *Atmosphere*.

The great Strength and Spring in the *Legs* of such as *leap*, is very notable; and so are the well-made *Feet* and strong *Talons* of such as *dig*.

27. That is, eight-eyed.
28. That is, six-eyed.
29. That is, many-eyed.
30. The dragonfly (suborder Anisoptera, order Odonata) is among the best-known insects.
31. "As blind as a beetle" was perhaps a fairly common proverb by the end of the seventeenth century. Erasmus used it as early as 1529 and again in *The Praise of Folly* (1549). English authors employed the proverb on at least nine occasions between 1544 and 1670.
32. In addition to the mechanoreceptors described, all insects also have chemoreceptors.
33. The water boatman (*Sigara atropodonta*) of the Corixidae family in the Hemiptera is an example.

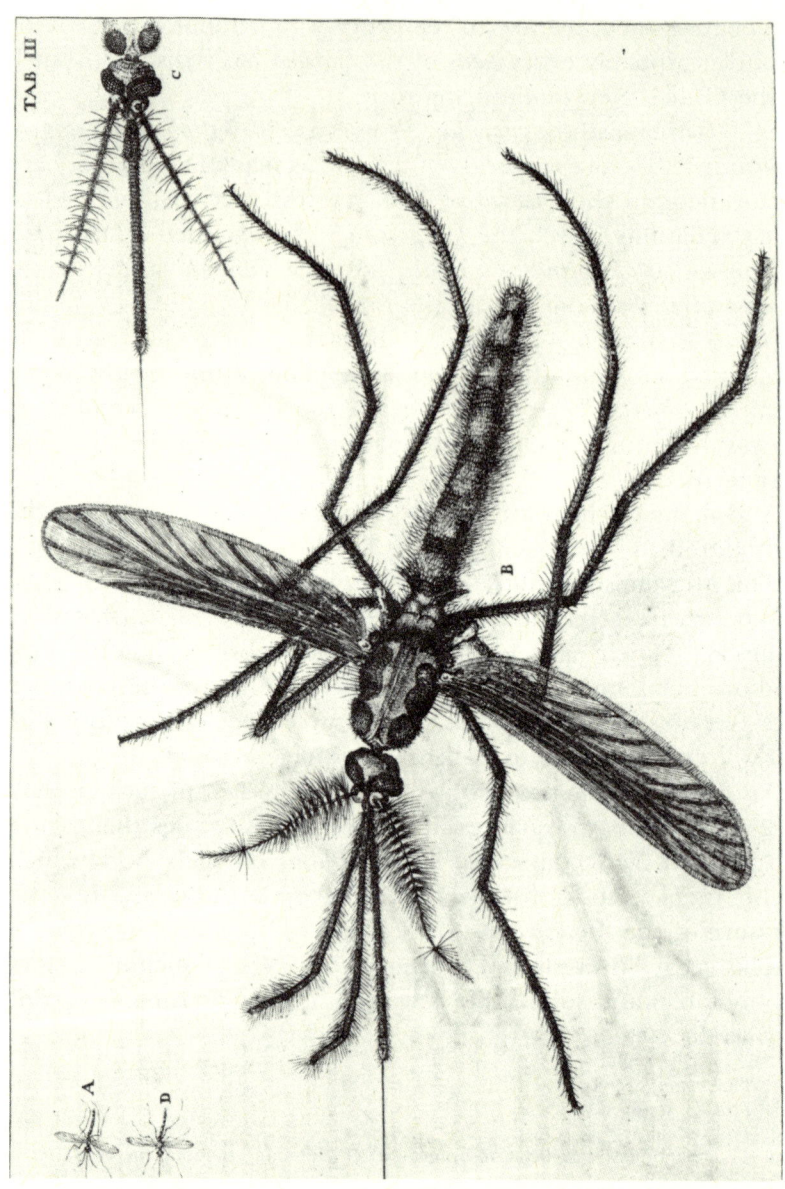

This plate is taken from Jan Swammerdam's *Historia insectorum generalis* (Utrecht 1669, 1682; Leiden, 1685). Figures A and D represent the male and the female, respectively, of the culex or gnat (mosquito). Figure B is a magnified image of the male. Figure C portrays the head of the female. Courtesy of the Rare Book and Special Collections Library, University of Illinois at Urbana-Champaign.

Essay 27. Of Insects

Admirable the Faculty of some that cannot fly, to convey themselves with Speed and Safety, by the help of their *Webs*, or some other Artifice that renders their Bodies lighter than the Air! How pleasantly do the *Spiders* dart out their *Webs*, and *sail* away by the help thereof; whereof Dr. *Lyster* and Dr. *Hulse*[34] were some of the first who made a discovery? There seems to be an hint of their *darting* in *Aristotle*, and in *Pliny;* but the Antients knew nothing of their *sailing*.[35] Some other little Animals may have their ways of *Conveyance* as unknown to us, as heretofore has been that of the *Spiders;* Creatures found in very new Pits, and Holes in the Tops of Houses, where they were never bred by any *equivocal Generation*. The *green Scum* on the Surface of stagnating Waters, which is nothing but prodigious Numbers of Animalcules; how come they there? And when gone, where do they go?

What can be better contriv'd than the *Legs* of *Insects*, most incomparably fitted for the intended Service? Or than their *Wings*, distended and strengthned with the finest *Bones,* and these cover'd with the lightest *Membranes*, whereof some are adorned with the most beautiful *Feathers;*[36] for the elegant Colours of *Moth* and *Butterflies* are owing to neat *Feathers* on their *Wings*, that are set in Rows with great Exactness and all the good Order imaginable? And some are provided with *Articulations*, for their *Wings* to be withdrawn, and folded up in Cases, and again readily spread abroad upon occasion: *Scarabs*[37] and other that have *Elytra,*[38] are thus accommodated. That their Body may be kept steady and upright, there is the admirable Artifice of *Pointels* and *Poises,*[39] under those who have no more than *two Wings*, (whereas the *four-wing'd* ones have no such things:) These *Poises* in the bipennated Insects are for the most part little *Balls,* that are set at the top of a slender Stalk, which they can move every way at pleasure to obviate *Vacillations*. If one of the *Poises* be cut off (or if the *four-winged* have

---

34. Probably Edward Hulse.
35. This flying is sometimes called "ballooning" or "parachuting."
36. The "feathers" are actually scales.
37. That is, the scarab beetle (Scarabaeidae) in Coleoptera, the largest order of the animal kingdom.
38. Elytra (sing. elytron) are the hardened or horny front wings in the Coleoptera order.
39. A pointel is a slender organ on the body of an animal, such as the horn of a snail. A poise is something that acts like a weight, such as the halteres or balancing organs on a fly.

lost one of their secondary or auxiliary Wings) the Insect will fly as if one side over-balanc'd the other, till it fall to the ground.

How *minute,* but how astonishingly *curious,* must be the Joints, the Muscles, the Tendons, and the Nerves, necessary to perform the Motions of these marvellous Creatures! These things concur even in the *smallest* Animalcules, and such as cannot be seen without our *Microscopes.*

When *Galen* had admired the Skill, *quod declarant Opifices cum in Corporibus parvis aliquid insculpant,*[40] instanced in the *Phaeton* in a *Ring,* where the Legs of the *Horses* were no bigger than those of a *Gnat,* he yet very justly cries out, their Make did not come up to those of a *Gnat: Major adhuc alia quædam esse videtur Artis ejus, qui Pulicem condidit, Vis atque Sapientia;* and is amazed that *Ars tanta in tam abjectis Animalibus appareat.*[41]

Among the celebrated Pieces of Human Art, there was the *Cup* that *Oswald Nerlinger*[42] made of a *Pepper-corn,* that held twelve hundred little Ivory Cups, all gilt on the Edges, and having each of them a Foot, and yet afforded room for four hundred more. But our *Derham* justly celebrates the more stupendous Art, which *plainly manifesteth the Power and Wisdom of the infinite Contriver of the inimitable Fineries* in the Bodies of our little *Insects;* they must have *Eyes,* a *Brain,* a *Mouth,* a *Stomach,* and *Entrails,* and other Parts of an Animal Body, as well as *Legs* and *Feet:* and all these must have their necessary *Nerves* and *Muscles;* all these are cover'd with an agreeable *Tegument,* whereof how neat the *Imbrications*[43] and other Fineries! All this Curiosity many times lying in a Body much smaller than the smallest *Grain* of Sand. A *Drop of Water* is a sort of an *Ocean* to them! Mr. *Derham* in a *Drop* of the *green Scum* upon Water, a *Drop* no bigger than a *Pin's-head,* sees no fewer than an hundred frisking about. How vastly many more in a *Drop* of *Pepper-water!*[44] How vastly many, many, many more, in a Drop of the *Leuenhoeckian* Examination![45] Dr. *Harris* affirms, that

40. "As when craftsmen carve something on small objects."
41. "Still greater it seems are other aspects of his art, [of him] who created the flea, namely, his power and wisdom.". . . "So great an art would appear in such lowly animals."
42. Oswald Nerlinger (fl. c. 1600) of Germany made this work of art for the duke of Bavaria. It was carried to Rome and shown to Pope Paul V.
43. That is, an overlapping as of tiles; a decorative pattern imitative of this.
44. That is, an infusion of black pepper, formerly used for microscopical observations of protozoa ("first animals").
45. That is, a microscopic examination.

Essay 27. Of Insects

not only in *black Pepper-water,* but also in Water wherein *Barley* and *Oats,* but especially *Wheat,* hath been steeped for about four or five Days, he hath seen prodigious Numbers of them. *Great GOD, we are amazed!*

The *Jews* have a foolish Notion, tho advanc'd by a Rabbi *Solomon,*[46] (upon the *Egyptian* Plague of *Lice*) *Quod Diabolus non dominatur super Creaturam, quæ Grano Hordei sit minor.*[47] Indeed a Man who by *Humility* shrinks himself into less than the *light Dust of the Balance,*[48] may take the comfort of the Notion. But then in *Philosophy* what a mighty Army of Animals less than a *Barley-corn* are found under the Dominion of the glorious GOD, who also has all the *Devils* as much under His Command as the least of these. I have read of a *Flea* in a *Chain, Beelzebub* is no more before the Almighty Maker of the *Flies,* and all the other *Insects.*[49]

The *Sagacity* observable in the generality of *Insects,* for their Provision against the Necessities of the *Winter,* is never enough to be admired.

Some having fed and bred themselves up to the Perfection of their *Vermicular State* in the Summer Months, then retire to a Place of Safety, and there throw off their *Nympha,* and put on their *Aurelia-state*[50] for all the Winter, in which they have no occasion for any Food at all; this is done by all the *Papilionaceous,*[51] as well as divers other Tribes.

Others, in their most perfect State, are able to subsist in a kind of *Torpitude,* without any Food at all; being at no *Action,* they are at no *Expence,* but can lie and sleep whole Months without any Sustenance. 'Tis remarkable that it is not any Stress of Weather which drives them into their intended Retirement, but they go to it in the *proper Season,* towards the end of Summer. 'Tis also

---

46. Rabbi Solomon ben Isaac.
47. "That the Devil does not have power over a creature which is smaller than a barleycorn."
48. Adapted from Isa. 40:15.
49. The last two sentences in this paragraph are based on Mark 3:22–26 (see also Matt. 12:24–28, Luke 11:15–20) where Christ's enemies accuse him of "casting out devils by Beelzebub," who is "the prince of devils" in the Gospels. In the Old Testament (2 Kings 1:3) "Baal-zebub" is the god of the Philistine city of Ekron. The name appears to mean "lord of flies" in Hebrew, and is so explained by ancient commentators. The Mishnah contains one reference to Beelzebub, understood as "fly-god."
50. The aurelia is the chrysalis or pupa of an insect.
51. The Papilionidae is a family in the order Lepidoptera. The common names of butterflies in the Papilionidae family are swallowtails, bird-wings, and parnassians. Mather may mean the Papilionoidea, a superfamily which includes all the families of butterflies.

remarkable, that every *Species* betakes itself to a convenient Receptacle, whereof there is a vast variety, where the *Frost* cannot come at them.

There are others who need *Food* in the Winter, and it is astonishing to see what a Foresight their glorious Creator has given them to lay up accordingly.

One of these Providers is the BEE,[52] reckon'd by *Aristotle* among the ζῷα πολιτικά, or *Civil People*.[53] Prepare now for a Scene of Wonders! Every Colony of *Bees* has a *King*, whereof *Pliny* gives this true Description: *Omnibus semper forma egregia, & duplo quam cæteris major, Pennæ breviores, Crura recta, Ingressus celsior, in Fronte macula quodam Diademate candicans, multum etiam Nitore a vulgo differunt.*[54] This majestical *Bee* has a *Sting*, which he can use without losing it;[55] but his Majesty rarely finds occasion for it. The *common Bees* (which have their four *Wings* and six *Legs*) are divided into *Bands*, which have their *Officers*, all working for the Good of the Whole, and as long as they live. But then there are *Drones*, which are bigger than they, and are Servants and Nurses under the *Honey-Bees*, in the hatching of their Brood.[56] A *Bee*, as *Rusden* observes, the first day of his flying abroad is an exquisite *Chymist*, or at least a diligent Purveyor and Collector of the *Honey-dews*, provided by Heaven for him on the Leaves of the Plants in the Field, which he lays up in convenient *Cells*, and there preserves it in a Covering of *Wax*, as foreseeing that a Winter is coming. How indefatigable the Pains of these industrious and marvellous Creatures! If they have *no King*, they pine, they die, they yield themselves a Prey to Robbers;[57] but they will not bear *two*. *Butler* observes, they abhor *Polyarchy*, as well as *Anarchy*. Their King oppresses none, is a Benefactor to all; so their Loyalty to him is inviolate. His Place of Abode makes a Court, a noble Retinue of *Bees* attends him.

52. All the bees together comprise the superfamily Apoidea (order Hymenoptera). The number of living species is immense; perhaps the best known is the honey bee (*Apis mellifera*). Throughout antiquity the queen bee was regarded as a king. Mather perpetuates this usage.

53. Another meaning is "social animals."

54. "All of them are always exceptionally well formed and twice as large as the others; their wings are shorter, their legs straight, their bearing more lofty, and they have a spot on their brow that shines white in a kind of fillet; they also differ from the common herd a great deal by their brilliant color."

55. The sting is actually an ovipositor or egg-laying organ. In using it the bee loses it.

56. Drones do *no* work in the hive. Their sole responsibility is to inseminate the queen bee.

57. Bee colonies can requeen after the old queen is killed or leaves the hive.

The Historie of Serpents.

vnto I will adde for a conclusion, that prouerbicall speech, of one Aspe borrowing poyson of another, out of *Tertullian* against the Hereticke *Marcion*, who gathereth many of his absurd impieties from the vnbeleeuing Iewes. *Desinat nunc hæreticus à Iudæo, aspis quod aiunt à vipera mutuari venenum*, that is, let the hæreticke now cease to borrow his venom of a Iew, as the Aspes doe borrow their poyson from Vipers. And true it is, that this prouerbe hath especiall vse, when one bad man is holpe or counselled by another; and therefore when *Diogenes* saw a company of women talking together, hee said merrily vnto thē, *Aspis par' echidnes pharmacon daneizetai*, that is, the Aspe borroweth venom of the Viper. Thus much of the Aspe.

⁑ Of the Description and differences of BEES.

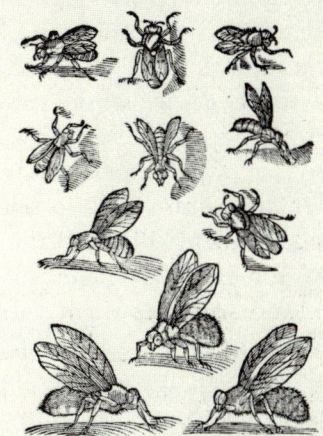

D: Bonham his discourse of Bees, wasps and Drones.

Amongst all the sorts of venomous Insects, (or cut-wasted creatures) the soueraigntie and preheminence is due to the Bees, who onely of all others of this kinde, are made for the nourishment of mankind, all other (cut-wasted) seruing onely for medicinall vse, the delight of the eyes, delectation of the eares, & the ornament, trimming, and setting forth of the body, which they performe at the full. They are called of the Hebrewes, *Deborah*. The Arabians terme them, *Albara*, *Nahalea*, and *Zabar*. The Illirians and Sclauonians, *Wezilla*. The Italians, *Ape*, *api*, *vna sticha*, *moscatella*, *ape* or *scoppa*, *pecchi*.
Names. The Spanyards, *Abeia*. Frenchmen, *Mousches au miel*. The Germaines, *Eenymbe*, *apen*. The Flemmings, *Bie*. The Polonians, *Pztzota*. The Irishmen, *Camilij*. In Wales a Bee is called *Gweniv*. Amongst the Græcians they haue purchased sundry names, according to the diuersitie of Nations, countries and places, but the most vulgar name is *Melissa*, & in *Hesiodus*, *Melie*. Othersome call a Bee *Plastis*, *à fingendo*, of framing. Some againe, *Anthedon*: and of their colour, *Zanthai*. Of their offices and charge, *Egemones*, *ab imperando*, from gouerning. *Sirenes*, *à suam cantu*, from their sweet voyce. The Latines call them by one generall name, *Apis* and *Apes*. *Varro* sometimes termes them *Aues*, but very improperly, for they might better be named *Volucres*, not *Aues*. So much for their names, now to the definition.

A Bee

This representation of bees is from Edward Topsell's *Historie of Serpents* (London, 1608). Topsell's historical value lies in his representation of prevailing zoological beliefs. His illustrations are reproductions from books by Konrad Gesner. Courtesy of the Rare Book and Special Collections Library, University of Illinois at Urbana-Champaign.

> *Rege incolumi Mens omnibus una est,*
> *Amisso rupere fidem.*
> *Ille operum custos; illum admirantur; & omnes*
> *Circumstant fremita denso, stipantque frequentes,*
> *Et sæpe attollunt humeris, & Corpora bello*
> *Objectant, pulchramque petunt per Vulnera mortem.*[58]

They have the *Orders* of their *King* for all the Work they do; and they never *swarm* without his *Orders.* The chief cause of their *Swarm* is the want of room. He usually goes himself with them, as in view of a more flourishing State, and leaves his decaying and unpleasant Kingdom, with the noisome old Combs, to such Successors as he has left alive. If the old one dies in his going forth, they return home to the Prince whom they had relinquish'd. And the King sometimes gives his Consent to a *second Swarm,* tho there be no lack of room, out of his respect to some of his Royal Lineage. In their *Hives* they are mighty just to one another, tho the fear of being robb'd makes them kill any Strangers that break in upon them. *Colonies* are sometimes engaged in Wars; the *King* usually orders the Battel, and animates them with his *Voice,*[59] and like a General, for whose Defence they unanimously expose themselves: They neither give nor take any *Quarter,* and they distinguish one another by their *smelling.* Spurt any thing among them that may make them *smell all alike,* and their *Hostility* ceaseth. The King is the only *Male* among the *Bees.* Each particular Cell in the *Honey-comb* is a Matrix. The King walks from one Cell to another, and injects a Seed[60] into each of them; the *Honey-Bees* mix with it a *generative Matter,* which they have lodg'd there, and add Water to it, and cover it with *Wax,* which is not opened till the young Bee opens its way out of it.[61] The *Drones* are also begotten by the King in like manner, but on a *generative Matter* something different, and in *deeper Cells.* The *Drones* are for no purpose, but only to lie at home close to the *Combs,* where the *young Bees* are breeding, and

---

58. "While the king is safe, all are of one mind; when he is lost, straightway they break their fealty. . . . He is the guardian of their toils; to him they do reverence; all stand round him in clamorous crowd and attend him in throngs. Often they lift him on their shoulders, for him expose their bodies to battle, and seek amid wounds a glorious death."

59. The queen bee animates the colony with varied pheromones, substances that serve as chemical signals between members of the same species.

60. Actually, an egg.

61. Worker bees feed grubs continuously during their larval stage.

hatch the young Brood, as a *Capon* does the Eggs assign'd to him. Hence the time for breeding the *Drones* is deferr'd till near the fall of the *Honey-Dews,* because they would have the use of them at as little charge of feeding as they can. But such is the Nature of the *Drones,* that if the *Bees* do not kill them, as they generally do, when they can be no further serviceable, they do by the Coldness of the Season in *September* die of themselves.

But now how many *moral Instructions* would the *Commonwealth of Bees* afford to a Mind willing to be *instructed* of God, by the Ministry of this *mysterious Insect!* Honest *Purchas*[62] has with an Imitation of it *gather'd* no less than three Centuries[63] of them; and yet these are but a few of the things which these *aculeated Preachers*[64] would advise us of: I will single out but this one peculiar Document from them for myself, which *Pliny* takes notice of: *Nullus Apibus, si per Cælum licuit, Otio perit Dies.*[65]

Another of these Providers is the ANT,[66] whereof the Wise-Man says, they are *exceeding wise; a People not strong, yet they prepare their Meat in the Summer.*[67]

Sir *Edward King* having been curious in examining their Generation, wonders to find them lying in Multitudes on their Eggs (which they industriously gather together) by way of Incubation. He wonders to see them in the Morning bringing them up towards the top of the Bank, and for the most part on the Southside of it; but at Night, especially if it be cool, or likely to rain, you may dig a Foot before you can find them. Indeed all is wonderful!

There is the *Field-Ant*[68] and the *Wood-Ant;*[69] the *Field-Ant* feeds upon small Seeds. They have their *Leaders* and *Rulers,* which they follow along their little *Paths* in exact Order, and return the same way; they all go out light, but all return home heavy laden, with

---

62. Samuel Purchas (the Younger) (d. c. 1658), not the author of the "Pilgrimes," was pastor of Sutton in Essex and author of *A Theatre of Politicall Flying-Insects* (London, 1657).
63. "Centuries" are groups of similar episodes or stories in a book, theoretically numbering one hundred. Books of the early modern period are often divided into "centuries." The "three centuries" are the major division of the second part of Purchas, *Theatre of... Insects.*
64. That is, bees (the Latin *aculeatus* means provided with prickles or stings).
65. "The bees lose not a single day in idleness when the weather grants permission."
66. Ants are members of the family Formicidae. Widely distributed throughout the world, they are regarded as the premier social insects. Most colonies contain queens, males, and workers.
67. Prov. 30:24–25.
68. Field ants are either *Pogonomyrmex* or *Pheidole,* both in the subfamily Myrmicinae.
69. Wood ants are of the genus *Atta.*

their *Burdens* on their *Backs*. The *Wood-Ant* feeds upon Leaves. You may see sometimes great *Paths* made by them, three and four Inches broad, and as beaten as the High-ways; they march stoutly under such Loads, that you cannot see their Bodies; a *Path* looks perfectly *green* with them.

In two Months of the Year they take *Wing,* and fly abroad in the warm Sun, to take their Pleasure, after the *Fatigue* of their *Labour* is over.[70]

And how unparallel'd the Tenderness, the Diligence, the Forecast of this little Animal, for the Safety of their *young ones!* A στοργή,[71] that filled *Squammerdam* with an unspeakable Pleasure at the view thereof; *Non sine Jucunditate spectabam!*[72] 'Tis very diverting to see how they carry about their *young ones,* and expose themselves to any *Dangers,* rather than leave their *young ones* exposed; and how they remove them from one place to another, as they find occasion.

Sometimes the *Ants* in the *Indies* will have Nests most artificially placed between the Limbs of huge *Trees,* and these Nests as big as an *Hogshead;* here is their *Winter Habitation.*

They will ransack strangely for Provisions, and in mighty Troops, which will follow wherever the foremost goes.

Excellently well Mr. *Derham* hereupon: "That the great *Wisdom* discernible in this little *Animal,* is owing to the Infusions of the great Conservator of the World, is evident; because either this *Wisdom* and *Forecast* is an Act of the Animal itself, or of a Being that hath *Wisdom:* but the Animal being *irrational,* 'tis impossible it can be its *own Act,* but it must be deriv'd or receiv'd from some *wise Being.* And who! What can that be, but an infinite LORD, and Conservator of the World!"

An *Ant-hill,* 'tis a Seat of a very curious Contrivance. *Johnston* makes it an Article of his *Thaumatography,* and says very truly, *Vix ullius Urbis artificiosior Structura.*[73] If you read the Description of the *quadrangular City,* four Foot long, and a Foot wide, the Streets wisely laid out, the convenient Granaries provided, the Civility of the Citizens to one another, as *Aldrovandus* has given it, you would see nothing in any *Strabo* more entertaining.

70. Flight is a distinctive attribute of most insects. Only the kings and queens of ants fly; this occurs during the mating process and the establishment of new colonies.
71. "Love," "affection."
72. "I observed them with much pleasure."
73. "[It is] a structure more skillfully wrought than almost any city."

I wonder not that the Wisdom of God sends me thither: *Go to the ant, thou Sluggard;* may I *learn her Ways and be wise!*[74]

But we are passing into a Theme, whereon there is *no end of the Wonders!* The Care of the *Insects* about their *Off-spring.*

Singular their Providence for their Young, in finding or making fit Receptacles for their *Eggs* or *Seed*, where they may enjoy a sufficient Incubation, and have ready an agreeable and sufficient Food for their Education.

They to whom *Flesh* is proper Food, lay their Sperm in *Flesh;* from which Nursery of *Maggots,* S. *Redi*[75] has for ever banish'd the old Whimsey of *anomalous Generation,*[76] by incontestable Experiments.

Others, to whom the Fruits or Leaves of the *Vegetables* are a *Food,* find a Repository there.

Some take this *Tree,* some take that *Herb;* and one Family still always the same.

If the *Cochineel*[77] were not accommodated with a Fruit like a *Prickle-Pear,* which opens after the Flower which protected it is by the Heat of the Sun scorched away, when the small *red Insects* are come to maturity, and would die and rot for want of more Food, if the *Indians* did not now come to shake them out; *Gentlemen,* where would you be supplied with your so much esteemed *Scarlet?*

Others require a greater degree of Warmth in their Lodging, and those look out the *Bodies* of *larger Animals,* that they may be lodged there. Many, if not most sorts of *Birds,* have their *Lice* in their Feathers; and several sorts of *Beasts* have peculiar *Lice* in their Hairs, all distinct from the two sorts wherewith *Man* is infested. It has been pretended that the *Ass* is free, and an odd reason assigned for it; but it has been rather supposed from a Passage in *Aristotle,* the *Chronology* whereof won't well suit with the *odd Reason* I refer to.[78]

Some work themselves into the very *Scales* of the *Fishes.* There *Lumbricus innascitur, qui debilitat;*[79] it was observed as long ago as

---

74. Adapted from Prov. 6:6.
75. That is, Signor Francesco Redi.
76. That is, spontaneous generation.
77. The cochineal insect (*Dactylopius coccus*) is a species of scale insect which feeds on the cactus plant (*Opuntia cochinillifera*) to produce the red dye called cochineal.
78. The odd reason assigned was "because our Saviour rode upon one, as some think."
79. "A worm is produced which makes them become weak." The "worm" was probably a crustacean rather than an insect.

the Days of the *Stagyrite*.⁸⁰ They find them even in the *Stomach* of *Cod-fish*.

55 The *Sheep* complains of them in his *Nose;* the *Kine* have them on their Backs; the *Horses* in their *Guts*.⁸¹

Those in the *Heads* of *Deer* are often mentioned by antient Writers.

Worms of many Yards long are bred in the *Legs* of *Men*, and in
60 other Parts of their Bodies; in their *Tongues,* their *Gums,* and their *Noses,* as 'tis reported in our *Philosophical Transactions;*⁸² in their *Eyes,* and their *Eyebrows,* as in the *German Ephemerides*.⁸³ *Mouffet* and *Tyson* will set before you what Worms the *Stomach* and *Bowels* of *Men* have often breeding in them. Lately in my Neighbourhood
65 a poor Man reaching to vomit, a monstrous *Worm* thrust up one end of itself, which the Man seizing on, fell to pulling of it, as a Fisherman hales up his Line, and pull'd till the *Worm* lay in an enormous heap; whence being drawn into its length and measured, the *Worm,* in the full extent of it, made about *one hundred and*
70 *fifty Foot long*. I may say, *Hisce ipse vidi Oculis*.⁸⁴ Yea, Dr. *Lyster* affirms true *Caterpillars* to have been vomited from thence. And Mr. *Jessop* affirms true *Hexapods* to have been also thrown up with a *Vomit*. Entertain unquestionable Accounts from *Germany,* and you will see *Toads,* and *Frogs,* and *Lizards,* cast up from an Human
75 *Stomach,* no doubt from the drinking of their Spawn. The *Livers* and *Kidneys* of Animals have had their *Worms;* yea, *Verzascha* has found them (without a Metaphor) in the *Brains* of Men; probably they were laid in the *Laminæ*⁸⁵ of the Nostrils, and gnawed their way into the *Brains* thro the *Os Cribriforme*.⁸⁶

80. That is, Aristotle, who was born in Stagira.
81. The sheep bot fly (*Oestrus ovis*) deposits its larvae in the nostrils of sheep. The cattle warble fly (*Hypoderma bovis*) lays its eggs on the legs of cattle. The larvae penetrate the skin and migrate to the back, where they develop in swellings or "warbles" under the skin. When full grown they escape through the skin and pupate on the ground. The larvae of the horse bot flies (family Gasterophilidae) infest the alimentary canal of horses.
82. That is, the *Philosophical Transactions* of the Royal Society of London. It carried several accounts of intestinal worms in human bodies starting with volume 8 (June 23, 1673). Most internal parasites are nematodes rather than worms or insects. Mather's references are probably to nematodes.
83. "German Ephemerides" was the popular English name given to a German counterpart of the *Philosophical Transactions* published by the Deutsche Akademie der Naturforscher and entitled *Miscellanea curiosa medico-physica academiae naturae curiosorum, sive ephemeridum medico-physicarum Germanicarum curiosarum*, 30 vols. in 25 (Leipzig, 1670–1706).
84. "I myself saw it with these eyes."
85. That is, a thin layer or plate, in this case of tissue.
86. Literally, a sievelike bone; the cribriform plate is part of the walls of the nasal

Wierus found them divers times in the *Gall-Bladder* of Persons whom he had opened. In divers *Fevers* the Blood has been found strangely *vermiculated*,[87] as *Kircher* and several others have upon Examination reported; [so *one Worm kills another!*] *Verminous* Collections are found in the *Small-Pox*, as *Lange*[88] and *Borellus* testify; and in *pocky Scabs* there are incredible Multitudes of them.

Others who make themselves *Nests* by Perforations in the *Earth*, or in some *Wood*, or in *Combs* of their own building; 'tis admirable to see how they lay in, and seal up the Provisions that will be necessary for their young ones there. So divers *Ichneumons*[89] carry in *Maggots*, which they take from the Leaves of Trees, which they sagaciously put up close into their Nests. *Aristotle* says they carry in *Spiders* too.

Their *Nidification*[90] is astonishing! When their *Eggs* are on the Leaves of Plants, or other Materials on the *Land*, how commodiously are they laid! Always carefully *glued* on, with one certain End lowermost, and handsome Juxta-positions.

When in the *Water*, in what beautiful *Rows!* In a *gelatine Matter* so fasten'd, as to prevent its Dissipation.

Single out but *Pliny's* Instance of the *Gnat*, a contemptible Animal, the Story of his Proceedings would give you a thousand Astonishments!

They who must perforate hard Bodies, to make their Lodgings there, have their Legs, Feet, Mouth, yea, their whole Bodies, very strangely accommodated to the Service.

But for them who build or spin their *Nests*, their Art, as Mr. *Derham* expresses it, *justly bids defiance to the most ingenious Artist among Men so much as tolerably to copy them*. The geometrical *Combs* of some, the terrestrial *Cells* of others; the *Webs*, the *Nets*, the *Cases*,

---

cavity; it is perforated by many openings for the exit of olfactory nerves. The "worm" was probably a hydatid cyst of the tapeworm (*Echinococcus granulosus*) which is common in cattle- and sheep-raising regions. This flatworm passes its adult life in dogs; the eggs may be passed to a human and enter the mouth. The larva develops into what is called a hydatid cyst and can be fatal.

87. That is, worm-eaten.

88. Probably Johannes Lange.

89. The ichneumon is any member of the insect family Ichneumonidae, order Hymenoptera. Although frequently called ichneumon flies, they are actually wasps. The female deposits her eggs within the body of the insect or spider host by means of her ovipositor. Ichneumons are beneficial in helping control overpopulation of harmful insects. Virtually every known species of insect is attacked by some species of parasite.

90. That is, the construction of a nest.

of divers. A Bishop of *Paris*[91] long since observed, *Nascitur Aranea cum Lege, Libro, & Lucerna;*[92] the very *Spider* knows its Lesson.

There is a *natural Glue* afforded by the Bodies of several to consolidate their Work. The *Wasps* have this, as well as the *Tinea Vestivora*[93] the *Cadew-worm,*[94] and several others; what *Goedart* also observes of his *Eruca*, this can be by some darted out at pleasure, and woven into silken Balls. Mr. *Boyle* mentions an *oval Case* of a *Silk-worm,*[95] which a Gentlewoman of his Acquaintance drawing out all the *silken Wire* that composed it, found it above *three hundred Yards,*[96] and yet weighed no more than *two Grains and an half*.

That wondrous Insect the *Silk-worm!* It has no *Eyes,*[97] but how fine its Performances. Let the *Historia Singularis* of them, written by *Libavius,* be perused, it will be found a Collection of Wonders. *Good God, shall thy Silk-worm adorn me, and shall he not instruct me too! There is another Worm, which would at least learn this of him, to spin out of his own Bowels, from his own Experience and his own Meditation, such things as may be useful to those to whom they shall be communicated.* But, O *vain person*, proud of the *silken Attire* that is rusling upon thee; is it possible that in a little *Worm* thy *Pride* should find a Nourishment!

There are others of these little Animals which make Nature itself serviceable to their Purpose, and make the Vegetation of *Trees* and *Herbs* the Means of building their little Habitations. They build in the *Galls* and *Balls* of the *Oak*, the *Willow*, the *Briar*, and other Vegetables, and are furnished with a *Piercer*, to prosecute their Business. Among these we will single out what the *Ichneumon-*

91. The reference is probably to Guillaume d'Auvergne or William of Auvergne (1180?-1249), but I have been unable to make a positive identification.

92. "The spider is born with the law, the book, and the light."

93. Several genera of small moths (Microlepidoptera) in the family Tineidae, the larvae of which are very destructive of cloth (*vestivoro* = cloth eaters), feathers, soft paper, decaying wood, stuffed birds, etc. The most common clothes moth is the webbing clothes moth, *Tineola bisselliella.* Next in importance among the clothes moths is the casemaking clothes moth, *Tinea pellionella.*

94. The cadew worm, also known as the caddis worm, is the larva of the caddis fly, any member of the order Trichoptera. Caddis fly larvae occur in various types of aquatic habitats—ponds, lakes, and streams. Casemaking larvae make little houses of leaves, twigs, sand, pebbles, or other materials which are fastened together with silk or cemented. Many casemakers are plant feeders. Some species spin silk nets from modified salivary glands and feed on the materials caught in the nets.

95. That is, a cocoon.

96. Each cocoon of the silkworm moth (*Bombyx mori*) is composed of a single thread about one thousand yards long.

97. The silkworm larvae have stemmata, which are lateral ocelli or simple eyes.

Fly[98] does to the Leaf of the *Nettle*. The Parent-Insect, with a stiff setacious[99] Tail, terebrates[100] the Rib of the Leaf when tender, and makes way for its *Egg* into the very Pith or Heart thereof, and probably lays in therewith some Juice of its Body, which will pervert the regular Vegetation of it. From this Wound arises a small *Excrescence,* which (when the *Egg* is hatch'd into a *Maggot*)[101] grows bigger as the *Maggot* increases, and swells on each side the Leaf, between the two Membranes, and extends itself into the parenchymous[102] part thereof, till it is grown as big as two Grains of *Wheat*. In this Mansion there lies a small, white, rough *Maggot,* which turns to an *Aurelia,* and afterwards to a very beautiful, green, small, *Ichneumon-Fly.*

A peculiar Artifice, and so far out of the reach of any mortal Understanding, that here must be, as Mr. *Derham* justly pauses upon it, *the Concurrence of some great and wise Being, that has from the beginning taken care for the Good of the Animal!* The Formation of these Cases is quite beyond the Cunning of the *Animal* itself, but it is the Act partly of the *Vegetable,* and partly of some *Virulency* in the Juice or Egg of the Animal reposited on the *Vegetable;* which *Malpighi,* in his Description of the *Fly* bred in *Oaken-Galls,* has notably confirm'd to us. *Erunt Plantarum Tumores, morbosæ Excrescentiæ, vi depositi Ovi a turbata Plantarum compage, & vitiato humorum Motu excitatæ, quibus inclusa Ova & Animalcula, velut in Utero foventur & augentur; donec manifestatis firmatisque propriis partibus, quasi exoriantur novam exoptantia auram.*[103]

It is a just Thought of one well skill'd in *Cosmology,*[104] That *Men* themselves, and much more *other Creatures,* may do many things which aptly serve to some *certain End* whereof they have no consideration. Creatures may be directed and constrained by a *strong Fancy* which they have of such and such Works, and of Actions that belong to them. Well, but who has imprinted it? It is the

98. The insect described is not an ichneumon, which is parasitic, but a gall wasp, a member of the subfamily Cynipinae of the family Cynipidae of Hymenoptera.
99. That is, like a bristle.
100. That is, to bore, pierce, perforate; to penetrate by boring.
101. Actually, a grub rather than a maggot.
102. That is, the fundamental tissue of leaves, pulp of fruits, etc.
103. "The tumors of plants are morbid excrescences, raised up by the force of the egg which has been deposited there, by the disturbed junction of the plants, and by the vitiated motion of the humors. In these tumors the enclosed eggs and little animals are nourished and augmented as if in a womb until with their parts manifested and strengthened, they come forth, desiring new air."
104. That is, Nehemiah Grew.

Great GOD, who will have *such Works* to be done. *Great GOD, shall we contrive what Service of thine thy nobler Creature MAN may thereby be helped to!* My excellent Philosopher[105] concludes: *The Divine Reason runs like a Golden Vein through the whole Leaden Mine of Brutal Nature.*

There is one thing more to be added: That the Numbers of *Insects* and *Vermin* may not be too offensive to us, Providence has ordained many Creatures, especially such as are in superior Orders, to make it their business to destroy them, especially when their Increase grows too numerous and enormous. As in the *Indies,* where they are sometimes exceedingly punished with *Ants,* there is the *Ursus Formicarius,*[106] whose very business is to devour them. Hideous Armies of *Worms*[107] do sometimes visit my Country, and carry whole *Fields* of Corn before them, and climbing up Trees, leave them as bare as the middle of *Winter.* Our *wild Pidgeons* make this the *Season* of their *Descent,* and in prodigious Flocks they fall upon these Robbers, and clear the Country of them.

The *Destruction* and *Death* of *Animals* does proclaim the *Fame* of the Divine *Wisdom* in adjusting of it!

The *Locusts,* that have sometimes proved so devouring a *Plague,* do also prove a *Dish* to the People that suffer from them. In *A Voyage round the World,*[108] I read That in the *East-Indies,* when these Creatures come in great Swarms to devour their Fruits and Herbs, the Natives take them with Nets, and parch them over the fire in an earthen Pan, on which their Wings and Legs would fall off, and their Head and Backs turn red, but their *Bodies* being full, would eat *moist* and *sweet* enough, and their *Head* a little crackle in one's Teeth; a Dish that People might subsist upon: tho the

105. Mather again refers to Grew.
106. "Anteater." According to Thomas Robinson, *A Vindication of the ... Mosaick System,* p. 73, Cardan named these creatures, which he said were numerous in the West Indies, *ursus formicarius.* This is a literal translation into neo-Latin of *oso hormiguero,* the Spanish common name of the lesser anteater (*Tamandua tetradactyla*), a tree- and ground-dwelling mammal found from Costa Rica to northern Argentina and in Trinidad. When frightened, it sits up like a bear, with outstretched arms. I thank Dr. John Bouseman of the Illinois Natural History Survey for help on this point.
107. The armyworm (*Pseudaletia unipuncta*) frequently does serious damage to corn and wheat. The larvae of these heavy–bodied moths (Lepidoptera: Noctuidae) frequently migrate in large numbers to new feeding grounds, which accounts for the insect's common name. The corn earworm (*Heliothis zea*) is another related pest which feeds on corn and other plants.
108. William Dampier, *A New Voyage Round the World.*

Condition of the *Acridophagi,* mentioned by *Diodorus,* and by *Strabo,*[109] would not encourage one to be confin'd to it.

Even the more noxious *Insects* and *Vermin* are such, that we may consider in them *the Finger of God!*[110] The *Sufferings* they inflict upon us, may be considered as the *Scourges* of God upon us for our Miscarriages, and be improved as Excitations to *Repentance.* I have read somewhere a Passage to this purpose: "I would carry on the Matter to so much of *Watchfulness,* in my apprehending Opportunities for *Thoughts of Repentance,* that the Provocations that may happen to be given to my *Bodily Senses* at any time, shall provoke such *Thoughts* in my Soul. If I happen to lodge where any *Insect* or *Vermin* assaults me, it shall *humble* me. I will think *I have been one among the Enemies of God in the World. These uneasy Creatures are part of the Armies which the Lord of Hosts employs, and with some Contempt, against his Enemies!* "

The *Worms* which, especially in places where the *salt* and *fresh* Water meets, do in such horrid Swarms eat into the Bottoms of our *Ships,* and render them even like *Honey-combs;*[111] the Coasts that are not infested with them, ought to acknowledge the Favour of Heaven in it; and the Merchant and Mariner that suffers by them, ought to consider *what Rebuke of Heaven upon their Dealings or Doings may lie at the bottom of such a Calamity!*

How wretched would our Condition be, if we were constantly infested with *Flies,* like the poor *winking People* of *New-Holland* in the *East-Indies!*[112] To be exempted from the Mischiefs which the *Justice* of God sometimes *inflicts* on People that do not acknowledge Him, 'tis what calls for our Acknowledgments of His *Goodness.*

If the *Lord of Hosts* please to single out from his *Armies,* whereof *there is no Number,*[113] no other *Legions* than those of *Insects,* even those *Velites*[114] commanded by Him, how would they *embitter,* and even *extinguish* our Lives! *Locusts* alone make whole Nations tremble: *Worms* have destroyed *Kings;* and *Flies* have scattered *Kingdoms.*

---

109. That is, a tribe or people neighboring the Ethiopians who lived on locusts and whose bodies were infected with parasites. They are described by the Greek historian Diodorus Siculus (late first century B.C.) in his world history (3.29) and in the *Geography* of Strabo (7.327).
110. Exod. 8:19.
111. These shipworms are not insects but wood-boring bivalves, greatly elongated clams (the *Teredo*).
112. These people kept their eyes half closed to protect them from insects. Mather's source refers to them as "blinking creatures."
113. Adapted from Job 25:3.
114. That is, lightly armed soldiers employed as skirmishers in the Roman armies.

But then the reverse: O *Cantharides*,[115] how many Millions of Lives are continually saved by your *epispastick*[116] Applications! *GOD is to be acknowledged in the Good which is done by a poor green Fly to the Children of Men!*

Honest Mr. *Terry* tells us, That among the *Persees* in the *East-Indies* they profess this Devotion: That the first Creature of *Sense* and of *Use* which they behold in a Morning, they employ still as a Remembrancer to them all the Day following, to draw up their Thoughts in *Thanksgiving* to the Almighty *God*, who hath made such a Creature for our Sevice.

My God, shall the Pagan rise up and condemn the Christian! If we should not from the View of thy Creatures have our Hearts drawn up to thy Praises, we should to our confusion find it so!

¶. "For what ENDS are all these little Creatures made? Most certainly for great ENDS, and for such as are worthy of a GOD!

"The exquisite Artifice which is conspicuous in the *Make* of these Creatures, does proclaim a marvellous and matchless *Wisdom* in the *Maker* of them; and Wisdom will *make nothing in vain.*[117]

"Tho the more *special Uses* of these Creatures be as yet unknown to us, the *only wise God*[118] sends to us this Advice concerning them: *What I do thou knowest not now, but thou shalt know hereafter.*[119]

"However, this we *know NOW;* for these and all Creatures this END is great enough, *that the Great God therein beholds with pleasure the various and curious Works of His Hands.* Behold a sufficient END, as well for a *World* as for a *Worm*, that the infinite God may with delight *behold His own Glories* in the Works which His Hands have wrought.[120] *My Readers*, let us come to a Comfort in the Doxology, *O Lord, thou hast created all things,*

---

57 Comfort/Confort

115. Cantharides is the plural of *cantharis*, the pharmacopoeial name of the dried bodies of the blister beetle, the Spanish fly (*Lytta vesicatoria*). It is used externally as a rubefacient and vesicant and internally as a diuretic and stimulant to the genitourinary organs. Improperly used, it can cause death.
116. That is, causing a blister or drawing out a serous discharge by producing inflammation.
117. Adapted from Ps. 104:24 and Isa. 45:18.
118. Jude 25.
119. John 13:7.
120. Gen. 1:1–31.

and for thy Pleasure they are and were created![121] The Great God has contrived a mighty *Engine,* of an Extent that cannot be measured, and there is in it a Contrivance of wondrous *Motions* that cannot be *numbred.* He is infinitely gratified with the View of this *Engine* in all its *Motions,* infinitely grateful to Him so glorious a Spectacle! when it becomes grateful to *us,* then *we* come into some Communion with Him. I will esteem it a sufficient END for the whole Creation of God, *that the Great Creator may have the Gratification of beholding His own admirable Workmanship.*[122] And I will esteem it a part of the Homage I owe to His Eternal Majesty, to be satisfied in such an END as this.

"I will transfer this *Meditation* to the Exercises which are to fill a *Life of Piety.* Have I not *Reason* enough, *Motive* enough, to abound in all the Exercises of a *pious Life,* even the most *secret* of them, and a Guard upon the *Frames* and *Thoughts* of my Heart within me? *The Great GOD is the Beholder of my whole Behaviour, He knows the way that I take; and I chuse the things that please Him in what I am now a doing.*"[123]

§. Finding myself now entred into the *Animal World,* I will take this opportunity to insert and pursue an Observation of the acute Dr. *Cheyne;* which is, That the *Production of Animals* is a thing altogether *inconsistent* with the *Laws of Mechanism:* from whence I infer, that it must be from something *superior* to them.

For first, the *Blood* is by the Force of the *Heart* squeezed from the *left Ventricle,* thro the *Arteries,* to the Extremities of the Body, and is thence returned by the *Veins* into the right Ventricle, thence by the *Arteria Pulmonalis* into the *Lungs;* from the Lungs by the *Vena Pulmonalis* again into the *right Ventricle.* The *Motion of the Heart* is caused by the *nervous Juices* mixing with the Blood, in the *muscular* part thereof; and these *nervous Juices* are both derived from the Blood, and forced into the *muscular* part of the Heart, by the Motion of the *Heart* itself, the Texture of the containing Vessels, and perhaps by the Pulsation of the *Arteries* upon the Nerves of the Brain. Here now, the *Heart* is the cause of the *Motion of the Blood* in the *Arteries;* and the *Motion of the Blood* in the *Arteries* urging their Juices thro the Nerves, is the cause of the *Motion of the Heart:* which is a plain Circulation of Mechanical Powers, a

---

121. Rev. 4:11.
122. Gen. 1:1–31.
123. Adapted from Ps. 139:1, Job 23:10, and Isa. 56:4.

95 *Perpetuum Mobile,* a thing unknown to Nature! An *Epicurean* cannot contrive a *Water Machine,* wherein the *Water* should move the *Machine,* and the *Machine* move the *Water,* and the same *Water* continually return in a *Circle* to move the *Machine.*[124]

99 Great GOD, it is thy immediate Influence on the Powers of Nature in me that keeps my Heart in motion. Oh! that I may love thee and serve thee with all my Heart! In thee I live!*[125]* To glorify thee, should be the Business of my Life!

Again, In all *Animals* how *small,* how *fine* the *Organs!* How indefinite the *Number* of them! *Sensation* is performed by the mediation of *Organs* arising from the *Brain,* and continued thro the part affected. Now there is not the least imaginable solid part of the *Vessels* or *Muscles* but what we find sensible; wherefore the Number of *Organs* that convey *Sensation* must be inconceivable! *Nutrition* is also performed by *Organs,* thro which a Supply is conveyed to the place to be nourished. Now there is *no Part* of the Body but what may be *increased* or *lessened;* so then in every *individual Point* of the Body there is the Termination of *Organs,* thro which a *Nourishment* may be conveyed. Furthermore, the *Canals* do all augment, and may all decay; and therefore every assignable part of these *Canals* must be the Termination of some *secretory Duct,* separating a *Fluid* fit for the repairing of their Losses, and these again must have others to repair their Losses; and how shall we conceive where to stop? Moreover, the most exquisite *Glasses* can discover nothing in the several parts of the Vessels and Muscles, but *Canals* amazingly slender; the better the Glasses, the more of these *capillary Pipes* are discovered. In short, all the *solid* Parts of the Body are nothing but either *Tubes* to convey some *Fluid,* or *Threads* in Bundles, tied by others that surround them, or going from one Fibre to another, or spread into thin Membranes; but each of these how *inconceivably* minute! the *Doctor* does not scruple to say, *infinitely!*

O infinitely Great GOD, I am astonished! I am astonished! For all those things hath my Hand made, saith the Lord.[126]

---

124. According to Epicureans, matter is in perpetual motion.
125. The first sentence is perhaps an allusion to Ps. 66:8–9; the next two sentences are adapted from Deut. 10:12 and Acts 17:28 respectively.
126. Isa. 66:2.

ESSAY XXVIII. *Of* Reptils.

LET us now handle the *Reptils,* which are a sort of *Animals* that rest one part of their Body on the Earth, while they advance the other forward.

In our way of doing it we shall *take up Serpents, and it shall not hurt us.*[1]

Concerning the meanest of these, namely, the *Earth-worm,* Dr. *Willis* makes this Remark: *Lumbricus terrestris, licet vile & contemptibile habeatur, Organa Vitalia, necnon & alia Viscera, & Membra Divino artificio admirabiliter fabrefacta sortitur.*[2]

And the *spiral Motion* of it is admired as well as described by Dr. *Tyson.*

The Motion of *Reptils* is extremely curious.

Their *Food* and their *Nest* lies in the next Clod, Plant, or Hole; or they can long bear Hunger and Hardship.[3]

So their *sinuous Motion,* perform'd with as much Art as what is in the *Legs* or *Wings* of other Creatures, and as curiously provided for, is found sufficient for the conveying of them.

There is abundance of *geometrical* Neatness and Niceness in the Motion of *Serpents;* their *annular Scales* lie cross their Belly, contrary to what those in the Back and the rest of the Body do: the Edges also of the *foremost Scales* lie over the Edges of the *following Scales;* and every Scale has a *distinct Muscle,* one end of which is tack'd to the middle of the Scale, the other to the upper Edge of the following Scale.

The *Snails*[4] have neither Feet nor Claws, but they creep with an undulating motion of their Body; on which Dr. *Lyster* has written: and by a *Slime* emitted from their Body, they adhere to all Kinds of Superficies.

---

1. Adapted from Mark 16:18. In the five-kingdom system of plant and animal classification, Reptilia is a class within the subphylum Vertebrata of the phylum Chordata. The common names of representative reptiles are crocodiles, lizards, turtles, and snakes.

2. "The earthworm, though it appears a vile and contemptible insect, has its vital organs and other viscera admirably framed by a skill which can be nothing less than divine." Earthworms are annelids (Annelida).

3. Aside from many turtles and very few species of lizards, reptiles are not plant eaters. Many reptiles can bear long periods without food owing to their low rate of metabolism.

4. Snails are in the class Gastropoda, phylum Mollusca.

This engraved frontispiece from Gerard Blasius's *Anatome animalium, terrestrium variorum, volatilium, aquatilium, serpentum, insectorum, ovorumque, structuram naturalem* (Amsterdam, 1681) suggests the range of comparative anatomy. Courtesy of the Rare Book and Special Collections Library, University of Illinois at Urbana-Champaign.

Essay 28. Of Reptiles

The *motive* Parts of *Caterpillars*[5] are admirably contrived, not only to serve their progression, but for gathering of their Food.

The *Spine,* and *Muscles* co-operating with the Spine, in such as have Bones;[6] and the *annular* and other muscles in such as have none; are incomparable Contrivances.

The *Magnitude* whereto some *Serpents* have grown, is prodigious. *Bochart* will astonish you with a Collection of Relations found in Antiquity concerning *Serpents,* and particularly *Dragons,* of a most enormous Magnitude. *Gesner* too will quote us Authors for some so big, that the little Book I am now writing will afford no room for them.

Yea, *Suetonius* affirms, that one was exposed by *Augustus,*[7] which was no less than fifty Cubits long. *Dio*[8] comes up with him, and affirms, that in *Hetruria*[9] there was one that was fourscore and five Foot long, which, after he had made fearful Devastations, was kill'd with a Thunderbolt. *Strabo* out-does him, and affirms, that in *Cælo-Syria*[10] there had been one which was an hundred Foot long, and so thick, that a couple of Men on horseback, on each side of him, could not see one another. Yea, one that was an hundred and twenty foot long, was kill'd near *Utica* by the Army of *Regulus*. Well might *Austin*[11] say of these dreadful Animals, *Majora non sunt super Terram.*[12]

Tho, if I might be allowed the Benefit of a *Metaphor,* I would say, *I have known where to find a greater than all of these!* But,

> *Ye Dragons, whose contagious Breath*
> *Peoples the dark Retreats of Death,*
> *Change your dire Hissings into heavenly Songs,*
> *And praise your Maker with your forked Tongues.*

'Tis what occurs in my Lord *Roscommon's* Paraphrase on Psalm cxlviii.

---

5. Caterpillars are insects.
6. Reptiles being vertebrates, all possess an internal skeleton of bone.
7. The Roman emperor Augustus (63 B.C.–A.D. 14)
8. That is, Dio Cassius or Dion Cassius.
9. That is, Etruria, a region in Italy now known as Tuscany.
10. That is, the Bekaa Valley in Lebanon.
11. That is, Augustine.
12. "Greater there are not on earth." The lengths of the snakes described are all exaggerated. The largest living snakes are the anaconda (*Eunectes murinus*) of South America and the reticulated python (*Python reticulatus*) of Malaya that are estimated to reach a maximum length of thirty-five to forty feet.

The *poisonous Tribes* have been made an Objection against the Divine Providence, as being destructive to the rest of the World.

The *Poison* of a *Viper* is found by Dr. *Mead,* on a microscopical Examination, *a parcel of small Salts, nimbly floating in the Liquor, but quickly changed, and shot out into Chrystals, of an incredible Tenuity and Sharpness, with something like to Knots here and there, from which they seemed to proceed:* it lies in a *Bag* in the *Gums,* at the upper-end of the *Teeth;* these Teeth are tubulated, for the conveyance of the *Poison* into the Wound which they make.[13] *Galen* says, Mountebanks did use to stop these Perforations of the *Teeth,* before they would let Spectators behold the *Vipers* to bite them.

Let it be considered, that the venomous Creatures have their great *medicinal Uses;* we see a *Treacle*[14] fetch'd out of a *Viper;* the *Viper's* Flesh cures *Leprosies,* and obstinate Maladies. The *Gall* of a *Rattle-snake* (which we take out of him in the more early Months of his yearly appearance, and work into *Troches*[15] with *Chalk* or *Meal*) is a rich *Cordial* and *Anodyne,* for which purpose I have often taken it, and given it: it invigorates the Blood into a mighty *Circulation,* when fatal Suppressions are upon it; it is highly *alexipharmick,*[16] and cures *Quartan-Agues.*[17] And yet this *Rattle-snake,* such a venomous Wretch, that if he bite the Edge of an *Axe,* we have seen the bit of *Steel* that has been bitten, come off immediately, as if it had been under a *Putrefaction.*

The very Steam of the *Serpents* in the famous *La Grotta delli Serpi,*[18] at *Sassa* in *Italy,* celebrated by *Wormius* from *Kircher,* and strangely discovered by a *Leper* happening to sleep there, does wondrous things.

Moreover, *ubi Virus, ibi Virtus;*[19] 'tis observed, the bruised *Flesh*

91 microscopical/microscopial

13. Snake venom readily crystallizes upon evaporation, but it is in fluid form when injected into a prey or adversary. The "bag" is the venom gland and lies well above the gums, behind the eye, and sends a duct forward beneath the eye to the upper opening of a hollow fang.
14. That is, a remedy for poison. Mather published anonymously a discourse entitled *A Treacle Fetch'd Out of a Viper* (Boston, 1707).
15. That is, a small, usually round, medicinal lozenge.
16. That is, preserving from the effects of poison; having the quality or nature of an antidote.
17. A type of malaria in which the paroxysms come every fourth day.
18. "The Grotto of the Serpents."
19. "Where there is a disease there is a remedy."

Essay 28. Of Reptiles

of the *venomous* Creatures applied to their *Bites,* cures the *Venom* of them.

But, as Mr. *Derham* observes, "There would be no Injustice in God for to make a Set of such noxious Creatures, as Rods and Scourges, to execute the Divine Chastizements on sinful Men." He adds, "I am apt to think, that the Nations which know not God are the most annoyed with those noxious *Reptils,* and other pernicious Creatures."

There is a strange Story related and asserted by *Franzius,* That *Anno Christi* 1564, vast Armies of *Serpents* appeared in *Hungary,* and occupied their Fields of Corn; and when the People were with a particular Contrivance by Fire going to destroy them, one who was bigger than the rest lifting up his Head, articulately cried out, *Nolite hoc facere, quia non nostro Arbitrio, sed a Deo huc missi sumus, ad perdendas Segetes.*[20] If the Story should be but a *Fable,* yet the *Moral* is wise and good.

It may be, they that have been thought *venomous* have not had in them so much *Venom* as has been thought. For Sir *Theodore Mayern* laughs at the Poison of a *Toad,* and says, 'tis no worse than a *Frog;* he had himself without mischief eaten several.[21]

There is one Mr. *Robinson* of *Cumberland,*[22] who offers it as a probable Conjecture, that the *venomous Creatures* lick up the Venom of the *Earth,* which, if it were diffused, might be more dangerous than their *Bite* or *Sting.*

The same Gentleman observes concerning the *crawling Worm,* which is despised, as the most useless among all the Creatures of God, that the Earth abounds with a gross, fat, luxuriant *Slime* at the time when these Vermin are engendred, and these Vermin then feed upon it; this, if it were not suck'd up, and contracted into the Bodies of these diminutive *Animals,* but were diffused thro the Grass and Herbage, would occasion *Murrains*[23] in Beasts, and perhaps *Diseases* in Men, whose Diet is much upon Herbage.

34 be,/be   34 *venomous/venomous,*
35 thought. For/thought for.

20. "We do not desire to do this, since not by our will, but by God we have been sent here to destroy your crops."
21. Frogs and toads are amphibians, not reptiles. Their poison is in their skin. It was a Mr. Pontaeus, a "Chymical Mountebank" according to Mayerne, and not Mayerne, who had eaten several toads. Mr. Pontaeus was likely to have suffered at least a stomachache.
22. That is, Thomas Robinson.
23. That is, a pestilence or plague affecting domestic animals or plants.

50   *A Worm now makes a pause, and adores the Divine Workmanship appearing in the Constitution of his Brethren!*

What amazing Effects follow on the Bite of the *Tarantula!*[24] The Patient is taken with an extreme difficulty of *breathing,* and heavy Anguish of *Heart,* a dismal Sadness of *Mind,* a *Voice* querulous
55 and sorrowful, and his *Eyes* very much disturbed. When the violent Symptoms which appear on the first Days are over, a continual *Melancholy* hangs about the Person, till by dancing, or singing, or change of Air, the poisonous Impressions are extirpated from the Blood, and the Fluid of the Nerves: but this is an Happiness that
60 rarely happens; nay, *Baglivi,* this wicked *Spider's* Countryman, says, *there is no Expectation of ever being perfectly cured.* Many of the Poisoned are never well but among the *Graves,* and in *solitary places;* and they lay themselves along upon a *Bier,* as if they themselves were *dead:* like People in despair, they will throw themselves
65 into a *Pit; Women,* otherwise chaste enough, will cast away all Modesty, and throw themselves into very exposing and indecent Postures; they love to be toss'd in the Air, but some will be mightily pleased with rolling themselves, like *Swine,* in the Dirt; and others cannot be pleased except they be soundly drubb'd on their hinder
70 Parts. There are some Colours agreeable to them, others offensive, especially *Black;* and if the Attendants have their Clothes of ungrateful Colours, they must retire out of their sight. The *Musick* with the *Dancing* which must be employ'd for their Cure, continues three or four Days; in this vigorous Exercise they *sigh,* they are
75 full of Complaints; like Persons in drink, they almost lose the right use of their Understanding; they distinguish not their very Parents from others in their treating of them, and scarce remember any thing that is past. Some during this Exercise are mightily pleased with *green Boughs,* of *Reeds* or *Vines,* and wave them with their
80 Hands in the Air, or dip them in the Water, or bind them about their Face or Neck; others love to be handling *red Cloths* and *naked Swords.* And there are those who, upon a little intermission of the *dancing,* fall a digging of Holes in the *Ground,* which they fill with *Water,* and then take a strange satisfaction in rolling there. When
85 they begin to *dance,* they call for *Swords,* and act the *Fencers;* sometimes they are for a *Looking-glass,* but then they fetch many

58 Air/Age

24. Spiders are arachnids, not reptiles.

a deep Sigh at the beholding of themselves. Their Fancy sometimes leads them to *rich Clothes,* to Necklaces, to Fineries, and a variety of *Ornaments;* and they are highly courteous to the By-standers that will gratify them with any of these things; they lay them very orderly about the place where the Exercise is performed, and in *dancing* please themselves with one or other of these things by turns, as their troubled Imagination directs them.

How miserable would be the Condition of Mankind, if these Animals were common in every Country! But our compassionate God has confined them to one little Corner of *Italy;* they are existing elsewhere, but no where thus venomous, except in *Apulia.*[25] *My God, I glorify thy Compassion to sinful Mankind, in thy Restraints upon the Poisons of the* Tarantula!

But who can behold the Dispositions of the poor *Tarantulates,* and not behold at the same time with Horror, a lively Exhibition of the *Follies* whereto *vicious People* are disposed? Perhaps the Thought well pursued would give such an Illustration of the *Venom* that befools, depraves, and enslaves *vicious People,* as to lead us into some very right Notions of the Methods, wherein the *evil Spirits,* to whose Conduct they have resign'd themselves, do, thro a just Judgment of God, operate upon them.

Vicious People, if you are not so *Tarantulated,* that it will fright you to look into a *Looking-glass,* bethink yourselves, and in the Condition of the Miserables that are stung with a *Tarantula,* behold as in a *Looking-glass* your own Behaviour and Confusion.

¶. "That the *least* and the *worst* of the Creatures may do *Man* the Service of leading him to God, a renowned Writer[26] has demonstrated, in singling out the Example of a *Toad.* A Gentleman saying, that in every one of the *Creatures* he could see Invitations to the *Praises* of GOD, one ask'd him, What! in a Toad? *Quomodo in Bufone potes laudare Deum?*[27] He made this good Answer, *This; That a good God has advanced me above the Baseness and Venom of that contemptible Animal!*

"The Bishops who in their Travel to the Council of *Constance,*[28] found a poor Country-man in the Tears of *Praises* to

---

25. That is, a region on the southeastern coast of Italy, opposite Albania.
26. That is, Johann H. Alsted.
27. "How are you able to praise God through the toad?"
28. The Council of Constance, one of the greatest assemblies of the medieval period, met from 1414 to 1417 for the purpose of restoring the unity of the Church, reforming it in head and members, and extirpating heresy, especially that of the Hussites.

God at the sight of a *Toad*, were struck into just Reflections, whereof this was one, *Surgunt Indocti, & rapiunt Cælum.*"²⁹

ESSAY XXIX. *Of the* FISHES.

*T*HE *Fishes of the Sea shall declare to thee!*¹
Let us become *Divers*, and visit the *watery World;* there we shall see, as Mr. *Derham* truly says, *a various, a glorious, an inexhaustible Scene of the Divine Power, Wisdom, and Goodness.*

The *Variety* of the Creatures that are the *Inhabitants of the Waters* is very considerable. *Pliny* in the eleventh Chapter of his thirty-second Book reckons up one hundred and seventy-six Kinds of them:² indeed he is very short in his Account. Our Christian *Pliny*, the excellent *Ray*, raises the Number of the *Fishes* to five hundred, excluding the *Shell-fish;* but of the *Shell-fish* more than six times the Number, and yet he thinks there may be but half the Species of the *Fishes* yet known to us.

If you'll believe *Pliny* and Company, *Vera est vulgi Opinio, Quicquid nascitur in parte Naturæ ulla, & in Mari esse, præterque multa quæ nusquam alibi.*³

Mr. *Willoughby* says *Aristotle*'s Division of the *Fishes* is the best, [better than *Rondelerius*'s] into three Kinds, the *cetacious*, the *cartilaginous*, and the *spinous*.

He gives us a Catalogue of *ninety-three* several sorts of our *English Fishes*.

The *Shape* of their Bodies, long and slender, or else very thin, is admirably accommodated to their Action of *swimming*, wherein they are to *divide the Waters.*⁴

The *Air-bladder*, wherewith most of the *Fishes* are furnished; this is what cannot be beheld without Astonishment! By this they poise their Bodies, and keep them equiponderant to the Water;

---

29. "The unlearned rise and seize the heavens."
1. Job 12:8.
2. Pliny actually counts seventy-four species of fishes, and another thirty of "those that have a hard covering." See *Natural History*, 9.16. The Pliny reference at 32.11 is to coral.
3. "So ratifying the common opinion that everything born in any department of nature exists also in the sea, as well as a number of things never found elsewhere."
4. Gen. 1:6.

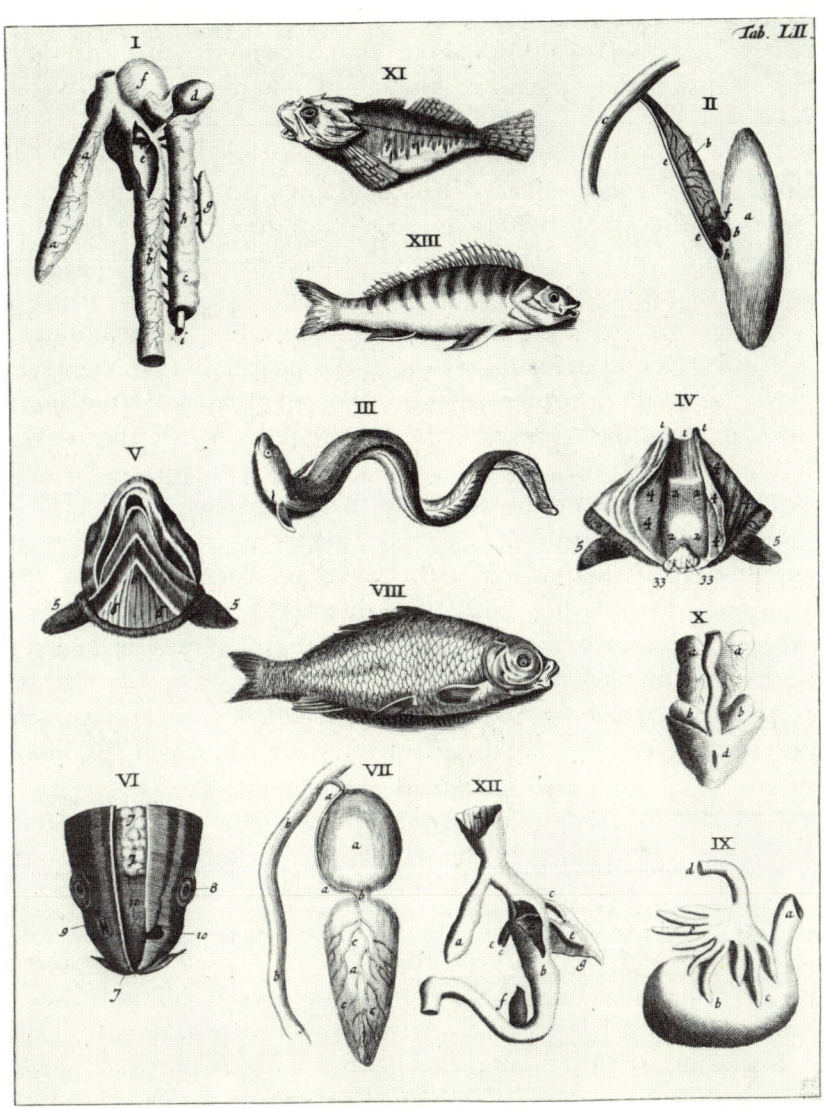

Blasius dealt with fishes as well as other creatures in his *Anatome animalium*. This illustration shows (center, top to bottom) what he calls a *cuculus* (fig. XI), a perch (fig. XIII), an eel (fig. III), and a carp (fig. VIII). Courtesy of the Rare Book and Special Collections Library, University of Illinois at Urbana-Champaign.

without it they would fall to the bottom, and lie groveling there, as it has been found, when that *Wind-bladder* has been broken. By *contracting* or *dilating* this Bladder, they are able to sink or to raise themselves at their pleasure, and continue in any depth of Water they please.

Fishes are *sensible* of *Sounds,* but whether they *hear,* or only *feel* the *Sounds,* is very much disputed. *Athanasius Kircher* observes, That tho the *Fishes* that have *Lungs* have also *Ears,* yet by what *Organs* the hearing of the rest is performed, *a nemine adhuc penitus exploratum est.*[5]

Their *Fins,* made of gristly *Spokes,* connected by Membranes, like our *Fans,* and furnished with *Muscles* for motion, these do partly serve them for progression, but chiefly to hold the Body upright: when these are cut off, as they were by *Borelli,* they waver to and fro, and when they die, their Belly turns upwards.

The great Strength, by which they dart themselves forward with an incredible Celerity, lies in their *Tails;* almost the whole musculous part of their Bodies is bestowed upon *them,* to assist the Vibration thereof. How *Fishes* row themselves by their *Tails,* and other Curiosities relating to *swimming,* you may read in *Borelli's* ingenious Discourse *de motu Animalium.*[6]

It is remarkable to see how *Fishes* have the *Center of Gravity* always placed in the fittest part of their Bodies, which is a Point of great Consideration in their fluid Element.

Consider the *Food* of these Animals; they neither *chew* their Meat in their Mouths, nor *grind* it in their Maws: but in their *Stomach* they are furnished with a *dissolvent Liquor,* which does corrode their Food, and reduce it, Skin and Bones and all, into a *Chylus* or *Cremor;*[7] and yet it is very marvellous, the Taste can perceive in this Liquor nothing of *Acidity:* it will manage Flesh as *Aquafortis*[8] does Metals, and yet no sensible *Sharpness* in it!

But where shall they find their Food? *Lord, these wait all upon thee, that thou mayst give them their Meat in due Season; what thou givest them, they gather; thou openest thine Hand, they are filled with*

---

5. "No one thus far has investigated it."

6. Giovanni Alfonso Borelli's *De motu animalium* was published posthumously in two parts at Rome in 1680 and 1681. A Leiden edition appeared in 1685.

7. Chylous consists of or is like chyle, which serves as the chief medium for the transfer of ingested fats to the blood. Cremor is a thick juice or concoction.

8. Literally, "strong water," but actually nitric acid, a strong oxidant used in metallurgy, etching, and engraving.

*Good.*⁹ How rich a *Promptuary*¹⁰ is this unlikely Element! From the largest *Leviathan* which *playeth in the Seas,*¹¹ to the smallest *Mite* in the Lakes and Ponds, all are plentifully provided for; as is manifest (which Mr. *Derham* notes) from *the Fatness of their Bodies, and the Gaiety of their Aspects and Actions.*

There is a Germination of divers *aquatick Plants* in the Waters; the Waters are also a sort of a *Matrix* to many Animals, particularly *Insects,* not only such as are peculiar to the Waters, but also many pertaining to the Air and the Land; who, by their near alliance to the Waters, delight in being about them, and so become a Prey to the Inhabitants thereof. Dr. *Schuyl* mentions the Horror of the *Water turned into Blood*¹² at *Leyden,* from nothing but the infinite Swarms of *Pulices*¹³ upon it; besides these, what mighty Shoals do we find of *lesser* Animals there, which the *greater* feed upon!

What a vast Supply of our Food have we in *sucking the Abundance of the Seas?*¹⁴ How many Millions of the *Fish* are every Year fetch'd out of their Element, and interr'd in the hungry Bowels of *Men?* Some of these very delicious, particularly the *White-fish,* whereof such infinite Shoals in the vast Lakes of the *North America,* which has this very singular Property, that all sorts of *Sauces* do but spoil it; it is always eaten, either boil'd or broil'd, without any manner of Seasoning.

You, *Gentlemen,* who think your own Country of *England* worth visiting with your *Travels,* as methinks you should before you go abroad, find the little River *Trent* in *Staffordshire* affording *thirty* several sorts of *Fishes;* you'll be ready to affirm of it, as the *Hungarians* do of their *Tibiscus,*¹⁵ two parts are *Water,* the third is *Fish.*

*My God, when in our Necessities we ask of our Father a Fish, our heavenly Father feeds us, how agreeably, how plentifully!*¹⁶

As the *smallest* Animals are bred in the Waters, witness those

---

9. Adapted from Ps. 104:27–28.
10. That is, a storehouse or repository. This meaning is now obsolete.
11. Adapted from Ps. 104:26.
12. Adapted from Rev. 11:6, 16:4.
13. That is, "fleas." The "little red animals" described by Schuyl were actually "waterfleas," a crustacean (*Daphnia pulex*) commonly encountered in Holland.
14. Deut. 33:19.
15. The Tibiscus or Tisza River, the longest Danube tributary in Eastern Europe, flows from western Ukraine through Hungary into Serbia, where it joins the Danube. It is known for its fisheries.
16. See Matt. 7:9–10 and Luke 11:11–12.

in *Pepper-water*,[17] so are the *largest;* those of the *cetaceous* Kind are there.

Pliny mentions the *Balænæ* of the *Indian Sea*, which were nine hundred and sixty Foot long; and he mentions *Whales* that were six hundred Foot long, and three hundred and sixty broad, which came into a River of *Arabia*. In the second Chapter of his ninth Book he offers a Reason why the *largest Animals* are bred in the Sea.

But I love to pass from him to a more trusty and modern *Pliny*, our industrious *Ray;* and we will now see something of his Remarks upon these *Belluæ Marinæ:*[18] The *Tail* in these has a different position from what it has in all other *Fishes;* it lies parallel to the Horizon in these, and it is perpendicular in the rest; hereby it supplies the use of the hinder pair of *Fins;* which these Creatures lack; and it serves both to raise and sink their Body at their pleasure. It is necessary that these Creatures frequently ascend to the top of the Water to *breathe,* and therefore they should be furnished with an Organ, by which their ascent and descent might be facilitated. The turning of their Bodies in the Water they perform like the *Birds,* by the motion of one of their *Fins,* while the other is quiescent. It is very remarkable that their whole Body is compass'd round with a copious *Fat,* which we call *the Blubber,* whereby their Bodies are poised, and rendred equiponderant to the Water, and the Water also is kept off at some distance from the *Blood,* the immediate Contact whereof might else have had some chilling force upon it; it serves likewise, as our Clothes do for us, to keep the *Fish warm,* in reflecting the hot Steams of their Body; and so redoubling the Heat thereof: hence they can abide the greatest Cold of the *Northern Seas,* to which they chiefly resort, not only for the Quiet which they enjoy there, but because the *Northern Air,* which is more fully charged with the Particles which we suppose to be *nitrous,* and that are the Aliment of Fire, is fittest of all to maintain their vital Heat in that Activity, which may be sufficient for to move such an unwieldy Bulk as theirs. The *stupendous Magnitude* of these Animals! Thou *Antitype* thereof, among the *Poets* which adorn our Age, describe them to us.

17. On pepper-water, see Essay 27, n. 44.
18. "Beasts of the sea."

> *While the vast Whale takes in the Deep his place,*
> *Prince of the Waters and the finny Race;*
55 *Rolling in Sport, the Billows he removes,*
> *And, like a floating Isle, the Ocean shoves:*
> *Now in his weedy Court he lies at ease,*
> *Now spouts against the Skies exhausted Seas.*

And yet one[19] says very well concerning him; he is *minima* 60 *quædam operum Dei, particula ac velut mica.*[20]

Let what I gave you of the *nine hundred and sixty Foot* pass for a *Plinyism;* and so what *Basil* in his *Hexaemeron* reports of *Whales* equal in bigness to the greatest Mountains, let the Censure of *Brierwood* pass upon it, as *an intolerable Hyperbole:* We will write 65 more sober things. Passing by what *Ælian* affirms of the *Whale* being five times beyond the largest *Elephant,* we find *Rondeletius* assigning him sometimes *thirty-six Cubits* of length, and *eight* of height. *Dion* is a grave as well as an old Writer, and he reports a *Whale* coming to Land out of the *German Ocean*[21] sixty Foot in 70 length, twenty in breadth. But *Gesner,* a later, affirms a *Whale* to have landed near our *Tinmouth-Haven,* in the Year 1532, which was *ninety Foot* in length, and the breadth of his Mouth six Yards and an half, and his Belly of such a compass, that one standing on the Fish, and slipping into his Belly, very narrowly escap'd being 75 drown'd there.

But then, if we may take *Hartenius*[22] for a Voucher, among the *twenty* several Kinds of *Whales* by him enumerated, he reckons one sort that is *thirty Ells*[23] long, and hath more than *seventy Teeth,* so large as to make Handles of Knives and other Instruments. He 80 reckons another sort that is *forty Ells* long, and overwhelms Vessels that come in his way. He proceeds to some *eighty Ells* long, and some of *ninety.*

All these proclaiming the *Grandeur* of their *Great Creator!*

Even in the *cold Sea* too, what a *Warmth* of Parental Affection

---

19. That is, Johann H. Alsted.
20. "A very small part of the works of God, a particle and a crumb, so to speak."
21. That is, Dio Cassius or Dion Cassius. The German Ocean is now known as the North Sea.
22. I am unable to identify Hartenius.
23. The ell was a measure of length which varied from country to country, with the range being from 22.257 to 45 English inches (565.3 to 1143 millimeters). Thus a whale of 30 ells would have ranged from 55.5 to 112.5 feet in length; one of 90 ells from over 166 to over 337 feet in length.

176 Animalium Mar. Ordo XII.

SEQVVNTVR CETE QVAEDAM EX OLAI MAGNI SEptentrionalis Oceani Europæi in Tabula Descriptione.

ICONVM quas subijciemus fides penes Olaum authorem esto. nos enim eas omnes ex Tabula ipsius depingendas curauimus. Apparet autem eum, ex narratione nautarum, nò ad uiuum, pleraque depinxisse. Vix probârim capita quorundam nimis ad terrestrium similitudinem esse ficta: ut neque pedes unguibus armatos, & fistulas binas (Rondeletius quidem Balænæ fistulam unicam tribuit) adeo prominentes, cum Balænarum, tum Pristis seu Physeteris, &c.

Harum belluarum nomina quædam confingemus, à similitudine terrestrium, ut Apri, Hyænæ, Monocerotis, Rhinocerotis, &c. Extare quidem in immensitate illa Oceani quàm plurimas diuersas & inusitatis formis belluas, quis dubitet? Et, si non temere est, quod uulgo dicitur: nomina etiam illa accolis Oceani Germanis & Gothis cognita, testari hoc possunt: qualia sunt, Wangwal/Andwal/Schwynwal/Rauëwal/Wittewal/Schiltwal/Hanerkeit/Monwarfrack/ Trolwal/Springwal/Herwal/Blotewal/Hill/Herill/Karckwal/Rußwal/Nachtwal/ Nordwal/Wintinger/Fischekeeke/Schellewyncke/Roze/Rostinger/Schlichtback/rc.

Volgend etliche Figuren auf der Tafel der beschreybung mittnächtischer landen des Olai Magni: wie wol vnd recht aber die selben conterfeetet syend/lassend wir den Olaum verantworten.

Balana erecta grandem nauem submergens. Videntur & alia quædam Cete ex eâdem Tabula Balænis adnumeranda, quæ ipse simpliciter Cete nominat, cum præter magnitudinem Balænis præcipuè conuenientem, nullam in se corporis partem raram aut monstrosam habeant: ut sunt quæ sequuntur aliqua.

Illustration from Konrad Gesner's *Historia animalium aquatilium in mari et dulcibus aquis degentium* (Zurich, 1560). Here Gesner, borrowing from Olaus Magnus, reports on monstrous whales that overturn passenger-laden boats in northern waters. Courtesy of the Rare Book and Special Collections Library, University of Illinois at Urbana-Champaign.

do the *old ones* express for their *young ones,* and how distinguishing! When the *Seals* are hundreds of thousands of them lying in a Bay coming out of the Sea, they bleat like Sheep for their Young; and tho they pass thro hundreds, yea, thousands of other young ones before they come to *their own,* yet they will suffer none but *their own* to suck them. *Even the Sea-Monsters draw out the Breast, they give suck to their young ones.*[24] *Monstrous Parents,* that are *without natural Affection!*[25] These Inhabitants of the *Sea* with open Mouth cry out against you.

¶. "I remember a *Crassus,* of whom 'tis reported, that he so tamed a *Fish* in his Pond, as to make him come to him at his calling him; verily, I shall have a Soul deserving *his Name,* and be more stupid than the *Fish,* if I do not hear the Calls which the *Fish* give to me to glorify the God that made them; and who has in their *Variety,* in their *Multitude,* in their *Structures,* their *Dispositions* and *Sagacities,* display'd his Glories. The *Papists* have a silly and foolish Legend of their St. *Anthony* preaching to the *Fishes;* it will be a Discretion in me to make the reverse of the *Fable,* and hear the *Fishes* preaching to me, which they do many Truths of no small importance. As *mute* as they are, they are *plain* and *loud* Preachers; I want nothing but an *Ear* to make me a profitable Hearer of them.

"It is a good Wish to be *in virtute Delphinus,*[26] to use the Dispatch of the quick *Dolphin* in all good Purposes.

"Tho 'tis the *way of the Sea* for the *greater to devour the lesser,* and the Wisdom of Heaven is conspicuous in it; yet I deprecate this *way of all the Earth:*[27] for indeed the *Fish,* who devour not those of their own particular Kind, therein condemn the cursed *Rapacity* too often seen among the Children of Men.

"To *catch Fish* is an *Employment* whereby many support themselves, a *Diversion* wherewith many refresh themselves; in managing this *Fishery* what an opportunity for many useful Reflections! In the *Means of Good* bestow'd upon us, the Glorious-One does *Retia Salutis pandere.*[28] How happy we, if taken in the *Nets of Salvation!* We are so when effectually persuaded to the embracing of our Saviour, and of his Religion.

24. Lam. 4:3.
25. Rom. 1:31 and 2 Tim. 3:3.
26. "To have the qualities of a dolphin."
27. Josh. 23:14 and 1 Kings 2:2.
28. "Spread the nets of salvation."

"Alas! the Ministers of the Gospel now *fish*,[29] not with *Nets*, but with *Rods;* and after long *angling*, and *baiting*, and *waiting*, how few are taken!

"In the *Temptations* to *Sin* and *Vice* which are offer'd to me, I see the *Hooks* with which the Destroyer proposes to *take* me, that I may be thrown into the *Perdition of ungodly Men*.[30] *My God, let not the Satanick Baits have any Power over me!*

"How *suddenly* is the *Fish* caught and killed, and with what a Surprize, when the poor Animal has not the least thought of such a Fate coming upon him! One moment *sporting, taken* the next; *he* pull'd away, *his Fellows* not at all regarding it! He was a wise Man who long since took notice of this; *Man knoweth not his Time: As the Fishes that are taken in an evil Net, so are the Sons of Men*.[31] *My God, help me to think seriously of Death every day, as not knowing but it may be my dying-day.*

"At our *Tables* we are now welcome to all the *Fish* we can fairly come at, whether they have any *Fins* or *Scales* or no; but methinks it gives a *special relish* to the Dish, '*tis a Dish which my admirable Saviour sometimes tasted of!*"[32]

ESSAY XXX. *Of the* FEATHERED.

THE BIRDS now invite us to *soar* and *sing* with them in the Praises of our God.

These ought immediately to follow the *Fishes,* not only for the *Order* of their *Creation,* but also because, as *Basil* notes, there is a συγγένεια τοῖς πετομένοις πρὸς τὰ νηκτά, *Volantibus Affinitas cum Natantibus*.[1]

These are either *Land-Fowl* or *Water-Fowl*. Of the *Land-Fowl* some have *crooked Beaks* and *Talons,* whereof some are *carnivorous,* called *Birds of Prey*.

40 admirable Saviour/admirableS aviour

29. Perhaps an allusion to Jer. 16:16, Matt. 4:19, Mark 1:17, and Luke 5:10.
30. 2 Pet. 3:7.
31. Adapted from Eccles. 9:12.
32. An allusion to Luke 24:42–43 and perhaps also to passages describing the feeding of the multitude with the loaves and fishes: Matt. 14:19–20, Mark 6:41–42, Luke 9:16–17, and John 6:9–11.
1. "A comparison of swimming things to flying things."

And some are also *frugivorous*,[2] called by the general Name of *Parrots*.

Others have their *Bills* and *Claws* more *streight*; of which there are some of a *larger* Size, which cannot fly at all.

Some are of a *middle* Size, and have either a *bigger* or *longer Bill*; some whereof do feed promiscuously, some only on *Fish*, some on *Insects*; or a *smaller* and *shorter Bill*, whereof some have a *whiter Flesh*, others a *blacker*.

Some are of *lesser Size*, called the *small Birds*; which are either the *soft-beak'd*, that feed mostly on *Worms* or *Flies*; or the *hard-beak'd*, that feed mostly on *Seeds*.

The *Water-Fowl* are either such as *frequent* the Waters for their Food, these are all *cloven-footed*, and generally have *long Legs*, and those *naked* for a good way above the Knees, that they may the more conveniently wade in the Waters; or they are such as do *swim* in the Waters, the most of these are *whole-footed*; some have but *three Toes* on a Foot, but most of them *four*; these either all connected by intervening *Membranes*, or more usually with the back *Toe loose*. Most *Water-Fowls* have a *short Tail*.

In *Birds* the Shape and Make of their Body is incomparably adapted to their *Flight; before* sharp, to pierce and make their way thro the Air, and then rising to their *full Bulk* by gentle degrees.

Their *Feathers*, how artificially placed, for facilitating the motion of their Body! Being placed any other way than what they are (as they would have been if meer *Chance* had placed them) they would have gathered *Air*, and been an Incumbrance to the Passage of their Body thro the *Air*; whereas in the neat *Order* wherein they are now placed, they are like a Boat new dress'd and clean'd, making its Passage thro the Waters. At the same time they have the Security of an admirable *Cloathing* in them, with a soft and warm *Down* next to their Body, but those next to the *Weather* of stronger Consistence, and closed most curiously. And then there is a most surprizing Accession to all this in the Art with which those Animals do *preen* and *dress* their *Feathers*, and the wondrous *Oil-bag* with which they are for this purpose accommodated. There is usually one *Gland*[3] (Mr. *Willoughby* sometimes found a couple)

2. That is, eating or feeding on fruit.

3. The uropygial gland, also known as the oil gland and the preen gland. Early speculations agreed that it serves as a water repellent and preserves the bird against wetting, but recent investigations indicate that it makes the feathers flexible and plays an important role in plumage hygiene.

in which there are divers little *Cells,* ending in two or three larger ones, which lie under the Nipple of the *Oil-bag;* this Nipple is perforated, and being press'd or drawn by the Bird's Bill or Head, emits a liquid *Oil* in some, an unctuous *Grease* in others, which being employ'd on their *Feathers,* contributes to their *nimble gliding* thro the Air.

How commodiously their *Wings* are placed! They that fly much, or have most occasion for their *Wings,* have them in the very best part imaginable, to balance their Body in the Air, and give them a swift progession. Alter their *Equipoise,* by cutting a *Wing,* or hanging a *Weight,* and how they reel! Such as have as much occasion for *swimming* as for *flying,* have their *Wings* therefore set a little out of the *Center* of their Body's Gravity; and for such as have more occasion for *diving* than for *flying,* these for that reason have their *Legs* more backward, and their *Wings* more forward.

The incomparable Curiosity of every *Feather!* The *Shaft,* hollow below, that it may be the stronger and the lighter; above a *Pith* filling it, which is also both strong and light; the *Strength* marvellous! The *Vanes,* how nicely gauged! broader on one side, narrower on the other, in both contributing to the progressive motion of the Fowl, and closeness of the *Wing.* The *Vanes* of the *Flag-feathers* of the Wing, the *Edges* of the exterior bending downwards, of the interior upwards, by which means they lie close to one another when the Wing spreads, and not one *Feather* misses its *full Impulse* on the Air; yea, the very *sloping* of the Tips of these Feathers is a *Nicety* to be wondred at.

Let an Eye assisted with Glasses view the *textrine*[4] Art of the *Plumage,* and, as Mr. *Derham,* who has given us a more particular Account of it, justly says, it will be found so exquisite, that *it cannot be viewed without Admiration!*

"My PEN, thou art fetch'd from the *Wing* of a Bird; thou wast one of the *Feathers,* which thou art now writing of! How surprizing an *Engine!* How surprizing, how extensive, how powerful thy *Operations* in the World! Never shall my *Pen* be employed in any thing but the Service of the glorious God, to whom I am indebted for it."

Admirable the *Apparatus* of the strong, but light *Bones* in the *Wings!* The *Joints* which move so as to answer all Occasions! The

---

4. That is, of or pertaining to weaving. The word is now obsolete.

Strength of the *pectoral Muscles* in *Birds* is greater than in any things not made for *flying*. *Borelli* observes, that the *pectoral Muscles* in *Men* are very small, and they don't come up to the fiftieth part of all the *Muscles;* but in *Birds* the *pectoral Muscles* are very *large*, & *equant, imo excedunt, & magis pendent, quam reliqui omnes Musculi ejusdem Avis simul sumpti*.[5] For which cause our *Willoughby* observes, that if Men would propose to prosper in their vain Project for *flying*, their *Wings* must be fastned not to their *Arms*, but their *Legs*, the *Muscles* being much stronger there.

The *Tail* of the Bird, which has been thought a sort of a *Rudder*, 'tis proved by *Borelli* that this is the *least use* of it; but it serves wonderfully to assist the *Ascent* and the *Descent* of the Bird in the Air, and obviate the *Vacillations* of the Body and Wings.

The *Flight* performed according to the strictest Rules of *Mechanism!* The untaught Artist gives a motion to his Wings, than which the acutest Mathematician could not give one more agreeable.

Blind Philosopher, canst thou see no GOD in all of this?

View next the *Feet* and *Legs*, which minister to their other motion.

Both of them very light, for their easier *Transportation* thro the Air.

In *Water-Fowl* how exactly do their *Feet* and *Legs* correspond to their way of living! Some of them have their *Legs* pretty long, that they may wade in the Waters, in this case their *Legs* are without Feathers a good way above their Knees, which is a Conveniency; their *Toes* also are all broad: and in the *Mudsuckers* two of the *Toes* are somewhat joined, that they may not easily sink in walking upon boggy places. Those that are *whole-footed*, or have their *Toes webbed* together, have their *Legs* generally short, which for *swimming* is most convenient; and it is pretty to see how artificially they gather up their *Toes* and *Feet* when they go to take their Stroke, and as artificially again extend or open their *Feet* when they drive themselves forward in the Waters.

Rapacious Birds, as they have *hooked Beaks*, thus they have strong, and sharp, and pointed *Talons*, fitted for the *Rapine* they are so intent upon, and for the tearing the *Flesh* that falls into them;

---

5. "And they are equal to, rather they exceed and weigh more than all of the remaining muscles of the same bird taken together."

and, as our *Willoughby* and *Ray* observe, they have robust and brawny *Thighs,* for striking down their Prey.

65 By the way; of this Kind there is a sort of *white Crows* (we must believe some who tell us this!) which they call *King-Carrion-Crows;*[6] and it is affirmed, that when a great number of *Crows* are assembled about a Carcase, if a *King-Carrion-Crow* be among them, he falls on first, and none of the rest will taste the least Morsel till he has 70 fill'd his Belly, and is withdrawn. I hope these *Crows* do no hurt by breaking in upon a Paragraph that is treating upon other Matters, especially if they effectually teach us, that the want of *good Manners* will never want a Condemnation.

Birds that climb, as the *Wood-pecker* Kind are, how fitted for 75 the purpose! Their *Thighs* very strong, their *Legs* very strong, but yet very short; their *Toes,* two forwards, two backwards, and so closely joined, that they may firmly lay hold on the Tree; an hard and a stiff *Tail,* bending downwards, on which they lean, and so bear themselves up in climbing.

80 How conveniently are the *Legs* of *Birds* curved, for their easy perching, and roosting, and rest! And to help them up upon their Wings in taking their Flight, and then to be so tuck'd up to the Body, as not to obstruct the Flight!

It is admirable that *Birds* as are *Fin-toed* are naturally directed 85 and carried to the *Water,* and fall to swimming there; thus *Ducklings,* tho hatch'd and led by an *Hen,* when they come near a Pond of *Water, in* they go, tho they never saw such a thing before, and tho the *Hen* clucks and calls, and is in a mighty Agony to keep them out, as *Pliny* expresses it, with *Lamenta circa* 90 *Piscinæ stagna, mergentibus se pullis, Natura duce.*[7]

There is a considerable Observation of *Aristotle,* πτηνὸν μόνον οὐδέν.[8] There is no *Flyer* but what has *Feet* as well as *Wings,* a power of *walking* or *creeping* on the Earth; 'tis because there is not always a sufficient Food to be had for them in the *Air,* nor 95 could the *Birds* take any rest, for without Feet they could not perch on the Trees; and if they lit on the ground, they could not

---

6. The carrion crow (*Corvus corone corone*) of western Europe and eastern Asia is one of the two races of the common crow (*Corvus corone*) found in Eurasia. The common crow of North America (*Corvus brachyrhynchos*) is similar. The carrion crow is black; there are no white ones.

7. "Her lamentations round the margin of the pond when the chicks, under the guidance of instinct, take to the water."

8. "Nothing is winged only."

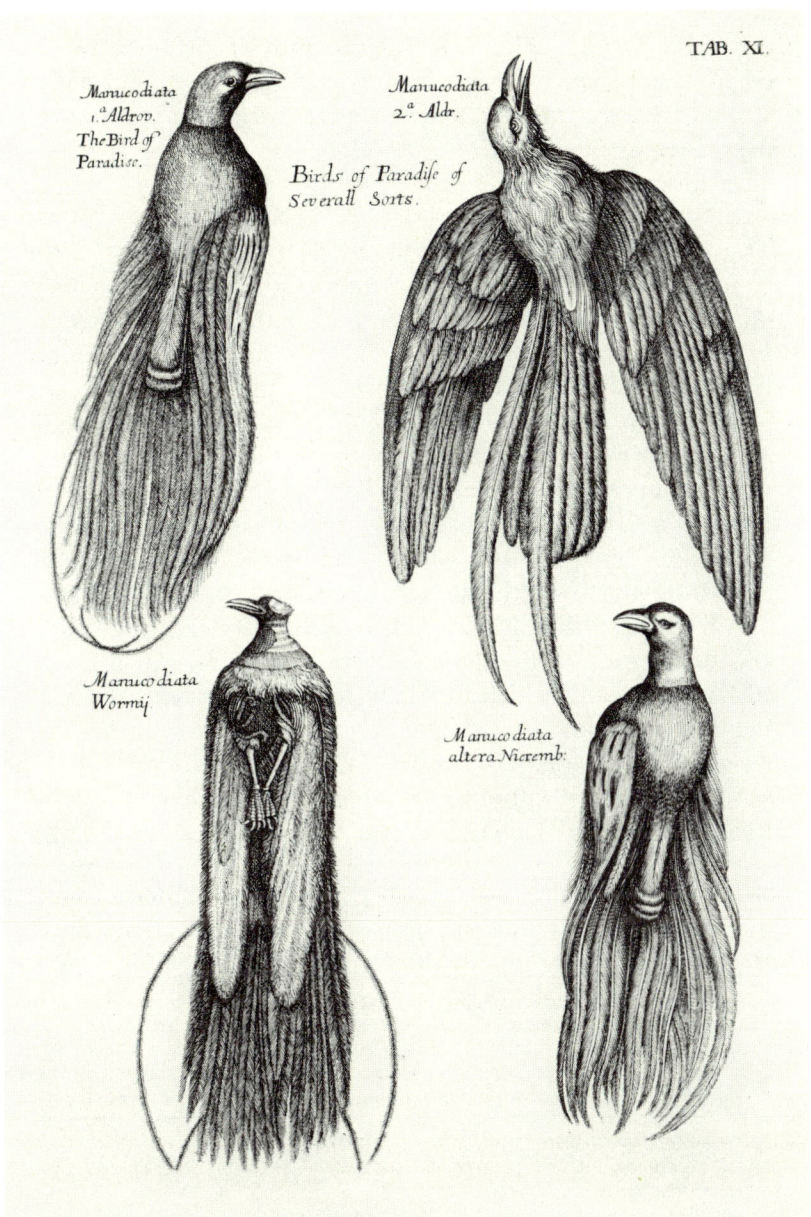

This illustration from Francis Willughby's *Ornithologiae libri tres* (London, 1676), which Mather never saw, demonstrates that the bird of paradise did indeed have feet. Courtesy of the Rare Book and Special Collections Library, University of Illinois at Urbana-Champaign.

again easily raise themselves; and where could they sit, hatch, and breed their Young? The Story of the *Bird of Paradise,* received even by the Learned in the former Age, is now found a *Fable;* that Bird has *Legs* and *Feet,* and those great and strong, and armed with *Talons,* as being a *Bird of Prey.*⁹

The *Bill* of Birds, how suited for gathering *Food,* and other Uses?

The *Eye,* how commodiously situated! (It is, by the way, a thing so remarkable, that nothing less than Astonishment can be the result of the Observation; that the *Fowls* in their Tribes have their *Centinels,* especially in the Night. The Watchfulness of the *Scart*[10] is true to a *Proverb:* One, by surprizing the *Centinel,* has caught three hundred in a Night.)

And the *Ear,* which would obstruct the Flight of it, were it like that of other Animals; the *inner Ear,* largely described by Mr. *Derham,* is a Contrivance that is a very amazing one.

Willis admires the Points wherein the *Brain* of *Birds* and *Fishes* agrees, differing from the *Brain* of *Man* and *Beasts.*

To *Steno* there appears *Elegans Artificis libere agentis indicium,*[11] in the Bifurcation of the *Aspera Arteria*[12] in *Birds,* which is not in other Animals, and which fits them for their *singing.*

In the *Swan* particularly, *Bartholin* celebrates it, as being *admirandæ Structuræ,*[13] by which means it may continue half an Hour under Water without any danger of choaking.

Read *Blasius* and *Coiter,* and admire the *Tongue* of the *Wood-*

---

9. The bird of paradise (any member of the family Paradisaeidae) occupies the tropical forests of New Guinea and neighboring islands, the Moluccas, and northeastern Australia. The males, among the most conspicuously decorated birds, are noted for their beautiful and varied plumage. These birds were introduced into Europe when the *Victoria,* the only surviving ship of Magellan's world circumnavigating expedition, reached Seville on 8 September 1522. Its cargo included several skins with feathers, but the skins lacked legs and feet. Legends quickly arose around these avian creatures, named birds of paradise by a Dutchman in about 1590. It was believed that birds of paradise possessed neither wings nor feet but passed all their time in the air, sustained on their spectacular plumes, resting at long intervals suspended on tree branches by wirelike tail feathers, drawing their food from the dews of heaven. In reality, they feed on berries, fruit, insects, and small animals.

10. Scart is another name for the cormorant. Mather probably refers to the common cormorant (*Phalacrocorax carbo*).

11. "An elegant proof of the contrivance of the Artificer."

12. Literally, "hard windpipe," now called the syrinx (pl. syringes). It is the organ of voice or song in birds. The syrinx is situated at or near the bifurcation of the windpipe (trachea) into two bronchi and typically comprises a resonating chamber (tympanum), vibrating membranes, and control structures of varying complexity, including cartilages and muscles.

13. "An admirable structure."

pecker, especially the sharp, horned, bearded *Point,* and the *glewy Matter* at the end of it, the better to stab and stick into the little *Maggots,* and to draw them out of the Wood.

The several ways the *Birds* have of purveying for their *Food,* call for our Consideration as we go along: but how can they be considered without some surprize of Pleasure at the view thereof. Among all these, that of the *Man-of-War Bird,* mention'd by *Dampier,*[14] is very singularly diverting. He sees a Bird called a *Booby,* and flying at him, gives him a *Blow,* which causes him immediately to disgorge the *Fish* he has in his *Crop;* and this he seizes on, perhaps before it can in its fall reach the Earth or Water. "'Tis in effect what *Men* do to one another, when the Justice of Heaven uses them to make *Seizures* on one another's Possessions. Have not the *French* in the late and long Wars,[15] been *Men-of-War Birds,* on our *English* Nation?"

Wonderful the Provision in the *Bill,* for the *judging* of the *Food!* It has peculiar *Nerves* for the purpose. These are smaller and less numerous in them that have the assistance of their *Eye:* but they are more numerous and thickly branched about, to the very end of the *Beak,* in such as hunt for their Food out of sight, in *Water,* in *Mud,* or under Ground. *Flat-billed Birds,* as Mr. *Clayton* and Dr. *Moulen* have observed, they that *grope* for their Meat have three Pair of *Nerves* that come into their Bills, whereby they accurately distinguish what may be proper for their Food.

Shall we stop a Moment, and consider how useful the *carnivorous* Birds of Prey become, even in prosecuting their voracious Inclinations? If the number of *lesser Birds* were not by their means lessened into such a Proportion, those *lesser Birds* would *overstock their feeding;* and then also, should those *lesser Birds,* which are so numerous, die of Age, they would leave their *Carcases* to rot upon

14. "Man-of-war bird" is the sailors' name for the frigate bird, the substantive name of the species of Freigatidae. There is a single genus *Fregata.* Frigate birds occur in tropic and subtropical oceanic areas; the magnificent frigate bird (*Fregata magnificens*) is found in the Caribbean, where Dampier observed them. Though skillful in forcing homecoming boobies to disgorge fish and then catching the food in midair, frigate birds are fully capable of fishing for themselves.

15. Mather refers to Queen Anne's War (1702–13), the American phase of the War of the Spanish Succession. During this conflict French and Indian warriors attacked the New England frontier, and New England troops aided by British forces captured Port Royal but failed to subdue Montreal. By the treaty of Utrecht (1713) Great Britain secured recognition of its claims in the Hudson's Bay country and the possession of Newfoundland and Acadia.

the Ground, and their *Stink* would corrupt the Air, and become insupportable.¹⁶

55 Dr. *Grew* observes, both *Birds* and *Beasts* having one common use of *Spittle,* are therefore furnish'd with the *parotid Glands,*¹⁷ which help to supply the Mouth with it; but the *Wood-Pecker,* and other *Birds* of that Kind, because they prey upon *Flies* which they catch with their *Tongue,* therefore in the room of the said *Glands,*
60 they have a couple of *Bags* filled with a *viscous Humour;* a sort of natural *Bird-lime,* which being by small *Canals,* like the *Salival,* brought into their Mouths, they dip their *Tongues* in it; and with the help thereof, they attack and master their Prey.

Pass from the *Mouth,* to its near Ally the *Stomach.* 'Tis admirable
65 in its *Duplicity;* one to *soften,* another to *digest!* Admirable in its *Variety,* suited unto a diverse *Diet: membranous* in some that are *carnivorous; musculous,* with a Strength agreeable, where *Grain* must undergo a Comminution!¹⁸

The *Gizzard* has a Faculty of *grinding;* to which purpose the
70 *Bird* swallows rough *Stones,* which when grown smooth, it throws up again as useless. Dr. *Harvey* says, this *grinding* may be heard in *Eagles* and some other *Fowls,* if you lay your Ear close to them when their *Stomachs* are empty.

In *Birds* there is no *Mastication* or *Comminution* of the Meat in
75 the Mouth; but in such as are not carnivorous, it is immediately swallowed into the *Crop* or *Craw,* or at least a kind of *Ante-Stomach,* (which Mr. *Ray* observed, especially in the *Piscivorous*)¹⁹ where it is moisten'd and mollified by proper Juices, from the *Glandules* there distilled in, then transferred from thence into the *Gizzard.*

80 Their *Lungs* adhere to the *Thorax,* and have little play; which is a good Provision for their *steady Flight.*

Wanting the *Diaphragm,* instead of it they have diverse *Bladders,* made of thin transparent *Membranes,* with pretty large Holes out of one into the other. These *Membranes* contain *Air* in them, and
85 are also Braces to the *Viscera.* The *Lungs* have large Perforations, thro which the *Air* has a Passage into the Belly. Doubtless the Body is hereby made more or less buoyant, and their *Ascent* or *Descent* facilitated.

16. Mather's explanation is almost but not quite Malthusian.
17. The parotid glands are a principal source of saliva for the oral cavity.
18. That is, breaking up into small fragments; pulverization.
19. That is, feeding on fishes.

Essay 30. Of the Feathered

Their *Necks,* how proportioned unto the *Length* of their *Legs!* Indeed, they that must search out their Food in the Waters, have them *longer* yet; and they have them so long, that when their *Heads* are extended in flight, they cause a due *Equipoise* and *Libration*[20] of the Body upon the *Wings.*

The Inspection of these Things would compel us to confess the glorious MAKER of them all!

Indeed what *Steno* says on a Description of a particular Subject, (the Myology[21] of the *Eagle*) may be more generally applied; *Non minus arida est Legentibus, quam Inspectantibus jucunda.*[22] For which reason I will not offer the Readers too many Particularities.

The *Nidification*[23] of *Birds;* a thing how full of Curiosity: They find out *secure* Places, and very *proper* ones; where their Young may lie safe and warm, and have their Growth promoted. But then, with what an *artificial Elegancy* are some of their *Nests* prepared? *Human Skill* could hardly imitate it. Among other Curiosities of *Nidification,* I will mention one that is observed in *Pidgeons* of my own Country. They build their *Nests* with little Sticks laid athwart one another, at such distances, that while they are so near together as to prevent the falling through of their *Eggs,* they are yet so far asunder, that the *cool Air* can come at their *Eggs.* And the REASON for this *Architecture* of their *Nests!* 'Tis this; their *Bodies* are much *hotter* than those of other *Birds;* and their *Eggs* would be perfectly addled by the *Heat* of their Bodies in the Incubation, if the *Nests* were not so built, that the *cool Air* might come at them to temper it.[24]

We have seen the *Nest* of an *Indian Bird* composed of the *Fibres* of certain *Roots,* which were so curiously interwoven, that it could not be beheld without Astonishment! These *Nests* they hang on the Ends of the Twigs of the Trees, over the Water, to secure their *Eggs* and *Young* from the Ravage of *Apes,* and other Beasts,

---

20. That is, the act of vibrating (as a balance does) before resting in equilibrium; the act of staying poised. Note the use of "cause." One might wonder as to how the necks *cause* balance.

21. The branch of anatomy dealing with the muscles.

22. "It is no less dry to those who read it, than it is pleasing to those who examine it."

23. The act of building a nest.

24. In reality, the body temperature of pigeons is no higher than that of most birds, and lower than that of many species that build thick, lined nests. Keeping eggs cool is not a problem for most birds. If eggs become too hot, birds simply stop incubating, or shade their eggs if they are exposed directly to the sun.

that else would prey upon them. They are justly enough called *subtle Jacks.*[25]

And what shall we say of the *Flamingo's*? They build their *Nests* in shallow *Ponds,* where there is much *Mud;* which they scrape together into little Hillocks, like *Islands,* appearing out of the Water about a Foot and a half high from the Bottom. They make the Foundation of these Hillocks broad, bringing them up tapering to the Top, where they leave a small hollow Pit, which they lay their *Eggs* in; and when they either lay or hatch their *Eggs,* they stand all the while, not on the *Hillock,* but close *by* it, with their *Legs* on the ground, and in the Water, resting themselves on the Hillock, and covering the hollow *Nest* upon it with their Bodies. Their *Legs* are very *long,* and building as they do upon the ground, they could neither draw their *Legs* conveniently into their *Nests,* nor *sit* down upon them otherwise than by resting their whole Bodies to the prejudice of their *Eggs* or Young, were it not for this rare Contrivance. [Psal. lxxxiv. 3.]

The *Incubation,* for which this Tribe of *Animals* is remarkable, opens a *new Scene* of Wonders unto us. The *Egg* with its crusty Coat is admirably fitted for it. Here we find one part provided for the *Formation* of the Body before 'tis grown to any considerable Dimensions, another for its *Nourishment* afterwards, till the *Bird* be able to shift for itself.

Willoughby confirms that Observation of *Pliny, Ipsum Animal ex albo Liquore Ovi corporatur: Cibus ejus in Luteo est.*[26]

But then the accurate *bracing* of these parts, by which they are kept in their due place, Mr. *Derham* observes, must be a *design'd,* as well as it is a *curious* piece of Workmanship. They are separated by *Membranes.* The *Chalazæ,*[27] (which because formerly thought the *Sperm* of the *Cock,* were called the *Treddles,*) are, as *Harvey* says, *As it were the Poles of this Microcosm, and the Connexions of the Membranes.* But as Mr. *Derham* observes, they serve only to keep one and the same part of the *Yolk* always uppermost, let the *Egg* be turned which way it will. The *Chalazæ,* it seems, are specifically

---

25. This bird is almost certainly the weaver, which is the substantive name of many species of Ploceidae (Passeriformes, suborder Oscines). They live in Africa south of the Sahara and in India; the ecological New World equivalents are Oropendolas (*Psarcolius* spp.) and Caciques (*Cacicus* spp.) in the American oriole family.

26. "The animal itself is formed out of the white of the egg, but its food is in the yolk."

27. A chalaza (pl. chalazas, chalazae) is either of the whitish spiral bands extending from the yolk to the lining membrane at each end of a bird's egg.

lighter than the *Whites* in which they swim; and being braced unto the Membrane of the *Yolk*, not exactly in the *Axis* of the *Yolk*, but somewhat out of it, it causes one side of the *Yolk* to be heavier than the other: so that the *Yolk* being by the *Chalazæ* made buoyant, and kept swimming in the midst of the two *Whites*, is by its own heavy side kept with the same side always uppermost, and probably this *uppermost side* is that on which lies the *Cicatricula*.[28]

It is affirmed, that our *Hens* once in every day of their Incubation *turn* their *Eggs*, without ever turning of one more than once, or leaving any one unturn'd. This is for a Service which they understand not themselves.

The Conveyance of what *Colours* we please to the Fowl that is hatching, by our painting of the *Eggs*, is a Curiosity.

That *Birds* must lay *Eggs,* is a sensible Argument of a *Divine Providence,* designing to preserve them, and secure them, that there might be a greater plenty of them, and that the *Destroyers* might not straiten their Generations. Had they been *viviparous,* if they had brought forth a *great number* at a time, the burden of their Womb would have rendred them so heavy, their *Wings* could not well have served them: or if they had brought forth but *one or two* at a time, they would have been troubled all the Year long with bearing or feeding their Young.[29] The Conveniency consulted in *oviparous* Animals, is one of Dr. *More*'s Triumphs over *Atheism.* Of these *Eggs* he makes an *Antidote* against that hellish Poison!

Dr. *Cheyne* will more particularly assure us, *We know* that the *Eggs* of *Animals* are only an *Uterus* for a little *Animal*, furnished with proper *Food*, and fenced from external Injuries: and *we know* likewise that all the Effects of *Incubation* are only to supply a proper degree of *Heat*, which may make the congealed *Fluids* to flow, and more easily pass into the nourishing Channels of the included *Animalcule*. On this occasion he goes on, *We are sure* that all the *Transformations* of *Insects* and other *Animals*, are nothing but the *Expansion* of their Parts, and the breaking of the *Membranes* that folded them up, by the Augmentation of these Parts; and all the several *Figures* they put on, are owing to the several *Membranes* in which they are involved. His Conclusion is what I was wishing

---

28. A cicatricle is the protoplasmic disc in the yolk of an egg from which the embryo develops.

29. The number of eggs laid in a clutch varies considerably, both between and often within species. The range is from one or two eggs, which is common, to about nineteen eggs.

for: *It is impossible duly to consider these things, without being wrapt into Admiration of the infinite Wisdom of the Divine Architect, and contemning the arrogant Pretences of the World-wrights, and much more the Production of Chance and justling Atoms.*

As Mr. *Derham* observes, what a *prodigious Instinct* is it, that *Birds,* and only *they,* should betake themselves to *this way* of *Generation!* How should they be aware that their *Eggs* contain their *Young,* and that they have in their power the *Production* of them? What should move them to betake themselves to their *Nests,* and there with *Delight* and *Patience* abide the *due number* of Days? And when their *Chickens* are *hatched,* how surprizing is their *Art,* and *Care,* and *Passion,* in bringing them on *until,* and only *until,* they are able to shift for themselves.

A Remark of our valuable *Ray* is worthy to be introduced here. It would be on many accounts inconvenient for *Birds* to *give suck;* and yet no less inconvenient, if not altogether destructive unto the *Chicken,* upon Exclusion all of a sudden, to make so great a change in its *Diet,* as to pass from a *Liquid* unto a *harder Food,* before the *Stomach* be consolidated, and by use habituated unto the concocting of it, and its tender and pappy Flesh fitted to be nourished by what shall be *strong* and *solid;* and before the *Bird* be by little and little accustomed to the using of his *Bill* in the gathering of it up, to which it comes not very readily: therefore there is a large *Yolk* provided in every *Egg,* a great part whereof remains after the *Chicken* is hatched, and is inclosed in its Belly, and by a *Channel* made on purpose, receiv'd by degrees into the *Guts,* and serves instead of *Milk,* to nourish the *Chicken* for a considerable time; which nevertheless in the mean time feeds itself by the *Mouth,* a little at a time, and gradually more and more, as it gets a more perfect Ability.

I will add a Curiosity relating to the *Pidgeons,*[30] which annually visit my own Country in their *Seasons,* in such incredible numbers, that they have commonly been sold for *Two-pence* a dozen; yea, one Man has at one time surprized no less than *two hundred dozen* in his Barn, into which they have come for Food, and by shutting

---

30. The reference is to the passenger pigeon (*Ectopistes migratorius*). Once the most abundant bird in North America, its population numbered millions of millions. These gregarious birds nested and migrated in enormous flocks; single flocks may have numbered in the tens or even hundreds of millions. In the century ending about 1914, the passenger pigeon was driven to extinction by hunting and habitat destruction. The last bird died on 1 September 1914 at the Cincinnati zoo.

This plate from Edward Howe Forbush's *Birds of Massachusetts and Other New England States,* part 2, *Land Birds from Bob-Whites to Grackles* (Boston, 1927) shows passenger pigeons, which were still common in eastern North America in Mather's day. An adult male and female sit on the top branch, with a young pigeon in juvenile plumage to the right on the same branch. Below are two mourning doves. The original painting is by Louis Agassiz Fuertes. Courtesy of the Massachusetts State Archives.

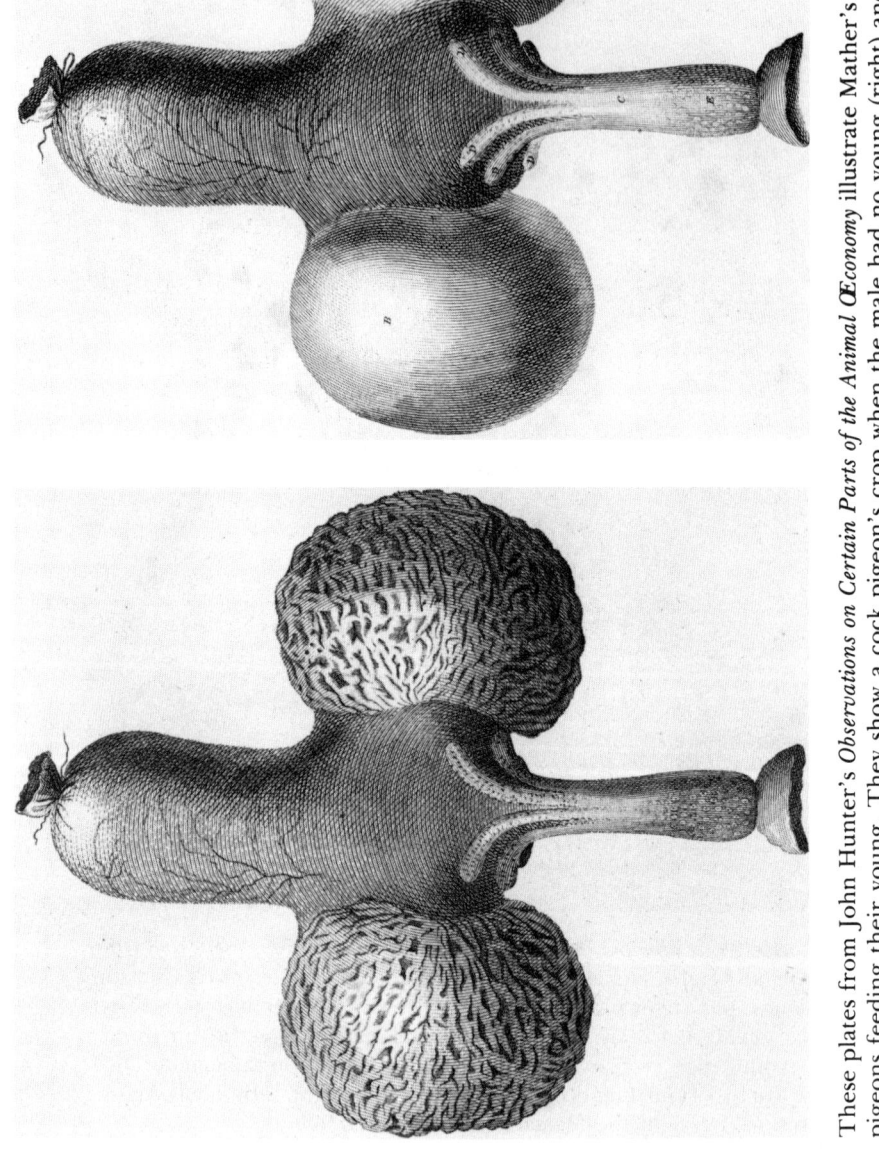

These plates from John Hunter's *Observations on Certain Parts of the Animal Œconomy* illustrate Mather's description of cock pigeons feeding their young. They show a cock pigeon's crop when the male had no young (right) and the thicker, more vascular crop of a cock pigeon with young. Courtesy of the Rare Book and Special Collections Library, University of Illinois at Urbana-Champaign.

the door, he has had them all. Among these *Pidgeons,* the *Cocks* take care of the *young* ones for one part of the day, and the *Hens* for the other. When they are taken, we generally take but *one Sex* at a time. In the Crops of the *Cocks,* we find about the quantity of half a Gill of a Substance like a tender *Cheese-Curd:* the *Hens* have it not. This *Curd* flows naturally into their *Crops,* as *Milk* does into the *Dugs* of other Creatures. The *Hens* could not keep their *young* ones alive when first hatched; but the *Cocks* do fetch up this *thickned Milk,* and throw it into the Bills of their *young* ones, which are so nourished with it, that they grow faster, and fly sooner than any other Bird among us. None but the *Cocks* which have young ones to care for, have this *Curd* found in their *Crops.* Kill one of those *Cocks,* and all the young ones pine away to death in the *Nest,* nowithstanding all that their Dams can do for them. See Sirs, and be instructed!

Masculus ipse fovet Fœtus, atque incubat Ovis;
Conjugii servat fœdera casta sui.[31]

All Birds lay a *certain number* of *Eggs,* or near that number, and then betake themselves to their *Incubation:* but if their *Eggs* be withdrawn, they will then lay more. When they have laid such a number of *Eggs,* as they can conveniently cover, and brood, and hatch, they give over, and begin to *sit.* This is not because they are necessarily *determined* to such a number: For *Hens,* for example, if you let their *Eggs* alone, when they have laid *fourteen* or *fifteen,* they will give over, and begin to *sit;* whereas if you daily take away their *Eggs,* they will go on to lay five times that number! This holds not only in *domestick* Birds; and so, as Mr. *Ray* observes, it can't be thought the effect of *Cicuration*[32] and *Institution:*[33] But the like was by Dr. *Lister* observed in *Swallows* too.

But altho almost the whole Tribe of *Birds,* do produce their Young by *Incubation,* there is a marvellous *Deviation* from it in some few Families which do it in a more *novercal way,*[34] and without any trouble at all, only by laying their *Eggs* in the *Sand,* exposed unto the *Heat* of the Sun. This Instinct of the *Ostrich*

---

31. "The male himself warms the young, and lies upon the eggs; He protects the pure agreement of their union."
32. Cicurate is a verb meaning to tame, to make mild or innoxious. The noun is now obsolete.
33. That is, training, instruction, teaching.
34. That is, befitting a stepmother.

particularly, *who leaveth her Eggs in the Earth, and warmeth them in the Dust,*[35] is ascribed unto GOD, who supplies the want of Concern in the *Parent-Animal* another way.

It is a surprizing thing, which the observing *Ray* has mentioned. Such *oviparous* Creatures as are long-lived, have *Eggs* enough at first conceived in them, to serve them for many years laying; probably for as many as they are to live: allowing such a proportion for every year, as will serve one or two *Incubations.* But *Insects* which are to breed but once, lay all their *Eggs* at once, have they ever so many. He says most justly, *Chance cannot govern it.*

The Scarcity of the *voracious* and *pernicious* Birds, and the Plenty of the *mansuete,*[36] and *useful,* and more desirable, is to go among the Matters of our Wonderment!

And so must the *swift Motion* of such whose Food is to be sought in distant Places, and in different Seasons; the *slow Motion* and short Flight of others more *domestick;* and the Awkwardness of some to *Flight,* whose *Food* is to be got near at hand, and without much *flying* for it.

It is amazing, *Who feeds the young Ravens when they cry!*[37] That Birds which feed their Young in the *Nest,* tho probably they cannot *count the Number*[38] of them, and tho they bring but *one Morsel* of Meat at a time, and tho they have not fewer it may be than seven or eight *young* in a *Nest* together, which at the return of their Dams do *all at once,* and with equal greediness, hold up their Heads and gape, yet they forget not one of them, they feed them all.[39] Our good *Ray* notes well, *'Tis beyond the possibility of a meer Machine to perform such a thing as this!*

With what an impetuous desire of *sitting* are the *Birds* inspired, while it is called for! After the Young are *hatch'd,* for some time they do almost constantly *brood* them under their Wings, lest they should suffer by any Inclemency of the Season; all this while how hard they labour to get them *Food! sparing* it out of their own Bellies, almost *pining* themselves to death rather than that their Young should want any thing! With what *Courage* are they inspired

---

35. Job 39:13–14.
36. That is, gentle, tame.
37. Adapted from Job 38:41, Ps. 147:9.
38. Rev. 13:18.
39. Some young receive no parental aid in food finding. The extent of parental involvement in feeding young birds who are active immediately after hatching varies substantially among various species.

in this time, to venture their very Lives in defence of them, and even fly in the Face of a *Man* that shall molest their *Young*, (as a *Hen* or a *Goose* will do) which they would never do in their own defence! These things are contrary to the Instinct of *Self-preservation*, and are eminent Pieces of *Self-denial*. Our good *Ray* says well, *They must needs be the Works of Providence for the upholding of the World!* These *Pains* are bestowed upon a thing which takes *no notice* of it, makes *no amends* for it, never acknowledges it with Thankfulness; and when the *young* one is grown *old* enough to shift for itself, the στοργή[40] is gone! The *old* one takes no further care of it, will beat it indifferently with such as it is not at all related to! The words of Mr. *Robinson*[41] on this Occasion are agreeable: "She does she *knows not what*, but yet it is what *ought to be done* by the most *exquisite Knowledge;* hence it is conclusive, that something else has *Knowledge* for her, even the *Creator* and *Contriver* of all things, who is the omniscient and omnipotent God." At the same time how remarkable to see that *Poultry* and *Partridge*, and other Birds, at the *first sight* know the *Birds of Prey*, and make a *Sign* of it with a peculiar *Note* to their *young* ones, who thereon hide themselves.

We celebrate the *Dove* of *Archytas*,[42] whereof *Gellius* tells us, *Simulachrum Columbæ e ligno ab Archyta, ratione quadam, disciplinaque mechanica factam, volasse;*[43] the same whom we find celebrated by *Horace*[44] for a noble *Geometrician*. This *Dove* surely had more Geometry in it than the πλαταγή, or *Childrens Rattle*, for which *Aristotle* celebrates him, as the Inventor of it. We are surprized at what *Ramus* tells us of the *Wooden-Eagle* and the *Iron-Fly*, made by *Regiomontanus;* the former of which flew forth of the City, met the Emperor a good way off, saluted him, and returned with him; the latter, at a Feast, whereto he invited his Friends, flew out of his Hand, fetch'd a round, and flew back to him again before the astonish'd Beholders. *Du Bartas* employ'd his *Poetry* on these Curiosities.

But what! No *Honours*, no *Praises* due to that infinite GOD,

---

40. "Love," "affection."
41. That is, Thomas Robinson.
42. Archytas of Tarentum (fl. first half of fourth century B.C.), allegedly the founder of mechanics, had a great reputation in antiquity. He built mechanical models.
43. "Archytas made a wooden model of a dove with such mechanical ingenuity and art that it flew."
44. Quintus Horatius Flaccus (65–8 B.C.), the Roman poet, author of odes, satires, and other verse.

who hath with so much Art contrived all the Variety of *Birds,* and accommodated every part of them within and without after so rare a manner, that there is not so much as a *Feather* misplaced, redundant, or defective! *Austin* says well, *Deus non solum Angelum & Hominem, sed nec exigui & contemptibilis animantis viscera, nec Avis pennulam, nec Herbæ flosculum, sine suarum partium convenientia dereliquit.*[45]

In the xivth of *Deuteronomy* there is a *Bird* called *Racham,*[46] which signifies *Mercy.* The *Talmudists*[47] have a Saying, That when this *Bird* appears, the *Mercy* of God and His *Messiah* is then coming to the World.[48] Verily, in every *Bird* that flies into our World, there is a display of the Divine *Goodness,* as well as *Power* and *Wisdom.* I wish that, in the reigning Dispositions of *Benignity* and *Compassion* among Mankind, *Racham* were making her Appearance!

Our excellent *Cosmologer*[49] makes his religious Remark upon it, That the *Birds* (and so the *Beasts*) which are *domestick,* or the most *useful,* are the most *prolifick;* there are more *Hens* than *Kites,* more *Geese* than *Swans.* A *Crane,* which is but scurvy Meat, hatches no more than *two Eggs* in a Year; several *Sea-Fowls* but one. The *Pheasant* and *Partridge,* excellent Meat, and easily come at, hatch

---

45. That is, Augustine. "God has not left angels and men, not even the entrails of the smallest and most contemptible animal, or the feather of a bird, or the little flower of a plant, without a mutual peace amongst all its parts."

46. Deut. 14 describes the killing of animals for food and lists the unclean creatures which it is forbidden to eat. Verse 17 places "the pelican, and the gier eagle, and the cormorant" in the latter category. The gier-eagle is the vulture. In Hebrew, the word for vulture is *racham,* but *racham* is also one of many Hebrew words for mercy.

47. The Talmudists are the authors of the Talmud, a Jewish compilation which embodies both the Mishnah and the Gemara. The two main forms of the Talmud, the Palestinian and the Babylonian, are similar in method and construction but by no means identical in content. The Palestinian Talmud reached final shape toward the end of the fourth or beginning of the fifth century, and the Babylonian Talmud about a century later. The Talmud has had an influence on the life of Judaism second only to that of the Hebrew Scriptures.

48. This saying is found in the rabbinic commentary (Gemara) on Hullin, the third tractate in Kodashim, the third order or division of the Talmud. Hullin deals with the slaughter of unconsecrated animals and with dietary regulations. In commenting on a passage in Hullin (3.6) which treats unclean birds, one rabbi said that the vulture is called *racham* because when the *racham* comes, mercy (*rachamim,* that is rain) comes to the world. Another rabbi added, "Provided it perches upon something and cries *sherak-rak.*" The latter was another Hebrew word for vulture. The commentary went on, "There is a tradition that if it [*racham,* the vulture] settles upon the ground and hisses, the Messiah will come at once, for it is said, *I will hiss for them* [an onomatopoetic Hebrew word is used here in imitation of the sound *sherakrak*] *and gather them*" (Zech. 10:8). See Herbert Danby, *The Mishnah* (Oxford: Clarendon Press, 1933), p. 518; *The Babylonian Talmud. Seder Kodashim. Hullin,* trans. Eli Cashdan, 2 vols. (London: Soncino Press, 1948), 1:343.

49. That is, Nehemiah Grew.

fifteen or twenty. The more valuable, which lay *fewer* at a time, sit the *oftner,* as the *Dove.* Thus, if it were not out of place to observe it here, there are more *Dogs* than *Foxes;* more *Cats* than *Lions.* The *Sheep* feeds and breeds in all Countries much alike.

Of *Wild-Fowl,* those which are the most *useful,* fly not singly, but are *gregarious,* which renders them the more *visible* and *audible* to us, and the more *plentiful* Game. And for our more quiet possession of things that are most *useful,* they are *naturally marked,* when there is occasion for it. *Wild-fowl,* and *Fishes,* and other Creatures, which are not fitted by Nature to be any Man's *Propriety,* have only such distinguishing Marks as belong to the whole *Species;* but of the *domestick,* as *Poultry, Horses, Dogs* and *Cats,* not only the *Species,* but the *Individuals* have their Marks. The *Sheep,* which are *proprietary,* if not so *marked,* it is compensated in this, that they do not *straggle.*

¶. "My Great Saviour has given me this Direction, Matth. vi. 26. *Consider the Fowls of the Air.*

"But is it possible to *consider* them without continual *Wonders* at the *Divine Workmanship* appearing in them! *Wonders* to be articulated and modulated into endless *Praises* of their Glorious Creator! Methinks the *sweet Notes* uttered by many *Tribes* of them invite me into a *Consort* with them.

"I know not what well to make of a Relation published a few Years ago, but so well attested, that a pious and worthy Man wrote a large Treatise upon it, entitled, *Vox Corvi!*[50] which affirms, That a *Raven* perching on the top of a Steeple, and thence turning towards a quarrelsome Neighbourhood, was heard very audibly and articulately to utter these Words, *Look into the third of the* Colossians, *and the sixteenth:* But this I know, *Ask the Fowls of the Air, and they shall tell thee.*[51] There needs no *Genius* to take a possession of our *Birds,* that we may hear from them the Admonitions of *Piety,* and Exhortations *to believe and adore an infinite* GOD intelligibly enough proceeding from them.

50. Alexander Clogie or Clogy (1614–98), learned divine and chaplain to Bishop William Bedell, was for forty years vicar of churches in Ireland and England. He published *Vox Corvi: Or, the Voice of a Raven* (London, 1694). The title page of Clogie's book indicates that Col. 3:15 is the proper biblical citation.
51. Adapted from Job 12:7 and Matt. 6:26.

"It was a celebrated Speech of the Philosopher,[52] *Si Luscinia essem, ut Luscinia canerem*;[53] I can *fly* much *higher* than they, and if I praise their Glorious Creator, I shall *sing* much *better* than they; *Homo sum, atque ut Homo canam colamque*.[54]

"The *Providence* of the Glorious GOD, in the Propagation and Sustentation of the *Fowls*, 'tis admirable; it extends to *Ravens*, to *Sparrows*; and shall I imagine *myself* excluded from the Care of that *Providence!* Holy Mr. *Dod* ventur'd upon the Difficulties and Contingences of a *married Life*, when he saw the *Hen* with her *Chickens* provided for. O *Unbelief*, I command an *eternal Silence* to thee! Shall the *Birds of Season* bring with them a Condemnation of my *Inadvertency*, to my fittest *Opportunities* for the doing and the getting of Good!

"There are the *Images* of many *Virtues* in *Birds*, (which have been called *Simulachra Virtutum*) of which I would endeavour an Imitation, and therein glorify the God that speaks to me by them; among these I would especially pitch upon two. Teach me, O *Stork*, how *gratefully* to treat my *Parent;* shew me, O *Dove*, how *lovingly* to treat my *Consort*."

Of such Reflections a famous Philosopher[55] says truly, *Rectis animis non poterunt non esse grata, licet perversis ridicula videantur*.[56]

The Man who learns all the Good which the *Birds* may mind him of, and then lives to the GOD, whose *Work* and whose *Voice* he discerns in the *Birds*, this *Man* shall be a *Phœnix*, and the Traditions of the Antients no longer a *Fable*.

ESSAY XXXI. *Of the* Four-Footed.

WE proceed to the *Animals* that are perfect, hairy, and *walking upon four.*

These Quadrupeds are either *hoofed* or *clawed*.

Of the *hoofed* or *ungulate;* some are *whole-hoofed*, whereof 'tis

---

52. That is, Epictetus (c. A.D. 55–135), a Greek slave who obtained his freedom and taught Stoic philosophy in Rome.
53. "If I were a nightingale, I would sing as a nightingale."
54. "I am a man, and as a man I will sing and worship."
55. That is, Johann H. Alsted.
56. "They are unable not to please those who have proper understanding, although they seem ridiculous to the perverse."

observ'd that none have Horns, nor have the Males any appearance of Breasts: there are four sorts of these.

Others are *cloven-footed;* of these there are two Divisions.

There is the *Bisulcate*[1] Kind, which is also subdivided.

There are the *Ruminant.*

Some of these have *perpetual Horns.*

Whereof there are six of the *Bull-kind,* five of the *Sheep-kind,* eleven of the *Goat-kind.*

Others have *deciduous*[2] *Horns;* these are the *Deer-kind,* whereof eight sorts have been reckon'd up.

Of those who do not *chew the Cud,* there is only the *Swine-kind,* whereof there are five sorts reckon'd up.

And then there is the Kind whose *Hoof* is cloven into *four Divisions;* we know five of these, but we know no *Rumination* in any of them.

Of the *clawed* or *digitate;* there is one sort whose Claws adhere to one another, cover'd with one common Skin, but with obtuse Nails, that stick out round the margin of the Foot; this is the *Elephant,* who must pass for *anomalous.*

There is another sort, which has only two Claws; namely, the *Camels,* which, tho they have no *Horns,* do ruminate, and have the *four Stomachs* of the *horned Ruminants.*

A third sort includes those which the *Greeks* call ἀνθρωπόμορφα,[3] whose Foot is divided into *many Claws,* with broad Nails on them: this is the *Ape-kind,* whereof there is a great variety; nine or ten Kinds have been described by the Naturalists.

A fourth sort is of those which have *many Claws,* yet they are not cover'd at the end with broad flat Nails, but have them rather like *Talons,* crooked and pointed; these had best be distinguish'd by their *Teeth.*

Some of these have many *cutting Teeth* in each of their Jaws; of these there is a *greater* sort, which either have a *short round Head,* as the *Cat-kind,* whereof there are *seven* sorts; and I hope the *Lion* will not be offended if *he* be reckon'd among them: or they have a *long Snout,* as the *Dog-kind,* whereof there are thirteen or fourteen sorts; and among these there are Varieties of *Mungrels,* and

1. That is, cloven-hoofed.
2. That is, shed at the end of the growing period.
3. "Of human form."

These plates from Edward Topsell's *Historie of Foure-Footed Beastes* (London, 1607) illustrate the mantichora (which had the body of a lion, the head of a man, the tail or sting of a scorpion, and the quills of a porcupine) and the lamia (a strange wild beast with a woman's face and large, comely breasts which she used to entice men). Mather drew on many authors who reflected the credulity of their age. Courtesy of the Rare Book and Special Collections Library, University of Illinois at Urbana-Champaign.

hebricious[4] Breeds: and there is also a *lesser* sort, which have a long and slender Body, with *short Legs;* these are the *Weasel-kind,* and there are about eight sorts of them.

Others of these have only *two* large remarkable *Teeth* in each of their Jaws; these are the *Hare-kind,* which live mainly on *Plants* and *Fruits;* and there are about half a score sorts of them.

To these Kinds of *Quadrupeds* there must be added several that are *anomalous.*

Some have a *long Snout,* with Feet which are divided into *many* Claws, and are furnish'd *with Teeth;* there are eight or nine sorts of these, whereof the *Hedge-hog* is in the Van.

Others of these are destitute of *Teeth,* and there are two sorts of these.

There are *Quadrupeds* that are *Flyers* too, as the *Bat-kind,* whereof there are different Forms.

There is one very odd Anomaly,[5] which has but three Claws on each of his *four Feet,* and has a Namesake too often among them that go *not upon four;* 'tis the *Ignavus,* a *Sloth* we call it: he takes eight or nine Minutes to move one of his Feet three or four Inches; and when he has grown fat and plump with eating all the Leaves on a Tree, he will be Skin and Bone before he reach another, which will be five or six Days, tho' it may be very near the former.

There are also viviparous and sanguineous[6] Quadrupeds, breathing with Lungs, but having only one Ventricle in their Hearts; to these we may add the *Tortoise,* whereof there are many Species, tho they be rather *oviparous.*

But then there are some *oviparous* Quadrupeds, which have a long Tail, horizontally stretched out; these are the *Lizard-kind,* and there be fourteen several sorts of them.

The *French* Gentleman[7] who writes *A Demonstration of the Existence of GOD from the Knowledge of Nature,* makes this Remark: "All the Animals owe their Birth to a certain Male and Female of their Species. All those different Species are preserved much

66 Anomaly/Anomale

4. That is, hybrid breeds.
5. That is, an anomalous creature.
6. This word, now obsolete as applied to an animal, means having blood or a circulatory system.
7. That is, François de Salignac de La Mothe Fénelon.

85 the same in all Ages. We do not find that for three thousand Years past any one has perished or ceased; neither do we find that any one multiplies to such an Excess, as to be a Nusance or Inconvenience to the rest."

And now since we are upon the *four-footed*, the Remarkables in
90 their *Legs* and *Feet* may be those which we may agreeably enough *begin* upon.

The *prone Posture of the Body* in the *Quadrupeds* is not only most beneficial to themselves, but also most advantageous to *Man;* they perform their own Actions the better for that Posture, and they
95 serve Man the better, both for *Carriage* and for *Tillage*.

But then it's observable how exactly their *Legs* are made conformable to this Posture.

It invites yet more Observation, how admirably their *Legs* and
99 *Feet* suit the Exercises of every Animal.

The *Elephant*, a Creature of prodigious Weight, has *Legs*, as *Pliny* notes, like *Pillars* rather than *Legs*.

The *Deer*, and the *Hare*, and other Creatures of a singular *Swiftness*, have their *Legs* accordingly slender; but they have
5 therewithal an incredible *Strength* adapted to their *Swiftness*.

Some have their *Feet* made only for *walking* and *running*, but some have them for *swimming* too.

The *Toes* on the Feet of the *Otter* are all conjoined with *Membranes*, and in *swimming*, when the *Foot* goes forward in the
10 Water, the *Toes* are close; but when backward, they are spread out; whereby they more forcibly strike the Water, and are driven forward. The *French Academists*[8] are surprized at the extraordinary Structure in the *Feet* of the *Bever:* their *hindmost* Feet, like those of a *Goose*, are more proper to *swim* than to *walk* with; but their
15 *foremost* are like *Hands* rather than *Feet*, and wondrously suit their Occasions.

Some, as the *Moles*, have their *Feet* for *walking*, and for *digging*.

Some, as the *Bats*, for *walking*, and for *flying* too.

In some the *Feet* are more lax and weak, for the plainer Lands;

---

8. The French Academists were members of the Royal Academy of Sciences led by Claude Perrault, who made anatomical dissections of animals. Perrault published a book on his findings in 1669 (2d ed., 1682). He also published *Mémoires pour servir à l'histoire naturelle des animaux* (Paris, 1671–76), which Professor Charles Singer called the crowning achievement of the Royal Academy and the foundation of comparative anatomy. This rare folio edition was published in English translation by Alexander Pitfeild as *Memoir's for a Natural History of Animals* (London, 1688). Mather also applies the term "Parisian Anatomists" to Perrault's group.

but others have them stiff, and less flexible; their Joints hardly discernible, as the *Elks,* and the *Goats,* which are to traverse the Ice, or to pass over the dangerous Precipices of the Mountains.

In some the *Feet* are shod with tough and hard *Hoofs,* (either whole or cleft, as there is most occasion) in others they have only a *callous Skin.*

And here 'tis admirable to see how their *Toes* are supplied, according to their several Conveniences.

The Structure of the *Bones* in *Quadrupeds* would be a mighty large Field for Curiosity and Admiration.

Galen remark'd a singular Provision of Nature for the *Strength* of the *Lion,* that his *Bones* are much more *solid* than those of other Animals.

Mr. *Ray* enquiring how so many Animals do to bear up against the extremest Rigor of the *Cold,* he notes, that the Extremities of their *Toes* are fenced with *Hoofs,* which in a good measure secure them: he adds, the main thing is, that the *Cold* is its own Antidote; for the Air being fully charged and sated with nitrous, or some other sort of Particles, (which are the great Efficients of *Cold,* and no less also the Pabulum for *Fire*) when it is inspired it causes a great Accension[9] in the Blood (as we see the *Fire* burns fiercely in such Weather) as enables it to a vigorous resistance of the *Cold.*

The *defensive Armour* given to some Creatures, with the Skill to use it, how admirable! The *Hedge-hog,* filled with sharp and strong Prickles, has also a Muscle given him on purpose, which enables him to contract himself into a *globular Figure,* and so inclose himself in his Thicket, that his rapacious Enemies cannot lay hold upon him. *Olaus Borrichius* is amazed at the wondrous Fabrick of that Muscle. The *Armadilla,* described by *Marcgrave,* is covered with a strong, hard, scaly Crust or Shell, of a boney Substance, with four transverse Commissures[10] in the middle of the Body, connected by tough Membranes. By a *peculiar Muscle* he brings his Tail to his Head, and so gathers himself into a round Ball, that there is nothing to be seen but his *Armature:* had such a Muscle been given to any Animal covered with soft Hair or Fur, there might have been a pretence to fancy that this was accidental and undesigned; but seeing there is not one Instance of this kind,

9. An archaic word meaning kindling, ignition, combustion.
10. That is, a joint or seam; the place where two parts of one body meet and unite.

Mr. *Ray* very justly says, *It must be great Stupidity to believe it, and Impudence to assert it.*

Let us pass to the *Head*. The Head of *Man* is of one singular Form. In the *Four-footed* the Form of the *Head* is almost as various as the Species, in some square and large, suitable to their Food, Motion, and Abode; in others more small, more sharp, and more slender, still to suit those purposes. How surprizingly is the *Head* and the *Neck* of the *Swine* adapted for his rooting in the *Earth!* How the Neck, Nose, Eyes and Ears of the *Mole*, adapted in the nicest manner to its way of subterraneous living! The strong Snout of the *Swine,* such that he may sufficiently thrust it into the Ground, where his Living lies, without hurting his Eyes; and of so sagacious a Scent, that we employ them to hunt for us; and even his *wallowing in the Mire,*[11] is a wise Contrivance for the Suffocation of troublesome Insects! The *Mole* so shaped, that our Doctor *More* makes this Creature a notable Ingredient in the Composition of his *Antidote against Atheism;* even his want of a *Tail* is a considerable Contrivance for his advantage.

The *Brain* of *Quadrupeds* obliges us to employ *ours* in a particular Contemplation of it; it is larger in *us* than in *them,* no doubt for the Accommodation of a nobler Guest, which we entertain in ours: but an exact Anatomist of that part, the famous Dr. *Willis,* has led us more particularly to contemplate the Situation of it. In *Man,* to whom God has given a *lofty Countenance,* with a Capacity to think on *heavenly things,* the *Brain* is placed above the *Cerebellum,*[12] and all the Sensories; in *Brutes,* whose *Brain* is incapable of Speculation, the *Cerebellum,* whose Business it is to minister to the Actions and Functions of the *Præcordia*[13] (the principal Office in those Creatures), is above the *Brain,* and the Eyes and Ears are placed at least equal to it: moreover, in the Head of *Man* the *Base* of the *Brain* and *Cerebell* is parallel to the *Horizon,* by which means there is less danger of their jogging or slipping out of their place; but in *Brutes,* whose Head hangs down, the *Base* of the Skull makes a *right Angle* with the Horizon; and yet lest the *Cerebell*

85 Creatures),/Creatures)

11. 2 Pet. 2:22.
12. The cerebellum is part of the brain in vertebrates and the major center for muscular coordination and body equilibrium. Mather calls it "cerebell" below.
13. The precordium is the part of the ventral surface of the body overlying the heart and stomach.

should be unsteady, and the frequent Concussions thereof should cause disorderly motions of the Spirits about the *Præcordia,* there is a sufficient provision made by the Artifice of Nature, by the *Dura Meninx*[14] closely encompassing of it; besides which, it has also in some a strong boney *Fence* about it.

The *carotid Arteries*[15] passing thro the Skull of *Quadrupeds,* and their branching into the *Rete mirabile,*[16] and some other such things, are particular Accommodations to their Circumstances, to prevent a too rapid Incursion of *Blood* into the *Brains* of Creatures that hang down so much.

At the great Aperture of the Shell in a *Tortoise,* there is at the top a raised Border, to grant a liberty to the *Neck* and Head, for the lifting of himself upwards; and this Inflection of the *Neck* is of great use to him, for without it he would be unable to turn himself when thrown upon his Back. The *French Academists* look'd upon the Contrivance as a surprizing one!

The Varieties in the inner and outer *Ear* of *Animals* entertained Dr. *Grew* with observable Curiosities. In an *Owl,* that perches *above,* and hearkens after her Prey *below,* it is produced *further* out above than it is below, that so the least Sound from that Quarter may be the more easily received; but in a *Fox* that scouts *underneath,* it is for the same reason produced further out *below.* In a *Polecat,* which hearkens directly forward, it is produced *behind,* for the taking of a forward Sound; but an *Hare,* which is very quick of hearing, and thinks of nothing but being pursued, has a *boney Tube,* a natural *Otacoustick,*[17] so directed *backward,* as to receive the smallest and farthest Sound that comes behind it; and in an *Horse,* which receives the Sound of the Driver behind, the Passage into the Ear is like that of the *Hare.*

It is remarkable that in *Quadrupeds* the *Necks* are commensurate to the *Legs;* the equality in the length of their *Necks* and their *Legs* is most remarkably seen in Beasts that feed constantly upon *Grass.*

14. That is, any of the three membranes enveloping the brain and the spinal cord.
15. The carotid arteries are the two main arteries that supply blood to the brain.
16. A complex network of arteries formed by the breaking up of a larger vessel into branches that usually reunite into one trunk. In Galenic theory, animal spirit, the motive force of the nervous system, was said to be manufactured in the *rete mirabile.* This network of blood vessels is found at the base of the skull in hoofed quadrupeds. It was supposed to exist in man as well, but Berengario da Carpi's most notable anti-Galenic heresy was denial of its existence in the human brain. In humans the region at the base of the skull is occupied by the circle of Willis.
17. An instrument to assist hearing; for example, an ear trumpet.

But that which is yet more surprizing, is, that in that sort of Creatures which must needs hold their *Heads* down in an *inclining Posture* for a considerable while together, which would be very painful to the *Muscles,* on each side the Ridge of the Vertebras of the *Neck,* Nature hath placed an ἀπονεύρωσις,[18] or *nervous Ligament,* very thick and strong, and apt to stretch, and shrink again, as need requires, and void of Sense, extending from the *Head* (to which and the next Vertebras of the *Neck* it is fastned at the end) to the middle Vertebras of the *Back* (to which it is knit at the other end) for the assisting of them to support the Head in that posture; it is by the Vulgar called *the Whitleather.*[19]

Indeed this Proportion is not kept in the *Elephant,* he has a *short Neck,* the excessive Weight of his Head and his Teeth to a *long Neck* would have been unsupportable; but then his *Proboscis! Tully* takes notice, *Manus data Elephantis, quia propter Magnitudinem Corporis, difficiles aditus habebant ad Pastum.*[20] He is provided with a *Trunk,* wherewith, as with an *Hand,* he takes up his Food, and his Drink, and brings it to his *Mouth;* a Member so admirably contrived, that Mr. *Derham* has just occasion to say, *'tis a manifest Instance of the Creator's Workmanship.*

Galen observing the *Necks* of Animals, how accommodated to their feeding, is not able to forbear his Acclamations of an *Opus Artificis Utilitatis memoris!*[21] He goes on with his Contemplation, and adds, as we cannot but also do, *Quo pacto non id etiam est admirandum!*[22]

On the mention of the *Elephant,* we will introduce a particular Curiosity relating to him; he has no *Epiglottis,* because there is no danger of any thing falling into his Lungs from eating or drinking, seeing there is in him no Communication between the *Oesophagus* and the Passage into the *Lungs;* the Passage to the Ventricle is thro the Tongue, an Hole near the Root of it is the beginning of the *Oesophagus,* and the Passage of the Air into the Mouth is quite stopped up; however, he is not sufficiently secured from small

27 Vertebras/Vertebres   31 Vertebras/Vertebres
32 Vertebras/Vertebres

18. "The end of the muscle."
19. The tough ligament in the neck of a quadruped; also known as paxwax.
20. "The elephant is even provided with a hand, because his body is so large that it was difficult for him to reach his food."
21. "The work of an Artificer mindful of the usefulness of the parts."
22. "How admirably is all this contrived."

Animals that may creep in and murder him; a *Mouse* creeping up his *Proboscis,* might get into his *Lungs,* and so stifle him: guess now the reason why an *Elephant* is so afraid of a *Mouse!* To avoid this danger, when he sleeps he keeps his *Proboscis* close to the ground, that nothing but *Air* could get in. Mr. *Ray* celebrates this as a rare Sagacity!

The *Stomach* of *Quadrupeds!* How adapted to the various Food intended for it! One kind of *Stomach* in the *Carnivorous,* another in the *Herbaceous!*

The peculiar Contrivance on the *Stomach* of the *Camel* deserves our Pause upon it; the words of the *Parisian Anatomists* upon it, are, At the top of the second of the four Ventricles there are several square Holes, which were the Orifices of about twenty Cavities, made like Sacks, placed between two Membranes, which do compose the Substance of this Ventricle; the view of these Sacks made us think that they might well be the Reservatories, where Pliny *saith that Camels do a long time keep the Water, which they drink in great abundance, to supply the want thereof in the dry Desarts.*

In some of the *Quadrupeds* the *Stomach* is fitted for a *Digestion* upon bare *Mastication;* but in others there is a whole Set of *Stomachs,* to digest with the help of *Rumination.* Mr. *Derham* is very sensibly affected with the curious Artifice of Nature here; but for the whole Business of *Rumination,* the learned *Peyerus* will give you a very affecting Entertainment in his *Merycologia, seu, de Ruminantibus & Ruminatione Commentarius.*

Dr. *Grew* observes, all *carnivorous Animals* have the *smallest* Ventricles, *Flesh* going farthest; those that feed on *Fruits* and *Roots* have them of a middle size; *Sheep* and *Oxen,* which feed on *Grass,* have the *greatest;* yet the *Horse,* tho graminivorous,[23] has comparatively but a little one, for that he is made for *Labour:* the same is to be said of the *Hare,* which is made for *Motion,* for which the most easy *Respiration* and the most free play of the *Diaphragm* is requisite, and that could not be if the *Stomach* were very big and cumbersome upon it.

There are *domestick Animals* which look up to me for their *Food,* sometimes for the *Crumbs* that fall from my *Table;*[24] I will consider myself as doing the part of a *Steward* for the Glorious GOD in feeding them; it shall be done with an *holy Delight,* and with such

23. That is, feeding on grass.
24. Adapted from Matt. 15:27 and Mark 7:28.

an Inference drawn from it as this: *And will not the Glorious GOD graciously and readily grant the Mercy which I look up to Him to bestow upon me!*

The Food of the *Castor*²⁵ is generally of *dry things,* and such as are hard of digestion; and now there is a wonderful provision made in the *Stomach* of that Creature, by a *digestive Juice,* lodg'd in the curious little Cells of it; the admirable Structure and Order thereof is described by *Blasius*²⁶ out of *Wepfer,* and then he adds, *Nimirum quia Castoris alimentum ex succum & coctu difficilimum est, sapientissimus & summe admirandus in suis Operibus rerum Conditor, D. O. M.*²⁷ *ipsi pulcherrima ista & affabrefacta Structura benignissime prospexit, ut nunquam deesset Fermentum, quod ad solvendum & comminuendum alimentum durum & asperum par foret.*²⁸

There is in the *Eye* of *Brutes* a *Periopthalmium,* or nictating Membrane, which the *Eye* of *Man* is a stranger to; the *Royal Academy* at *Paris* have been very curious and punctual in the description of it: their Opinion of it is, that this *Membrane* serves to clean the *Cornea,* and to hinder, that by *drying* it grow not less transparent. *Man* and the *Ape,* which are the only Animals wherein this Membrane is not found, have not wanted this provision for the cleaning of their Eyes, because they have Hands, with which they may, by rubbing their *Eyelids,* express the Humidity contain'd in them, which they let out thro the *Ductus Lachrymalis;*²⁹ as is known by Experience, when the *Light* is darkned, or when the *Eyes* are pained, or itching, these Accidents do cease upon the *rubbing of the Eyes.*

In the *Heart* of *Quadrupeds* there is an excellent provision for the living of those Creatures.

The *Foramen Ovale*³⁰ in some (that which in a *Fœtus* makes the

---

25. That is, the beaver.
26. Gerardus Blasius or Gerard Bläes is quoting Wepfer.
27. Mather changes the "T.O.M." of his source to "D.O.M." His probable meaning is *Deus optimus maximus,* "the great work of God."
28. "For as the aliment of the castor [beaver] is of a dry nature and of difficult digestion, the most wise and wonderful Author of Nature hath so constructed the stomach of the animal, that it is perpetually supplied with a fermenting juice which is fitted for dissolving and grinding down that hard and rugged aliment."
29. That is, the tear duct.
30. The oval opening in the septum secundam of the fetal heart that provides communication between the atria. In the fetus the circulation of the blood must bypass the lungs, since these are not operational during intrauterine life. The bypass is accomplished by the foramen ovale, which connects the left and right atria.

Anastomosis,[31] by the means whereof the *Blood* goes from the *Cava* into the *Aorta,* without passing thro the Lungs) is an Accession to the Wonders.

This Passage between the *Arteria Venosa*[32] and the *Vena Cava*[33] is kept open in *Amphibious Quadrupeds;* this maintains a degree of Heat and Motion in the Blood, which may be sufficient for them while they are under Water.

The *Epiglottis* in such Creatures is also larger and stiffer than it is in others, that so when they are feeding under Water, the Water may not break in upon their Lungs.

I confess Mr. *Cheselden* is of the Opinion, that it is not the *Foramen Ovale,* but the *Ostium Venarum Coronoriarum,*[34] which being very near it, may easily be mistaken for it, that the Anatomists have made their curious Remark upon; however the provision is admirable!

The *Heart* in *Beasts* is near the *middle* of the whole Body, in *Man* it is nearer the *Head;* this *Aristotle* observes: but Mr. *Lower,* who has been a most curious Anatomist of this Part, gives us a reason for it; the Trajection and the Distribution of the *Blood* wholly depending on the *Systole* of the *Heart,* and so either the *Heart* must have been stronger in Man, or the *Head* would have wanted its due Proportion of *Blood,* if it had not been so near to the *Heart;* whereas in *Beasts,* whose *Heads* hang down, the *Blood* goes a plainer way, and often a steep one.

There are also peculiar *Nerves* reaching to the Heart of *Beasts,* besides the *sixth Pair,* as in *Man,* a Relief provided by Nature, lest their *prone Heads* might fail of imparting Animal Spirits copiously to it.

The *Cone* of the *Pericardium*[35] in *Quadrupeds* is loose from the *Diaphragm,* whereas in *Man* it is fastned to it; thus the motion of the *Midriff,* in the necessary Act of *Respiration,* is notably assisted in the posture of both. Dr. *Tyson*'s Remark upon it is, *This must needs be the Effect of Wisdom and Design,* and it is plain was intended in *Man* to walk upright, and not upon all four, like the *Quadrupeds.*

31. An act of joining; the union of artery and vein.
32. That is, the pulmonary vein.
33. That is, a vein cavity or chyle vessels.
34. That is, the mouth (opening) of the coronary veins. I thank Mr. Robert J. Adelsperger, Curator of Special Collections, Library of Health Sciences, University of Illinois at Chicago, for help in defining this term.
35. The sac of membrane that encloses the heart and the roots of the great blood vessel of vertebrates.

In the *Four-footed* there is not that Communication between the *Head* and the *Heart* which there is in a *Man*, especially by the Branches of the *intercostal*[36] Pair of *Nerves*, which are sent from the *cervical Plexus*[37] to the *Heart*, and the *Præcordia*, a thing which Mr. *Derham* cannot behold without calling it *a prodigious Care of Nature;* thus the *Head* and *Heart* of Man have a more *intimate Concern* with each other, and a greater and quicker Correspondence, than what is in other Creatures: *Brutes* are more simple Machines; but in *Man*, by the Commerce of the *cervical Plexus*, the Conceptions of the *Brain* presently affect the *Heart*, and agitate its Vessels, and the whole Appendage thereof, together with the *Diaphragm;* whence the Alteration in the motion of the Blood, the Pulse, and Respiration: and when any thing affects or alters the *Heart*, the Impressions are not only retorted by the same Duct of the *Nerves*, but also the *Blood* itself, with a changed Course, flies to the *Brain*, and there agitating the Animal Spirits with diverse Impulses, produces various Conceptions in the Mind. This is Dr. *Willis*'s Observation; who adds, that the Antients therefore made the *Heart* the Seat of *Wisdom;* and certainly the Works of *Wisdom* and *Virtue* do very much depend upon the Commerce which is between the *Heart* and the *Brain*. This eminent Person dissecting a *Fool*, found, besides the Smallness of his *Brain*, the principal difference between him and a Man of Sense to be, that the *Nervi-Intercostalis Plexus, in hoc Stulto valde exilis, & minorum Nervorum Satellitio stipatus fuerit.*[38] The want of the *intercostal Commerce* with the *Heart* in Brutes, is truly an admirable thing! MAN, ponder upon this, and say, *Where is God my Maker, who teaches us more than the Beasts of the Earth!*[39]

I cannot here forbear to introduce a good Observation of a Gentleman who writes *Christian Religion's Appeal*,[40] which he thus expresses; "That God should endow us with *Reason*, and make us differ from the *Brutes*, only that we may rule *them*, and not

88 *Religion's/Religious*

36. That is, situated between the ribs.
37. That is, a network of interlacing nerves in the neck.
38. "The plexus of the intercostal nerve was in the idiot very small, and furnished with much less appendage of nerves."
39. Job 35:10–11.
40. John Smith (fl. 1675–83), rector of St. Mary's Colchester, was the author of *Christian Religion's Appeal from the Groundless Prejudices of the Skeptick, to the Bar of Common Reason* (London, 1675).

ourselves, and put a *golden Mattock* in our Hands, only to dig *Dunghills;* has not the least Congruity with the *Decorum* observed by Him in all His Works, which are framed in Weight, Number, and Order."

95 *Lactantius,* do thou pass a Censure on the *Men like the Brutes that perish,*[41] who do not from the *Beasts* learn the Being and the Glory of a GOD! *Illos qui nullum omnino Deum esse dixerunt, non modo non Philosophos, sed ne Homines quidem fuisse dixerim; qui*
99 *mutis simillimi, ex solo Corpore constiterunt, nihil videntes animo.*[42] [lib. 7. c. 9.]

Galen gives us a notable Relation of a *Kid,* which he took alive out of the Belly of the Dam, and brought it up; the *Embrio* presently fell to walking, as if he had *heard,* says *Galen,* that *Legs*
5 were given him for that purpose; then he smelt into all the things that were set in the Room, and refusing them all, only supped up the *Milk:* after two Months the tender Sprouts of *Shrubs* and *Plants* appeared, and then refusing the rest, he kept to those which are the peculiar Food of *Goats.* But that which to *Galen* appeared
10 most admirable of all, was, that a while after it began to *chew the Cud;* whereupon says he, θεασάμενοι πάντες ἀνεβόησαν ἐκπλαγέντες ἐπὶ ταῖς τῶν ζῴων δυνάμεσιν, All that saw cried out with Admiration, being astonished at the natural Faculties of Animals. He complains thereupon that many neglect such *Works of Nature,* and admire
15 none but μόνα τὰ ξένα θεάματα, unusual Spectacles. Mr. *Ray* notes, *One may fill a Volume with Comments on this pleasant Story.*

The *Sagacity* of some *Quadrupeds,* tho so far short of *Man*'s, yet is a matter of Astonishment to *Man;* and *Man*'s will be short of *theirs,* if it see not the glorious GOD of Nature operating in it.

20 Indeed there was Humour enough in *Rorarius,* who upon hearing a learned Man prefer such a Wretch as *Frederick Barberossa,* before that great Emperor *Charles* V. was thereby so provoked, that he wrote his two Books to prove *that Beasts often have more Use of Reason than Men.* The Consequence of the *absurd Reasoning*
25 he found among *Men* was this with him, *Itaque in Mentem mihi venit Animalia Bruta sæpe Ratione uti melius Homine.*[43] But the

41. Adapted from Ps. 49:10, 12, 20.
42. "[For] as to those who have altogether denied the existence of God, I should not only refuse to call them philosophers, but even deny them the name of men, who, with a close resemblance to dumb animals, consisted of body only, discerning nothing with their mind."
43. "And so it occurred to me that brute beasts often make better use of reason than do men."

Consequence of his own *absurd Reasoning* will soon be found such as will carry thousands of *Terrors* with it.

It is enough that what of *Reason* appears in the *Brutal Tribes*, is an immediate Effect of the Providence exerted by the all-wise Creator, and applied for the *Preservation* of His Creatures. *O Lord, thou preservest not only Man, but Beast also!*[44]

The Words of the excellent Sir *Richard Blackmore*, in his Essay on *the Immortality of the Soul*, are worthy to be transcribed and pondered on this Occasion. "I must acknowledge that I look upon the *Souls* of *Brute* Creatures as *immaterial*, for I cannot conceive how an internal Principle of *sensitive Perception* and *local Motion* can be framed of *Matter*, tho ever so subtile and refin'd, and modified with the most artful Contrivance; yet they are plainly of a base and low Nature, and destitute of those *intellectual Faculties* and that *free Choice* that should make them Subjects of *Moral Government*, enable them to discern the Obligation of *Laws*, and the Distinction of *Virtue* and *Vice*, and understand the Notion of being an *accountable Creature*, and receiving *Rewards* and *Punishments*. Whether the *Animal Souls* in a State of Separation remain *stupid* and *asleep*, or whether they are *dispersed* thro the Creation, and employ'd to animate *other Beings*, or return to *one common Element*, whence they were at first deriv'd, is unrevealed; but this is certain, the *Souls* of *Brutes* are not design'd by the Great Creator for such a Life of *Pleasure* and *Happiness*, as that of *Human Souls* in a State of *Immortality* and *Perfection*, for the Enjoyment of which they have no Dispositions and Capacities."

The Opinion of *Descartes*, and *Gassendus*, and *Willis*, and others, That the *Soul* of Brutes is *material*, and the whole Animal a meer Machine, is clogg'd with insuperable Difficulties.

Our excellent *Ray* bespeaks a *lower degree of Reason* for them, and his Argument is fetch'd from some of their Actions, which, without allowing some *Argumentation* in *them*, can hardly be accounted for; he singles out the *Dog*, the *Dog* running before his Master, will stop at a divarication of the way, till he see which way his *Master* will take. Again, when the *Dog* has got a *Prey*, which he fears his Master will take from him, he runs away to hide it, and afterwards returns to it. Once more, if a *Dog* be to leap upon a Table which he sees too high for him to reach at

---

44. Adapted from Ps. 36:6.

once, let there be a *Stool* or *Chair* near it, he will first mount *that*, and so the *Table,* yea, tho the *Stool* stand so that the Creature takes not a *direct Leap* towards the place finally intended; if he were a meer Piece of *Clockwork,* and this Motion caused by the striking of a *Spring,* there can be no reason imagin'd why the Spring being set on work, should not carry the Machine in a *direct Line* towards the Object that put it in motion, as well when 'tis on an *high* Table as when 'tis on a *low.*

They that have written *de Canum Fidelitate & Sagacitate,*[45] have entertained us with Stories full of Wonders. The Observers have thought themselves obliged sometimes to suspect that the *Dogs* might have a *Spirit of Python*[46] in them, *Camerarius* in his *Horæ Subcesivæ* has collected surprizing, but credible Relations, of such as we may call *reasonable Dogs.*

A well-known King, who dealt much in them, at a famous *Act* in one of our Universities, very publickly determin'd it, *that they could make Syllogisms,* and so 'tis no longer to be disputed. The Authority is as great as that of *Jacobus Micyllus,* who wrote an *Elogium Canis,* which is thought a very elegant Epigram.

There is a surprizing thing related of the *Sea-Tortoises,* both *Aristotle* and *Pliny* have remark'd it; That when *Tortoises* have been a long time upon the Water, during a Calm, their Shells will be so dried with the *Sun,* that they are easily taken by the Fishermen, because being become too light, they cannot plunge into the Water nimbly enough. The *French Academists* do not refer this easiness to be now taken, merely to the *Lightness* of the Creature's Body, for he could easily let *Air* enough out of the Lungs to render his Body heavier than the Water, upon which he would sink immediately, but to a *Sagacity* of the cautious Animal, which is truly marvellous. The *Tortoise* is always careful to keep himself in his *Equilibrium,* and therefore he dares not let the Air out of his Lungs, to acquire a Weight which would make him to sink immediately; for he fears lest the wetting of his Shell should render it so heavy, that being sunk to the bottom of the Water, he might never afterwards have the power of re-ascending. What *Foresight* here! What a degree of *Argumentation* too!

---

45. "On the Faithfulness and Sagacity of the Dog."
46. The Spirit of Python is a familiar spirit, the demon which possesses a soothsayer. The term also refers to one possessed by such a spirit.

They that have written *de Sollertia Animalium* (as many besides *Plutarch* have done)⁴⁷ have reported such Essays and Shadows of *Reason* in many of them as are diverting.

The *Fox* is often catch'd in Tricks, which afford as pleasant Stories as any in that old Volume, *The delectable History of Reynard.*⁴⁸ His way to get rid of his *Fleas* is notorious.⁴⁹

What notable *Architects* are our *Bevers!* They lay their *Logs,* and build their *Dams,* and form their *Chambers,* with a marvellous Artifice. A Nation of *Indians* do sometimes in scarce any thing but their Speech *out-man* a Nation of *Bevers.*

Elephants, what *reasonable,* but what *prodigious* things have been related of them! Things that almost have *Religion* in them. The Story of *Hanno*⁵⁰ is an amazing one, *Pierius* is our Author for it. Well may I *write* of them that have themselves been so susceptible of Discipline as to *write* whole Sentences; 'tis affirm'd that *Elephants* have done so. *Alsted* spends two whole Pages together, in his concise way, enumerating but the Heads of the *strange things* which this *tractable,* and almost *rational* Quadruped arrives to!

What a notable, docile, tractable Animal the *Horse!* The *Horse,* of whom the admirable *Buchanan* sings,

*Equus ad cunctos se accommodat usus.*⁵¹

Read *Solinus,* and see what Approaches the *Horse* makes to *Reason!* One would question which had most, *Caligula* or *Incitatus.*⁵² Dr. *Grew* admires him, as being *swift* and *strong,* above most other Animals, and yet strangely *obedient;* both comely and clean; he breeds no *Vermin* of any sort; his Breath, his Foam, his Excrements and Sweat, all sweet and useful; fitted every way for Service or Pleasure, for the meanest or the greatest Master. There are antient

3 *Sollertia / Solertia*

47. Plutarch's dialogue *De sollertia animalium,* which discusses whether land or sea animals are more clever, makes the point that all animals are rational.
48. The source of the story of Reynard the Fox is Aesop's fables. Latin collections of animal fables based on Aesop were common during the Middle Ages. William Caxton's English translation, *The History of Reynard the Fox,* printed in 1481, was very popular in England.
49. The fox rids himself of fleas by lowering himself into water with a bundle of hay. The fleas gather on the hay-bundle, and the fox dives into the water.
50. Hanno was the elephant of Emanuel, sometimes known as Manuel I, called the Great (1469–1521), king of Portugal during that country's golden age.
51. "The horse adapts itself to all uses."
52. Caligula, Roman emperor from A.D. 37 to 41, built a marble stable for his horse Incitatus.

Topsell's *Beastes* offered readers exhaustive accounts and illustrations of real animals as well as mythological creatures, and the elephant is one of the most popular. Courtesy of the Rare Book and Special Collections Library, University of Illinois at Urbana-Champaign.

Examples of other *Horses* besides *Bucephalus*[53] and *Lethargus*,[54] that have been honour'd with stately *Funerals* and *Sepulchres* at their *Deaths,* as well as their Masters; yea, tho the Epitaph of *Adrian* be lost, his *Horse*'s is preserved to this day.[55] The *Riders* of *Horses,* who in their *Lives* will submit to no *Bridles,* nor do any *Service* for Him that made them, deserve at their *Deaths* to pass away no better esteem'd than their *Horses,* but will have a worse Fate than they.[56] The Gentleman, who going home with his Head full of the *sickly Fumes* from the *Healths* of the Evening's Debauch, could not compel his *Horse* to drink an *Health* which at the next Brook he proposed to him, had so much *Reason* left him (and a very little might serve) as to make that Reflection, *That the Man in the Saddle was the greater Beast of the two.*

How innumerable are the *Appearances* of Nature, which are above the Powers of *Mechanism?* 'Tis religiously and most reasonably observed by Dr. *Cheyne,* that all these are so many undeniable Proofs for the Being of a GOD; there must be a *Power superior* to those of *Mechanism,* and this must lead us to Him, *who alone does great and marvellous things.*[57]

How often have I heard this, and how plainly seen it; this Power belongeth to God![58]

After all, do we see something in these, and other, and all *Creatures,* that appears *defective* to us? A wise Remark made by the Marquis of *Pianezza*[59] shall be introduced upon it; his remarkable words are these: "The *limited Perfections,* and the seeming *Irregularities* of the World, rather afford us occasion to acknowledge and glorify the *Providence* of GOD, which not

53. Bucephalus was the favorite horse of Alexander the Great.
54. Lethargus has the distinction of gaining Mather's attention without leaving sufficient traces to be identified. The name, from the Greek word meaning "morbid drowsiness," does not seem to fit a horse which merits the honors described, certainly not a race horse. Is it possible that Mather's memory played a trick on him? A sepulchral epigram by Pisander of Rhodes in the *Anthologia Palatina* ("Palatine Anthology") (7.304) relates that a man named Hippaemon, a Thessalian from Crete who perished fighting in the front ranks, had a horse named Podargos and a dog named Lethargos. I am indebted to Professor George Pesely for this reference.
55. Hadrian, Roman emperor from 117 to 138, built a tomb with a stele and an inscription for his horse Borysthenes.
56. This sentence alludes to Zech. 10:5 and Amos 2:15.
57. Adapted from Ps. 136:4.
58. Ps. 62:11.
59. Carlo Emanuel Filiberto Giacinto de Simiana, Marquis de Pianezza (1608–77), Italian soldier, diplomat, and man of learning, abandoned worldly employments to become a monk. He published *La Christiana essere la sola religione verace* (Turin, 1700), which was translated as *The Truth of the Christian Religion* (London, 1703).

only declares, that all the Creatures are too *imperfect* to deserve to be *worshipped* as *Deities*, but also amidst their *Imperfections* obliges them to confess, as it were with their own Mouths, *one infinitely perfect Deity; a Deity* that would not have *Man* fix on *them* as the Objects of his Love and Admiration, but that from them he should pass on to the Love and Esteem of his only true GOD."

There is one very surprizing thing, and without acknowledging a Superintendency of a *Divine Providence* there can be no accounting for it. The *Mansuete*[60] *Creatures* bring forth no more than one or two at a time, the *Beasts of Prey* bring forth as often, and seven, or nine, or eleven at a Litter; and yet! what inexpressible Multitudes of the *Mansuete* have we to serve us! What vast Herds of *Beeves*! What vast Flocks of *Sheep*! Whereas they that live upon *Prey* appear in very little Numbers. How rarely is a *Wolf* met withal, tho a Price be set upon his Head! What Rarities are *Lions*, and *Tygers*, and *Ounces*! To be caged in the *Tower* for *Spectacles*!

And then the Liberty given us to *butcher* our useful Creatures at our pleasure; 'tis observed by Mr. *Robinson*,[61] that this will be found a *Kindness*, rather than a *Cruelty* to the Creatures; if we kill them for our *Food*, their Dispatch is quick, and much less dolorous, than that they should be torn to pieces by such cruel Masters as the *Lion*, and the *Tyger*, and *Bear*, who would not give them time to *die*, but even eat their Flesh from their Bones *alive*; and if they should live to the tedious Condition and Melancholy of *Old Age*, it would, after many Tortures, kill them, and leave their Carcases rotting, stinking, and useless upon the ground.

The *short Life* of a *Beast*, compared with the *Life of Man*, deserves to have some Remark made upon it; this at least: *Man, do not lead* the *Life of a Beast*, if thou wouldst not be condemned and confined to the *short Life* of a *Beast*, nor come under the Execution of that Sentence, *The Days of the Wicked shall be shortned*.[62] There is a way of *living*, by some called *living apace*; it is indeed not *living* at all, but rather *dying apace*; a *beastly* Life ought to be a *shortned* one.

What useful *Instructions* would the Properties of the several *Animals* yield to the *Christian Philosopher*, would he be duly and

60. That is, gentle, tame, not wild or fierce.
61. That is, Thomas Robinson.
62. Adapted from Prov. 10:27.

wisely attentive to them! *Franzius,* and *Simpson,* and others, have cultivated this Theme, not unusefully; 'tis capable of a much more vast Cultivation: *Christian,* hearken to the Voice of the many *Preachers* thou hast about thee, lest thou *mourn at the last,*[63] and say, *I have not obeyed the Voice of my Teachers, nor inclined mine Ear to them that instructed me!*[64]

I remember one Observation of *Seneca,*[65] which a little exemplifies a *moral Remark* on the Properties of some *Four-footed; Omnia quæ Natura fera ac rabida sunt, consternantur ad Vana. Idem inquietis & stolidis Ingeniis evenit, rerum suspicione feriuntur.*[66] I thought this worth mentioning, but not because I do not think a *Christian* of a *good Understanding* might easily produce ten thousand more.

The Account which honest *Leguat* gives of the *solitary Bird,*[67] which he and his Companions observed on the Isle of *Rodrigo,* is as admirable as unquestionable; the *Bird* has *Wings,* but so small that it cannot fly with them, they serve to flutter with a mighty noise when they call one another; they never lay but *one Egg,* which is bigger than that of a *Goose;* the *Male* and *Female* sit upon it in their turns, and all the while they are hatching it, or bringing it to provide for itself, (which is divers Months) they will not suffer any other Bird of their own Species to come within two hundred Yards round of the place: but this is very singular, the *Males* will never drive away the approaching *Females,* but call for their own *Females* to do it; the *Female* does the like, and upon the Approach of any other *Males,* call their own *Males* to chase them away. After these *Birds* have raised their *young one,* and left it to itself, "we have often observed (says my ingenious Traveller) that some days after the *young one* leaves the *Nest,* a Company of thirty or forty brings another *young one* to it, and the new-fledg'd Bird, with its Father and Mother joining with the Band, march to some by-place; we frequently followed them, and found that

63. Prov. 5:11.
64. Prov. 5:13.
65. That is, Seneca the Younger.
66. "All creatures by nature wild and savage are alarmed by trifles. The same is true of men, whether they are by nature restless or inert. They are smitten with suspicions."
67. Rodriguez Solitaire (*Pezophaps solitaria*) was one species of the family Raphidae, which included the dodo (*Raphus cucullatus*) and was found only on the Mascarene Islands. Rodriguez Solitaire was about the size of a turkey. Since the islands never had any land connections, the bird must have arisen from flying ancestors, with flightlessness and gigantism evolving from absence of ground predators. This species became extinct in the late eighteenth century.

afterwards the old ones went each their way alone, or in couples, and left the two *young ones* together, which we call'd a *Marriage.*" My religious Traveller does give all possible Assurance for the Truth of this Relation, and adds, *I could not forbear to entertain my Mind with several Reflections on this Occasion. I sent Mankind to learn of the Beasts.*

It is an Observation made by one of the most refin'd Philosophers[68] by whom our Age has been illuminated; "Most Creatures have some *Quality,* whereby they admonish us of what is BEST. Of *Neatness,* all *Birds* which love to be perpetually pruning of themselves; and *Cats,* which commonly cover their Excrements, and wipe their *Mouths* after Dinner. Foul Water will breed the Pip[69] in *Hens,* and Nastiness Lice and Scabs in *Kine;* and all Creatures, even *Swine* themselves, which love Dirt, yet thrive best when kept clean. Of *Forecast,* the *Sitta*[70] and the *Ant,* which lay up Nuts and other Seeds in their Granaries, that serve them in the Winter. Of *Modesty,* the *Elephants,* the *Dromedaries,* and the *Deer,* which always conceal their Venereal Acts. Of *mature Marriage,* all Animals which beget their best Breed at their full Growth. Of *Conjugal Chastity,* the *Doves* and *Partridges,* which keep to one Husband and Wife. Of *Conjugal Love,* the *Rook,* the Male helping the Female to make her Nest, feeding her while she sits, and often sitting in his turn. Of *Maternal Love,* the domestick *Hen,* gentle by Nature, and unarmed, yet, in defence of her Chickens, bold and fierce; and the *Tyger* herself, the fiercest of Beasts, yet is infinitely fond of her Whelps."

The same excellent *Fellow of the* ROYAL SOCIETY carries on his Observation; "The most odious or noxious things do serve for Food or Physick, or some Manufacture, or other good use; neither are they of less use to *amend our Minds,* by teaching us *Care,* and *Diligence,* and more *Wit:* and so much the more, the worse the things are, we see and should avoid. *Weasels,* and *Kites,* and other mischievous Animals, induce us to Watchfulness;

---

68. That is, Nehemiah Grew. Mather describes him at the beginning of the next paragraph as a fellow of the Royal Society.

69. That is, a disease of poultry and other birds, characterized by the secretion of a thick mucus in the mouth and throat, often with the formation of a white scale on the tip of the tongue (hence sometimes applied to this scale itself).

70. The genus name of true nuthatches, a subfamily of the nuthatches.

Thistles and *Moles* to good Husbandry; *Lice* oblige us to *Cleanliness* in our Bodies, *Spiders* in our Houses, and the *Moth* in our *Clothes:* the Deformity and Filthiness of *Swine* makes them the *Beauty-spot* of the Animal Creation, and the Emblem of all *Vice;* and the *Obscenity* of *Dogs* shews how much more beastly it is in *Men:* the *Fox* teaches us to beware of the *Thief,* and the *Vipers* and *Scorpions* those more noxious Creatures, which carry their Venom in their *Tongues* or their *Tails."*

I will prosecute this Observation of my Brother,[71] with only observing so much further upon it; that no little part of the *Homage* we owe to the glorious *Creator* of all these things, is to learn those *Virtues,* and those *decent* and *honest* things, whereof, if the Faculties of our Minds be awake, we shall easily perceive His *Creatures* to be the *Monitors.*

In writing these things I cannot but call to mind the expressive Words of *Theodorus Gaza* in his Preface to *Aristotle's* Books *de Animalibus; In contemplandis Animalium Moribus, Exempla suppetunt omnium Officiorum, & Effigies offeruntur Virtutum summa cum Authoritate Naturæ, omnium Parentis, non simulatæ, non inconstantes, sed vere ingenuæ atque perpetuæ.*[72] He goes on to shew how powerfully the Kindness of the Brutes to those of their own Kind, rebukes the *unbrotherly Carriage* too often found in Mankind; and adds a variety of Admonitions, which, *my Reader,* thou art not unable to discover by thy own Ingenuity.

¶. "One of the most valuable Writers that ever was in the World,[73] brings from the glorious Creator of the *Beasts* this Voice to Man; *Sic utere illis, ut Exempla Virtutum quæ in illis apparent, observes, & omnibus Viribus coneris illa longo intervallo superare, ut ne Bestialem Animam reperiam in tuo Corpore Humano.*[74]

"It would not be a *Fancy* destitute of *Judgment,* if I should

---

71. Mather, believing himself a fellow of the Royal Society at the time he wrote this book, refers to Grew, also an F.R.S., as "my Brother."

72. "In contemplating the behavior of animals, [we see that] they supply examples of all their duties, and pictures of virtue are supplied with the highest authority of nature, the parent of all; the pictures are not simulated, not inconstant, but genuine and enduring."

73. That is, Johann H. Alsted.

74. "Use them in such a way that you observe the examples of the virtues which appear in them and that you attempt with all your might to surpass them by such a great distance that I may not find in your human body a beastly soul."

set before me the *Tabella Hieroglyphica*, wherewith *Alsted* has obliged us.⁷⁵

"But of all the Tribes that graze in the Field, there is none that I would more chuse for an *Emblem* than the *Sheep;* the clean, patient, innocent Creature, which has nothing belonging to it but what is of a celebrated *Usefulness. O thou most honourable Creature, what a Dignity has the Son of God Himself put upon thee!*

"I see so much of GOD in the Circumstances of the *Brutal Tribes*, as obliges me to *look upwards* in a way too high for them.

"At the same time, tho I would by no means fall into *Pythagorean* and *Mahometan* Superstitions,⁷⁶ yet I would abhor to treat any of the *Brutes* with barbarous Cruelties, *Immanities*⁷⁷ and *Inhumanities; cruelly* to delight in their *Miseries,* or to be *unmerciful* to them, is an Offence to God, and what a *righteous Man* would not be guilty of; *unknown Punishments* may be reserved for it.

"*Great GOD, if I do not acknowledge Thee, I am condemned by the Ox, which knows his Owner, and by the Ass, which knows his Master's Crib!*"⁷⁸

*Luther*⁷⁹ seeing the *Cattel* go in the Fields, used this Expression; *Behold, there go our Preachers, our Milk-bearers, and Wool-bearers, which daily preach to us Faith towards GOD, that we trust in Him as our loving Father, who will maintain and nourish us.*

It is very certain our *Dominion* over the Creatures is very much impair'd by our Fall from God. Those Creatures do now either *fly from* us, or *fly at* us, which, if we had been faithful to our God, would not have done so. Honest *Egardus* propounds two *Admonitions of Piety* on these Occasions; the one, *Fuga Animalium a te, moneat te de tua fuga a Deo per peccatum.*⁸⁰ The other, *Animalium in te ad lædendum impetus hostilis, moneat te de Odio & Furore Diaboli, & Mundi, adversus te immani.*⁸¹

---

75. The name of a table in Alsted (pp. 514–17), which lists animals and their temperaments.
76. Mather's statement is based on a source which refers to the Pythagorean pity toward animals and the Turkish belief that grace is obtained by feeding and caring for animals. At the time he wrote "Turkish" signified "Muslim."
77. Immanity is an obsolete word meaning monstrosity or inhuman.
78. Adapted from Isa. 1:3.
79. That is, Martin Luther (1483–1546).
80. "Let the flight of the animals from you warn you concerning your flight from God through sin."
81. "Let the hostile attack of the animals against you to do harm warn you concerning the great hatred and wrath of the devil and the world towards you."

I conclude with an Observation of Dr. *Grew*'s; "As the *essence* of every thing, and its relation, in being fitted, beyond any Emendation, for its *Actions* and *Uses*, evidently proceeds from a Mind of the *highest Understanding,* so the nature of these *Actions* and *Uses,* in as much as they are not any way destructive or troublesome; no, but each thing tends apart, and all conspire together to conserve, cherish, and gratify: this is an Evidence of their proceeding from the *greatest Goodness.* There are many who are very *cunning* and *subtile* in the Invention of *Evil*, and *Engines* have been fitted, with much Contrivance, for the tormenting of Men; how easy had it been for the Creator of the Universe to have stock'd it with Creatures that should never have moved so much as one Limb without *Pain*, or have had the least Sensation without a mixture of horrible *Torment,* or have entertain'd the least Imagination, but what should have had *Horror* in it? But behold, our good God has ordered it, that whatever is *natural* is *delightful,* and has a tendency to Good; He has employ'd His transcendent *Wisdom* and *Power,* that He might make way for His *Benignity.*"

Great GOD, Thou art Good, and Thou dost Good; Oh teach me Thy Statutes![82] So sings the Poet:[83]

O Deus, O Mundi solus qui flectis habenas,
Ut tua nunc Bonitas oculis est obvia cunctis![84]

ESSAY XXXII. *Of* MAN.

AND now let the *Lord of this lower World* be introduced, MAN, who is to do the Part of a *Priest* for the rest of the Creation, and offer up to God the *Praises* which are owing from and for them all.

In Libro Creaturarum continetur Homo (as one of the School-

---

82. Ps. 119:68.
83. "The Poet" is apparently the psalmist; the Latin verse that follows paraphrases Ps. 8:1, 9.
84. "O God, you who alone turn the reins of the world so that now your goodness is exposed to all eyes!"

Divines happens to express it well) & *est principalior Litera ipsius Libri.*[1]

It was most reasonably done of thee, Father *Austin*,[2] to tax the Folly of them who admired the *Wonders* in the other Parts of the Creation abroad, & *relinquunt seipsos, nec mirantur,*[3] but see nothing in *themselves* to be *wondred* at. It is not for nothing that *Mankind* is in the *Gospel* called *every Creature;*[4] he that beholds *Man,* may therein behold what is most wonderful in *every Creature.*

It is well express'd in a Treatise entitled, *Schola & Scala Naturæ!*[5] "Nature doth not lead thee towards GOD by a far-fetch'd and winding Compass, but in a short and strait Line. The *Sun* waits upon the *Rain,* the *Rain* upon the *Grass,* the *Grass* serves the *Cattel,* the *Cattel* serve *thee,* and if *thou* serve GOD, then thou makest good the highest Link in that *golden Chain,* whereby *Heaven* is joined to *Earth;* then thou standest where thou oughtest to stand, in the *uppermost Round* of the *Divine Ladder,* next to the most High; then thou approvest thyself to be indeed what thou wert designed by God to be, *the High-Priest and Orator of the Universe;* because thou alone, amongst all the Creatures here below, art endued with Understanding to know Him, and Speech to express thy Knowledge of Him, in thy Praises and Prayers to Him."

I may now say with honest *Stigelius,*

Jam vocat ad pulchros nos Fabrica Corporis Artus,
Quæ mira Authorem monstrat in Arte Deum.[6]

The BODY of MAN being *most obvious* to our view, is that which *we will first* begin with; a *Machine* of a most astonishing Workmanship and Contrivance! *My God, I will praise Thee, for I am strangely and wonderfully made!*[7]

"But is it possible for me to consider this BODY as any other

---

1. "Man is included in the book of creatures and is the principal letter of that book." The "School-Divine" is Raymundus de Sebonde (Sebonde has many variant spellings) or Raymond of Sebonde.
2. That is, Augustine.
3. "And yet pass themselves by, nor wonder."
4. Mark 16:15. "Go ye into all the world, and preach the gospel to every creature."
5. *Theologia ruris, sive schola et scala naturae: Or, The Book of Nature* (London, 1686). See the Endnotes for 41:45–52.
6. "Now the body's fabrication, which in marvelous fashion shows God to be its author, summons us to beautiful parts."
7. Adapted from Ps. 139:14.

than a Temple of GOD! A *Vitruvius*⁸ will teach us that the most exquisite and accurate Figure for a *Temple* will be found in a Conformity to an *Human Body;* indeed an *Human Body* ought for ever to be beheld and employed, as designed for an *holy Temple;* for me to apply any Part of such a *Body* to any Action forbidden by God, would be a very *criminal Prostitution.*

"By *using* my *Body* in and for the Service of God, and by *praising* the Glorious-One, who has formed every Part of my Body, *and clothed me with Skin and Flesh, and fenced me with Bones and Sinews,*⁹ I desire to assure my share in an happy *Resurrection* of this *Body* from the *Grave,* into which it is falling: for *tho a Man die, he is to live again;* an *appointed Time* will come, when *Thou, O my God, wilt call, and I shall answer thee, and thou wilt have a desire to see the Work of thine Hands* revived and restored."¹⁰

The *erect Posture* of Man, the *Os sublime,*¹¹ how commodious for a *rational Creature,* who must have *Dominion* over those which are not so, and must invent and practise things useful and curious! *Tully*¹² admires the *Providence of Nature,* as he calls it, adding the reason for it; *Sunt enim e Terra Homines, non ut Incolæ atque Habitatores, sed quasi Spectatores superarum rerum, atque Cælestium, quarum Spectaculum ad nullum aliud Genus Animantium pertinet.*¹³ By this posture Man has the use of his *Hands,* which, as *Galen* observes, are, *Organa sapienti Animali convenientia;*¹⁴ and his *Eyes,* which as they have the glorious *Hemisphere* of the Heavens above him, so they have the *Horizon* of three Miles on a perfect Globe about them, when they are six Foot high, and by the Refractions of the *Atmosphere* they have much more than so: his *Head* is also sustained, which is heavy, and how painful to be carried in another Posture!

The provision made for this Posture is very surprising; what *Ligaments?* especially that of the *Pericardium* to the *Diaphragm,*

---

8. Vitruvius Pollio (early first century–c. 25 B.C.), Roman military engineer and architect, was the author of *De architectura,* which touches on topics beyond architecture.
9. Job 10:11.
10. Adapted from Job 14:14–15.
11. "Uplifted face."
12. That is, Marcus Tullius Cicero.
13. "For men are sprung from the earth not as its inhabitants and denizens, but to be as it were the spectators of things supernal and heavenly, in the contemplation whereof no other species of animals participates."
14. "The instruments most suitable for an intelligent animal."

which, as *Vesalius* and *Blancardius* note, is peculiar to *Man?* The *Bones,* how artificially placed and braced? Most remarkably the *Vertebræ* of the Back-bone? The *Feet,* how exquisitely accommodated! For the rare Mechanism whereof, a *Cheselden* may be consulted; yea, every Writer of *Anatomy* will offer enough to *trample Atheism under foot.* To all add the Ministry of the *Muscles,* which answer all Motions, and yet with easy and ready Touches, keeping the *Line of Innixion*[15] and the *Center of Gravity* where it ought to be! Yea, all the Parts of the Body so disposed as to *poise* it! All in a nice *Equipoise!* With a prodigious variety of *Muscles* placed throughout the Body for the Service! *Borelli* observes, "'tis worthy of Admiration, that in so great a variety of *Motions* Nature's Law of *Equilibration* should always be observed; so that if it be transgressed or neglected, the Body necessarily and immediately tumbles down."

Every thing does conspire to assure us, that the Maker of Man intended Man for such a *Posture.*

The most indigent Condition wherein *Man* is born into the World, but the plentiful Provision which he finds made by a gracious and merciful God for him in the World, this invites *Man* to return to God, and to taste His *Love,* in all the Creatures that accommodate him, and rely upon His *Care* for ever, for the Supply of all his Wants. And, as Mr. *Arndt* expresses it, *Homo Dei Amorem in omnibus rebus eo intimius degustaret, in caducis Creaturis Deum immortalem inveniens discret, quod immortalis Deus melius possit exhilarare, consolari, corroborare, ac conservare hominem, quam omnes omnino Creaturæ fluxæ & cito perituræ.*[16]

A Comparison between the *Macrocosm* and the *Microcosm* would afford a very edifying and acceptable Entertainment to a contemplative Mind; the excellent *Alsted* will therewith entertain the Gentlemen that will visit his *Theologia Naturalis.*

Indeed he that speaks to MAN, speaks to *every Creature;* and *Man* is therefore the more *concerned,* as well as *capable* to hear *every Creature* speaking to him.

'Tis what calls for a deep Consideration with us, that in the *Body* of Man there is nothing deficient, nothing superfluous, an

---

15. That is, a leaning, pressing, or resting upon something.
16. "Man experiences the love of God in all things so much more profoundly; finding the immortal God in fallen creatures he learns that the immortal God is better able to cheer, console, strengthen, and preserve man than all of the transient and short-lived creatures."

End and *Use* for every thing. *Natura non abundat in superfluis, nec deficit in necessariis.*[17] There is no Part that we can well spare, nor any that can say to the rest, *I have no need of you!* The *Belly* and the *Members* cannot quarrel with one another.[18] Even the *Paps* in Men, besides their adorning of the *Breast,* and their defending of the *Heart,* sometimes contain *Milk,* as in a *Danish* Family mention'd by *Bartholinus.*[19] A Man mention'd by *Beccone,* upon the Death of his Wife, suckled the Infant himself. He concludes, that since, according to *Malpighius* and others, the *Paps* of Men have the same Vessels with those of Women, 'tis intended that, if need requires, the *Young* should be suckled at them, who, upon a little pulling, soon fetch *Milk* into them.

What should we do with a *Bavarian Poke*[20] under our Chins?

Our pious *Ray* makes this Remark, That if we consider no more than the very *Nails* at our Fingers ends, we must *be very sottish if we can conceive that any other than an infinitely good and wise God was our Author and Former.* And there was an honourable Person[21] who long before him said, *An non videmus in singulis summis Digitis, Artificium Dei? Estne unguis aliquis qui non reddat Testimonium Deum esse Opificem eximium?*[22]

No sign of *Chance* in the whole Structure of our Body. It is remarkable, in Bodies of different Animals there is an *Agreement of the Parts,* as far as their *Occasions* and *Offices* agree; but a *difference* of those where there is a *difference* of these. Dr. *Dowglass* will tell you what *Muscles* are in a *Man* that are not in a *Dog,* what in a *Dog* that are not in a *Man.* The Matter, the Texture, the Figure, the Strength, with the necessary Accoutrements of every Part, how amazingly commodious! How often does the *Ars, Pro-*

---

17. "Nature does not abound in superfluities, nor does she lack the necessities." Aristotle had formulated the maxim, "Nature does nothing in vain" (*De partibus animalium* 4.11.691b).

18. Adapted from 1 Cor. 12:14–26. The italicized phrase is quoted nearly verbatim from 1 Cor. 12:21. Note that certain parts of the body, such as kidneys, may be taken for transplants. Tonsils have been removed as superfluous in the recent past, but they are now thought to relate to the body's immune system and are normally left in place.

19. That is, Thomas Bartholin.

20. That is, a morbid baglike swelling on the neck, or the goiter. This chronic enlargement of the thyroid gland, usually secondary to decreased intake of iodine, is also called a Bavarian Poke.

21. The reference is to Philipp Camerarius.

22. "Or do we not see in the ends of the individual fingers the handiwork of God? Is there any nail which does not give testimony that God is an extraordinary artisan?"

videntia, & Sapientia CONDITORIS,²³ appear to the Pagan *Galen* upon the Contemplation!

In the *Body* of Man the *Lodgment* of the Parts is as admirable as the *Parts* themselves. Where could the *Eye*, the *Ear*, the *Tongue*, be so commodiously placed as in the *upper Apartments* assigned for them? *Tully* says truly, *Mirifice ad usus necessarios collocati sunt!*²⁴ And for the other Parts, he notes, *Recte in illis Corporunt partibus collocata sunt.*²⁵ Four of the *five Senses,* how commodiously lodged, near the *Brain,* the common Sensory, and a place well guarded; *Galen* celebrated this wondrously agreeable Situation! And how could the *fifth Sense,* that of the *Touch,* be more agreeably lodged, than with a Dispersion into all Parts of the Body! Where should the *Hand,* the *Feet,* the *Legs* be, but just where they are! Where the *Heart,* the *Sol Microcosmi,*²⁶ which is to labour about the whole Mass of Blood, but in the *Center* of the Body? Where can the *Viscera* discharge their Offices better, than in the place assigned to them? Where could the *Bones* and the *Muscles* be better disposed of? And what better *Covering* were it possible for the whole Body to have, than the *Skin;* whereof the *Microscopical Views* given by *Cowper* in his *Anatomy,* must give a vast Surprize to us!

What can be more *ornamental,* than that those Members which are *Pairs,* do stand by one another in an *equal Altitude.*

The Provision made in the Body of Man to *stave off Evils,* is very admirable. The *Secretions* made by the *Glands,* whereof *Cockburn, Keil,*²⁷ *Moreland,* and others, give us notable Accounts, are such as cannot be considered without some Amazement. How many Parts of the Body stand ready to do what belongs to faithful *Centinels!* The principal and more essential Instruments of Life and Sense, how well *barricado'd* are they? Of how many Parts are we supplied with *Pairs,* to make up a Defect which may happen in any of them? The *Pairs* of *Nerves,* and the Ramifications of the *Veins* and *Arteries* in the fleshly Parts, what *Cases of Disaster* are answered in them? Mr. *Derham* here justly adores *the infinite Contriver!* Dr. *Sloane* justly admires the Contrivance of our *Blood,* which on some Occasions, as soon as any thing destructive to the

---

23. "Skill, foresight, and wisdom of the Founder." Mather changes Galen's *Creator* ("Creator") to *Conditor* ("Founder").
24. "[The senses] are marvelously adapted to their necessary services."
25. "[The ears] are rightly placed in the upper part of the body."
26. "Sun of the microcosm."
27. That is, James Keill.

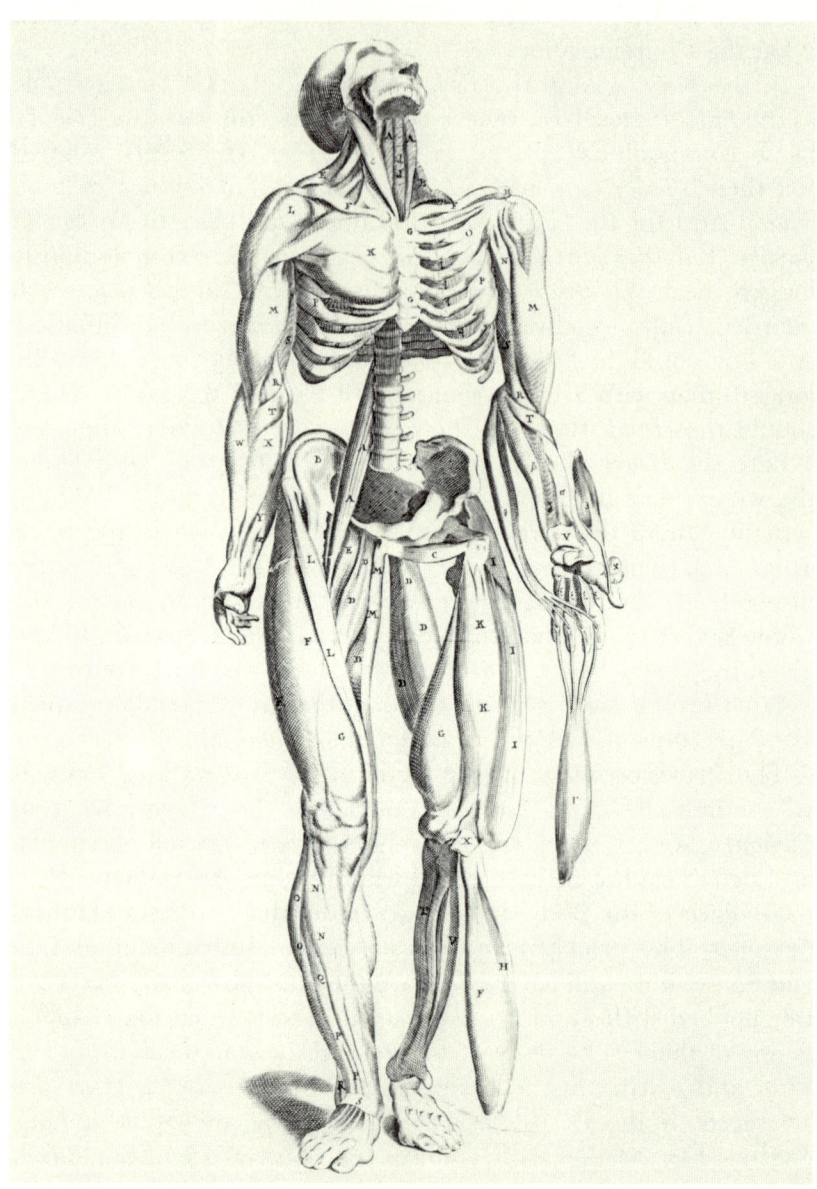

This illustration from Thomas Bartholin's *Anatome ex omnium veterum recentiorumque observationibus* (Leiden, 1673), the tenth edition of his father Caspar's *Anatomicae institutionis corporis humani* (Wittenberg, 1611), shows the muscles of the anterior part of the body. Mather draws on the Bartholins as leading anatomical authorities. Courtesy of the Rare Book and Special Collections Library, University of Illinois at Urbana-Champaign.

Constitution of it comes into it, immediately by an *intestine Commotion* endeavours to thrust it forth, and so 'tis not only freed from the new Guest, but sometimes what likewise might long have lain lurking there.

What *Emunctories*[28] has the Body, and what surprizing *Passages*, to carry off Mischiefs, which we foolishly bring upon our selves! And how astonishing the *Methods* and *Efforts* of *Nature* to set all things to rights. *Valsalva* discovered Passages into the Region of the *Ear-drum*, which are of mighty use to discharge morbifick Matter from the *Head*. *Hippocrates* in his Book *de Alimentis*[29] makes his Remarks upon the *Sagacity of Nature*, to find out Passages for the discharging of things offensive to the Body; and indeed they who confess no Wonders in it, are *Hippocraticis Vinculis alligandi*.[30] Modern Stories of what Nature has done for this, occurring in the *German Ephemerides*,[31] and elsewhere, would scarce be credible, were not the Fidelity of the Relators unreproachable. Dr. *Grew* bestows his just Remarks upon it, that in most *Wounds,* if kept clean and from the *Air,* the *Flesh* will glue together with a *native Balm* of its own; and that *broken Bones* are cemented with a *Callus*,[32] which they themselves help to make: yea, *Diseases* themselves are not useless, for the *Blood* in a *Fever,* if well govern'd, like Wine upon the *fret,* will discharge itself of all heterogeneous Mixtures. But the Philosopher last quoted observes, *Nothing can be more admirable than the many ways Nature hath provided for preventing or curing of Fevers.* Yea, Mr. *Boyle* and others have entertained us with surprizing Relations, how the Senses of *Seeing* and *Hearing* have been restored and strangely quickned by *acute Fevers* befalling those that wanted them.

The *Harmony* and *Sympathy* between the Members of the Body, made by the Commerce of the *Nerves,* and their most curious *Ramifications* thro the whole Body, is, as Mr. *Derham* observes, a most *admirable thing,* and such as greatly sets forth the Wisdom

28. A cleansing organ or canal; a term applied to the excretory ducts and organs of the body.

29. The treatise *De alimentis* ("Nutriment") applies the Heraclitean theory of permanent change to the assimilation of food by a living organism. In this work, which shows the influence of the Sophist school and may have been written by a follower of Hippocrates, the philosophic element predominates over the scientific.

30. "Bound by the Hippocratean chains."

31. An annual publication dealing with science. See Essay 27, n. 83.

32. That is, generally a hardened or thickened part of tissue which is normally soft; specifically it can mean the bony material thrown out around and between the two ends of a fractured bone during the process of healing.

and Benignity of the Great Creator; to see how *God hath so tempered the Body together, that the Members should have the same care one for another, and if one Member suffer, all the Members suffer with it!*³³

One Instance is by Mr. *Derham* singled out; there is one *Conjugation of the Nerves*, which is branched into the *Ball*, and the *Muscles*, and the *Glands* of the *Eye;* to the *Ear*, to the *Jaws*, and the *Gums*, and the *Teeth;* to the *Muscles* of the *Lips*, to the *Tonsils*, the *Palate*, the *Tongue*, and the Parts of the *Mouth;* to the *Præcordia*³⁴ too; and lastly, to the *Muscles* of the *Face*, and very particularly those of the *Cheeks*. Hence 'tis that a *gustable* thing, seen or smelt, excites the *Appetite*, and affects the *Glands* and Parts of the *Mouth*. A *shameful* thing seen or heard affects the *Cheeks*. If the *Fancy* be pleased, the *Præcordia* are affected, and the Muscles of the *Mouth* and *Face* are put into the Motions of *Laughter*. When *Sadness* is caused, it exerts itself upon the *Præcordia,* and the *Glands* of the *Eyes* emit their *Tears;* wherein also, as was long since noted, *Fletus ærumnas levat,*³⁵ and the Muscles of the *Face* put on a sorrowful Aspect. Hence also the *torvous*³⁶ *Look*, produced by *Anger* and *Hatred;* and a *gay Countenance* accompanies *Love*, and *Hope*, and *Joy*. Finally, hence 'tis that, as *Pliny* notes, the *Face* in *Man* alone is the *Index of all the Passions.*

It is an inexplicable Sympathy which there is between the Diseases of the *Belly* and those of the *Skin;* whence very stubborn *Diarrhœa's* cured by *Diaphoreticks.*³⁷

What a Sympathy between the *Feet* and the *Bowels!* The Priests walking *barefoot* on the Pavement of the Temple, were often afflicted, as the *Talmuds* tell us, with Diseases in their *Bowels*. The Physician of the *Temple* was called a *Bowel-Doctor. Belly-achs* occasion'd by walking on a cold Floor, are cured by applying *hot Bricks* to the Soles of the *Feet*.

A glorious Providence of God is to be seen in three *remarkable Dissimilitudes* between *Men* and Men, *Faces, Voices,* and *Writings*.

First, Such is the variety of Lineaments in the *Faces* of Men, that tho *Valerius Maximus,* and some others, gives us Examples of

---

33. Adapted from 1 Cor. 12:14–26.
34. That part of the ventral surface of the body lying over the heart and stomach; the epigastrium and anterior surface of the lower thorax.
35. "Weeping lightens woe."
36. That is, stern, grim, fierce.
37. A diaphoretic has the property of inducing perspiration.

Men that have been very *like* one another, yet there are no *two Faces* in all things alike. Had Nature been a blind Architect (as our curious *Ray* well observes) the *Faces* of several Men might have been as like as *Eggs* laid by the same Hen, or *Bullets* cast in one Mould. It was one of *Pliny*'s Wonders, *In Facie Vultuque nostro, cum sint decem aut paulo plura membra, nullas duas in tot millibus Hominum indiscretas Effigies existere.*[38] Now, as my modern and better *Pliny* proceeds upon it, "should there be an indiscernible Similitude between *divers Men,* what Confusion and Disturbance would necessarily follow? What *Uncertainty* in all Conveyances, Bargains and Contracts? What *Frauds* and *Cheats,* and suborning of *Witnesses*? What a Subversion of all *Trade* and *Commerce*? What Hazard in all *judicial Proceedings*? In Assaults and Batteries, in Murders and Assassinations, in Thefts and Robberies, what *Security* would there be to Malefactors? How many other Inconveniences?"

Secondly, The *Voices* of Men differ too; not only divers Countries pronounce in ways peculiar to themselves, but in the same Country how many Dialects? *Britain* as well as *Greece* exemplifies this variety; thus *Gileadites* can discover *Ephraimites.*[39] *A-Lapide* tells us how the *Flemings* discover a *Frenchman;* and *Fuller,* what way they took in *England* long since to discover a *Dutchman:* yea, some have demonstrated that *Voices* do distinguish *Individuals* as much as *Faces,* and in some Cases more; for this way the Discovery is made in the *Dark,* and by the *Blind* also.

Thirdly, Dr. *Cockburn* shall supply us with one *Dissimilitude* more: "To no other Cause than the wise Providence of God can be referr'd the no less strange variety of *Hand-writings.* Common Experience shews, that tho Hundreds and Thousands were taught by one Master, and one and the same Form of Writing, yet they all *write differently;* there is something *peculiar* in every one's *Writing,* which distinguishes it; some indeed can counterfeit another's Character and Subscription, but the Instances are rare, nor is it done without Pains and Trouble: nay, the most

---

38. "Though our physiognomy contains ten features or only a few more, to think that among all the thousands of human beings there exist no two countenances that are not distinct."

39. The people of Gilead, an area in ancient Israel, had an old grudge against the Ephraimites, who were the descendants of Ephraim, the second son of Joseph. The Gileadites identified the Ephraimites by their pronunciation of "Shibboleth" (they said "Sibboleth"), and slew 42,000 men. See Judg. 12:4–6.

Expert and Skilful cannot *write much* so exactly like, that it cannot be known whether it be genuine or counterfeit; and if the Providence of God did not so order it, what Cheats and Forgeries too would be daily committed, which would run all into Confusion? The diversity of *Hand-writing* is of mighty great Use to the Peace of the World; and what is so very useful is not the Effect of any Human Concert; Men did not of themselves agree to it, they are only carried to it by the secret Providence of God."

My God, let me never do any thing that may be to the Damage of that which thou proclaimest thyself so very tender of! HUMAN SOCIETY, Mankind associated.

The *Variety* of the *Parts* whereof the Body is composed cannot but oblige our Admiration, cannot but compel our admiring Souls to acknowledge our glorious Maker!

The *Bones* in a Skeleton are two hundred and forty-five, besides the *Ossa Sesamoidæa*,[40] which are forty-eight.

The *Muscles* of the Body are four hundred and forty-six.[41]

The *Nerves* which come immediately out of the Skull, from the *Medulla oblongata*,[42] are ten Pair.

The *Nerves* which come out between the *Vertebræ* are thirty Pair.[43]

The *Scarf-skin*[44] examin'd with a Microscope, appears made up of Lays of exceeding small *Scales,* which cover one another more or less, according to the different Thickness of the *Scarf-skin* in the several Parts of the Body; but in the Lips they only in a manner touch one another. *Leuenhoeck* reckons that in one cuticular Scale there may be five hundred *excretory Channels,* and that one Grain of *Sand* will cover two hundred and fifty Scales; wherefore one Grain of Sand will cover one hundred and twenty-five thousand *Orifices,* thro which we are *daily perspiring.* What a prodigious number of *Glands* must there now be on the Surface

---

40. That is, sesamoid bones, or certain small bones and cartilages found in tendinous structures. The actual number of bones in the skeleton is 206. The number of sesamoid bones varies with the person and the person's age. In infants there are no sesamoid bones. The patella (kneecap) is the largest sesamoid bone.

41. We now know that there are over six hundred muscles in the body.

42. That is, a part of the vertebrate brain which is continuous with the spinal cord and contains nuclei that link spinal with higher centers and various centers mediating the control of involuntary vital functions. Twelve pair of nerves are connected with the brain.

43. Actually, thirty-one pairs of nerves come out of the vertebrae.

44. The outer layer of the skin; the epidermis; cuticle.

of the whole Body! Into every one of these *Glands* there enters an *Artery,* a *Vein,* and a *Nerve.* How many *Organs* now in all the Body!

Look upon thy *Skin,* O Man, and say, *Great God, how wondrously hast thou clothed me!*[45]

Daily perspiring, I said. The Sum of all the Particles that are strained thro the *cuticular Glands,* is reckon'd by *Sanctorius* to amount to about *fifty Ounces* in a day; so that supposing a Man's Body to weigh *one hundred and sixty Pounds,* in *fifty one Days* a Quantity equal to the whole Body is perspired. The *Medicina Statica* will multiply the Calls to us to glorify the God who *so upholds our Souls in Life.*[46]

But then the *multitude* of *Intentions* which our Creator has in the Formation of our several Parts, and the *Qualifications* they require to fit them for their various Uses, this also calls for our Wonders. Dr. *Wilkins* takes notice of it, that according to *Galen* there are in an Human Body above six hundred several Muscles, and there are no less than ten several Intentions to be observed in each of these; about the *Muscles* alone there are at least six thousand several Ends or Aims to be attended to. They reckon the *Bones* to be two hundred and eighty-four, the distinct Intentions of each of these are no fewer than forty; the whole Number of Scopes for the *Bones* arise to an hundred thousand: thus it is in proportion with all the other Parts, the *Skin, Ligaments, Vessels, Glandules, Humours,*[47] but more peculiarly with the several Members of the Body, which do in regard of the *multitude* of Intentions or Qualifications required to them, very much exceed the *homogeneous* Parts; a failing in any one of these would cause an Irregularity in the Body, and in many of them, as the Doctor notes, it would be such as would be very notorious. *My Friend,* contemplate the Figures of *Spigelius,* and *Bidloe,* and *Lyserus,* if thou canst without Astonishment! Who can behold a Machine composed of so many Parts, to the right Form, and Order, and

---

45. Adapted from Job 10:11 and perhaps Ps. 139:14.
46. Adapted from Ps. 66:8–9.
47. The ancients recognized four cardinal humors or fluids of the body responsible for one's health and disposition—blood (sanguine, cheerful), phlegm (phlegmatic, sluggish), yellow bile (choleric, quick-tempered), and black bile (melancholic, gloomy). About the time Mather wrote, three general humors of the body were distinguished—blood, lympha, and the nervous juice—along with several particular humors—chyle, bile, spittle, etc. Three humors of the eye were also noted—aqueous, crystalline, and vitrious or glassy.

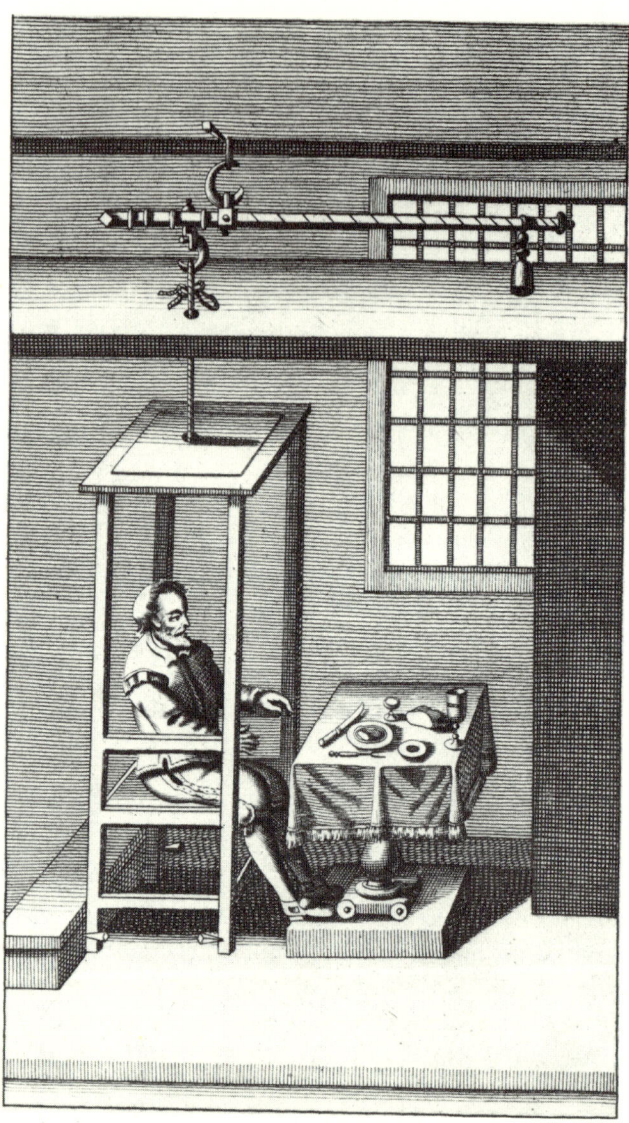

The frontispiece of Santorio Santorio's *De statica medicina sectionibus aphorismorum septem comprehensa* (Venice, 1614), translated by John Quincy as *Medicina Statica: Being the Aphorisms of Sanctorius* (London, 1718), shows Santorio weighing himself. He made quantitative experiments with a chair scale, checking daily variations in body weight and discovering that a large part of excretion takes place invisibly through the skin and lungs. Courtesy of the Rare Book and Special Collections Library, University of Illinois at Urbana-Champaign.

Motion whereof there are such an infinite number of Intentions required, without crying out, *Who can be compared to the Lord!*⁴⁸

The *variety of Offices* which sometimes *one Part* performs, will here come into Consideration. Thus the *Tongue,* it serves not only for tasting, but also for the *Mastication* and the *Deglutition*⁴⁹ of our Food; and then for the Formation of our Words in *speaking,* the use of it is admirable! The *Diaphragm,* with the Muscles of the *Abdomen,* are of use, not only in *Respiration,* but also for the compression of the *Intestines,* that the *Chyle* may be forced into the *Lacteal* Veins, and out of them into the *Thoracick* Channel; and no doubt the comminution of the Meat in the *Stomach* is likewise hereby assisted. The *muscular Contraction* of the *Heart,* in the Pulse of it, serves not only for the *Circulation* of the *Blood,* but also for the more perfect *Mixture* of it, by which it is preserved in its due Crasis⁵⁰ and Fluidity, and it incorporates the *Chyle* and other Juices it receives with it.

Even *Pain* itself, however afflictive it be, yet is of *Use* to us; it quickens us to seek for Help, and makes us careful to avoid what may be for our *Hurt;* it is, as Mr. *Ray* calls it, a πολύχρηστον⁵¹ in the Government of the World.

The mention of *Pain* leads one to think on *Sleep; Sleep,* a thing so necessary to repair the great Expence of Spirits we make in the day-time, thro the constant exercise of our *Senses* and motion of our *Muscles.* 'Tis a little surprizing, that tho we lie long on one side, we have no sense of *Pain* during our *Sleep,* no, nor when we awake. One would think the whole Weight of the Body pressing the Side on which we lie, should be very burdensome and uneasy, and create a grievous *Pain* to us; and if we lie *long awake* we really find it so. Our ingenious *Ray* supposes that our Ease in this case may be owing to an *Inflation of the Muscles,* whereby they become soft, and yet renitent,⁵² like so many Pillows, dissipating the force of the Pressure, and so the feeling of the Pain. Hence when we rest in our *Clothes* we loose our Garters, our Buckles, and other

---

48. Ps. 89:6.
49. That is, the act of swallowing.
50. That is, a blend or combination of constituents. This meaning is now obsolete.
51. That is, a polychrest, a drug or medicine useful as a remedy in several diseases.
52. That is, resisting pressure, restraint, or compulsion. Another form of the word is used below (281:80).

Ligatures, to give the Spirits[53] a free Passage, else these Parts will be *pained*, which when we are awake are not so. The reason of this ἀναλγησία[54] during and after a long Sleep on one side, is by Dr. *Lyster* and by Dr. *Jones* attributed to the *Relaxation* of the *Nerves* and Muscles in the time of Sleep; or *Pain* while we lie awake, is owing to the *Tension* of them.

O merciful God, thou makest my Bed for me!

Let more particular Parts of our *Body* come into Consideration with us; 'tis impossible for them to do so without coming into Admiration too!

The *Head* ought certainly to be first considered. The *Head*, because it must contain a large *Brain*, is made of a most capacious Figure, as near as may be to a *Spherical*.

What an infinite number of *Glands* in the *cortical* part, and of beginning *Nerves* in the *medullar part;* an hundred whereof exceed not one *single Hair*.

Upon the *Head* grows the *Hair*, which is of great use, not only to quench the Stroke of a Blow at the Skull, but also to cherish the *Brain;* it serves also to disburthen the *Brain* of a superfluous Moisture, wherewith it abounds. *Marchetti* finds that *Baldness* comes from the *Dryness* of the *Brain*, and the shrinking of it from the Skull; he found an *empty Space* between the *Brain* and the *Skull* in the *Bald*. The *Hair* is likewise a graceful Ornament, else, as Mr. *Ray* observes, *the present Age would not bestow so much Money upon Periwigs.*

How commodiously are the *Nerves*, wherewith four of the *Senses* are served, as well as all the *superiour Parts*, all sent out the shortest and safest ways, thro proper Holes in the *Head*. And those that serve the *Inferiour*, carried down in a *Bony Channel*. And as Dr. *Cheyne* remarks, it is very remarkable, that the *Veins* do not pass out at the same Holes the *Arteries* enter; for if they did, then upon any violent Motion of the Blood, or any greater Quantity thereof than ordinary, lodged in the *Arteries*, their *Dilatation* and *Pulsation* would compress the *Veins* against the *Bony* Sides of their Passage, and so occasion a *Stagnation* and

---

53. That is, the vital or natural spirits. The reference assumes a distinction, long accepted at the time Mather wrote, between two kinds of spirits in an animal body—animal spirits in the brain and vital spirits in the mass of the blood. Vital spirits were considered the most subtle parts of the blood. They were believed to be actuated and fermented in the blood, making it fit for nourishment.

54. That is, analgesia, insensitivity to pain without loss of consciousness.

Extravasation[55] of the Blood in the *Brain,* to the Destruction of the whole *Machine,* which by these different *Entries* and *Exits* of these Vessels is prevented.

The *Brain,* the *cortical* Parts thereof, serve to make the *Animal Spirits;*[56] that is, to separate them from the Blood: The *Medullary* Parts to receive them, and convey them from thence into the *Nerves.*

The inner *Meninx,*[57] by its *Constriction,* upon occasion, causes a more vigorous Efflux of the Spirits, and thereby the better Irradiation of the *Organs* of Motion and Sense. By the frequent Repetition of this *Constriction* all the day, being tired, as all other *Muscles* are by continual Action, it is anon relaxed, or suspended from Action. Hereupon, the Efflux of the Spirits into the said *Organs,* being made more slowly, we fall *asleep.*

A great Philosopher[58] observes and affirms, that the Clearness of our *Fancy* depends on the *regular* Structure of the *Brain;* by which it is fitted for the receiving and compounding of all Impressions with the more Regularity. In *Fools* the *Brain* is deformed. The Deformity is not easily noted in other People: But, no doubt, a smaller Difference than can be imagined, may alter the Symmetry of the Brain, and so the Perspicuity of the Fancy.

Gracious God! how much ought I to adore the Goodness of thy superintending Providence, which gave my Brain that Conformation, that enables me now to see and write thy Praises.

The *Head* has wonderful things to show: But can any thing in the World be shown so curious and marvellous as the EYE! Our excellent *Ray* says truly, *Not the least Curiosity can be added to it.* What *Rhetorick* what *Poetry* can sufficiently celebrate the Glories of this admirable *Organ!* How perverted the Eye, which is not ever

---

55. That is, the escape of an organic fluid (e.g., blood, sap) from its proper vessels into the surrounding tissues. It is largely true that veins do not pass out of the same holes the arteries enter, but there are three or four exceptions to this rule in the skull. Arteries and veins also pass through the same orifices in other parts of the anatomy.

56. Animal spirits were long considered to be the motive force of the nervous system. They were believed to be a very thin liquor manufactured in the *rete mirabile,* that is, distilled from the blood in the cortical substance of the brain, from which they were sent into the nerves and spinal marrow to perform all the actions of sense and motion.

57. The meninges (sing., meninx) are the three membranes that envelop the brain and spinal cord.

58. That is, Nehemiah Grew.

unto the Lord, the glorious Maker of it! There was much Discourse all over *Europe* a while ago, concerning a Child, in whose *Right Eye* there were very apparent and legible, those *Latin* Capitals, *DEUS MEUS;*[59] and in whose Left Eye, those *Hebrew* Letters, אדני, *My Lord.* This we may justly say, No rational Beholder can look upon the *Eye,* without seeing Reason in the wondrous Workmanship thereof, to make this Confession, *The Maker of this Organ is for ever to be adored, as MY God and MY Lord.*

The Place of the *Eye,* even in the *Head,* how agreeable! 'Tis here not only near to the *Brain,* but also advantaged for the *better View* of Objects, and better defended and secured. How unhappy were the People, if there were any such as *Pliny* tells of, *Oculis Pectori affixis,*[60] and *Oculos in humeris habentes;*[61] from whom our famous Romancer *Mandeville,*[62] doubtless, took hints for some of his Fables. *Galen* would satisfy us, if we wanted any Satisfaction, that the *Eye* in the *Hand* would have had many inconvenient Circumstances.

The *spherical Form* of the *Eye,* how commodious! To lodge the Humours, and also take in the Objects, and likewise to befriend the Motions! The Parts of the *Eye* being made convex, especially the *chrystalline,* which is of a lenticular Figure, convex on both sides; by the Refractions there made, there is a direction of many Rays coming from one point in the Object, namely, as many as the Pupil can receive, to one point answerable in the bottom of the Eye, without which the Sense would be obscure and confused. The difference between a Picture that is received on a *white Paper* in a *dark Room,* thro an open or empty Hole, and the same received thro an Hole furnished with an exactly polished lenticular Chrystal, is brought by Mr. *Ray* to illustrate this.

The *Membranes* and *Humours* of the *Eye* are all purely *transparent,* purely *pellucid;* thus none of the *Rays* let in are suffocated before they reach the bottom of the *Eye,* nor are they *sophisticated* with the Tincture of any *Colour,* by which that Colour might be refunded on the Object, and the Soul deceived.

4 *chrystalline/chrystaline*

59. "My God."
60. "Their eyes attached to their chests."
61. "Having their eyes in their shoulders."
62. Sir John Mandeville (fl. fourteenth century) is the name assumed by the author-compiler of a book of travels which describes the Holy Land as well as nearly all of Asia, combining geography with romance.

The *uveous Coat* or *Iris* of the *Eye* has a musculous Power, and can contract or dilate the Pupil; the former is to preserve the *Eye* from Injury, by too lucid an Object that may be too near to it; the latter is to apprehend a remoter Object, or one placed in a fainter Light: all, as 'tis justly said by *Scheiner, Tam miro Artificio, quam munifica Naturæ largitate.*[63] There are some Animals which can so close the *Pupil* as to admit of, one may say, one single Ray of *Light,* and by throwing all open again they can take in the faintest Rays; 'tis an incomparable provision for them who must watch for their *Prey* in the *Night.* These have also another astonishing provision for their business, which is a *Radiation of the Eyes,* from the shining of the *Retina* about the *Optick Nerve.* Man has not this provision, because he has no occasion; and yet there have been Instances of some whose *Iris* has had the Faculty so to dart out Rays of *Light,* that they could see in the Dark. *Willis* and *Briggs* mention divers Instances; and *Pliny* tells us, 'twas reported of *Tiberius Cæsar,* that *Expergefactus noctu paulisper, haud alio modo quam luce clara, contueretur omnia.*[64]

The *uveous Coat* and the inside of the *Choroides*[65] are wonderfully blackened; this is, that the Rays may be suppressed there, and not so reflected backwards as to confound the Sight: if any be reflected by the *retiform*[66] *Coat,* they are soon choak'd in the black inside of the *Uvea;* were they to and fro reflected, there could be no distinct Vision; as the *Light* admitted into a dark Room would obliterate the Species, which before were seen upon white Paper, by the Light let in thro an Orifice in the Wall; Dr. *Briggs* adds this reason for it, *Quo Radii in Visione superflui, qui ab Objectis lateralibus proveniunt, hoc ritu absorbeantur.*[67]

Dr. *Grew* makes a just Exclamation: What more wonderful than to see *two Humours* of *equal Use* to true Vision, bred so near together as to be contained within one common Coat, and yet one of them as *clear* as Chrystal, the other as *black* as Ink!

Since the *Rays* from an Object nearer to us, or farther from

63. "As much by marvelous craftsmanship as by the munificent liberality of nature."
64. "If he awakened in the night for a short time he could survey everything just as in bright daylight." Tiberius Julius Caesar Augustus (42 B.C.–A.D. 37), adopted by the emperor Augustus when the latter married his mother, led the Roman army in many campaigns in Germany, Dalmatia, and Pannonia, and became the second emperor of Rome (A.D. 14–37).
65. A choroide is a membranous structure which serves as a coat of the eye.
66. That is, composed of crossing lines and interstices, netlike.
67. "That the superfluous rays which proceed from lateral objects may be absorbed."

us, don't meet just in the same distance behind the *chrystalline Humour*,⁶⁸ therefore the *ciliary Processes*,⁶⁹ or the Ligaments observed in the inside of the *sclerotick Tunicles*⁷⁰ of the *Eye*, do serve instead of a Muscle, by their contraction, to alter the Figure of the *Eye*, and make it broader; and consequently draw the *Retina* nearer to the *chrystalline Humour*, and by the relaxation thereof suffer it to return to its natural distance, according to the Exigency of the Object, in respect of distance more or less. Dr. *Grew* ascribes to the *Ligamentum Ciliare*⁷¹ a power of making the *Chrystalline* more convex, as well as of moving it either to or from the *Retina;* and indeed by the Laws of *Opticks* there must be something of this necessary to distinct Vision.

The *chrystalline Humour*, when dried, appears manifestly to be made up of many very thin *Spherical Scales*, lying one upon another; *Leuenhoeck* reckons there may be two thousand of them in one *Chrystalline*, from the outermost to the Center: every one of these wonderful Scales is made up of one single Fibre, or the finest Thread imaginable, wound in a stupendous manner this way and that way, so as to run several Courses, and meet in as many Centers, and yet not in any one place to interfere or cross one another. Some ingenious Men have question'd this, but Mr. *Derham* silences them with, *It is what I myself have seen, and can shew to any body with the help of a good Microscope.*

Peter Herigon has observed a remarkable thing about the Insertion of the *Optick Nerve* into the Bulb of the *Eye*. The Situation of it is not just *behind* the *Eye*, but on *one side*, lest that part of the Image which falls upon the Hole of the *Optick Nerve* should want its Picture. But Mr. *Ray* will rather have the reason to be, because if the *Optick Axis* fall upon such a Center, as it would were the *Nerve* seated just behind the *Eye*, this great Inconvenience would follow, that the middle point of every Object we view'd would be invisible, or there would a dark Spot appear in the midst of it. Behold, a Situation of a *Nerve*, which any one would at first have thought inconvenient, now evidently found to be assign'd by a most admirable Wisdom!

68. A fluid in the eye with a structure resembling crystal.
69. The vascular folds on the inner surface of the ciliary body that give attachment to the suspensory ligament of the lens in the eye.
70. A hard-covering membrane or integument.
71. A muscle situated in the ciliary body of the eye and serving as the chief agent in accommodation of the eye.

And then, what a wise Contrivance, particularly about the motion of the *Eye*, in uniting into one that Pair which are called *the motory Nerves?* Each of these do send their Branches in each Muscle of each *Eye;* this would cause a *Distortion* of the *Eyes:* but being united near their Insertion, they cause *both Eyes* to have but one motion; when one Eye is moved this or that way, the other is turned the same way with it. But what shall we say concerning this? There is a decussation[72] of the *Rays* in the *Pupil* of the *Eye*, the Image of the Object in the *Retina,* or bottom of the *Eye,* is *inverted;* whence does it come to pass that it appears not so, but in its *natural Posture?* Why the *visual Rays* coming in strait Lines by those Points of the Sensory, or the *Retina,* which they touch, affect the common Sense or Soul,[73] according to their direction; they signify to it, that the several Parts of the Object, from whence they proceed, lie in strait Lines (Point for Point) drawn thro the *Pupil,* to the several Points of the *Sensory,* where they terminate, and which they press upon: Whereupon the Soul must needs conceive the Object in its true Posture. The *Nerves* are naturally made, for to inform the Soul, not only of the external Objects, which do press thereupon, but also of their *Situation.* Hence the Objects will appear double, if the *Eyes* be distorted. This is *Des Carte's* way of accounting for this Mystery: *Notitia illius ex nulla Imagine pendet, nec ex ulla Actione ab objectis veniente, sed ex solo situ exiguarum partium cerebri, e quibus Nervi expullulant.*[74] Mr. *Molyneux* contents himself with this Account: *The Eye is only the Organ or Instrument, it is the Soul that sees by means of the EYE. To enquire how the Soul perceives the Object erect, by an inverted Image, is to enquire into the Soul's Faculties.*

Even the *aqueous Humour*[75] is not an useless one: It sustains the *Uvea Tunica,*[76] which else would fall flat upon the *Chrystalline.*

72. Crossing so as to form a figure like the letter X; intersection.

73. That is, the rational soul. Mather inherited the Platonic theory that three souls ruled and yet used the body. These three, the rational, irascible, and concupiscible or vegetative soul, seated in the brain, heart, and liver respectively, were regarded as different aspects of the one soul inhabiting the body. The rational soul presides over reasoning thought and provides sensation and motion.

74. "This knowledge does not depend on any image, nor on any action which proceeds from the objects, but only on the position of the small points of the brain whence the nerves originate."

75. A nearly pure water contained in the space between the cornea and the lens of the eye.

76. The membrane covering the iris and ciliary body together with the choroid coat.

Because the *outermost Coat* of the *Eye* might chance to be wounded or pricked, and this fluid Humour be let out, there is therefore a Provision made, speedily to repair it, by the help of certain *Water-Pipes,* or *Lymphaducts,* inserted into the Bulb of the *Eye,* proceeding from *Glandules* designed by Nature to separate this Water from the Blood for that Use. *Antonius Nuck* found, that if the *Eye* of an Animal be pricked, and the *aqueous Humour* squeezed out, in the space of ten hours the Humour and Sight would be restored unto the *Eye,* at least if the Creature be kept in the Dark. *Verzascha* gives divers Examples, both antient and modern, of *Sight* strangely recovered, by the Reparation of the *aqueous Humour,* after it had been let out at very dangerous Wounds.

It is remarkable, that the *horny Coat* of the *Eye* does not lie in the same Superficies with the White of the *Eye;* but it rises up, as it were on Hillock, above its Convexity, and is of an *Hyperbolical* or *Parabolical* Figure. Tho' the *Eye* seems to be perfectly round, in reality it is not so; but the *Iris* thereof is protuberant above the *White:* and the Reason is, because if the *Cornea Tunica,*[77] or *Chrystalline Humour,* had been concentrical to the *Sclerodes,*[78] the *Eye* could not have admitted a whole Hemisphere at one View; and as by *Sheiner* noted upon it, *Sic Animalis Incolumitati in multis rebus minus cautum esset.*[79]

Dr. *More* has now a Remark, That the Eye being thus perfect, the Reason of Man would easily have rested here, and admired the Contrivance. Being able to move himself every way, he might have thought himself every way sufficiently provided for. But, behold! An Addition to this Perfection! There are *Muscles* also added unto the *Eyes!* For we have occasion, particularly in *reading,* to move our *Eyes,* without moving our *Head.* The *Organ* is therefore furnished with no less than *six Muscles,* to move it upwards, downwards, to the right, to the left, obliquely, and round about.

And now, for the Security of this wonderful *Organ,* the *Eyes* are sunk in a convenient Valley, where, as *Tully* says, *Latent*

77. The horny or translucent tissue constituting the outer coat of the eye.
78. The white of the eye, a fibrous coat forming the outer envelope of the eye, except for the cornea.
79. "Thus in many things there would not have been sufficient care taken for the animal's safety."

utiliter;[80] and they are encompassed round with Eminencies, as within a Rampart: *Excelsis undique partibus sepiuntur.* This defends them from the Strokes of any flat or broad Bodies. Above stand the *Eye-brows*, to keep off any thing from running down upon them, says the same Orator, *Superiora Superciliis obducta, sudorem a Capite & Fronte defluentem repellunt;* the *Eye-lids* then fence them from sudden and lesser Stripes: whereas the *Fishes*, who have no occasion for a Defensative against Dust and Motes, are destitute of *Eye-lids!* The *nictating Membrane*[81] is an abundant Provision for all their Occasions! These *Eye-lids,* also round the Edges, are fortified with *Bristles*, like Palisadoes, to keep off the Incursions of troublesome Insects. 'Tis remarkable, that these Hairs grow to a *determinate,* but a most *commodious* Length, and need no *cutting,* as many other Hairs of the Body do; and that their Points do stand out of the way, bending *upwards* in the upper Lid, in the lower *downwards.* But then *Sleep* is necessary for us. This would be disturbed, if the *Windows* were always open to the Light. Here are *Curtains* then to be drawn, for the keeping of it out. Yet more: The outward Coat of the *Eye* must be kept *pellucid.* This would anon dry and shrink, and lose its *Diaphaneity,* if the Eyes were always open. The *Eye-lids* are therefore so contrived, as often to wink. Thus they varnish the *Eyes* with their *Moisture* over again: They have *Glandules,* on purpose to separate an Humour for that use, and withal wipe off whatever Dust or Filth may stick to them. And lest the Sight should be hinder'd, they do it, with what Celerity! *Cicero* adds, they are *Mollissimæ tactu, ne læderent Aciem:*[82] And I will add, *Man,* who is a *sociable Creature,* and should exhibit *social Affections* by some *visible Tokens,* is here furnished with *Tears* for that purpose, beyond any other Animal.

My God, let me ever employ them, on the just Occasions for them.

59 nictating / nictitating

80. The three Latin quotations that follow are part of one passage from Cicero that reads as follows (the translation of Mather's material is in italics): "The eyes are in an *advantageously retired* position, and *shielded on all sides by surrounding prominences; . . . the parts above them are covered by the eyebrows which prevent sweat from flowing down from the scalp and forehead.*"

81. A thin membrane which is found in many animals (but not human beings) at the inner angle or beneath the lower lid of the eye and is capable of extending across the eyeball.

82. "Very soft to the touch, so as not to hurt the pupil."

It is a Passage which drops from the Pen of *a Person of Quality,*[83] in a Treatise, entitled, *A View of the Soul:* "It does not seem wonderful to behold a Distillation from the Eyes, 'tis to be found in *Beast,* as well as in *Man,* upon an *offensive Touch* thereof: But when there is no such Cause to be alledged, to have the *Body,* as it were, *melted* on a sudden, send forth its Streams thro that unusual Channel, makes it seem to me no less than the quick and violent Agitation of some *Divine Flame,* thawing all the *vital Parts,* and drawing the Moisture thro the chief and clearest Organ of the Body, the *Eye,* and not to be caused by any thing, which is part of itself."

This brings to my mind an antient Problem: *Cur Deus Oculos fletus instrumentum esse valuit.* And the Answer to it, *Ut quo sordes peccatorum hauriuntur, eodem per lachrymas diluantur.*[84]

And then the *Ball* of the *Eye* has the exterior Coat made so thick, so tough, so strong, that it is a very hard matter for to make a Rupture in it. But because the *Eye* must be exposed at all Seasons, and in all Weathers, there is provided for it an hot Bed of *Fat,* which fills up the Interstices of the Muscles; nor is it so sensible of Cold, as other Parts of our Body. 'Tis a strange thing, which the *French* Academists found by Experience! The *Aqueous Humour* of the *Eye* will not *freeze.* Admirable! It has the Fluidity and Perspicuity of common Water, nothing singular to be discovered in the Taste or Smell of it. Of what *Ethereal Nature* must we imagine it?

Shall we, on this occasion, *look back* on the *Eyes* of other *Animals,* and compare *ours* with *theirs?* The *Chrystalline Humour,* in the *Eyes* of the *Fishes,* is much nearer to a *Sphere,* than that of *Land-Animals.* 'Tis because the Light has a different *Refraction* in the *Water,* from what it has in the *Air:* That *Convexity,* which would unite the Rays of Light in the *Air,* will not in the *Water.* In those *Animals,* that gather their Food from the Ground, the *Pupil* is *Oval* or *Elliptical,* the greater *Diameter* going *transversely* from Side to Side. In those that seek their Food on higher Places, the greater *Diameter* is the *Perpendicular.* These two Figures are

---

83. Richard Saunders or Sanders (1613–87?) practiced astrology and chiromancy and wrote on these subjects. He also published anonymously a work entitled *A View of the Soul* (London, 1682).

84. "Why does God wish the eyes to be the instrument of weeping?" . . . "So that the filth of sins may be washed away by the same instrument in which it is absorbed in the first place."

wonderfully fitted unto their different Necessities. Those *Animals*, that have no Motion of their *Neck*, have a Cluster of *Semisphærical Eye-balls*, which send in the Pictures of Objects all round about them; and they that seek their Food in the dark, have a *Retina* coloured *white*, which reflects the Light, and enables them to see best, when they have least of it. An acute Philosopher[85] says justly, "These are wonderful and surprizing Instances of *Foresight* and *Counsel*, in the Being who framed those *Organs*."

But why don't we *see double* with our *two Eyes*? *Galen*, and others after him, took this to be from a Coalition, or Decussation of the *Optick Nerves*. I pass by the Assertion of the *Bartholines*,[86] that they are united, not by any Intersection, *sed per totam Substantiæ Confusionem*.[87] Dr. *Gibson* says there is the closest *Conjunction*, but no *Confusion* of the Fibres. Others apprehend only a *Sympathy* between the *Optick Nerves*. Mr. *Briggs* thinks that the *Optick Nerves* of each Eye consist of *homologous Fibres*, and that these *Fibrillæ* have the same Tension and other Circumstances in both Eyes; and so when an Image is painted on the same corresponding and sympathizing Parts of the *Retina*, the same Effects are produced, the same Notice or Information is carried to the *Thalamus Nervorum Opticorum*,[88] and so imported and imparted to the Soul, that is to judge of all. Our great Sir *Isaac Newton* says, *Are not the Species of Objects seen with both Eyes, united where the Optick Nerves meet, before they come into the Brain, the Fibres on the right side of both Nerves uniting there?* Monsieur *Tauvry*, in his *Rational Anatomy*, thinks this Answer to be enough: "When we see the same Body with *two Organs*, we judge it to be one, because we see it still in *one place*, and refer it to *one place*; for every Point of the seen Object is directed upon *one place*, by the perpendicular Rays of each of the two Cones; this is what we call *the Direction of the Optick Axis*. 'Tis a natural Consequence from this Explication, that certain Distortions of the Eye will make the Object

---

85. That is, George Cheyne.
86. "The Bartholines" is a reference to Thomas Bartholin's edition of the medical works of his father Caspar Bartholin (1585–1629), entitled *Institutiones anatomicae* (Leiden, 1641, and later eds.).
87. "But through the whole combination of the substance."
88. "Chamber of the optical nerve." The reference is to the two bodies which protrude at the base of the brain and give rise to the optic nerves. Thomas Willis, Jean Riolan, and other seventeenth-century anatomists credited Galen with naming this body.

appear double, because we then direct the same Point of the Object to two different places."

We might go on and resume our Enquiry, Why Objects which are *inverted* in the bottom of the *Eye* do not appear so, but in a direct Position? *Tauvry* thinks it enough to say, "We do not judge of the Situation of the Bodies by the *Part* which is affected in the *Eye*, but by the *Manner* in which it is affected; the Soul judges of the Object by the *Manner* in which the *Organ* is affected."

To conclude our Observations of the *Eye*; Mr. *Derham* very justly says, *None less than GOD could contrive, order, and provide an Organ as magnificent and curious as the Sense is useful.* And *Sturmius* had reason enough to say, he was fully persuaded, that no Man who survey'd the *Eye* could abandon himself to any *speculative Atheism.* And *Cheyne* passes a most equal Sentence, when he says, *He certainly deserves not to enjoy the Blessings of his Eyesight, whose Mind is so depraved as not to acknowledge the Bounty and Wisdom of the Author of his Nature, in the ravishing and astonishing Structure of this noble Organ!*

"*Good God!* How *unreasonable* am I, if the *Eyes* made by Him should not *be ever to the Lord!*[89]

"An *envious Eye* is an *abused* one; an *haughty Eye* is a *distorted* one; an *unchaste Eye*, how ignominiously misapplied! It has *Dirt* thrown into it. *Gracious God, let not my Eyes be Port-holes of Wickedness. Let no Death get into my soul by those Windows.*[90]

"A *pitiful Eye* a *bountiful Eye*, and the *Eye* on the *Book* that will feed it well, how much to be wish'd for! And an *Eye* upon a CHRIST at His Table, *evidently set forth as crucified* before it.[91]

"'Tis an odd Question in *Tympius*, Why the *Eyes* are the *last* things *quickned*, and the *first* that are decayed? It is answered, *Ut quo magis est ipsorum Periculum, eo minus sit nocendi Spatium?*"[92]

The EAR is what falls next under our Consideration; *double*, not only to provide against the *Loss* of one, but also for the more commodious *hearing*.

'Tis astonishing to see the Sagacity of some *deaf* Persons, who

---

89. Adapted from Ps. 25:15.
90. Adapted from Jer. 9:21.
91. Adapted from Gal. 3:1.
92. "Just as they are in greater danger, so they have less time to do harm."

85 come to understand things that are spoken, only by *seeing* the motion of the Lips in the Speaker; but the Instances of this are so rare, that they abate nothing of our Obligations to our glorious Maker for bestowing the noble Sense of *Hearing* upon us.

The Situation of the *Ear* is where it may give the most *speedy*
90 *Information,* and where it will *occasion* and also *encounter* the least Annoyance.

The *outward Ear* is most nicely adjusted to the peculiar Circumstances of every Animal. Dr. *Grew* celebrates the marvellous *Varieties* in the *Ears* of several Animals for the reception of Sound,
95 according to their several Exigences. And Mr. *Derham* challenges our Confession of a *notable Prospect of the Handy-work of God even in so inconsiderable a Part as this.* In Man the *Form* of it is of all the most agreeable to the *erect Posture* of his Body. 'Tis pity the
99 most eminent of our modern *Anatomists* cannot yet agree whether it has any *Muscles* belonging to it.

What a surprizing Spectacle the *Helix,*[93] which in its tortuous Cavities collects the *sonorous Undulations,* and gives them a gentle *Circulation,* with some *Refraction,* and conveys them to the *Concha,*
5 that large and round Cell at the entrance of the *Ear!* Then to bridle the Evagation[94] of the *Sound* when arrived thus far, but at the same time avoid any Confusion thereof by *Repercussions,* what a curious provision is there made by those little Protuberances called the *Tragus*[95] and the *Antitragus*[96] of the *outward Ear,* softer
10 *than the Helix,* and blunting the Sound without repelling it! Monsieur *Dionis* observes, they that have this *Ear* cut off have *but a confused way of hearing.*

That the Substance of the *outward Ear* should be *cartilaginous;* this is an admirable Contrivance of the most wise Creator. Dr.
15 *Gibson* observes, if it had been *Bone,* it would have been troublesome, and might by many Accidents have been broken off; if it had been *Flesh,* it would have been subject to Contusion, yea, we may add, it would not then have remained so well expanded, nor have so kindly received *Sounds,* but have absorbed them, and
20 retarded them; whereas now the Sounds have their agreeable

---

93. The incurved rim of the external ear.
94. That is, the act of wandering.
95. The prominence in front of the external opening of the ear.
96. The prominence on the lower posterior portion of the concha of the external ear opposite the tragus. The tragus and the antitragus are vestigial aspects of the ear that have nothing to do with the sense of hearing.

Volutations,[97] as in well-built Arches, and the *Whispering-places,* whereof the World has had many famous ones.

How artfully tunnell'd the *auditory Passage!* But then, because the Passage must be always open, therefore to prevent the invasion of *noxious things,* which love to retreat into every little Hole, behold, the Passage secured with a bitter and nauseous Excrement, afforded from *Glands* appointed for that purpose! Where the *Meatus auditorius*[98] is long enough to afford harbour to any Insects, there this *Ear-wax* is constantly to be found; but *Birds,* whose Ears are cover'd with Feathers, and where the *Tympanum*[99] lies but a little way within the Skull, have none of it. *Schelhammer* confutes the old Anatomists, who make this *Ear-wax* an Excrement of the *Brain,* and justly says, *Nil absurdius!*[100] Dr. *Drake* has given us an handsome Cut of the *Glandulæ Ceruminosæ.*[101] *Pliny* ascribes a great medical Virtue to the *Ear-wax,* the *Sordes ex Auribus,*[102] as curing the *Bites* of *Men,* (which he says, *inter asperrimos numerantur*)[103] and of *Scorpions* and *Serpents.* And Mr. *Derham* asserts he had found it a good *Balsam* in his own Experience.

The Notion of an *innate Air* in the Ear, is by *Schelhammer* found but a Fancy; the Passage into the *inner Ear* from the *Throat* confutes it: but in this Passage there is a wise provision, as he notes, that no *Air* might pass in thither but what shall be changed and warmed, and so rendred harmless: *Imo fortassis non facile alius, nisi ex Pulmonibus.*[104]

The Passage from the *Ear* to the *Palate* (the *Tuba Eustachiana*)[105] accurately described by *Valsalva;* this is to give way to the *inner*

---

97. A volutation is the action of rolling; a revolution involving progression.

98. The auditory meatus or auditory canal which leads from the outer to the middle ear.

99. The tympanic membrane is a triple membrane separating the outer and middle ear.

100. "Nothing is more absurd."

101. "Ceruminous glands." The ceruminous gland is one of the modified sweat glands that produces earwax.

102. Literally, "dirt out of the ears," that is, earwax.

103. "[Are] considered the most severe."

104. Mather borrows only part of a Latin passage. It will promote clarity to translate the entire passage, with Mather's phrase in italics: "So that the external air cannot immediately find entrance, but receives a change and temperature from the heat of the body, *and must, perhaps, be expired from the lungs before it gets into the ear.*"

105. The Eustachian tube is a bony and cartilaginous tube connecting the cavity of the middle ear with the nasopharynx. The tube opens with swallowing and yawning, which are the two most common ways of equalizing air pressure on both sides of the tympanic membrane.

Air upon every motion of the *Membrana Tympani*,[106] the *Malleus*, the *Incus*, and the *Stapes*;[107] and if this be shut up, *Deafness* ensues.

And then the *Os Petrosum*,[108] that *Bone* which contains the rest, this has a remarkable Texture and Hardness above the other Bones of the Body, and so it serves not only as a very substantial *Guard* to the Sensory, but also, as Dr. *Vieussens* observes, to oppose the Impulses of the *æthereal Matter*, that there be no loss of Sound, and no confusion in it, but that the *auditory Nerves* may have it regularly convey'd to them.

The *Membrana Tympani*, as long ago as *Hippocrates*'s Time, had some notice taken of it, whether it has any disengaged part, by which it is not fastned to the *boney Circle*, in which it is enchased, as Monsieur *Dionis* affirms, is disputed. Mr. *Derham* could not find it. But then Dr. *Vieussens* discover'd a further inner Membrane, *Tenuissimæ raræque admodum Texturæ*,[109] whereof the Uses are to keep the Gate of the *Labyrinth*, lest the *thick Air* abroad hurt the *pure Air* within, and that a due Heat may be preserved in the Basis of the *Labyrinth*.

But now the astonishing *four little Bones*, and *three little Muscles* about them, to move them, and adjust the whole Compages[110] to the several Purposes of *Hearing*, and for all manner of *Sounds!*

These were wholly unknown to the antient Anatomists. *Jacobus Carpensis* was he by whom the *Malleus* and the *Incus* were first of all discovered; the Gentleman who was indeed the first Restorer of the *Anatomick Art*, which *Vesalius* afterwards carried on. The *Stapes* was found out by *Johannes ab Ingrassia*, a learned *Sicilian*. The fourth was what *Francis Sylvius* first lit upon.

In *Man*, and in the *Four-footed*, they are *four*, curiously inarticulated with one another, with an external and internal Muscle, to draw or work them in extending or in relaxing of the Drum.

106. The eardrum.
107. The malleus is the outermost of the three auditory ossicles of mammals, part of which is fastened to the tympanic membrane and part of which articulates with the head of the incus. The incus is the middle of the chain; also called the anvil. The stapes is the innermost of the chain and has the form of a stirrup, a base that occupies the fenestra vestibuli of the tympanum, and a head that is connected with the incus.
108. "Petrosal bone." The petrosal bone is the petrous portion of the temporal bone, a compound bone on the side of the human skull that houses the inner ear.
109. "Of a very thin and rare texture."
110. That is, a complex structure. Note that above (47–48) Mather refers to three bones in the ear: the malleus, incus, and stapes.

In *Fowls* Dr. *Moulen* could never find any more than *one Bone* and a *Cartilage,* making a Joint with it, that was easily moveable.

It is a probable Thought of *Rohault,* That for us to *give Attention,* is nothing else but for us, by extending or relaxing the *Tympanum* of the *Ear,* to put it into that position, *in qua tremulum aeris externi motum excipere possit,*[111] wherein it shall be most sensible of the motion of the *external Air.* The Benefit which *deaf* Persons receive by *loud Noises,* enabling to hear what shall be spoken to them in the midst thereof, helps to clear this Matter. Dr. *Willis* tells of one who hired a Servant who was a *Drummer,* on purpose that his *deaf Wife* might hear his Discourses, which, while the *Drum* was beating, she was able to do.

In *Birds* the *auditory Nerve* is affected from the impression made on the Membrane, only by the intermediation of the *Columella;*[112] but in *Man* it is done by the intervention of the *four little Bones,* with the Muscles acting upon them, his *Hearing* being to be adjusted to all kinds of Sounds or Impressions made upon the *Membrana Tympani;* the Impressions are thus made upon the *auditory Nerve,* they first act upon the *Membrane* and the *Malleus,* the *Malleus* upon the *Incus,* the *Incus* upon the *Os orbiculare*[113] and *Stapes,* and the *Stapes* upon the *auditory Nerve,* the Base of the *Stapes* not only covering the *Fenestra ovalis,*[114] wherein the *auditory Nerve* lies, but also having a part of the *auditory Nerve* spread upon it. Our valuable *Derham,* upon a diligent Examination, found this to be the *Process of Hearing.*

How will the *Wonders* grow upon us, if we pass now to the *Labyrinth!* And there survey the wonderful Structure of the *Vestibulum*[115] and the *Cochlea,*[116] and yet more particularly the *semicircular Canals!* These last are three, and of three different Sizes. *Valsalva* thinks, that as a part of the *auditory Nerve* is lodg'd in these Canals, thus they are of *three Sizes,* the better to suit all

---

111. "In which it will best receive the impression and motion which the sound gives to the external air."

112. The bony or partly cartilaginous rod connecting the tympanic membrane with the internal ear in birds and many reptiles and amphibians.

113. A knob at the tip of the long limb of the incus which articulates with the stapes.

114. An oval opening between the middle ear and the vestibule having the base of the stapes or columella attached to its membrane.

115. The central cavity of the bony labyrinth of the ear.

116. A spiral canal in the petrous part of the temporal bone in which lies a smaller spiral passage that communicates with the sacculus at the base of the spiral, ends blindly near its apex, and contains the organ of Corti.

the variety of Tones; and tho there be some difference as to the Length and Size of the Canals in different Persons, yet lest there should be *Discord* in the *auditory Organs* of one and the same Man, those *Canals* have always in the same Man a most exact Conformity to one another.

Shall we take notice of one Curiosity more! There is one of the *auditory Nerves,* whose Branches do spread partly to the Muscles of the *Ear,* partly to the *Eye,* partly to the *Tongue* and Instruments of *Speech,* and inosculated[117] with the *Nerves,* to go to the *Heart* and *Breast;* by means hereof there is an useful and wondrous *Consent* between these Parts of the Body. It is natural for most Animals, upon the hearing of any uncouth sound, presently to erect their *Ears,* and prepare them for the catching of every *Sound,* and therewithal open their *Eyes,* to stand as faithful Centinels upon the Watch, and be ready with the *Mouth* to call out, or utter, according to the Dictates of the present Occasion; when surprized with any frightful Noise, they give a Shriek immediately.

Dr. *Willis* observes another great Use of this *nervous Commerce* between the *Ear* and the *Mouth; Usum alium insigniorem præstat:*[118] that is, that the *Voice* may correspond with the *Hearing,* and be a kind of *Echo* to it; that what is *heard* with one of the two *Nerves,* may be readily expressed with the *Voice,* by the help of the other.

SOUND is the Object of this admirable Sense; the intricate nature of it has puzzled the best of Naturalists.

How many *sounding Instruments* have yet been contrived by the Wit of Man, whereby *Sounds* have been augmented, and conveyed, and rendred serviceable!

The biggest Bell in *Europe* is reckon'd to be at *Erfurt* in *Germany,* which may be heard, they say, *four and twenty Miles.*

It is reported that *Alexander* the Great had a *Tube,* which might be heard an hundred *Stadia,* whereof the Figure is preserved in the *Vatican.* It is a little strange that no one should hit upon the like Invention, till *Athanasius Kircher,* in our Days, and soon after him Sir *Samuel Moreland,* whose *Tuba Stentorophonica* was publish'd in 1672.

117. United by apposition or contact; united or joined so as to become or be made as if one. Mather's reasoning here is physiologically inaccurate as applied to human beings. The nerves to the eye, nose, and tongue go through the same opening in the skull and travel together until they part in the mass of the parotid gland, but they have nothing to do with one another. See also n. 139 below.

118. "It shows another more remarkable use."

Caves have out-done *Tubes* for bellowing. *Olaus Magnus* describes a *Cave* in *Finland*, into which if a Dog or any other Animal be cast, it sends forth so dreadful a Sound as to knock down every one that is near it; and they have therefore guarded it with high Walls to prevent such a Mischief. *Peter Martyr* informs us of a *Cave* in *Hispaniola*, which with a small Weight cast into it, will with its hideous noise at five-Miles distance endanger Deafness. *Kircher* in his *Phonurgia* finds a Pit in the *Cucumer Mountains*[119] of *Switzerland*, that sends out a fearful *Noise*, and *Wind* accompanying of it; and a Well in that Country, a noise in which is equal to that of a great Gun.

Olaus Magnus mentioning the vast high Mountains of *Angermannia*,[120] tells, that the Waves of the Sea striking at the bottom thereof, make such a terrible noise, as not only to *deafen* the Mariners, but also to *sicken* them, and even to fright them out of their Wits, if they dare approach them. *Habent Bases illorum Montium in Fluctuum ingressu & egressu tortuosas rimas, sive scissuras, satis stupendo Naturæ Opificio fabricatas, in quibus longa Voragine formidabilis ille sonitus, quasi subterraneum tonitru generatur.*[121]

The prodigious *Cataract* of *Niagara*, whereof *Hennepin* has given some relation, produces a *Noise* which perhaps nothing on Earth has equall'd; a *Noise* which it might well nigh deafen one to think upon.

What is the Matter of *Sound?* The *Atmosphere* in gross? Or the *ethereal part* of it? Or some soniferous *Je-ne-sçay-quoy*[122] Particles of Bodies?

That the *Air* is the Medium of *Sound*, is manifest from Experiment. In an *unexhausted Receiver* a small Bell may be heard at the distance of several Paces; but when it is exhausted, it can scarce be heard at the nearest distance: if the *Air* be compressed,

55–56 Angermannia / Augermannia

119. Athanasius Kircher, *Phonurgia nova: sive conjugium mechanico-physicum artis et naturae paranympha phonosophia concinnatum* ([Kempten [sic], 1673], rpt., New York: Broude Brothers, 1966). I am unable to identify the Cucumer Mountains.
120. That is, Angermanland, a historic province in northeastern Sweden, on the Gulf of Bothnia.
121. "The bottoms of these mountains have winding creeks at the entering and going out of the waters, or cliffs made by the wonderful work of Nature, wherein by reason of the long cavities, that formidable noise is made like to thunder under the ground."
122. "I do not know what."

a *Sound* will be louder, proportionably to the Compression, or the Quantity of *Air* crouded in; the Experiment succeeds, not only in *forced* Rarefactions and Condensations of the Air, but also in such as are *natural*. The Story of the *Pistols* discharged by *Frœdlichius* on the *Carpathean* Mountains, related by *Varenius*,[123] gives an Instance how the *Sound* was diminished, by the rarity of the Air, at the great Ascent up to the Atmosphere; but how magnified by the *Polyphonisms*, or the Repercussions of the Rocks and Caverns, and other *phonocaptick*[124] *Objects* in the Mount below!

The *Water* also is capable of transmitting a *Sound*; the *Sound* of a *Bell* struck under Water is heard, tho as much more dull, and not so loud: Judges in *musical Notes* pronounce it about a *fourth* deeper.

Divers at the bottom of the Sea can hear *Noises* made above, but confusedly; those above cannot hear the *Divers* below at all.

Dr. *Hearn*[125] tells of Guns fired at *Stockholm*, which were heard an hundred and fourscore *English* Miles. In the *Dutch* War, Guns were heard above two hundred Miles. If we go more Southward, Guns at *Florence* are heard at *Leghorn*, which is sixty-five Miles. When the *French* bombarded *Genoa*, they were heard at *Leghorn*, which is ninety Miles. In the Insurrection at *Messina* they were heard at *Syracuse*, which is an hundred Miles. This inclines Mr. *Derham* to think that *Sounds* fly near as far in the Southern as in the Northern Regions, tho the *Mercury* in the *Barometer* does rise higher without the *Tropicks* than within the *Tropicks*; and the more Northerly, still the higher, which may increase the *Sounds*.

Celebrated Authors differ about the Velocity of *Sounds*. Mr. *Derham* has by nice Experiments determined, that there is a small difference in *Sounds* before the *Wind* and against it, and this a little abated or augmented, according to the Strength of the *Wind*; but nothing else in the World will affect it: and there is one motion to all kinds of *Sounds*, whether loud or low; and they all fly *equal Spaces* in *equal Times*; and lastly, the *Mean* of their Flight is at the

---

123. David Froelich (fl. 1644) of Kesmarck in Upper Hungary published *Bibliotheca, seu cynosura peregrinantium, hoc est, Viatorium . . . in duas partes digestum* (Ulm, 1643-44).

124. The word *phonocamptick* is probably intended. It means, having the property of reflecting sound, or producing an echo.

125. Urban Hearne (fl. 1705), a medical doctor, published an article on Lake Vatter, the second largest lake in Sweden, entitled "Memorabilia nonulla Lacus Vetteri," *Phil. Trans.*, 24 (Apr. 1705), 1938-46.

rate of a Mile in 9¼ half Seconds, or 1142 Feet in one *Second* of Time.

The Power of *musical Sounds* over the *Spirits* of *Men*, yea, and over their *Bodies* too, is very surprizing. What could the famous *Timothy* the Musician[126] do upon *Alexander?* What another upon *Ericus?*[127] *Athanasius Kircher* in his *Phonurgia,* and *Isaac Vossius* writing *de Poematum Cantu & Rythmi Viribus,* report strange things of the Power which *Musick* has over the Affections.

The *German Ephemerides* mentions those, who at some Notes of *Musick* are unable to hold their *Water. Morhoff* tells us of those who would break *Romer Glasses*[128] with their *Voice:* Great Sea-Commanders have observed, that their *wounded Men,* with broken Limbs, undergo much Pain at the Enemies Discharges. 'Tis well known that *Seats* will sometimes tremble at the Sound of *Organs.*

The Force of *Musick* on Persons poisoned with the *Tarantula,* is altogether astonishing!

Ismenias the *Theban,* by playing on the *Flute* or *Harp,* cured the *Sciatica.*[129] In the late *French* History of the *Academy of Sciences,*[130] there is a Man cured of a *Fever* and *Frenzy* by *proper Tunes* play'd to him.

But after all, who but a God infinitely wise could contrive such a *fine Body,* so susceptible of every Impression that the Sense of *Hearing* has occasion for; and thus empower Animals to express their *Sense of things* to one another?

Mr. *Derham* thus justly concludes his Discourse on the Sense of *Hearing;* "Who can survey all this admirable Work, and not as readily own it to be the Work of an omnipotent and infinitely wise and good God, as the most *artful Melodies* we hear, are the Voice or Performances of a living Creature!"

---

126. Timotheus the Milesian (fl. 336–332 B.C.), son of Thersander, was the musician who added the tenth and eleventh strings to the harp. It is said that the Phrygian sound of the music of Timothy could excite Alexander the Great to arms.

127. The story is that "a certain musician" could drive Eric, king of Denmark, to such fury as to kill some of his best and most trusty servants. From the eleventh to the fifteenth century several kings of Denmark were named Eric.

128. The Römer was a popular German glass for drinking white Rhine wine. Typically it has a somewhat spherical bowl, a wide hollow stem decorated with prunts, and a flared bowl decorated with trailing.

129. Ismenias was a celebrated musician of Thebes, but it was Aesculapius (Latinized form of Asclepius), the god of the medical art, who is said to have prescribed the blowing of the trumpet to heal sciatica.

130. The reference is probably to *Histoire de l'Académie Royale des Sciences,* 2 pts. (Amsterdam, 1706–9).

Great God, let me ever use my Ear to learn what thou wouldst have me to know, and shut my Ear upon those things, wherewithal to be unacquainted is a learned Ignorance!

"May I have the Happiness of that Experience, *Faith comes by hearing.*"[131]

I will add one Remark: Many have been born destitute of *Seeing;* many born destitute of *Hearing;* exposed unto many Inconveniences by the want of the *Sense* whereof they were destitute; however capable of being provided for. I could never learn, that any Child of Man was born destitute of *both Senses;* one destitute of both could not be in any Capacity of being provided for. *My God, I behold thy Compassion, and I adore it!*

What a Provision has our Glorious Creator made for our Smelling? The Apertures of our *Nostrils,* which are cartilaginous, and accommodated with proper and curious *Muscles,* have, as our *Derham* notes, *all the Signatures of Accuracy.* And long before him, *Tully; Nares, eo quod omnis odor ad superiora fertur, recte sursum sunt, & quod Cibi & Potionis Judicium magnum earum est, non sine causa vicinitatem Oris secutæ sunt.*[132] Here the *olfactory Nerves* receive the odoriferous Effluvia of Bodies; and because the odorant Particles are drawn in by Breathing, the upper part of the Nose is barricadoed with *Laminæ,*[133] which fence out noxious Bodies from entering the breathing Passages; (for which purpose the *Vibrisci,* or Hairs placed at the entrance of the Nostrils, are a notable Contrivance) and they receive also the Divarications of the *olfactory Nerves,* which are here spread very thick, and thus meet the *Scents* which enter by the *Breath,* and strike upon them. The more accurate the Sense of *Smelling* is in any Animal, the longer these *Laminæ* are, and the more in Number, folded and crouded with the more nervous Filaments, to detain and fetter the *odorous Particles.* There are Animals, the *chief Acts* of whose Lives are performed by the Ministry of this *wonderful Sense,* and these have certain *Points of Provision,* which are not in *Man;* but, I will not say, are *wanting* in him: For he has enough; and he has utterly lost all *Sagacity,* if he be not sensible of *enough,* to oblige his Praises of the God that made him.

131. Rom. 10:17.
132. "The nostrils are rightly placed high inasmuch as all smells travel upwards, but also, because they have much to do with discriminating food and drink, they have with good reason been brought into the neighborhood of the mouth."
133. A lamina is a thin plate or scale.

Our *Tasting* is as well provided for.

75 For the Causes of *Tastes,* and their Diversities, Dr. *Grew* will give us a more accurate Account than *Theophrastus.*[134]

Concerning the *Organ* of *Tasting,* we will not recite the various Opinions of *Bauhin,*[135] and *Bartholin,* and *Laurentius,* and our *Wharton.* Our *Willis* determines, *Præcipuum & fere solum gustatus* 80 *Organon est Lingua.*[136] Our *Derham* inclines to that of *Malpighi,* that since the outward Covering of the *Tongue* is perforated, and under this there lie the *Papillary Parts,* whereof Mr. *Cowper* has given us Cuts full of Elegancy, the Taste probably lies in these: *Occurrunt Papillaria Corpora, probabilius est in his ultimo, ex sub-* 85 *intrante sapido Humore, Titillationem & Mordicationem quandam fieri, quæ Gustum efficiat.*[137]

There are *Nerves* curiously divaricated about the *Tongue* and *Mouth,* to receive the Impressions of every *Gust,* and these Nerves guarded with a firm and proper Tegument, which defends them 90 from Harms, but so perforated in the *Papillary Eminences,* that the *Tastes* of all things are freely admitted there.

Admirable the Situation of the *Taste* with the *Smell,* for the Discharge of their Offices, at the first Entrance into the way to the grand Receptacle of our Nourishment: that they may therefore 95 judge what is nourishing, and what unsavoury and pernicious.

The *Taste: Qui sentire eorum quibus vescimur genera debet;* as *Tully* long since observed, *Habitat in ea parte Oris, qua esculentis, & poculeatis iter Natura patefecit.*[138]

99 Our most wise Creator has established a great Consent between the *Eye,* and the *Nose,* and the *Tongue,* by ordering the Branches of the *same Nerves* to each of those three Parts. Hereby there is all the Guard that can be against Food that may hurt us; it is to

---

134. Nehemiah Grew's paper, "A Discourse of the Diversities and Causes of Tast[e]s Chiefly in Plants," read before the Royal Society on 25 March 1675, was published in Grew's *Anatomy of Plants* ([London], 1682), pp. 279–96.

135. That is, Gaspard Bauhin.

136. "The principal and almost the sole organ of taste is the tongue."

137. "[There are] certain papillary parts; it is probable that in these lies the sense of tasting, which is produced by the sapid humour entering, and slightly stimulating these parts." A papilla (pl. papillae) is any small nipplelike projection or outgrowth. Papillae of the tongue, also known as taste buds, are of three kinds: vallate, fungiform, and filliform.

138. "Which has the function of distinguishing the flavors of our various viands, is situated in that part of the face where nature has made an aperture for the passage of food and drink."

Essay 32. Of Man 271

undergo the Scrutiny of *three Senses, before it goes into the Stomach.*[139]

But if the other Senses have their *peculiar Seats,* there is one, to wit, *Feeling,* that is dispersed thro the whole Body, both without and within. *Every Part* needs to be *sensible* of what may be for its own Safety, and therefore our most wise Creator has admirably lodged the Sense of *Feeling* in *every part.* It was *Tully's* Remark, *Toto Corpore æqualibiter fusus est, ut omnes Ictus, omnesque nimios & Frigoris & Caloris appulsus sentire possumus.*[140] *Pliny* adds, *Tactus sensus omnibus est, etiam quibus nullus alius.*[141]

The *Organ* of this wonderful Sense, is the *Nerves;* which are, in a most curious, astonishing, incomparable manner, scattered throughout the whole Body.

Malpighi, upon many Observations, has determin'd, that as *Tasting* is performed by the *Papillæ* in the *Tongue,* so *Feeling* is performed by the like *Papillæ* under the *Skin.* That these *Papillæ Pyramidales,*[142] thrusting their Heads up to terminate in the outer Skin, are those by which we *feel;* he speaks of an *Animus abunde certior redditus.*[143] Our diligent *Cowper* has confirmed this, and given us elegant Cuts of these *Papillæ,* from the Informations of the Microscope.

Dr. *Cheyne* observes, the apt proportioning of that Sense, our *Feeling,* unto the Actions and Impulses of the Bodies among which we live, is wonderful! Had the Sense been ten times as exquisite as it is, we should have been in perpetual Torment. Had it been many times duller and more callous than it is, we should have lost many of our most agreeable Delights, and we should have had our tenderest Parts consumed without Knowledge or Concern. This nice Adjustment!

We were but now pretty near the *Teeth;* of these the *Numbers* are *thirty two.* But, Oh! how many more the *Wonders! Galen* observes, we commend the Skill and Sense of him that shall well

---

139. Mather's emphasis on design in nature again leads him to explain, erroneously, that the nerves to the eye, nose, and tongue are part of the same branch. The sense of smell is actually controlled by cranial nerve number one; the movement of the eye by numbers three, four, and six; the taste of the tongue by numbers seven, nine, and part of ten; and the motor nerves to the tongue by number twelve.

140. "[The sense of touch] is evenly diffused over all the body, to enable us to perceive all sorts of contacts and even the minutest impacts of both cold and heat."

141. "All creatures have the sense of touch, even those that have none of the others."

142. That is, papillae in the shape of pyramids.

143. "[A] mind made much more certain."

marshal a Company of *thirty two:* and shall we not admire him who hath so admirably disposed these *thirty two?*

We will here single out eight or nine things, that are very remarkable: The *Teeth* continue to *grow* in their Length as long as we live, as appears by the unsightly Length of a *Tooth,* when the opposite happens to be pulled out. Thus Providence repairs the waste that is daily made of the *Teeth,* by the frequent Attrition in Mastication. That part of the *Teeth,* which is above the Gums, is not invested with the sensible Membrane, called *Periostium,* with which the other Bones are covered; but then the *Teeth* are of a *closer* and *harder* Substance than the rest of the Bones, that they may not be so soon worn down by grinding the Food. For the *nourishing* of these necessary Bones, the Glorious Creator has wonderously contrived an *Unseen Cavity* in each side of the *Jaw-Bone,* in which are lodged an *Artery,* a *Vein,* and a *Nerve,* which thro lesser Gutters do send their Twigs to each particular Tooth. But because *Infants* are to feed a considerable while upon *Milk,* and lest their Teeth should hurt the tender Nipples of the *Nurse,* Nature defers the Production of them for many Months; whereas divers Animals, which must *seek betimes* a Food that needs Mastication, are born with them. The different Figure of the *Teeth,* how surprizing! The *Foreteeth,* called *Incisores,* broad, with a thin and sharp Edge, to cut off a Morsel from any solid Food. The *Eye-Teeth,* called *Canini,* stronger, deeper, and more able to tear the resisting sort of Aliments. The *Jaw-Teeth,* called *Molares,* flat, broad, uneven, accommodated with little Knobs, to hold, and grind, and mix the Aliments.

Because the Operations, to be performed by the *Teeth,* sometimes require a considerable Strength, what strong *Muscles* is the lower Jaw provided withal! And evey *Tooth* is placed in a strong, a close, a deep Socket; and the *Teeth* are furnished with *Holdfasts,* that are suitable to the stress, which in their different Offices they may be put unto. The *Fore-teeth* and the *Eye-teeth* have usually but *one Root,* which, in the latter, is very long; but the *Grinders,* that must manage hard Bodies, have *three Roots,* and in the upper Jaw often *four,* because these are pendulous, and the Jaw something softer. How convenient the Situation of the *Teeth!* The *Grinders,* nearest the Center of their Motion, because the greatest Force is required in them, the *Cutters,* where they may readily cut off what is to be transmitted to the *Grinders.* Finally, the *Jaw,* that is furnished with

75 *Grinders,* has an oblique or transverse Motion, which is necessary for the Comminution of the Meat: But this Motion is not in the Jaw of Animals, which have not such *Teeth* belonging to them.

"*Temperance* in *Feeding,* is one special Article of the Homage we owe to the Glorious One, who has, in our *Teeth,* so display'd
80 his admirable Workmanship!"

And we are now not far from the *Tongue,* the *Uses* whereof are, how various! how marvellous! and the *Texture* how much to be wonder'd at! You were in the right of it, *Vesalius,* when you told us, *That no Mortal had ever yet thorowly consider'd all the Wonders*
85 *of it.*

This is the main Organ of *Tasting;* it helps also in the *Mastication,* and the *Deglutition* of the Food.

Here the *Spittle* has its Vent; which, tho commonly taken for an *Excrement,* is indeed an *Humour* wonderfully serviceable; because
90 a great part of our Food is dry, there are provided several *Glandules,* to separate this Juice from the Blood, and no less than four pair of *Channels* to convey it into the Mouth, which are lately found out, and called the *Ductus Salivales;*[144] and through which the *Saliva* continually distilling, serves to macerate our Food, and,
95 by tempering of it, render it fit for chewing and swallowing. And hereby also the *Concoction* in the Stomach is not a little promoted.

But the grand *Glory* of the *Tongue,* is, that it is the main Instrument of *speaking; and therewith we bless God, even our*
99 *Father!*[145] This is a Faculty peculiar to *Man:* It was never known that a *Beast* could attain to any thing of it. A *Bird* indeed has been taught now and then a few words, and with no little difficulty; but then he *understands not* the meaning of his few words, nor does he use them for Signs of things conceived by him. The most
5 that can be pretended, is, that a *Parrot* being used unto such for such Enjoyments or Afflictions, at the Prolation[146] of certain *words,* may express his Passions by the noise of these *words.* The *Jewish* Rabbins[147] were not so very absurd in defining a Man, *Animal loquens,* a Creature that speaks. By the way, "you that are

---

144. "Saliva ducts." There are six separate salivary glands, but only two pair of ducts—one pair from the parotid gland and one pair from the submandibular gland.
145. James 3:9.
146. That is, utterance; bringing forth of words.
147. Rabbin means rabbi, but the word was used mainly in the plural to designate the chief Jewish authorities on matters of law and doctrine, the most important of whom flourished between the second and thirteenth centuries of the Christian era.

Stammerers[148] ought exceedingly to humble yourselves before the Holy God, under his Rebuke upon you, in an *Organ*, which, well employed, would be your *Glory*. Our Saviour, seeing a Man that had *an Impediment in his Speech*,[149] he *sighed* upon it; no doubt it grieved him to see a Man so *marked* by the Displeasure of God, in a most sensible Wound upon so distinguishing a Faculty. *My Friends*, learn to *speak deliberately*. This Expedient alone would help you wonderfully: For in *Singing* there is no *Stammering*. Speak but *little*, don't affect a *Loquacity*; a Folly *your Tribe* are often subject to! tho 'tis more burdensome and ungrateful in *them*, than in other People. What *little* you speak, let it be very *wise*, very *good;* such as may bespeak some respectful Regard for what you say. Then be not altogether discouraged under your Calamity: A MOSES, a PAUL, and a BOYLE, will make a noble *Triumvirate* of Companions for you,[150] under your uneasy Infirmity." I go on: The necessity of the *Tongue* for *Speech* will remain generally to be asserted, notwithstanding the Tricks of the *Ventriloqui*,[151] taking advantage of the Duplicature of the *Mediastinum*,[152] to form various Voices; and notwithstanding the rare Instance reported by *Roland*, in his *Aglossostomagraphia, sive Descriptio Oris sine Lingua, quod perfecte loquitur, & reliquas suas functiones naturaliter exercet.*

What the Emperor *Justinian* himself asserts in his Rescripts; [*Vidimus venerabiles Viros, qui abscissis radicitus Linguis;*] that he himself saw *venerable Men*, who when their *Tongues* were cut out, at the very Root, yet continued plainly speaking the Truth of Christianity against the *Arians;*[153] a Fact whereof many Witnesses are subpœna'd by *Cujacius*: it looks miraculous!

My God, thou hast made Man's Mouth! Make thou the Speech of mine what it ought to be. A pure Language![154] I have said, I will take

---

148. Mather had stuttered in his youth, but he learned how to overcome his handicap.

149. See Mark 7:32.

150. Exod. 4:10 relates that Moses was "slow of speech, and of a slow tongue." There is no good biblical evidence that Paul stammered, but Mather may have inferred as much from 1 Cor. 2:1–5 and 2 Cor. 10:10.

151. That is, ventriloquists.

152. That is, the midpoint of the chest. The mediastinum is the membrane that divides the lungs and other viscera of the chest into three parts—two lungs and the heart and great vessels.

153. The Arians, followers of Arius (c. 250–c.336), denied the divinity of Jesus Christ.

154. Zeph. 3:9.

heed, *that I do not sin with my Tongue.*[155] *Assist me to keep such a Resolution, and abhor all rotten or faulty Communication.*[156] *I resolve my Mouth shall speak the Praise of the Lord:*[157] *Oh that my Tongue may be like choice Silver,*[158] *for the good Use and Worth of what is thereby articulated, and as a Tree of Life,*[159] *in all my Conversation!*

If we pass down from the *Mouth,* we are quickly entertained with a *Wind-Pipe,* which is all made up of *Wonder!* A continual *Respiration* is necessary for the Support of our Lives; it is therefore made with *annulary Cartilages,* to keep it constantly open, and that the Sides of it may not flag and fall together. And lest, when we swallow, our Meat or Drink should fall in to do mischief there, it hath a strong Valve, an *Epiglottis,* to cover it when we swallow. For the more convenient bending of our Necks, it is not made of one *continued Cartilage,* but of many *annular* ones, which are joined by strong Membranes; and these Membranes are *muscular,* compounded of strait and circular Fibres, for the more effectual Contraction of the *Wind-pipe,* in any violent Breathing or Coughing. And that the *Asperity* of the Cartilages may not hurt the *Gullet,* which is of a tender and skinny Substance, or hinder our swallowing of our Food, these annulary Gristles are not entire Circles; but where the *Wind-pipe* touches the *Gullet,* there the Circles are fitted up with only a soft membrane, which may easily give way to the Dilatation of the *Gullet.* But now to proclaim a plain Design in this Conformation, as soon as the *Wind-pipe* enters the *Lungs,* its *Cartilages* are no longer *deficient,* but perfect *Circles;* it was no longer necessary they should be deficient, it was more convenient they should be *perfect.* And then, to finish the Collection which our excellent *Ray* has made (for I have him now before me) of these Curiosities; for the various Modulation of the *Voice,* the upper end of the *Wind-pipe,* is endued with several *Cartilages* and *Muscles,* to contract or dilate it, as we would have our *Voice* flat or sharp; and the whole is continually moistened, with a *glutinous Humour* issuing out of the small *Glandules,* that are upon its inner Coat: so 'tis fenced, that neither the *Air* fetched in, nor the *Breath* going out, may hurt it; yet it is of so quick a Sense,

---

155. Adapted from Ps. 39:1.
156. Adapted from Col. 3:8 and Eph. 4:29.
157. Ps. 145:21.
158. Adapted from Prov. 10:20.
159. Adapted from Prov. 15:4.

that it is provoked easily to cast out, by *coughing*, whatever may be offensive to it.

Caspar Bartholin has further observed, that where the *Gullet* perforates the *Midriff,* the carneous Fibres of that muscular Part are inflected and arcuate,[160] as a *Sphincter* embracing it, and closing it fast; which is a sensible Providence, lest, in the perpetual Motion of the said *Midriff,* the upper Orifice of the Stomach should gape and cast out the Food as fast as it received it.

Dr. *Grew* observes, that the Variation of the *Wind-pipe* is observable in every Creature, according as it is necessary for that of the *Voice;* and the *Rings* of the *Wind-pipe* are fitted for the Modulation of the *Voice.*

The Faculty of the *Glottis,*[161] in so exquisitely *contracting,* or *dilating* of itself, as to form all *Notes,* is, as Mr. *Derham* says, *prodigious!* For, as Dr. *Keil* notes, if you suppose the greatest Distance of the two sides of the *Glottis,* to be one *tenth part* of an *Inch,* in sounding *twelve Notes,* to which the Voice easily reaches, the Line must be divided into *twelve Parts,* each of which gives the Aperture that is requisite for such a *Note* with a certain Strength. But if we consider the *Subdivision of Notes,* into which the *Voice* can run, the Motion of the Sides of the *Glottis* will be still vastly nicer. A *Voice* can divide a *Note,* at least into an *hundred Parts,* which a *just Ear* can perceive; but then it follows, that the different Apertures of the *Glottis* actually divide the *tenth Part* of an *Inch* into *twelve hundred Parts,* and a *good Ear* will be sensible of the Alteration. But because each side of the *Glottis* moves just equally, therefore the *Divisions* are double, the Sides of the *Glottis,* by their Motion, do actually divide one *tenth part* of an *Inch,* we must say, into *two thousand and four hundred Parts.*

My God, I desire that never any evil Word may have my leave to go thro so curious a Passage, and that the Dispositions of my Mind may not be so vicious and odious, as to render so elegant a Passage, the vent of an open Sepulchre.[162] *"'Tis fit that nothing but Confessions of God, and Kindnesses to Men, should have such an exquisite Passage found for them."*

We cannot leave these Parts, without considering *Respiration.*

---

160. That is, bent or curved in the form of a bow.
161. The structure that surrounds the space between the vocal fold and arytenoid cartilage of one side of the larynx and those of the other side.
162. The last clause is adapted from Ps. 5:9.

Essay 32. Of Man

A Faculty of such importance to *Life*, that in the sacred Oracles, and indeed in our common Phrase also, *Breath* and *Life* are so concomitant, as to be equivalent: *Lord, thou takest away their Breath, and they die.*[163]

The Uses of *Respiration* were but indifferently assigned, until *Malpighi's* Discoveries. *Willis*, and *Mayow*, and others, do mention Uses thereof that are not contemptible; but our *Thurston*[164] rejects the Opinion of their being the *principal*, and thinks, 'tis principally to move, or pass the *Blood*, from the right to the left Ventricle of the Heart. Experiments made, by divers ingenious Men, on strangled Animals, have demonstrated his Opinion: For which cause the learned *Etmuller* also espoused it, who having reckoned up no less than *thirteen* Uses of *Respiration*, which are of great consequence, but conduce rather to the *Well-being*, than the *Being* of the living Creature, he concludes with a *fourteenth*, as the chief of all, which is, *For the passing of the Blood thro the Lungs, that is thrown into them by the Heart.* Anon comes Dr. *Drake*, and he not only establishes this Notion of *Respiration*, but also carries it further, and makes it the true Cause of the *Diastole* of the Heart; which neither *Borelli*, nor *Lower*, nor *Cowper*, much less any before those eminent Persons, have well accounted for. Dr. *Lower* has proved, that the *Heart* is a *Muscle*. The Motion of all *Muscles* does consist in *Constriction*. This accounts for the *Systole:* but the *Heart* has no *Antagonist Muscle*. What shall we now do for the *Diastole?* Great Wits have been puzzled here. But now Dr. *Drake* makes the weight of the incumbent *Atmosphere* to be the true *Antagonist* for all the *Muscles;* which serve both for the Constriction of the Heart, and for ordinary *Respiration*.

Dr. *Cheyne* adds yet one Use more for this great Faculty and Action; that is, to form the *Elastick Globules,* of which the *Blood* does principally consist, and without which there would be a general Obstruction in all the *capillary Arteries*.

Dr. *Wainwright* observes, the *Air* can't remain in the *Lungs*, without being much heated, and thereby having the Spring of it unbent, and so become specifically lighter than the external *Air:* For which reason it will, by a known Principle in *Mechanicks*, give place to it, and rise to such an height, as till it meet with *Air* of

---

163. Ps. 104:29.
164. Malachi Thruston (fl. 1665–81) was the author of *De respirationis usu primario, diatriba* (London, 1670).

its own Weight, and there it will remain. But then the Sides of the *Blood-Vessels*, which by the Inflation of the *Lungs* were drawn asunder, now, when the *Lungs* are crouded on an Heap, will be forced together, and so the *Blood* contained in them will be broken into innumerable Parts, exceeding small, and thereby rendered the fitter to pass the several *Strainers* of the Body.

Great God! thou hast in thy Hand my Breath and all my ways;[165] *I resolve to serve thee as long as I breathe; I resolve to look on thy Service as the end for which thou dost continue my Breath; I resolve to employ my Breath in thy Service to the last: I will praise thee as well as I can to and in my last Breath; and when I have no Breath, I shall do it better.*[166]

Behold now the *Lungs*, a most surprising Piece of Workmanship! Consult the Description of them given by *Malpighi*, who first of all discovered their *Vesiculæ;* and by *Willis*, who, writing after him, has proceeded upon it yet more accurately, and by *Cowper* in his admirable Tables. Then stand and *admire the Work of God*. You can do no otherwise! We will not meddle with the Controversy between *Etmuller* and *Willis*, whether the *Vesiculæ* of the *Lungs* have any muscular Fibres, or no. We will content ourselves with *Galen's* Conclusion upon the Parts ministring to *Respiration*, that *admirabilem Sapientiam testantur.*[167]

While the *Fœtus* is yet in the *Womb* (as Dr. *Keil* observes) the Vesicles of the *Lungs* lying flat upon one another, compress all the *capillary Blood-Vessels*, which are spread upon them. As soon as we are born, the *Air*, by its Gravity and Elasticity, rushes into the empty Branches of the *Trachea Arteria*,[168] and blows up the Vessels into Spheres: by which means the Compression being taken off from the *Blood-Vessels*, and they equally expanded with the *Lungs*, all the *Blood* has a free Passage thro the *Pulmonary Artery*. But when the *Air* is thrust out again, by a Contraction of the Cavity of the *Thorax*, it being a fluid Body, compresses the Vesicles and *Blood-Vessels* upon them, every where equally. By this Compression, the red Globules of the *Blood*, which thro their languid Motion, in the Veins, were grown too big to circulate in the fine *capillary Vessels*, are broken, and again divided in the *Serum*, and

---

165. Adapted from Dan. 5:23.
166. These lines are adapted from Ps. 150:6.
167. "They bear witness to an admirable wisdom."
168. That is, the trachea, the cartilaginous and membranous tube descending from the larynx to the bronchi, the air passages within the lungs.

the *Blood* is made fit for Nutrition and Secretion. This Pressure of the *Air* on the *Blood-Vessels,* Dr. *Keil* says, is equal to an hundred pound weight. It is also probable, he thinks, that Particles of the *Air* must enter the *Blood-Vessels,* and mix with the *Blood* in the *Lungs.*

The Divine Workmanship about the HEART, who, that has any *Heart,* can forbear admiring of it, with most sensible Acknowledgments! This is that admirable Bowel, which with its incessant Motion distributes the *Blood,* the Vehicle of Life, throughout the whole Body. From this *Fountain of Life*[169] and *Heat,* there are *Conduit-Pipes* even to the least, yea, and most remote Parts of the Body. 'Tis the Machine, which receives the *Blood* from the *Veins,* and forces it out by the *Arteries,* thro the whole Body. The force with which the *Heart* squeezes out the Blood into the *Arteries,* is, in *Borelli's* Reckoning, equal to the force of *three thousand Pound weight.* For this important Use it is most exquisitely contrived. Being a *muscular* Part, the Sides of it are composed of two Orders of *Fibres,* running circularly or spirally from the Base to Tip, contrarily the one to the other; and so being drawn contrary ways, do violently constringe and straiten the *Ventricles,* and strongly force out the *Blood.* And then the Vessels, we call *Arteries,* which carry from the *Heart* to the several Parts, have their *Valves,* which open *outwards* like Trap-doors, and give the *Blood* a free Passage out of the *Heart,* but will not suffer any Return of it thither; and the *Veins,* which bring it back from the several Members to the Heart, have their *Valves,* or Trap-doors, which open *inwards,* and give way for the running of the Blood into the *Heart,* but prevent its running that way back again. Moreover, the *Arteries* consist of a *Quadruple Coat,*[170] the third of which is made up of annular, or orbicular, carneous *Fibres,* to a good Thickness, and is of a *muscular* Nature, (which was first observed by Dr. *Willis*) and this, after every Pulse of the *Heart,* serves to contract the Vessel successively with incredible Celerity, so by a kind of *peristaltick Motion,* forcibly and very swiftly impelling the *Blood* onwards to the *capillary Extremities,* and thro the *Muscles;* wherefore the Pulse of the *Arteries* is not caused only by the Pulsation of the *Heart,* which drives the Blood thro them after the manner of a Wave, as many

169. Ps. 36:9 and Prov. 13:14, 14:27.
170. The arteries actually consist of three coats. From the outside in they are the adventitia, the media, and the intima.

would have it, but also by the *Coats* of the *Arteries* themselves, as it has been confirmed by the Experiments of many modern Physicians, yea, and of *Galen* also. We may add one thing more, that the *Heart* and the *Brain* do notably enable one another to work; for the *Brain* cannot live unless it receive continual Supplies of Blood from the *Heart*, much less can it perform its Functions of preparing and of dispensing the *Animal Spirits;* nor can the *Heart* afford a *Pulse*, unless it receive Spirits or something descending from the *Brain* by the *Nerves:* do but cut asunder the *Nerves* that go from the *Brain* to the *Heart*, the *Motion* thereof ceaseth immediately.

For the Motion of the *Heart*, Monsieur *Tauvry* flies to a *subtile Matter* managing the *Fibres* of it, but seems to acknowledge it a *Matter which no Mortal has traced yet to Satisfaction*. In fine, the *Heart* is a compound *Muscle*, and each Ventricle of it will (as Dr. *Keil* observes) contain an *Ounce* of Blood. We may well suppose the *Heart* throws into the *Aorta* an *Ounce* of Blood every time it contracts; the *Heart* contracts four thousand times in one Hour, sometimes more, sometimes less; hence there passes thro the *Heart* every Hour *four thousand Ounces of Blood*, that is to say, three hundred and fifty Pound.[171] Now the whole Mass of Blood is no more than twenty-five Pound, so that a Quantity of Blood equal to the whole Mass passes thro the *Heart fourteen times in one Hour*, which is about once in every four Minutes; not the *whole Mass* itself: we don't suppose that the *Blood* which goes to the Extremities, can return to the *Heart* as soon as the *Blood* which goes only to the *Kidneys* or the *Liver*.

"Without making any fanciful Excursions upon *Metaphors* drawn from the *Figure* and *Office* of the *Heart*, I am sure 'tis infinitely reasonable that I should behold this *Bowel* with a most hearty and lively Sense of my Obligations *to give thee my Heart, O my God, and love thee with all my Heart!*"[172]

The *Stomach* has in it how many things that are truly admirable! The greatest Philosophers have cried out, "How great a Comprehension of Nature did it require to make a *Menstruum* that should corrode all sorts of *Flesh* coming into the *Stomach*, and yet not the *Stomach* itself, which is also *Flesh!*" 'Tis *membranous*, and capable of being *dilated* or *contracted*, according to the Quantity

---

171. The correct figure should obviously be two hundred and fifty.
172. Adapted from Prov. 23:26, Deut. 6:5, Matt. 22:37, and Mark 12:30.

of Meat contained in it; the Situation of it under the *Liver*, accommodates with an *Heat*, that carries on the *Concoction;* when it has gone thro with the *Concoction*, it can shrink itself, and cast out the Food. But, *Concoction*, how performed? Inform us, Dr. *Drake!* There is in Bodies a *Principle of Dissolution*, which upon the Extinction of their vital and vegetative Faculty, begins to exert itself towards the *Destruction* of the Subject. This *Principle of Corruption* is, perhaps, the same that in a State of *Circulation* and *Vegetation* was the *Principle of Life*, but now being denied that Passage which it had before, it makes its way *irregularly*, and so destroys the Continuity of the *Solids*, in which it is included, and introduces that Change in the whole Mass which is called *Corruption*. This *active Principle* is a sort of Air, which is mixed in a considerable Quantity with all sorts of *Fluids;* this (tho its natural and essential Motion be expansive or *quaquaversum*)[173] when it is introduced into Bodies, has two kinds of motion, one *expansive*, by which it communicates that *intestine Motion* which all Juices have, and by which the containing Parts are gradually extended, and have their Growth; but the other *progressive*, and indeed *circulatory*, which is occasioned by the Renitency of solid Parts, and obliges its taking that Course which is most open and free. This *Motion* being stopt, the *expansive* still remains, and continues to act, till by degrees it hath so far overcome the Resistance of the including Bodies, as to bring itself into an equal degree of *Expansion* with the *external Air*, which cannot be done without a *Destruction* on the Texture and Continuity, or specifick degree of Cohæsion of the *Solids;* and this is called *a State of Corruption*. This *destructive Quality* of the *Air* in Bodies may be promoted, either by *weakening* the Tone of them, and the Cohæsion of the Parts, and so facilitating the Work of the *Air*, as it is done when *Fruit* is bruised; or by intending the *expansive Force* of the *Air* itself with *Heat*, or other co-operating Circumstances. The former is done in *Mastications*, the latter is done by the *Heat* of the *Stomach*, which forcibly rarefying the *Air*, enables it to rend the including Bodies to pieces the sooner, and so to let loose the Fluids, and perhaps likewise produce a Comminution upon several parts of the *Solids*, so as to make them sustainable in the *Liquor;* which latter is the Operation that compleats the *Digestion in the Stomach*. In *stewing*, tho the *Heat* be unspeakably short of what is in *roasting*

173. "To all sides."

and in *boiling,* the Operation is of all the quickest, because it is performed in a pretty close Vessel, and full, by which means the *Succussions* are more often repeated, and more strongly reverberated. The Operation of the *Stomach* is mightily resembled by the *Digestor* of Monsieur *Papin;* in this the *Meat* is put, together with so much *Water* as exactly fills the *Engine,* the Lid is then skrewed on so close as to admit of no external *Air,* and with two or three lighted *Charcoal,* or the *Flame* of a *Lamp,* it is reduced into 'a perfect *Pulp,* or indeed a *Liquor,* in a very few Minutes, in six, or eight, or ten, or twelve, or sixteen, according to the Toughness of the Matter to be digested, or the Augmentation of this little Fire; this way even the *hardest Bones* are presently dissolved. Thus the *Stomach* naturally closes on the Aliments, which descend to it; it strictly embraces them when it is full; by keeping out extraneous Air, it fortifies and invigorates the *Succussions*[174] of that which is contained in the Aliments, and this is enabled hereby to break and resolve the Bodies which included it, into Particles that may be small enough to enter the *Lacteals.* When all the *Chyme* and *Chyle* is pressed out, the *Stomach,* which follows the motion of its Contents, is again by means of its *muscular Coat* reduced into a State of Contraction, and the inner is brought thereby to lie in Folds, and by means of the *Peristaltick Motion* rubbing lightly upon one another, produce that Sense of a *Vellication*[175] which we call *Hunger:* this being felt first in the upper Orifice, which is first evacuated, begins first therefore to prompt us to replenishing; but as by degrees the remainder of the Contents are expelled, this Friction of the Membranes upon each other, spreads gradually over the whole *Stomach,* and renders our *Hunger* more impatient.

Great God, I bless thee for all my Food. My gracious Feeder, I bless thee that I have not known the terrible Famine. I will take no Food without looking up to thee for thy Blessing, by which alone I live!

The *Intestines;* these receive the *Chyle* from the *Pylorus;*[176] these further digest it, prepare it, separate it: these by their peristaltick Motion drive it into the Lacteals: but the excrementitious Parts they send off elsewhere, from whence there is no regress, unless

---

174. The action of shaking or condition of being shaken, especially with violence.
175. A rare or obscure word meaning the twitching or irritation of a muscle or other part of the body.
176. The muscular opening from the stomach into the intestine.

upon a Relaxation or Laceration befalling the Valve of the *Colon.* Can you behold the Structure of the *Intestines,* as reported by *Kerkringius,* by *Glisson,* by *Willis,* and *Peyer,* and others, without Astonishment!

The *intestines,* 'tis wonderful, they are six times as long as the *Body* to which they appertain; and now that they should keep their *tone,* and their *Site,* and hold on doing their *Office,* and give an undisturb'd Passage to what every day passes thro them, and this for some Scores of Years together, 'tis impossible for me to consider it without falling down before the glorious God, and making that Acclamation, *What hast thou done in me, O thou Preserver of Men!*[177] *How much do I depend upon thee for my Preservation from grievous Diseases!*

The *Liver* does admirable things, in continually separating the *Choler* from the Blood, and emptying it into the *Intestines,* where it is useful, not only to provoke Dejection, but also to attenuate the *Chyle,* and render it so subtile and fluid, that it may enter at the Orifices of the *Lacteals.*

The *Bladder* is an admirable Vessel! The Substance is *membranous,* and extremely dilatable, for the receiving and containing of the *Urine,* till a convenient opportunity of emptying it; it hath also Shuts for the Ends of the *Ureters,* which are so artificially and marvellously contrived, as to give the *Urine* a free entrance, but stop all passage backward: the *Wind* itself cannot be transmitted thro the Shuts, tho never so strongly forced upon them!

In the *Kidneys,* how admirable the innumerable *Siphons,* the little and curious *Tubes,* conveying the urinous Particles into the *Ureters!* discovered first by *Bellini,* afterwards illustrated by *Malpighi.*

Leuenhoeck has discovered Vessels in an Human Body, the *Diameters* whereof are more than *seventy-nine thousand* times less than an *Inch;* and, as Dr. *Wainright* observes, at least so small must be the Diameters of the *Lacteals.* My God, how exquisite, how curious are thy Works! But then how much do I depend upon thee to keep all the Vessels of my Body, doing their Office in their order![178] That so

---

55 dilatable/dilateable

177. Adapted from Num. 23:11 and Job 7:20.
178. Adapted from 1 Chron. 6:32.

fine an Engine is not ruin'd a thousand times in *a day,* but holds on in its motion for *twenty-five thousand five hundred and sixty-seven Days!*[179]

75 All the *Glands* of the Body, each of them an admirable *Congeries* of many Vessels, in a stupendous Variety, curled, complicated, circumgyrated, and marvellously woven into one another; these give the *Blood* an opportunity to stop a little, and separate thro the Pores of the *capillary* Vessels into the *secretory* ones, which after all exonerate into one common *Ductus.* Read *Wharton,* and
80 *Bartholin,* and *Bilsius,* and others; but prepare always for a Field of *Wonders,* equal to any *in the Field of Zoan!*[180] But then consider too the Variety of *Humours* that are separated by the *Glands;* all different in Colour, in Taste, in Smell, and in other Qualities.

The *Bones,* how admirable in their Circumstances! The *Back-*
85 *bone* is contrived with an Artifice truly astonishing! It is divided into many *Vertebras,* for the commodious *bending;* one entire and rigid *Bone* of that length would have been often in danger of snapping; it is *tapering,* in the form of a *Pillar,* the lower *Vertebras* being the broadest and largest, the superior in order lesser and
90 lesser, that so the Trunk of the Body may have the greater Stability: but the several *Vertebras* are so elegantly compacted and united, that they are as firm and strong as if they were but one single Bone; they are all perforated in the middle, with a large Hole for the *Spinal Marrow* (that wondrous *Pith!*) to pass along, and each
95 of them hath an Hole on each of their sides, to transmit the *Nerves* to the *Muscles* of the Body, and thereby convey both Sense and Motion. By the close Connection of the *Vertebras,* the *Back-bone* is formed so as to admit of no great Flexure and Recess from a right
99 Line; it also admits no *angular,* nor any but a moderate *circular* bending, lest the *Spinal Marrow* should be compressed, and so the Passage of the Spirits to and fro meet with some Obstruction.

Dr. *Grew* observes, that in *Trees* there is a new Ring added every Year out of the Bark to the Wood; so too in *Animals,* while they
5 grow, there is a new *Periostium*[181] added from time to time out of

86 *Vertebras/Vertebres*   88 *Vertebras/Vertebres*
91 *Vertebras/Vertebres*   97 *Vertebras/Vertebres*

179. That is, seventy years (including leap years)—the span of a human life.
180. Ps. 78:12–43 names the field of Zoan as the place where the miracles associated with the deliverance from Egypt took place.
181. A thin membrane that encloses all the bones in the body except at the articular surfaces.

the *muscular Membranes* to the Bones: *The sweet Harmony with the vast Variety in the Works of God!*

Admirable the Provision that is made for the more easy and expedite Motions of the *Bones* in their Articulations: a twofold *Liquor* is prepared, by the Inunction[182] whereof their *Heads* or *Ends* enjoy some Lubrification;[183] first, there is an *oily* one, furnish'd by the *Marrow;* and then there is a *mucilaginous* one, furnished by certain *Glandules,* that are seated in their Articulations; both of these together make up the most *proper Mixture* for this purpose that can possibly be thought upon; both of the Ingredients are *lubricating*. But more than this, from their Composition they mutually improve one another; the *Mucilage* adds to the smoothing Efficacy of the *Oil,* and the *Oil* preserves the *Mucilage* from Inspissation,[184] and from contracting the Consistency of a Jelly. Hereby the *Motion* of the *Bones* is facilitated; for if they were dry, they would not readily obey the Pulls of the *motory Muscles,* which we find in the Wheels of our Clocks; the ends of the *Bones* are hereby also kept from an inconvenient *Incalescency,*[185] which, if they were dry, being so hard, a swift and long Motion would necessarily give to them; and thus the Wheels of our Coaches must be besmeared with a Mixture of *Grease* and *Tar* (an Imitation of *ours!*) that they may not be set on fire. What a *slothful World* must we have had, and how confined to Deliberation, if this Care had not been taken of our *Bones!* And finally, a great Mischief is now prevented, the *Ends* of our *Bones* are not *worn down,* by a grievous Attrition in their motion rubbing against one another; 'tis indeed a strange thing that this proves a sufficient Preservative to prevent the Consumption of the *Bones,* when we see the tops of *Teeth,* which are harder, worn off by *Mastication,* and brought so low, that the very *Nerve* lies bare, and for meer Pain they can be used no more. The ingenious Mr. *Havers,* who makes these Remarks in his *Osteology,* makes this Conclusion: *Here we cannot avoid the notice of the visible Footsteps of an infinite Reason, and we can never sufficiently admire the Wisdom and Providence of our great Creator!*

182. That is, an act of applying an oil or ointment. The account that follows on lubrication of the joints is erroneous. There is only one synovial fluid made in the joints of the human body, and that is the mucilaginous one, which is supplied by the synovial membrane itself.
183. An archaic word for lubrication.
184. That is, thickened in consistency; made thick, heavy.
185. A state of being warm.

40 We may add, wonderful the *Construction* of the *Bones,* that are to support the Body, or bear heavy Burdens, or be employed in difficult Exercises; they are made *hollow,* this wonderfully accommodates them for both *Lightness* and *Stiffness;* an *hollow* Body is more inflexible than a *solid* one, of the same Substance and Weight: but the *Ribs,* which do not carry *Loads,* nor do any thing wherein so much Strength is required, but are only to fence the Breast, these have no *Cavity* in them, and these, towards the fore part of them, are broad and thin, so that they may give way, without much danger of any Fracture; and when they are bent, they do by their *elastick Property* again return to their Figure: and yet the *Hollow* of the *Bones* is not useless, but it contains the *Marrow,* which supplies an *Oil,* for the Maintaining and Inunction of the *Bones,* and of the *Ligaments,* and facilitating their Motion, and to secure them from Disruption, to which they would by any sudden Contortions be otherwise obnoxious. The mention of the *Ribs* will bring on one Observation more; That altho the *Breast* is encompassed with *Ribs,* the *Belly* is left free; this is, that it may give way to the motion of the *Midriff* in *Respiration,* and to the necessary reception of our Food, and to the convenient bending of our Body. The *Females* also find the Benefit of it in the time of their Pregnancy. *Great God, all my Bones must say, who is like to thee!*[186] *I bless thee for that thou dost not chasten the multitude of my Bones with strong Pain!*[187]

It cannot be without Admiration looked upon, that all the *Bones,* and all the *Muscles,* and all the *Vessels* of the Body, should be so contrived, so adapted and compacted, for their several Motions and Uses! All according to the strictest Rules of the *Mathematicks!* If you attempt an Innovation or Alteration, you *mar* all instead of *mending* any thing. In the *Muscles* alone there is more *Geometry* than in all the artificial Engines in the World; the greatest Mathematicians have not found a nobler Subject for their Disquisitions and Contemplations than *de Motu Animalium.*[188] The Essays of *Croon,* and *Steno,* and *Borelli,* on that Subject, have been very curious.

Dr. *Grew* observes, that no less than forty or fifty *Muscles,* besides many other subservient Parts, go to execute that one *Act of Laugh-*

---

186. Adapted from Ps. 35:10.
187. Adapted from Job 33:19.
188. Giovanni Alfonso Borelli published *De motu animalium,* 2 pts. (Paris, 1680–81), a work on muscular motion.

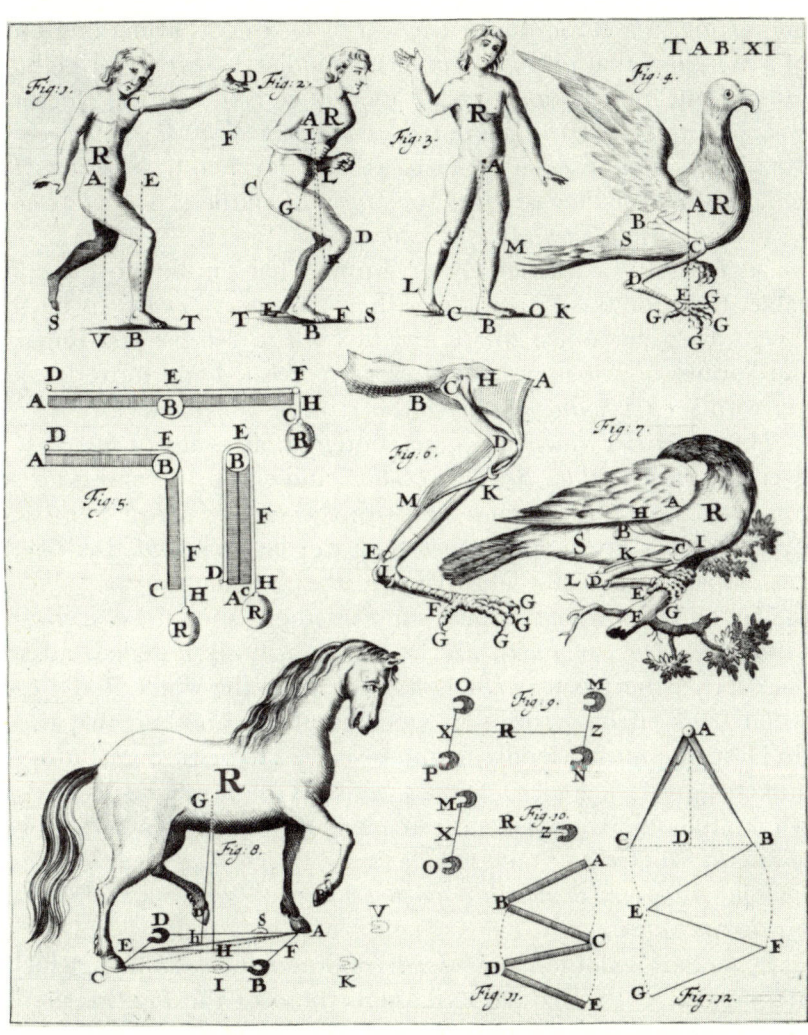

This illustration from the Leiden (1685) edition of Giovanni Alfonso Borelli's *De motu animalium* shows the functioning of the muscles and limbs mechanically along Cartesian lines. Courtesy of the Rare Book and Special Collections Library, University of Illinois at Urbana-Champaign.

ing, certainly then *laughing for nothing* may be indicted for an *Act of Folly!* He goes on with his Observation, That in some Cases we cannot execute *one single Thought* without such a Retinue. Suppose one sitting in a Room has a *Thought* of looking at something out of a Window, that one *Thought* has immediately seventy or eighty *Muscles* put into motion to wait upon it; *so that,* says the Doctor, *there is not a Monarch upon Earth served with such Majesty as every Man is within the Territory of his own Body:* But then how *reasonable* is it, O Man, for thee to serve the *Maker* of all these! *Glorious God, I will do it with all my Muscles, with all my Powers!*

Dr. *Grew* has a further Observation; What can be more admirable, than for the Principles of the *Fibres* of a *Tendon* to be so mixed as to make it a soft Body, fit both to receive and impart the Species of *Sense,* and to be easily nourished and moved, and yet with such a *Softness* to have the *Strength* of *Iron!*

Those *Muscles* which appear as contemptible as any of the Body, even the *Muscles of the Belly,* tho *Galen,* and other Anatomists after him, have contented themselves with reckoning four or five *Uses* of them, they are indeed more than can be reckoned. Dr. *Grew* has employed almost a large Page in the Enumeration.

'Tis admirable that under our *Skin* there should be such an unknown *variety of Parts,* and so very variously mingled, all so pack'd that there is no *unnecessary Vacuity* in the whole Body, yet so far from clashing with and hindring of one another, that they do all in the most friendly manner *conspire* to assist one another, and concur in the general Design, which is the Preservation of the whole. Behold, *Arguments* (as our pious *Ray* well notes hereupon) *of infinite Wisdom and Counsel! He must be worse than mad, that can find in his Heart to imagine all not provided by a most wise and intelligent Cause!*

Every Part is clothed, joined, corroborated by *Membranes,* which are capable of prodigious Extension; those of the *Peritonæum*[189] are a particular Instance of *that,* out of which alone, in *hydropical Persons,*[190] there have been drawn forty Gallons of Water, by a *Paracentesis.*[191] The undoubted Authorities of *Tulpius* and *Blasius,* and other Physicians, oblige us to believe surprizing things of this Importance.

189. The membrane that lines the cavity of the abdomen.
190. That is, swollen with water or edematous.
191. A perforation of some cavity in the body for the removal of a fluid or gas.

It is notable, that all our Organs are involved in *Coats,* one or more, consisting of tough or muscular Fibres, intended not only to *protect* them, as has been commonly thought, but also by a due Constriction to assist them in *straining* off their several Contents.

These Parts which at first appear to be of no more use than to fill up empty Spaces, will upon Examination be found exceeding serviceable. The *Fat* serves to cherish the Body, and keep it warm; yea, will maintain it for some time, when *Food* is wanting, and be as a sulphurous *Pabulum,* to preserve the *Heat* of the *Blood.* By what Vessels the *Fat* comes to be separated from the *Blood,* is a Point of curious Enquiry; the collection of it more on some certain Parts (as the *Caul*[192] and the *Reins*)[193] than on others, appears to be for the cherishing of those Parts with *Warmth;* the *Caul* is like an Apron of *Woollen Cloth* to the lower Belly. The *Gladiator,* whose *Caul* was cut out by *Galen,* felt so much *Cold,* that he was forced constantly to keep his Belly covered with *Wooll.* The *Intestines* containing much Food, there to undergo its last Concoction, and Vessels of Blood not flowing thither, need such a *Covering* to defend them; doubtless a constant *Heat* is required about the *Reins,* for the Separation of the *Urine* from the Blood: for we see if the Blood be chilled, the *Secretion* of the *Urine* will be sensibly stopt, and the *Serum* cast upon the *Glandules* of the Mouth and Throat.

Monsieur *Bernoulli,*[194] in a curious Meditation about *muscular Motion,* has observed another thing, that must not be pretermitted; that in *muscular Motion* the Expence of *Animal Spirits* is not in proportion to the Labour which the Animal is at: and so a Man reduc'd to hard Labour, is not reduc'd to the necessity of having twice or four times as much Victuals as one that is under no such necessity of working. Now the *Spirits* are the most precious things in all the *Animal Body,* we live by them; so needful and useful a Substance was to be saved by all the Means that were possible. And behold, as Dr. *Cheyne* expresses it, *we see the wise Author of Nature has taken wonderful Care that no Expences should be made that could be avoided.*

It has been observed by some, that to provide Matter for the

---

192. The greater omentum, a layer of fat which hangs down from the stomach and the large bowel. Often called the apron, it is referred to by physicians as the watchdog of the abdomen. The omentum becomes attached to anything that is inflamed and can help wall off a ruptured appendix.
193. That is, kidneys, or region of the kidneys.
194. That is, Johann Bernoulli.

generation of *Spirits* in *Man,* a vast Quantity of *Blood* is prepared, far exceeding what is found in *other Animals.* The *Blood* for the Body of *Man* bears the Proportion to his Weight, of *one to ten;*[195] in other *Animals* 'tis but *one to twenty.* And for the fetching of *Spirits* out of this Matter, there is the *Laboratory* of the *Brain,* which in a *Man* is twice as much as in a *Beast* four times as big.

It is Dr. *Cheyne's* Proposition, That the *Strength* of *Animals* is in a *triplicate* Proportion to the Quantity of *Blood* running in the Vessels.

The *Lympha* of the Blood is a marvellous thing; a Liquor separated in the *Membranes* and *Glandules,* which is the Medium whereby the *serous* and *fibrous* Parts of the *Blood* are united, and the *Bones* and *membranous* Parts of the Body are nourished. But how marvellous the *Lymphatick Vessels,* which convey this *exquisite Liquor!* They disappear when the *Animal* dies; their number is unaccountable: they were first of all discovered by *Thomas Bartholin* and *Olaus Rudbeck,* in the Years 1650 and 1651. *Pecker* made a progess in the discovery of them; and their Valves were demonstrated by *Frederick Ruysch,* which permit this transparent Liquor to pass thro them towards the *Heart,* but are like shut *Floodgates* upon the returning; they rise in all Parts of the Body. The *Glands* that separate the *Lympha* are of the smallest kind, and scarce visible by the finest *Microscopes;* but the *Lymphaducts* unite with one another, and grow larger as they approach the *Heart;* and yet they do not, like the *Veins,* open into one common Channel. The whole Contrivance of these *fine Vessels,* who can behold without Amazement!

About the *Blood,* this is admirable; the Branches which go off at any small distance from the Trunk of an *Artery,* unite their Channels into one Trunk again, whose Branches likewise communicate with one another, and with others; whence it comes to pass (as Dr. *Keil* observes) that when any small *Artery* is obstructed, the Blood is brought by the communicating Branches to the Parts below the Obstruction, which must otherwise have been deprived of their Nourishment. And in the *Veins* there is the like Provision, that so justly surprizes us in the *Arteries.*

The *Viscidity* of the *Blood* is increased by the *Heat* in a *Fever;* if we apply a much less degree of *Heat* than will boil Water, it will turn the *Serum* into a Jelly; the *Heat* of the *Skin,* where the Pulse

---

195. According to Mather's source, the blood in the human body bears the proportion of one to twenty of the body's weight.

will beat sixty Strokes in a Minute, is to the *Heat* of *boiling Water* as 16 to 52; *boiling Water* is but little more than three times as *hot* as the *Blood* of an healthy Man. If the *Heat* of the *Blood* increase in proportion to the Beat of the Pulse (as it must, if it beat with the same Strength it did) a Man whose Pulse beats 195 Strokes in a Minute, would be as hot as *boiling Water;* now 120 Strokes is common. Behold whence the *Siziness*[196] of the Blood in inflammatory Distempers!

"Why should I sinfully *over-heat* my *Blood?* But since my Life depends on the good Constitution of this *red Liquor,* which is yet so easily depraved, so easily disturbed, so easily overturned; *O God of my Life, I wonder that I live! I desire to live as a dying Man! But I live, because thou art the God of my Life!*"[197]

But at last the *Instrument* all this while employed in writing these things, that ὄργανον ὀργάνων,[198] demands of me that it be not forgotten; the HAND, the HAND, whereof I need no *Cicero* to be my Monitor, *Quam aptas, quamque multarum Artium Ministras, MANUS, Natura homini dedit!*[199] It is divided into four Fingers, bending *forwards,* and one stronger than any of them that bends *backwards,* to join with them; 'tis fitted thus to *lay hold* on Objects of any shape, or size, or quantity; and sometimes *one Finger* alone can discharge many Offices: the *Fingers* are strengthned with several *Bones,* jointed for motion, furnished with *Muscles* and *Tendons,* to bend them circularly forwards; how convenient this for the holding and griping of any Object! The *Fingers* also have their *Muscles,* to extend and open the *Hand,* and move them to the right and left; and thus the *whole Hand* may be employed, as all of a piece.

But then how notable is it, how wonderful! That the *Tendons* bending the *middle Joint* of the Fingers are so *perforated,* as to give passage to the *Tendons* of the *Muscles* which draw the uppermost Joints, and all bound close down to the *Bone* with strong *Fillets,* like so many Bow-strings, lest they should start up, and hinder the Hand in its Operations: finally, the *Ends* of the *Fingers* are fortified with *Nails,* which indeed *adorn* them as well as *defend* them; yea,

---

196. An archaic word meaning viscousness.
197. Adapted from Ps. 42:8.
198. "Organ of organs."
199. "Then what clever servants for a great variety of arts are the hands which nature has bestowed on man."

and have their further *Uses* too, if what *Camillus*²⁰⁰ writes in his Treatise upon the forming of *Judicia Medica*²⁰¹ from the Inspection of the *Nails,* may be relied upon: and how thin the *Skin,* and how exquisite the *Sense* at our *Fingers-ends,* by which we may judge of what we have there to be handled! We know who considered this Question, and how long ago; *Num eam omnino Constitutionem habeat* Manus, *qua meliorem aliam habere non potuit?*²⁰²

The *Uses* of this *astonishing Instrument* cannot be reckoned up; a whole *Book* written by *it,* might be easily filled with an Enumeration of its *Uses. Aristotle* says well, They *do ill* who complain that Man is *worse* dealt with than other Creatures, who are born with *natural Weapons* to defend themselves, and offend their Enemies; an *Hand,* with *Reason* to use it, abundantly supplies the Uses of all those *natural Weapons;* 'tis an *Horn,* an *Hoof,* a *Claw,* a *Tusk,* and all! Dr. *Grew* says very truly, *Never was there made an Instrument so curious!* The sixteen several general *Motions* of it are the *Elements of Operation,* as the *Letters* are of *Speech;* how infinitely to be diversified! What shall we call this but *the Handy-work of our God!*

Galen having described the Parts of the Fingers, and their Motion, cries out, *Considera hic mirabilem CREATORIS Sapientiam!*²⁰³

"When I apply *my Hand* to *any Action* which could not be done without it, I have *my Mind* invited to such a Thought upon it; *Great God, I bless thee for arming me with so curious and so adapted an Instrument! May I never ungratefully put forth my Hand to an evil Action.*

"Such a Thought often rolling in my *Mind,* and ruling of my *Hand,* would be better *Token for Good*²⁰⁴ to me, than the most promising *Lines* of any silly *Chiromancy.*"²⁰⁵

Voluntary Motion should not be left unconsidered; whereof Dr. *Cheyne* observes, the only Conception we can form, is, that the *Mind,* like a skilful Musician, strikes on that *Nerve* which conveys

---

36 *Tusk/Tush*

200. That is, Camillus Baldus or Camillo Baldi.
201. "Medical judgment."
202. The reference is to Galen, whose statement is made fully clear in context (the portion which Mather quotes is in italics): "[Let us investigate the hand], examining it to determine not simply whether it is useful or whether it is suitable for an intelligent animal, but *whether [the hand] is in every respect so constituted that it would not have been better had it been made differently.*"
203. "Consider, then, the marvelous wisdom of the Creator!"
204. Adapted from Ps. 86:17.
205. Divination by examination of the hand; palmistry.

Animal Spirits to the *Muscle* that is to be contracted, and adds a greater force than the *natural* to the *nervous Juice*, whereby it opens its passage into the Vesicles of which the *muscular Fibres* consist; but this Action of the *Mind* or *Will* on these *Animal Spirits*, is altogether *unaccountable* from the Laws of Motion. *My God, in thee I move!*[206] The astonishing *Power of spontaneous Motion* is what thou hast given me! Oh! may I never employ it in any Acts of Rebellion against Him that gave it.

Certainly Men may do well also to consider, whether the very *Configurations* of several Parts, may not afford good and great Admonitions of *Morality* to them. I need not explain my self, when I offer an Hint I have somewhere met withal: *Ponder, O Man, what Parts of thy Body have Bridles of Nature upon them!*

Some Consideration is also due to the astonishing *Strength* with which the *Bones* of Men have been sometimes endured. The *Strength* for which a *Samson* has been so famous, was indeed owing to a Possession and Assistance of a *Spirit* entring into him from above;[207] but the *ordinary Strength* of our *Nerves*, exerted in *moving* and *lifting*, is truly admirable; the Force of the *nervous Fluid!* And the Ability of the *little Fibres*, to sustain what it puts them on! And there are now and then, since the Days of *Milo* the *Ox-carrier*,[208] Examples of *Strength*, which will yet more *strongly* call for our growing Admiration; it would swell my Essay so big, that it would require a Man of such *Strength* to carry it, if on this and other Occasions I should insert all that has occurred to our purpose, in *Valerius Maximus*, in *Cælius Rhodiginus*, in *Zuinger*, in *Camerarius*, in *Hakewell*, in *Wanly*, and in other Collectors; however, a touch or two may not be unacceptable.

The Tyrant *Maximus* would with his Hands draw loaden *Carts* and *Wains*, break the Bones of *Horses*, and cleave *Trees* asunder. *Marius*, who of a *Cutler* became an *Emperor*, could with his fourth Finger stop a *Cart* that was drawn with Horses, and force it backwards; and a *Fillip* of his Finger (which they also report of *Tiberius*)[209] would knock a Man down like a Blow of an *Hammer*.

---

206. Adapted from Acts 17:28.
207. The reference is to Judg. 13:24–16:31. Judg. 14:19 relates that "the Spirit of the Lord came upon [Samson]" and gave him strength.
208. Milo or Milon of Crotona (fl. 511 B.C.), a celebrated athlete, was often victor in wrestling at the Olympic and Pythian games. Many stories were told of his marvelous feats of strength.
209. Tiberius Julius Caesar Augustus. See n. 64 above.

One *Salvius*,²¹⁰ mentioned by *Pliny*, having an *hundred-Pounds* weight at his *Feet*, and as many in his *Hands*, with twice as much on his *Shoulders*, could go up a pair of Stairs. *George Castriot* with his massy *Scimeter* did amazing Executions, he cut the *Turks* to pieces, *Barletius* affirms, three thousand of them with his own Hands, and scorn'd ever to throw away more than *one Blow* upon an Object; he could cleave *Helmet* and *Harness*, as if they were but Straw before him. *Cardan* saw one dancing with two in his Arms, two on his Shoulders, and one hanging about his Neck. A Baron of *Mindelheim*²¹¹ would with his middle Finger do things that surpass Imagination; he would shove a *Cannon* where he pleased; he would break *Horse-shoes* with his Hands like Potsherds; (which is a Circumstance they also relate of *Pocova*,²¹² a *Polish* Gentleman.) Little *Venetianello*²¹³ would with his Hands wreath great Pins of *Iron*, as if they were softned with the Fire, and carry on his Shoulders an erect *Beam* of twenty foot long and a foot thick, and shift it without the use of *Hands* from one Shoulder to another. A Provost at *Misna*²¹⁴ would make nothing with his bare *Hands* to fetch a Pipe of Wine out of a Cellar, and lay it on a Cart. *Mayolus*²¹⁵ affirms he saw a Man who took a Pillar of Marble three foot long, and one foot in diameter, which he cast up very high into the Air, and received it again in his Arms, and play'd with it as a little Ball; and another who would break a *Cable* as big as a Man's Arm, as easily as if it were a Thread of *Twine*. *Froisard*, a faithful Historian, tells of a Man who would make nothing to carry a great *Ass*, with all his *Load*, upon his Back. The Stories we have of the mighty Burdens carried by some of our *Cornish* Men, related by Mr. *Carew*, and others, are truly wonderful.

Can we now do any other than fall down before the glorious GOD, who has given such *Strength* to the Children of Men, as if their *Strength* were *the Strength of Stones*, or *their Flesh were Brass*;²¹⁶

---

210. Fufius Salvius was a strong man described by Pliny, *Natural History* (7.20), quoting Marcus Varro.
211. George of Froansberg, baron of Mindleheim (1473–1527).
212. Potocova, a Polish colonel of cossacks, was beheaded at Warsaw by permission of the Polish king at the urging of the Turkish ambassador.
213. Venetianello ("Little Venetian"), a Venetian by birth, though short of stature was of enormous strength. He was a famous funambulist.
214. Nicholas Klunder (fl. c. 1529) was provost of the church at Misnia in Thuringia.
215. Simone Maiolo (1520–97), bishop of Volturara, Italy, published three interesting books. His *Dies caniculares* (Rome, 1597), containing theologically oriented dialogues on aspects of nature, went through several editions.
216. Adapted from Job 6:12.

(and yet, when God pleases, *crush'd before the Moth!*)[217] with the antient Adoration, *O Lord God of Hosts, who is a strong God like to thee!*[218]

I conclude with the pathetical Words of an outlandish Doctor of Philosophy;[219] *O Deus, si totius Corporis mei Membra verterentur in Linguas, Nominis tui magnificentiam enarrare non possem.*[220]

But in MAN, must that have the *last* Consideration, the *State* whereof, alas, is that which too commonly is the *last* considered! The SOUL, which has mustered the many *Thoughts* wherewith our *Christian Philosopher* has fill'd his Pages, must now be thought upon. But oh! How much is *the Father of Spirits*[221] to be herewith acknowledged and glorified! Even the Pagan Orator shall be our Monitor; *Jam vero Animum ipsum, Mentemque Hominis, Rationem, Consilium, Prudentiam, qui non Divina Cura perfecta esse perspicit, is his ipsis Rebus mihi videtur carere.*[222]

'Tis high time for us now to take the SOUL of Man into our Contemplation. The SOUL, whereof *Juvenal*,[223]

> *Sensum a cœlesit demissum traximus arce,*
> *Cujus egent prona, & terram spectantia.*[224]

The SOUL, whereof *Claudian*,

> *Hæc sola manet, bustoque superstes*
> *Evolat.*[225]

And if our *Philosophy* terminate in *Theology*, the surprizing Words of a Pagan Physician will be proper to be introduced on the Occasion: O *Galen*, we Professors of *Christianity* will be thy surprized Hearers, while thou speakest at this rate to us: *Si quis nulli Sectæ*

217. Adapted from Job 4:19.
218. Ps. 89:8.
219. That is, Johann H. Alsted.
220. "O God, if the parts of my entire body were turned into tongues, I would not be able to describe the magnificence of your name."
221. Adapted from Heb. 12:9.
222. That is, Cicero: "Coming now to the actual mind and intelligence of man, his reason, wisdom and foresight, one who cannot see that these owe their perfection to divine providence must in my view himself be devoid of these very faculties."
223. Decimus Junius Juvenalis (c. A.D. 50/65-c. 127), last great Roman satiric poet, wrote his verse to protest the corruptions of Roman society.
224. "That we have drawn from on high that gift of feeling, / which is lacking to the beasts that grovel with eyes upon the ground."
225. Mather quotes only the italicized portion of the following lines by Claudian: "These [two souls] fail and perish with the body; / *the first* [man's spirit] *alone remains, survives the pyre and flies away.*"

*addictus, sed libera sententia rerum Considerationem interit, conspicatus in tanta Carnium & Succorum colluvie tantam Mentem habitare, (omnia enim declarant Opificis Sapientiam,) perfectissimæ Theologia verum principium constituet; quæ Theologia multo est major atque præstantior tota Medicina.*²²⁶ [De usu Part. lib. 17. c. 1.] Wonderful Words from a *Pagan Physician!*

The stupendous Faculties of the SOUL!

The *Wisdom,* with which a SOUL may perform wonderful things. 'Tis the *Wisdom* that *God puts into the Heart* of a *Solomon.*²²⁷

The Performances of that reaching *Philosophy,* which we have seen sagacious Minds endued withal, they have been amazing ones!

The Performances of the *Politician,* have sometimes been as amazing as those of the *Philosopher.*

Men of a *Great Soul,* what astonishing things have they arrrived unto!

And yet, I will venture to say, the *Love of GOD* in the Soul, or a *Principle of Grace* infused into it, is a *Divine Workmanship,* that is more *noble* than all its other Faculties, and will unspeakably *enoble* them all.

I have read, in the *Asceticks,*²²⁸ of a Servant of God, a Passage of this Importance: "I am not unable to write in *seven Languages;* I feast myself with the Sweets of all the *Sciences,* which the more polite part of Mankind ordinarily pretend unto. I am entertained with all kinds of *Histories,* antient and modern. I am no Stranger to the *Curiosities,* which by all sorts of Learning are brought to the Curious. Nevertheless, it appears unto me more valuable than all of this, it appears more delectable, it is a thing of a superiour Character, with a true *Spirit of Charity,* to relieve a

---

226. Mather condenses Galen, and to make the quotation clear it is given in full, with Mather's portion in italics. "*When anyone looking at the facts with an open mind sees that in such a slime of fleshes and juices there is yet an indwelling intelligence* and sees too the structure of any animal whatsoever—*for they all give evidence of a wise Creator*—he will understand the excellence of the intelligence in the heavens. Then a work on the usefulness of the parts, which at first seemed to him a thing of scant importance, *will be reckoned truly to be the source of a perfect theology, which is a thing far greater and far nobler than all of medicine.*"

227. The reference is to 1 Kings 4:29–30, 5:12.

228. Mather's reference is to the sober, righteous, godly life in the days of primitive Christianity, a way of life which he hoped to revive in his own day. This reference suggests a book. Rev. Edward Stephens (d. 1706), who wrote pamphlets on political and theological topics and devoted his later years to theological controversy, published *Asceticks: Or the Heroick Piety and Virtue of the Ancient Christian Anchorets and Coenobites* (London, 1696). There is no évidence that Mather knew this work, and the quoted passage is not found in it.

poor, mean, miserable Neighbour; much more to do any extensive Service for the Redress of those *Epidemical Miseries*, under which *Mankind* in general is languishing, and to advance the *Kingdom of God* in the World."

REASON, what is it, but a *Faculty* formed by GOD, in the Mind of Man, enabling him to discern certain *Maxims of Truth*, which God himself has established, and to make true *Inferences* from them! In all the Dictates of *Reason*, there is *the Voice of God*. Whenever any *reasonable thing* is offered, I have GOD speaking to me. Behold a Method in which a *Man*, (who will *shew himself a Man*,²²⁹ and *hearken to Reason*) may fill his Life with *Acts of Obedience* to GOD! Whatever I see to be *Reason*, I will comply with it, from this Consideration, *'tis what GOD calls me to! Reason* extends to Points of *Morality*, with as much Evidence as to those of *Mathematicks*. 'Tis as evident, *that* GOD, *my Maker, is to be glorified;* and, *that I am to do as I would be done unto;*²³⁰ as it is, *that three and four makes seven;* or, *that a Square is double to a Triangle, of equal Base and Height.* May the *Fear of* GOD for ever preserve me from doing any thing, whereof I may say, *it seems to me unreasonable.*²³¹

The prodigious *Learning*, wherewith some great Literators have been enriched! *Ideas*, like the *Sands on the Sea-shore*,²³² for the vast *variety* of them! There have been Men of so extensive a *Genius*, that they have been worthy to have a Celebration of their Obsequies, in as many Languages as were those of *Peireskius:*²³³ A Collection whereof, entitled *Panglossia*,²³⁴ had in it no fewer than *forty Languages*.

We see sometimes a much richer Soul than that of *Tostatus;* of whom yet *Bellarmine* says,

*Hic stupor est mundi, qui scibile discutit omne.*²³⁵

What a Character could *Vives* give of his *Budæus? Casaubon* reports of *Joseph Scaliger, There is nothing that any Man could desire to learn,*

---

229. Adapted from 1 Kings 1:52, 2:2.
230. Adapted from Matt. 7:12 and Luke 6:31.
231. Adapted from Acts 25:27.
232. Adapted from Gen. 22:17, Josh. 11:4, 1 Sam. 13:5, 1 Kings 4:29, and Heb. 11:12.
233. Nicolas Claude Fabri de Peiresc (1580–1637), French astronomer and celebrated patron of science.
234. The *Panglossie* was a collection of verses in forty languages by members of the Académie romaine des Humoristes in honor of Peiresc. It was edited by Jean J. Bouchard and published as *Monumentum romanum Nicolao Claudio Fabricio Perescio . . . factum* (Rome, 1638).
235. "This man is a wonder of the world, for he knows all that can be known."

but that he was able to teach: He had read nothing, (and yet what had he not read?) but what he did readily remember. Salmasius *gives a Report, little short of this, concerning* Casaubon. *Voetius* and *Vossius*,[236] how do they celebrate the vast Erudition of our *Usher!* Others will or may do as much for theirs. *Bochart* is rarely mentioned without the Epithet of *the incomparable. Grotius* was no *little Man, Selden* was not much smaller than he, both concluding their Lives with Testimonies to the Preference of *real Piety,* before all their Skill in *Languages* and *Sciences.*

My dear *Witsius,* lately dead, must for ever *live* in the Catalogue of *wonderful Men;* and Mr. *Baxter* too!

Of these two, and of some others, what *Amberachius*[237] writes of *Zuinger,* may be the consummate Elogy; *Cujus magna fuit Doctrina, sed exigua, si cum Pietate conferatur.*[238] Such was *he,* of whom I am going to repeat what I have heretofore asserted; had I Learning enough to manage a Cause of that nature, I should be ready to maintain, that there never was known under the Cope of Heaven a more learned Man than the incomparable *ALSTEDIUS;*[239] he has written on every one of the Subjects in the whole *Circle of Learning* as accurately and as exquisitely as those Men who have spent all their Lives in cultivating but any one of the Subjects. The reason why many of his Composures are not more esteemed, is the *Pleonasm*[240] of their Worth, and their deserving so much Esteem. To hear some silly and flashy Men, with a scornful Sneer, talk as if they had sufficiently done his Business, by a foolish Pun, of *All's-tedious,* is to see the ungrateful and exalted Folly of the World; for *Conciseness* is one of his peculiar Excellencies: they might more justly call him any thing than *tedious.*

The early Attainments and Achievements of some have been the just Admiration of the World. Mr. *Baillet* has drawn up a curious List of *illustrious Youths.*

When I see such *Men,* and their *Works,* I must for ever *look off,* and *look up* to the glorious God, and acknowledge, *Great God, thou*

27 not/no   34 Achievements/Atchievements

236. That is, Gerardus Vossius.
237. That is, Basilius Amerbach.
238. "Whose erudition was great, but slight, if compared with his piety."
239. That is, Johann H. Alsted.
240. The use of more words than are necessary for the expression of an idea; redundancy.

*art the Father of these Lights!*²⁴¹ *These had nothing but what they received from thee!*²⁴² And if such *Perfections* may be found in frail and weak Men, what, Oh! what are the Excellencies of the infinite God, before whom all these Men are but *as the Drop of the Bucket, and the light Dust of the Balance!*²⁴³ But when I consider how far the sinful Children of Men may come to have the *Chambers of their Souls* filled with *precious things,*²⁴⁴ it leads me to think, *What is that MAN, who is more than a meer MAN! That MAN who is the Son of God! O God, the Heavens do praise thy Wonder!*²⁴⁵ BOOKS which have in them vast Amassments of most *valuable Treasures,* cannot well be laid out of our Hands without such Thoughts as these.

But what shall we say when we see the vast *Performances* and *Capacities* of some SOULS, from which the want of *Bodily Senses* would have prohibited all our Expectations of any thing that should be considerable. *My God, I know that thou canst do every thing; all Souls are thine,*²⁴⁶ *and thou canst make them do what thou pleasest!*

The *Jews* tell us of a Professor in their Academy of *Sura,*²⁴⁷ who was called *Sagi Nahor,*²⁴⁸ or *Joseph of great Light;* he was *blind,* but it seems he had a Soul full of Knowledge.

We have had eminent *Preachers* who were *blind* Men, and educated for and serviceable in the Evangelical Ministry; Mr. *Cheesman* of *East-garston* was one, who lost his Eyes by the Small-Pox before he was four Years old: thus Mr. *Francis Tailor,* and Mr. *Homer Jackson.*²⁴⁹

But then that they should prove *Writers* too, learned, acute, polite *Writers!*

The Books of Mr. *John Troughton* are valuable things; his *Luth-*

48 Amassments/Amazements   55 *Sura/Sora*

241. Adapted from James 1:17.
242. Adapted from John 3:27 and 1 Cor. 4:7.
243. Adapted from Isa. 40:15.
244. Adapted from Prov. 24:4.
245. Adapted from Ps. 89:5.
246. Adapted from Job 42:2 and Ezek. 18:4.
247. The reference is probably to Joseph ben Hiyya (d. 333), head of a Jewish academy, who possessed exceptionally comprehensive knowledge, was distinguished in biblical exegesis, and had many pupils who transmitted statements in his name. A severe illness was doubtless the cause of his blindness.
248. *Saggi Nehor* is a euphemism for blindness; literally, "abundant in light." Joseph is not called *saggi nehor* in the Talmud, but the term is applied to him in modern Hebrew books.
249. These blind preachers resist closer identification. East-Garston, Berkshire, is due west of London, not far from the city.

erus Redivivus could be writ by none but a *Seer,* and an Eagle-ey'd one.

But if *many blind Men* have done learnedly, thou, Mr. *William Jameson,* hast *excelled them all!* That miraculous Man, a Professor of History in the famous University of *Glasgow,* tho blind from his Nativity, has published a variety of Books, and these in the *Latin* as well as the *English* Tongue, and full of *Quotation,* full of *Criticism,* full of accurate and exquisite Explanations on the nicest Controversies: when I read such things I cannot but see, and say, *the Finger of God!*[250]

That one Faculty of the Soul, the MEMORY, how amazing the Powers of it, how stupendous the Performances! The Account *Seneca*[251] gives of himself, if half of it be true! *Nam & duo millia Nominum recitata, quo ordine erant dicta, reddebam.*[252] Of his *very dear Companion,* as he calls *Latro Porcius,*[253] he affirms, that he retained in his *Memory* all the Declamations he had ever spoken, and never had his *Memory* failing him so much as in one single word. *Pliny* will give us more Examples of what the *Memory* of Man has done; a *Cyrus,*[254] who could call all the Soldiers in his Army by Name; a *Mithridates,* who could speak to twenty-two several Nations in their own Languages; a *Carneades,* who *Quæ quis exegerat in Volumina in Bibliothecis, Legentis modo representavit.*[255] Such was the *Memory* of Dr. *John Rainolds,* that he was called a *living Library,* and a *third University. Lipsius* had all *Tacitus* exactly in his Memory, and *Suarez* had all *Austin.*[256] *Homer's Iliads* have thirty-one thousand six hundred and seventy Verses, his *Odysses* no less; and yet the younger *Scaliger*[257] committed all *Homer* to his *Memory* in one and twenty Days. The *Memory* of our famous *Jewel* would perform Wonders, he would readily and exactly repeat any thing he had written, after once reading of it, and would have done it if the Auditors had been shouting, or fighting, and given

---

250. Exod. 8:19.
251. Lucius Aeneas Seneca the Elder (c. 55 B.C.–A.D. 37/41), Spanish-born writer on rhetoric, possessed a phenomenal memory.
252. "When two thousand names had been reeled off I would repeat them in the same order."
253. Marcus Porcius Latro (d. c. A.D. 4), Spanish-born friend of Seneca the Elder; an Augustan rhetor.
254. Cyrus I (d. 529 B.C.), founder of the Achaemenid Persian empire.
255. "He recited the contents of any volumes in libraries that anyone asked him to quote, just as if he were reading them."
256. That is, Augustine.
257. That is, Joseph Scaliger.

Essay 32. Of Man

him the greatest Occasions of Confusions; even Scores of barbarous Words, after once reading, he would repeat forwards and backwards, without hesitation. *Zuinger* mentions many strange Examples of a strong Memory, among which that of *Christopher Longolius* is very memorable; scarce any Length of Time was able to dislodge any thing he had once lodged in his *Memory!* But then how unaccountable the Instances of a *Læsa Memoria*,[258] reported by *Zuinger*, and *Forestus*,[259] and *Schenkius*, and others, especially when an *Apoplexy* has left a Man *Memory* enough to *write* Volumes, but unable to *read* a Syllable! The various *Inclinations* of the SOUL are a most admirably wise Provision of our good God, that the *Business of the World* may be all transacted, and with Satisfaction:

Diversis gaudet Natura ministris.[260]

We find *Homer*[261] sometimes admiring this Variety; and *Horace*[262] entertains us with a *Sunt quos Curriculo*,[263] which might have been extended to a Volume; for as one says, "there may be found a *Sunt quos* for every thing under the Sun."

Tho *Solomon* declares truly, *that much Study is a Weariness to the Flesh*,[264] yet with what Assiduity do many apply themsleves to it, and how delightfully! There have been other *hard Students* besides *Cato*, of whom *Tully* says, *Erat in eo inexhausta aviditas legendi, nec satiari poterat.*[265]

The *Jews* have done well to place this among their *Beracoth*;[266]

14 Sun."/Sun.

258. "Damaged memory"; that is, loss of memory.
259. Petrus Forest or Pieter van Foreest (1522–97), a Dutch physician who practiced in Delft for many decades, published histories of various maladies.
260. "Nature makes glad by different ministries."
261. The reference is to the *Odyssey,* 8.167–68, and to the *Iliad,* 13.730–33.
262. Quintus Horatius Flaccus (65–8 B.C.).
263. In the ode from which Mather quotes, Horace describes different walks of life, including his own as a poet. The bracketed passage from Horace, which Mather omits, will make his meaning clear: "Some there are [whose one delight it is to gather Olympic dust] upon the racing car."
264. Eccles. 12:12.
265. "He possessed a voracious appetite for reading, and could never have enough of it."
266. Berakoth, which in Hebrew means benedictions or blessings, is the name of the first tractate in the first of the six divisions of the Talmud. This division contains the precepts that apply to agriculture, upon which depends the life of all creatures. The Berakoth deals with three major areas of prayer—the obligation of daily reciting the Shema, Judaism's confession of faith, which begins, "Hear, O Israel: The Lord our God is one Lord"; public prayer in the congregation; and blessings at meals.

Deus facit ut unicuique suum Opificium placeat.[267] The *blessed* God is to be acknowledged in it. There is an Instance which Dr. *Edwards* has pitch'd upon: Would a *Gentleman* brought up a *Scholar,* and one very nice, neat, and curious, visit sick Persons whenever they call him, and leave his own Bed that he may give his Visits to them in theirs, and enter into Rooms that are filled with the most ungrateful Steam and Stench, and all his Days converse with Excrements, continue situated *inter Stercus & Urinam?*[268] One would think this were a Degradation to the *Velvet Cap* and *Scarlet Robe;* to go in Silk and Plush to the most squalid and nasty Chambers, looks a little strange; to suck in the Air of a Room which the Breath of the Diseased has infected, for this to be done by Persons of an honourable Character, and for them to undergo patiently and cheerfully more servile things than what are undergone in the basest and most servile Callings! But,

"*Behold, I have created the Smith, who blows the Coals in the Fire!*[269] so saith our God: and he is to be seen in the disposition to *profess every honest Trade for necessary Uses!*[270] When I behold any Man cheerfully following the Business of his *Calling,* I would upon the Invitation say, *Glorious God, it is well that thou hast so disposed the Mind of this my Neighbour!*"

They who have written *de Morbis Artificum,*[271] have mentioned no Case more deplorable than this, *for a Man to be sick of his Calling.*

Our Great GOD is to be seen, confessed, adored, in that admirable *Variety* of Matters which the *Invention of Man* has reach'd to! And the admirable *Sagacity* that prosecutes them! When such inventive Wits as *Helmont* and *Wallis* have taught the *Deaf* and the *Dumb* to *read* and *speak,* methoughts I have seen that *Sagacity* notably exemplified.

"Glorious GOD, my Soul with all possible Prostration before thee receives thy *faithful Sayings,* wherein thou hast instructed me: *Every good Gift comes down from the Father of Lights!*[272] And *the Lord giveth Wisdom!*[273] Not only of the *four Children*[274] that

---

267. "God causes his work to be pleasing to each person."
268. "Among dung and urine."
269. Isa. 54:16.
270. Adapted from Titus 3:14.
271. Bernardino Ramazzini, *De morbis artificium diatriba* (Modena, 1700).
272. Adapted from James 1:17.
273. Adapted from Dan. 2:21.
274. Daniel, Shadrach, Meshach, and Abednego, the wise children taken to Babylon's court.

had it, but of all that have ever had any thing of it, it must be own'd *God gave them Knowledge and Skill in all Learning and Wisdom:*[275] If a *Bazaleel*[276] have it, *O Spirit of God,* thou art He who givest him *Knowledge in all manner of Workmanship.*"[277]

But then there is another thing wherein the Superintendence of the Glorious Creator and Governor of the World is most conspicuous; and that is, the *Progress* which the *Invention* of Man has made: things of *greater* use were *sooner* invented, things of a *lesser* use *later,* every thing in the *Time* wherein our Great God has had his excellent Purposes to be served with it; things *equally plain* with such as have been formerly discovered, and as much desired, have been lock'd up from Human Understanding, till the God, *in whose hand are our Times,*[278] is pleased wisely to make them understood by the Children of Men. 'Tis not from your *fortuitous Concourse of Atoms,* ye foolish *Epicureans!* Why must *Printing* be withheld from the Service of Mankind till the Year 1430, when the *First-born of printed Books* was by the Hand of *Laurence Koster* midwifed into the World, and the Skill immediately improved by *Faust* and *Schoeffer?* Why must Mankind have no *Telescopes* till the Year 1609, when one whom *Syrturus*[279] would suspect almost an *Angel* in the Shape of a *Dutchman,* instructed *Lippersein* at *Middleburgh* to proceed upon them? To mention a Subject which my *Christian Philosopher* has very much liv'd upon, What is the Anatomy of *Mundinus,* if compared to our *modern?* (tho *Cardan,* and other learned Men, have so much cried it up with their Elogies and Commentaries.) *Baglivi* says truly, *'tis as far short of it as a Flea is of an Elephant.* We will pass to another Instance: The *Romans* had not so much as a *Sun-dial* till the second *Punick* War, and when they had one, they had no more than that one, in the *Forum,* above an hundred Years, tho *Pliny* says it never went right in all this time. Our King *Alfred*[280] had no better shift than this for measuring his Hours, the burning of a *Candle,* marked into *twelve*

---

275. Dan. 1:17.
276. Bezaleel was a craftsman of the tribe of Judah charged with construction of the tabernacle in the wilderness and its equipment. His skill as a versatile craftsman is attributed to his being filled with the Spirit of God.
277. Exod. 31:2, 3.
278. Adapted from Ps. 31:15.
279. Hieronymus Syrturus or Girolamo Sirtori published *Telescopium: sive ars perficiendi novum illud Galilaei visorium instrumentum ad sydera* (Frankfurt, 1618).
280. Alfred the Great (849–99).

parts, for which a *Lanthorn*[281] was needful to secure it from the Winds of the *Windows,* for *Glazing* was not yet in fashion. Dr. *Grew* observes, the first Conceit which tended to a *Watch,* was a *Draw-well;* first, People found the drawing of Water with a *Hand-cord* and a *Pitcher* troublesome, so they thought of a *Draught-wheel;* by and by they conceived such a Movement applicable to a *Spit,* if the motion of the Weight could be made slow enough, this was done by adding more *Wheels* and a *Flyer,* which made a *Jack:* by and by Men began to see, that if the motion were yet slower, it would serve to *measure Time* also, then instead of a *Flyer* they put a *Balance,* and thus made a *Clock;* this being so useful, Men considered how it might be made portable, by something answerable to a *Weight,* and so instead of that they put the *Spring* and the *Fuse-wheel,* which make a *Watch.* Here is the Pedigree of the noble Engine. But to what an astonishing Perfection is *Clock-work* and *Watch-work* now arrived! We will hardly allow a Gentleman of such Antiquity as *Boethius* to be the Inventor of the *Clock-work,* that hath been so mightily improved; no, *Regiomontanus,* thou shalt have the Honour of being the Instrument employed by God for the rare Invention, not more than between two and three hundred Years ago. The curious Performances of *Clock-work* cannot be related without our finding a Surprize of Pleasure in the Relations; how many *Motions* produced! How many *Designs* answer'd! The Gentlemen who writes *The Artificial Clock-maker,*[282] has with his Calculations made provision for a marvellous variety of them. What *Heylin*[283] in his *Cosmography* reports of the Clock at *Lunden* in *Denmark,* what *Gaffarel* in his *unheard-of Curiosities* reports that he himself *saw* in a *Clock* at *Leghorn,* and the *Clock* which every day diverts the Spectators at *Haarlem,* are notable Instances among many others. The *Repeating-Clocks* are now *common* on thousands of Tables, but how *curious!* At length Mr. *Huygens* has invented the way of applying *Pendulums* to *Watch-work.* If *Galilæo* entertained a Thought of such a thing, yet he never brought it to Perfection. We must not let Mr. *Huygens* be robb'd of his Claim,

14 *Leghorn / Ligorn*   15 *Haarlem / Harlem*

281. That is, lantern.
282. William Derham, *The Artificial Clock-maker: A Treatise of Watch and Clock-work* (London, 1714).
283. Peter Heylyn (1600–1662), champion of the Anglican Church against the Puritans, and author of *Cosmographie* (London, 1669).

either by *Becher* or the Academy *Del Cimento*.²⁸⁴ The first that was made in *England* was in the Year 1662. The Uses of these *Pendulum-Watches* cannot be sufficiently celebrated.

But useful indeed will be these *Measurers of Time,* if they teach and help us to be the more wise *Redeemers*²⁸⁵ of it.

It was thought, that he, who when Patents for *Monopolies* were granting in *France,* begg'd for one *to demand a Shilling from every Man who wore a Watch, but had no care how he spent his Time,* ask'd for what would have afforded a Revenue too rich for a Subject!

If the *Mathematicks,* which have in the two last Centuries had such wonderful Improvements, do for two hundred Years more improve in proportion to the former, who can tell what Mankind may come to! We may believe, without having *Seneca*²⁸⁶ our Author for it, *Multa venientis ævi populus ignota nobis sciet.*²⁸⁷

The Union between the SOUL and the BODY is altogether inexplicable, the *Soul* not having any *Surface* to touch the *Body,* and the *Body* not having any *Sentiment* as the *Soul.* The *Union* of the *Soul* and *Body* does consist, as Monsieur *Tauvry* expresses it, in the *Conformity* of our *Thoughts* to our *Corporeal* Actions; *but,* says he, *for the Explication of this Conformity, we must have recourse to a superior Power.* Truly, *Sirs,* do what you can, you must quickly come to *that!*

Our *nervous Parts* are very sensible. *Objects* do affect our *Senses,* and make Impressions on them; the *Senses* receiving such Impressions, the *Modifications of the Organs* produced by them terminate in the *Brain;* if they do not so, the *Soul* is unconcerned in them; but there is a *Law* given to the Soul by the glorious *God, who forms the Spirit of Man within him,*²⁸⁸ that in their doing so there shall be such and such *Thoughts* produced in the *Soul.*

"O *my Soul,* what a wondrous Being art thou! How capable of astonishing *Improvements!* How worthy to be cultivated with the best *Improvements!* How worthy to have all possible Endeavours used for thy *Recovery* from the *Depravations* which thy *Fall from God* has brought upon thee! How worthy to be *kept*

---

284. L'Accademia del Cimento or the Academy of Experiment, founded in Florence in 1657 under the patronage of Leopold, a liberal and intelligent prince, sought to encourage the development of scientific studies, especially physics. It dissolved in 1667 under some ecclesiastical pressure.
285. An allusion to "redeem the time"; see Eph. 5:16 and Col. 4:5.
286. That is, Seneca the Younger.
287. "Many things that are unknown to us the people of a coming age will know."
288. Adapted from Zech. 12:1.

*with all Diligence*²⁸⁹ from every thing that will bring any more *Wounds* upon thee! What *reason* is there that thou shouldst be filled with the *Love* of God, and acted by the *Faith* of thy only Saviour! And if the *Image* of the glorious God, which has been impaired by *Satanick* Impressions on thee, be revived and restored in thee, what marvellous, and even eternal *Felicities*, art thou sure of arriving to!"

But, O MAN, wilt thou stop here, and know nothing *above thy self*? Among the antient *Jews* there was a sort of *natural Philosophers*, who are by the *Rabbins* called חַכְמֵי הַסְּבְקָר, *Sapientes Inquisitionis*, or *Sapientes Scrutationis*,²⁹⁰ from their *enquiring* after *natural Causes;* perhaps our Apostle may mean these, when he says, 1 Cor. 1. 20. *Where is the Enquirer of this World? Jerome's* Version favours it.²⁹¹ Now of these Gentlemen it is reported, that they denied the Existence of *superior Intelligences;* our *Christian Philosopher* will not be guilty of such a Stupidity.

We are now soaring into the *invisible World*, a World of *intellectual Beings*, but invisible to such *Eyes* as ours. I do here in the first place most religiously affirm, that even *my Senses* have been convinced of such a World, by as clear, plain, full *Proofs* as ever any Man's have had of what is most obvious in the *sensible World;* *Proofs* which I am ready to offer in the most convenient Season. But then, *how glorious art thou, O God, in thy innumerable Company of the holy Angels, and in thy Government over those also that have made themselves evil ones!*²⁹² All the Wonders we have hitherto seen in the *visible Creation*, what are they, compared to those that are out of sight, those that are found among the *Angels that excel in Powers*, the Hosts of the infinite GOD, *the Ministers which do His Pleasure!*²⁹³

There is a *Scale of Nature*, wherein we pass regularly and proportionably from a *Stone* to a *Man*, the Faculties of the Creatures in their *various Classes* growing still brighter and brighter, and more capacious, till we arrive to those noble ones which are found

---

289. Adapted from Prov. 4:23.
290. "Wise at investigation"; "wise at examination."
291. 1 Cor. 1:20. The AV reads: "Where is the disputer of this world?" The RSV reads: "Where is the debater of this age?" The Vulgate, for the most part the work of Jerome, reads: "Ubi inquisitor huius saeculi?" ("Where is the enquirer of this age?")
292. This passage is in part an allusion to Heb. 12:22.
293. Adapted from Ps. 103:20–21.

in the *Soul* of MAN; and yet MAN is, as one[294] well expresses it, *but the Equator of the Universe.*

90 It is a just View which Dr. *Grew* had of *the World,* when he came to this Determination: "As there are several Orders of *animated Body* before we come to *Intellect,* so it must needs be that there are several Orders of *imbodied Intellect* before we come to *pure Mind.*"

95 It is likely that the Transition from *Human* to *perfect* MIND is made by a *gradual Ascent;* there may be *Angels* whose Faculties may be as much superior to *ours,* as ours may be to those of a *Snail* or a *Worm.*

99 By and by we may arrive to *Minds* divested of all *Body,* excellent *Minds,* which may enjoy the Knowledge of Things by a more *immediate Intuition,* as well as without any Inclination to any *moral Evil.*

The highest Perfection that any *created Mind* can arise to, is
5 that in the *Soul* of our admirable Saviour, which is indeed *embodied;* but it is the *Soul* of the *Man* who is personally united to the SON of GOD.

Anon we see an infinite GOD; but *canst thou by searching find out GOD? Canst thou find out the Almighty to Perfection?*[295]

10 It is a good Thought, and well expressed of an honest Writer on *the Knowledge of God from the Works of Creation.*[296] "It is true there are some *Footsteps* of a *Deity* in all the Works of Nature, but we should ascend by these *Footsteps* as by a *Footstool* to the *God* of the World, as *Solomon* by several Steps ascended to his
15 Throne, and by the *Scale of Nature* ascend to the *God* of *Nature.*"

This is what we shall now, tho in a more *summary way,* a little more distinctly proceed to.

No *Dominion over the Creatures*[297] can be more acceptably, more delightfully exercised with me than this; for me to *employ them* as
20 often as I please in *leading me to GOD,* and so in serving that which I propose as the chief END for which I *live,* and *move,* and have my *Being;*[298] which is, *to glorify GOD, and acknowledge Him.* When

---

294. That is, Nehemiah Grew.
295. Job 11:7.
296. Matthew Barker, *Natural Theology, or the Knowledge of God from the Works of Creation* (London, 1674).
297. An allusion to Gen. 1:26, 28 and Ps. 8:6.
298. Adapted from Acts 17:28.

the *Creatures* were brought to our *Protoplast*,²⁹⁹ *to see what he would call them*,³⁰⁰ he did not exercise a more desirable *Dominion* over them, in giving what *Name* he pleased to each of them, than I shall do in having them all brought to me, that I may read the *Name* of God, so far as it is to be seen in them, and be assisted in my *Acknowledgments* of the Glorious-ONE.

¶. *Hear now the Conclusion of the Matter*.³⁰¹ To enkindle the *Dispositions* and the *Resolutions* of PIETY in my Brethren, is the *Intention* of all my ESSAYS, and must be the *Conclusion* of them.

Atheism is now for ever chased and hissed out of the World, every thing in the World concurs to a Sentence of *Banishment* upon it. *Fly, thou Monster, and hide, and let not the darkest Recesses of Africa itself be able to cherish thee; never dare to shew thyself in a World where every thing stands ready to overwhelm thee!* A BEING that must be *superior* to *Matter*, even the *Creator* and *Governor* of all *Matter*, is every where so conspicuous, that there can be nothing more *monstrous* than *to deny the God that is above*.³⁰² No *System* of *Atheism* has ever yet been offered among the Children of Men, but what may presently be convinced of such *Inconsistences*, that a Man must ridiculously believe *nothing certain* before he can imagine them; it must be a *System* of *Things which cannot stand together!* A Bundle of *Contradictions* to themselves, and to all *common Sense*. I doubt it has been an *inconsiderate* thing to pay so much of a Compliment to *Atheism,* as to bestow solemn *Treatises* full of learned *Arguments* for the Refutation of a *delirious Phrenzy*, which ought rather to be put out of countenance with the most *contemptuous Indignation*. And I fear such Writers as have been at the pains to put the *Objections* of *Atheism* into the most plausible Terms, that they may have the honour of *laying a Devil when they have raised him*,³⁰³ have therein done too *unadvisedly*. However, to so much notice of the raving *Atheist* we may condescend while we go along, as to tell him, that for a Man to question the *Being* of a GOD, who requires from us an *Homage* of *Affection*, and *Wonderment*, and Obedience to Himself, and a perpetual Concern for the Welfare of the *Human Society,* for which He has in our *Formation* evidently

---

299. That is, Adam.
300. Gen. 2:19.
301. Adapted from Eccles. 12:13.
302. Adapted from Job 31:28.
303. That is, settling strife that has been stirred up. An adaptation of the proverb, "it is easier to raise the devil than to lay him."

suited us, would be an *exalted Folly,* which undergoes especially two Condemnations; it is first condemned by this, that every Part of the *Universe* is continually *pouring in* something for the *confuting* of it; there is not a Corner of the whole World but what supplies a *Stone* towards the Infliction of such a *Death* upon the *Blasphemy* as justly belongs to it: and it has also this condemning of it, that Men would soon become *Canibals* to one another by embracing it; Men being utterly destitute of any Principle to keep them *honest in the Dark,* there would be no *Integrity* left in the World, but they would be as the *Fishes of the Sea to one another,* and worse than *the creeping Things, that have no Ruler over them.*[304] Indeed from every thing in the World there is this Voice more audible than the loudest Thunder to us; *God hath spoken,*[305] *and these two things have I heard!* First, *Believe and adore a glorious GOD, who has made all these Things,*[306] *and know thou that He will bring thee into Judgment!*[307] And then *be careful to do nothing but what shall be for the Good of the Community which the glorious GOD has made thee a Member of.*[308] Were what God *hath spoken* duly regarded, and were these *two things* duly complied with, the World would be soon revived into a desirable *Garden of God,*[309] and Mankind would be fetch'd up into very comfortable Circumstances; till *then* the World continues in a wretched Condition, *full of doleful Creatures,* with *wild Beasts crying* in its *desolate Houses, Dragons* in its most *pleasant Palaces.*[310] And now declare, *O every thing that is reasonable,* declare and pronounce upon it whether it be possible that *Maxims* absolutely *necessary* to the *Subsistence* and *Happiness* of Mankind, can be *Falsities?* There is no possibility for this, that *Cheats* and *Lyes* must be so *necessary,* that the *Ends* which alone are worthy of a glorious GOD, cannot be attain'd without having *them* imposed upon us!

Having dispatch'd the *Atheist,* with bestowing on him *not many* Thoughts, yet *more* than could be deserved by such an *Idiot;* I will proceed now to propose two general Strokes of *Piety,* which will appear to a *Christian Philosopher* as unexceptionable as any Proposals that ever were made to him.

304. Adapted from Hab. 1:14.
305. Ps. 62:11.
306. Adapted from Jer. 14:22.
307. Eccles. 11:9.
308. The community is the body of Christ. See Rom. 12:4–5 and 1 Cor. 12:12–27.
309. That is, Eden. See Ezek. 28:13 and 31:8, 9.
310. Adapted from Isa. 13:21–22.

First, the Works of the glorious God exhibited to our View, 'tis most certain they do *bespeak,* and they should *excite* our *Acknowledgments of His Glories* appearing in them: the Great GOD is infinitely *gratified* in beholding the Displays of His own infinite *Power,* and *Wisdom,* and *Goodness,* in the Works which He has made; but it is also a most acceptable Gratification to Him, when such of His Works as are the *rational Beholders* of themselves, and of the rest, shall with devout Minds *acknowledge* His Perfections, which they see shining there. Never does one endued with *Reason* do any thing more evidently *reasonable,* than when he makes every thing that occurs to him in the vast Fabrick of the World, an *Incentive* to some agreeable Efforts and Salleys of *Religion.* What can any Man living object against the *Piety* of a Mind awaken'd by the sight of God in His Works, to such Thoughts as these: *Verily, there is a glorious GOD! Verily, the GOD who does these things is worthy to be feared, worthy to be loved, worthy to be relied on! Verily, all possible Obedience is due to such a GOD; and most abominable, most inexcusable is the Wickedness of all Rebellion against Him!* A Mind kept under the Impression of such Thoughts as these, is an *holy* and a *noble* Mind, a *Temple* of God, a *Temple filled with the Glory of God.*[311] There is nothing but what will afford an *Occasion* for the *Thoughts;* the oftner a Man improves the *Occasion,* the more does he *glorify* GOD, and answer the *chief End of Man;*[312] and why should he not *seek occasion* for it, by visiting for this purpose the several *Classes* of the Creatures (for *Discipulus in hac Schola erit Peripateticus*)[313] as he may have opportunity for so generous an Exercise! But since the horrid Evil of all *Sin* is to be inferred from this; *it is a Rebellion against the Laws of the glorious GOD, who is the Maker and the Ruler of all Worlds; and it is a disturbance of the good Order wherein the glorious Maker and Ruler of all Things has placed them all;*[314] how much ought a quickned *Horror of Sin* to accompany this Contemplation, and produce this most agreeable Resolution, *My God, I will for ever fear to offend thy glorious Majesty!* Nor is this all the *Improvement* which we are to make of what we see in the *Works of God* in our *improving* of them, we are to accept of the *Rebuke* which they give to our *Presumption,* in pretending to criticize upon the

---

311. An allusion to Phil. 2:5; 1 Cor. 6:19.
312. This echoes the Westminster Confession. See the Introduction, n. 4.
313. "A student in this school must be a Peripatetic."
314. An allusion to Rom. 1:18–32.

dark things which occur in the Dispensations of His *Providence;* there is not any one of all the *Creatures* but what has those *fine things* in the *Texture* of it, which have never yet been reached by our *Searches,* and we are as much at a loss about the *Intent* as about the *Texture* of them; *as yet* we know not what the glorious God *intends* in His forming of those *Creatures,* nor what *He has to do* in them, and with them; He therein proclaims this Expectation, *Surely they will fear me, and receive Instruction.*[315] And the Point wherein we are now instructed is this: "What! Shall I be so vain as to be *dissatisfied* because I do not *understand* what is done by the glorious GOD in the Works of His *Providence!" O my Soul, hast thou not known, hast thou not heard concerning the everlasting God, the Lord, the Creator of the Ends of the Earth, that there is no searching of His Understanding?*[316]

And then, secondly, the CHRIST of God must not be forgotten, who is *the Lord of all. I am not ashamed of the Gospel of CHRIST,*[317] of which I will *affirm constantly,*[318] that if the *Philosopher* do not call it in, he *paganizes,* and leaves the finest and brightest Part of his Work *unfinished.* Let *Colerus*[319] persuade us if he can, that in the Time of *John Frederick* the Elector of *Saxony*[320] there was dug up a *Stone,* on which there was a Representation of our *crucified Saviour;* but I cannot forbear saying, there is not a *Stone* any where which would not look *black* upon me, and *speak* my Condemnation, if my *Philosophy* should be so *vain* as to make me lay aside my Thoughts of my *enthroned Saviour.* Let *Lambecius,* if he please, employ his Learning upon the Name of our Saviour CHRIST, found in Letters naturally engraven at the bottom of a large *Agate-Cup,* which is to be seen among the Emperor's Curiosities; I have never drank in that *Cup,* however I can more easily believe it than I can the *Crucifixus ex Radice Crambes enatus,*[321] or the *Imago Virginis cum Filiolo in Minera Ferri expressa,*[322] and several more such things,

58 Crambes/Crambres

315. Adapted from Zeph. 3:7.
316. Adapted from Isa. 40:28.
317. Rom. 1:16.
318. Titus 3:8.
319. Probably either Johann Colerus (d. 1639), author of books on rural economy, or Christoph Coler (d. 1604), professor of history and politics at Altdorf.
320. John Frederick, called the Magnanimous (1503–54), elector of Saxony (1532–47) and a leader of the League of Schmalkalden.
321. "Crucifix sprouted from the root of a cabbage."
322. "Image of the Virgin and Child represented in iron ore."

which the Publishers of the *German Ephemerides* have mingled with their better Entertainments: but I will assert, that a glorious CHRIST is more to be considered in the *Works of Nature* than the *Philosopher* is generally aware of; and my CHRISTIAN *Philosopher* has not fully done his Part, till He who is *the First-born of every Creature*[323] be come into Consideration with him. *Alsted* mentions a *Siclus Judæo-Christianus*,[324] which had on one side the Name *JESUS*, with the Face of our *Saviour,* and on the other the Words that signify *the King Messiah comes with Peace, and God becomes a Man;* and *Leusden* says he had a couple of these *Coins* in his possession. I have nothing to say on the behalf of the *Zeal* in those *Christianized Jews,* who probably were the Authors of these *Coins,* a *Zeal* that *boil'd* into so needless an Expression of an Homage, that indeed cannot be too much expressed in the *instituted ways* of it to a Redeemer, whose *Kingdom is not of this World:*[325] but this I will say, *all the Creatures in this World are part of His Kingdom;*[326] there are no *Creatures* but what are His *Medals,* on every one of them the Name of JESUS is to be found inscribed. Celebrate, O *Danhaver,* thy *Granadilla,*[327] the *Peruvian Plant,* on which a strong Imagination finds a Representation of the *Instruments* employed in the *Sufferings* of our Saviour, and especially the *bloody Sweat* of His Agonies; were the Representation as really and lively made as has been imagined, I would subscribe to the Epigram upon it, which concludes:

> *Flos hic ita forma vincit omnes Flosculos,*
> *Ut totus optet esse Spectator Oculus.*[328]

But I will, with the Exercise of the most *solid Reason,* by every part of the World, as well as the *Vegetables,* be led to my Saviour.

78 *Granadilla / Granatilla*

323. Col. 1:15.
324. The reference is to a Jewish-Christian shekel or coin.
325. John 18:36.
326. An allusion to Gen. 1.
327. The Passion Flower (*Passiflora incarnata*), also known as the Granadilla. It was named by seventeenth-century Jesuit missionaries who discovered it in South America and related parts of the flower to the Passion of Christ. The spiky purple crown above the petals resembles the Crown of Thorns, while other parts resemble instruments of the crucifixion. The plant is native to the New and the Old World, but principally to tropical America. Many species are grown as ornamentals, but several are important economically for their edible fruits.
328. "This flower so exceeds all other blossoms in beauty, that a viewer would wish to be all eye."

A *View of the Creation*³²⁹ is to be taken, with suitable Acknowledgments of the glorious CHRIST, in whom the *eternal Son of God*³³⁰ has personally united Himself to ONE of His *Creatures*, and becomes on *his* account propitious to *all the rest;* our *Piety* indeed will not be *Christianity* if HE be left unthought upon.

This is HE, of whom we are instructed, *Col.* 1. 16, 17. *All things were created by Him, and for Him; and He is before all things, and by Him all things consist.* It is no contemptible Thought wherewith *De Sabunde* has entertained us: *Productio Mundi a Deo facta de Nihilo, arguit aliam productionem, summam, occultam, & æternam in Deo, quæ est de sua propria Natura, in qua producitur Deus de Deo, & per quam ostenditur summa Trinitas in Deo.*³³¹ And certainly he that as a *Father* does produce a *Son*, but as an *Artist* only produce an *House*, has a Value for the *Son* which he has not for the *House*; yea, we may say, if GOD had not first, and from Eternity, been a *Father* to our *Saviour*, He would never have exerted Himself as an *Artist* in that *Fabrick*, which he has built *by the Might of His Power, and for the Honour of His Majesty!*³³²

The Great Sir *Francis Bacon* has a notable Passage in his *Confession of Faith: I believe that God is so holy, as that it is impossible for Him to be pleased in any Creature, tho the Work of his own Hands, without beholding of the same in the Face of a Mediator; without which it was impossible for Him to have descended to any Work of Creation, but He should have enjoyed the blessed and individual Society of three Persons in the Godhead for ever; but out of His eternal and infinite Goodness and Love purposing to become a Creature, and communicate with His Creatures, He ordained in His eternal Counsel that one Person of the God-head should be united to one Nature, and to one particular of His Creatures; that so in the Person of the Mediator the true Ladder might be fixed, whereby God might descend to His Creatures, and His Creatures ascend to Him.*

It was an high Flight of *Origen*,³³³ who urges, that our *High-*

---

329. *A View of the Creation* is probably the subtitle of a work on physicotheology, but I am unable to identify the author and title.

330. An allusion to 1 John 5:20.

331. "The creation of the world, which was accomplished by God out of nothing, declares another creation, supreme, secret, and eternal in God, which is of his own nature, in which God is created from God, and by which is shown the Trinity in God."

332. Adapted from Dan. 4:30.

333. Origenes Adamantius (c. 185–c. 254), an Alexandrian Christian noted for extreme asceticism and profound learning, was the principal apologist of the early church.

Priest's having *tasted of Death*,³³⁴ ὑπὲρ παντός, FOR ALL, is to be extended even to the very *Stars,* which would otherwise have been *impure* in the sight of God; and thus are ALL THINGS restored to the *Kingdom* of the Father. Our Apostle *Paul* in a famous Passage to the *Colossians* [i. 19, 20.] may seem highly to favour this Flight. One says upon it, "If this be so, we need not break the Glasses of *Galilæo,* the *Spots* may be washed out of the *Sun,* and *total Nature* sanctified to God that made it."

Yea, the sacred Scriptures plainly and often invite us to a Conception, which Dr. *Goodwin* has chosen to deliver in such Terms as these: "The *Son of God* personally and actually existing as the Son of God with God, afore the World or any Creature was made, *He* undertaking and covenanting with God to become a *Man,* yea, *that Man* which He hath now taken up into one Person with Himself, as well for *this End,* as for *other Ends* more glorious; God did in the Fore-knowledge of *that,* and in the Assurance of that *Covenant* of His, proceed to the *creating* of all things which He hath made: and without the Intuition of *this,* or having *this* in His Eye, He would not have made any thing which He hath made."

O CHRISTIAN, *lift up now thine Eyes, and look from the place where thou art*³³⁵ to all Points of the Compass, and concerning *whatever thou seest,*³³⁶ allow that all these things were formed *for the Sake* of that Glorious-One, who is now *God manifest in the Flesh*³³⁷ of our JESUS; 'tis on *His* Account that the eternal Godhead has the *Delight* in all these things, which preserves them in their Being, and grants them the *Help,* in the *obtaining* whereof they *continue to this day.*³³⁸

But were they not all made *by the Hand,*³³⁹ as well as *for the Sake* of that Glorious ONE? They were verily so. *O my JESUS, it was that Son of God who now dwells in thee, in and by whom the Godhead exerted the Power, which could be exerted by none but an all-powerful GOD, in the creating of the World!* He is that WORD of GOD *by*

---

334. An allusion to Heb. 2:9.
335. Gen. 13:14.
336. Adapted from Rev. 1:11.
337. Adapted from 1 Tim. 3:16.
338. Adapted from Acts 26:22.
339. The notion that the hand of God made all things is frequently expressed in the Bible. It is clearly stated in Acts 7:50.

whom all things were made, and without whom was not any thing made that was made.[340]

This is not all that we have to think upon; we see an incomparable *Wisdom* of GOD in His *Creatures;* one cannot but presently infer, *What an incomprehensible Wisdom then in the Methods and Affairs of that Redemption, whereof the glorious GOD has laid the Plan in our JESUS!* Things which the *Angels desire to look into.*[341] But, O *evangelized Mind,* go on, mount up, soar higher, think at this rate; *the infinite Wisdom which formed all these things is peculiarly seated in the Son of God;* He is that *reflexive Wisdom* of the eternal *Father,* and that *Image of the invisible God, by whom all things were created;*[342] in *Him* there is after a peculiar manner the original *Idea* and *Archetype* of every thing that offers the infinite *Wisdom* of God to our Admiration. Wherever we see the *Wisdom* of God admirably shining before us, we are invited to such a Thought as this; *this Glory is originally to be found in thee, O our Immanuel!*[343] 'Tis in Him *transcendently.* But then 'tis impossible to stop without adding, *How glorious, how wondrous, how lovely art thou, O our Saviour!*

Nor may we lay aside a grateful Sense of this, that as the *Son of God* is *the Upholder of all Things in all Worlds,*[344] thus, that it is owing to his potent *Intercession* that the *Sin of Man* has made no more havock on this *our World.* This *our World* has been by the *Sin of Man* so perverted from the *true Ends* of it, and rendred full of such loathsome and hateful Regions, and such *Scelerata Castra,*[345] that the Revenges of God would have long since rendred it as a *fiery Oven,* if our blessed JESUS had not *interceded for it.* O my Saviour, what would have become of me, and of all that comforts me, if thy Interposition had not preserved us!

We will add one thing more: Tho the one GOD in His *three Subsistences* be the *Governor* as well as the *Creator* of the World, and so the *Son* of God ever had what we call the *natural Government* of the World, yet upon the *Fall* of Mankind there is a *mediatory Kingdom* that becomes expedient, that so *guilty Man,* and that which was *lost,* may be brought to God; and the singular Honour of this

---

340. An allusion to John 1:1–5; the last two lines are adapted from John 1:3.
341. Adapted from 1 Pet. 1:12.
342. Adapted from Col. 1:15–16.
343. Perhaps an allusion to Isa. 7:14, 8:8 and Matt. 1:23.
344. Perhaps an allusion to Heb. 1:3.
345. "Accursed regions." Suetonius used the phrase *castra scelerata* to relate that Drusus, the father of Claudius, died in his summer camp, which was thereafter called "accursed." See *Life of the Deified Claudius,* 5.1.

mediatory Kingdom is more *immediately* and most *agreeably* assign'd to the *Son of God,* who assumes the Man JESUS into His own Person, and has *all Power in Heaven and Earth given to Him;*³⁴⁶ all things are now commanded and ordered by the *Son of God* in the *Man upon the Throne,* and this *to the Glory of the Father,* by whom the *mediatory Kingdom* is erected, and so conferred. This *peculiar Kingdom* thus managed by the *Son of God* in our JESUS, will cease when the illustrious Ends of it are all accomplished, and *then* the *Son of God* no longer having such a *distinct Kingdom* of His own, shall return to those eternal Circumstances, wherein He shall reign with the *Father* and the *Holy Spirit,* one God, blessed for ever. In the mean time, what Creatures can we behold without being obliged to some such Doxology as this; *O Son of God, incarnate and enthroned in my JESUS, this is part of thy Dominion! What a great King art thou,*³⁴⁷ *and what a Name hast thou above every Name,*³⁴⁸ *and how vastly extended is thy Dominion!*³⁴⁹ *Dominion and Fear is with thee, and there is no Number of thine Armies!*³⁵⁰ *All the inhabitants of the Earth, and their most puissant Emperors, are to be reputed as nothing before thee!*³⁵¹

But then at last I am losing myself in such Thoughts as these: *Who can tell* what *Uses* our Saviour will put all these *Creatures* to at the *Restitution of all things,*³⁵² when He comes to rescue them from the *Vanity* which as yet captivates them and incumbers them; and His raised People in the *new Heavens* will make their Visits to a *new Earth,*³⁵³ which they shall find flourishing in *Paradisiack* Regularities? *Lord, what thou meanest in them, I know not now, but I shall know hereafter!*³⁵⁴ I go on, *Who can tell* how sweetly our Saviour may *feast* His *chosen People* in the *Future State,* with Exhibitions of all these *Creatures,* in their various *Natures,* and their curious *Beauties* to them? *Lord, I hope for an eternally progressive*

13 Paradisiack/Paradisaick

346. Adapted from Matt. 28:18.
347. Adapted from Pss. 47:2, 95:3.
348. Adapted from Phil. 2:9.
349. Perhaps an allusion to Dan. 4:34.
350. Adapted from Job 25:2–3.
351. Adapted from Dan. 4:35.
352. Adapted from Acts 3:21.
353. Adapted from 2 Pet. 3:13 and Isa. 65:17, 66:22.
354. Adapted from John 13:7.

Knowledge, from the Lamb of God successively leading me to the Fountains of it!

I recover out of my more *conjectural Prognostications,* with resolving what may *at present* yield to a serious Mind a *Satisfaction,* to which this World knows none superior: When in a way of *occasional Reflection* I employ the *Creatures* as my *Teachers,* I will by the *Truths* wherein those ready *Monitors* instruct me, be led to my glorious JESUS; I will consider the *Truths as they are in JESUS,*[355] and count my *Asceticks*[356] deficient, till I have some Thoughts of HIM and of His *Glories* awakened in me. To conclude, It is a good Passage which a little Treatise entitled, *Theologia Ruris,* or, *The Book of Nature,*[357] breaks off withal, and I might make it my Conclusion: "If we mind *Heaven* whilst we live here upon *Earth,* this *Earth* will serve to conduct us to *Heaven,* thro the Merits and Mediation of the *Son of God,* who was made the *Son of Man,* and came thence on purpose into this lower World to convey us up thither."

I will finish with a Speculation, which my most valuable Dr. *Cheyne* has a little more largely prosecuted and cultivated.

All *intelligent compound Beings* have their whole Entertainment in these three Principles, the DESIRE, the OBJECT, and the SENSATION arising from the *Congruity* between them; this *Analogy* is preserved full and clear thro the *Spiritual World,* yea, and thro the *material* also; so *universal* and *perpetual* an *Analogy* can arise from nothing but its *Pattern* and *Archetype* in the infinite God our Maker; and could we carry it up to the Source of it, we should find the TRINITY of Persons in the eternal GODHEAD admirably exhibited to us. In the GODHEAD we may first apprehend a *Desire,* an infinitely active, ardent, powerful *Thought,* proposing of *Satisfaction;* let this represent GOD the FATHER: but it is not possible for any Object but God Himself to *satisfy Himself,* and fill His *Desire* of Happiness; therefore HE Himself *reflected* in upon Himself, and contemplating His own infinite Perfections, even the *Brightness of His Glory,* and the *express Image of His Person,*[358] must answer this glorious Intention; and this may represent to us GOD the SON.

43 our/or

355. Adapted from Eph. 4:21.
356. See n. 228 above.
357. *Theologia ruris, sive schola et scala naturae: Or, The Book of Nature* (London, 1686).
358. Heb. 1:3.

Upon this Contemplation, wherein GOD Himself does behold, and possess, and enjoy Himself, there cannot but arise a *Love,* a *Joy,* an *Acquiescence* of God Himself within Himself, and worthy of a God; this may shadow out to us the third and the last of the Principles in this *mysterious Ternary,*[359] that is to say, the Holy SPIRIT. Tho these *three Relations* of the Godhead in itself, when derived analogically down to Creatures, may appear but *Modifications* of a *real Subsistence,* yet in the supreme Infinitude of the Divine Nature, they must be infinitely *real* and *living* Principles. Those which are but *Relations,* when transferred to *created Beings,* are glorious *Realities* in the infinite God. And in this View of the Holy Trinity, low as it is, it is impossible the SON should be without the FATHER, or the FATHER without the SON, or both without the Holy SPIRIT; it is impossible the SON should not be necessarily and eternally begotten of the FATHER, or that the Holy SPIRIT should not necessarily and eternally proceed both from Him and from the SON. Thus from what occurs throughout the whole Creation, *Reason* forms an imperfect Idea of this incomprehensible Mystery.

But it is time to stop here, and indeed how can we go any further!

F I N I S.

---

64 *Realities / Relatives*

359. That is, a set or group of three. The word is now obsolete.

ENDNOTES

INTRODUCTION

7:14–15 The Westminster Catechism was prepared by the Westminster Assembly of Divines simultaneously with the Confession of Faith, the classic English expression of Calvinist thought. The Massachusetts churches accepted the doctrinal part of the Confession at the Cambridge Synod in 1648.

7:19–22 Plato's *Timaeus* gave expression to the notion that the visible, living world is patterned on the universal idea of the living thing, a product of intelligent design and beneficent purpose. Plutarch reinforced and popularized the Platonic teaching. These authors complemented the Judeo-Christian doctrine of God as creator of the universe. On the distinction between Creator and Architect, see Pierre Janet, *Final Causes,* trans. William Affleck (Edinburgh, 1878), p. 332.

9:28–30 Mather's source is an annotation on Ps. 29 in Sebastian Münster, *Hebraica Biblia,* 2 vols. (Basel, 1534–35). The Latin quotation, from the notes to a commentary by Rabbi David Kimchi, is in the 2d ed. (Basel, 1546), 2:1182.

9:40–42 The source is Augustine, *Contra Julianum Pelagianum,* 4.15, 75 (PL, 44:776). Mather omits three words from the original. The translation is from *The Fathers of the Church,* vol. 35: *Saint Augustine: Against Julian,* trans. Matthew A. Schumacher (New York: Fathers of the Church, 1957), p. 231. I am indebted to J. van Sint Feijth of the *Thesaurus Linguae Augustinianae,* Eindhoven, The Netherlands, and Robert P. Russell, O.S.A., of the Augustinian Institute at Villanova University, Villanova, Pennsylvania, for the location of this passage.

9:44–50 Mather loosely quotes Jean de La Bruyère, *The Characters, Or the Manners of the Age* (London, 1699), p. 324.

10:54 The source is Pliny, *Natural History,* preface, sect. 21. My translation.

10:62–63 See Martial, *Epigrams,* 1.29, 38, 53, 72 for the charges of plagiarism.

10:73 The Jansenists were a group of seventeenth-century Roman Catholics who defended Augustinian and Thomist teachings on original sin and the role of divine grace in effecting salvation in opposition to the more optimistic view of human nature advanced by the Jesuits. They took their name from Cornelius Jansen, a Fleming whose *Augustinus* was published at Louvain in 1640. Their center was a former convent at Port-Royal des Champs. Both Louis XIV and the papacy opposed Jansenism, and in 1665 authorities dispersed the community.

10:84–11:85 On Mather's American communications see Raymond P. Stearns, *Science in the British Colonies of America* (Urbana: University of Illinois Press, 1970), pp. 156, 405–26; Otho M. Beall, Jr., "Cotton Mather's Early 'Curiosa Americana' and the Boston Philosophical Society of 1683," *William and Mary Quarterly,* 3d ser., 18 (July 1961), 360–72; and Kittredge.

11:85 On this subject see Walter E. Houghton, Jr., "The English Virtuoso in the Seventeenth Century," *Journal of the History of Ideas,* 3 (Jan.–Apr. 1942), 51–73, 190–219. Robert Boyle, *The Christian Virtuoso: Shewing, that by Being Addicted to Experimental Philosophy, A Man is Rather Assisted, Than Indisposed, to Be a Good Christian* ([London], 1690), went far in equating the word *virtuoso* with the word *scientist* (although the latter word first came into use in the nineteenth century). The library of the Mathers had a copy of Boyle's work. Mather's manuscript of the present book originally was entitled "The Christian Virtuoso."

11:92–93 On the early literature dealing with "the works of the six days," see Frank E. Robbins, *The Hexaemeral Literature: A Study of the Greek and Latin Commentaries on Genesis* (Chicago: University of Chicago Press, 1912).

11:97–98 Mather quotes from a commentary by Rabbi David Kimchi in the second edition of Münster's *Hebraica Biblia,* 2:1185. My translation.

11:4–5 I am unable to identify the source of the quotation. My translation.

11:7–12:39 Ibn Tufayl's philosophical romance described a religion of nature before the notion became familiar to Christians. The hero of the book, cast on a desert island as a child, grows up by merely contemplating the physical creation to discover the sciences, the existence of the soul, and God. When in mature life he comes across the revelation in the Koran, he is willing to accept it as merely confirming what he has already learned by the light of nature. The younger Edward Pococke (1648–1727) published the complete Arabic text together with a Latin translation entitled *Philosophus autodidactus: sive epistola Abi Jaafar Ebn Tophail de Hai Ebn Yokdhan* (Oxford, 1671). Mather used an anonymous translation by George Ashwell of Pococke's Latin version entitled *The History of Hai Eb'n Yockdan, An Indian Prince: Or, the Self-Taught Philosopher* (London, 1686). This was the first English translation. Mather takes his material from pp. 85, 86–87, 111–12, slightly modifying the original and adding italics. The idea of the self-improvement of man in the state of nature was congenial to Western thought in the late seventeenth and eighteenth centuries, and other translations of Ibn Tufayl's book followed in

English, Dutch, German, and Hebrew. Simon Ockley's translation from the original Arabic was published as *The Improvement of Human Reason, Exhibited in the Life of Hai Ebn Yokdhan* (London, 1708). The *Risala*, as the book is known in Arabic, has been regarded as the parent of Daniel Defoe's *Robinson Crusoe*, the first part of which appeared in 1719.

12:39–40 Based on Ausonius of Bordeaux (d. c. 395), *Gratiarum actio ad Gratianum imperatorem pro consulatu*, 4 (Loeb ed., 2:228).

12:44 The source is Alsted, p. 25. My translation.

12:48–13:51 Mather apparently quotes Minucius Felix, *Octavius*, 17.6, but he does not closely follow any of the four English translations published between 1636 and 1708.

Religio Philosophica

17:16–18 The source is *Theophilus of Antioch ad Autolycum*, text and trans. by Robert M. Grant (Oxford: Clarendon Press, 1970; originally published Oxford, 1684), 2.12. Grant's translation reads "ten thousand tongues" and "ten thousand years."

17:21–18:24 John Chrysostom wrote that God instructs man by means of both the Scriptures and the creation. See *The Homilies of S. John Chrysostom, Archbishop of Constantinople, on the Statutes, Or, To the People of Antioch* (Oxford, 1856), Homily 9 (pp. 161–63); Homily 10 (p. 175), Homily 12 (pp. 203–4, 208–13).

18:26–27 Mather's reference is from Matthew Barker, *Natural Theology: Or, the Knowledge of God, from the Works of Creation* (London, 1674), p. 17. Barker is probably based on two classical accounts which relate that a "certain philosopher" asked Anthony of Egypt how he endured being deprived of the comfort of books, and the latter answered, "My book is the nature of things that are made." See *Vitae Patrum: sive historiae Eremeticae libri decem*, 6.4, 16 (PL, 73:1018), with an English translation by Helen Waddell, *The Desert Fathers* (New York: Henry Holt and Co., 1936), p. 176, and Socrates (c. 380–450) "Scholasticus," *Historia ecclesiastica*, 4.23 (PG, 67:518), with an English translation by A. C. Zenos, *Church History from A.D. 305–439*, in *A Select Library of Nicene and Post-Nicene Fathers of the Christian Church*, 2d ser. (New York, 1890), 2:107. A variant account is in Athanasius, *The Life of Saint Anthony*, trans. Robert T. Meyer, in *Ancient Christian Writers: The Works of the Fathers in Translation*, ed. Johannes Quasten and Joseph C. Plumpe, no. 10 (Westminster, Md.: Newman Press, 1950), pp. 80–81.

18:31–33 The source is Paul Egard, Γνωθι Σεαυτον: *sive Tractatus utilissimus de vera microcosmi cognitione tum naturali tum supernaturali: vel de scientia illa divina . . . qua homo seipsumcognoscit secundam tum naturam gratiam vel tum in Adamo, tum in Christo* (Hamburg, 1621), p. 15. My translation. Mather greatly admired Egard and used the latter's writings as "bedbooks" out of which to read to his wife for entertainment and profit before rising in the morning. See Mather, *Diary,* 2:395, 396, 438, 440.

1. OF THE LIGHT

18:44–19:58 Mather's source is Cheyne, chap. 1, pp. 7, 16, 20, 55. Cheyne cites both Newton's *Principia Mathematica* (1687) and *Opticks* (1704).

19:65 Mather probably takes the quotation from Derham 2, p. 29. The ultimate source is Seneca, *Natural Questions,* bk. 1, preface, 13.

19:66–68 Mather's point is stressed in Arndt, bk. 4, pt. 1 (2:317–25).

19:71–75 Mather's source is Harris, *LT,* 2:s.v. "Light." Harris wrote that light "is the inworking of the Diaphanous Body." The original is in Robert Hooke, "Lectures on Light," in *The Posthumous Works of Robert Hooke* (London, 1705), pp. 75–76. The source in Aristotle is *On the Soul,* 2.7.418b.9–10. In translation Aristotle reads: "Now light is the activity of this transparent substance *qua* transparent."

19:76–20:84 Mather apparently condenses this paragraph from Molyneux, *Dioptrica nova,* pp. 198–200. Molyneux cites Newton's *Principia.*

20:85–21:42 These eight paragraphs paraphrase Cheyne, chap. 1, pp. 72–95 passim. Cheyne draws on Newton and cites Christiaan Huygens, *Cosmotheoros: sive de terris coelestibus, earumque ornatu, conjecturae* (The Hague, 1698), translated as *The Celestial Worlds Discover'd: Or, Conjectures Concerning the Inhabitants, Plants and Productions of the Worlds in the Planets* (London, 1698). The data on Römer (21:29–33) is probably based on Molyneux, *Dioptrica nova,* p. 199.

21:44–47 Mather apparently takes his information on Dee and Cardan from John Dee, *The Elements of Geometrie of the Most Auncient Philosopher Euclide,* trans. by H. Billingsley (London, [1570]), p. ciiii verso.

21:47–52 Stevin had written about a machine called the "Almighty" (Pancration) because, as he put it, it was "of infinite force, that is to say virtually, not actually." This instrument contained a wheel and an axle which was driven by a system of cog wheels powered by a crank. Stevin had determined how far the earth could be moved, assuming a place on which to stand for the purpose, "by turning the crank during ten years 4,000 times a minute." E. J. Dijksterhuis, *Simon Stevin: Science in the Netherlands around 1600* (The Hague: Martinus Nijhoff, 1970), pp. 60–61.

21:55 The ultimate source is Cardan, *De rerum varietate* (*The Variety of Things*), in *Opera omnia,* 10 vols. (Leiden, 1663), 3:186. Mather substitutes "*quietes*" for "*quae res*" in the original.

22:56–78 Mather quotes Richard Blackmore, *Creation: A Philosophical Poem* (London, 1712), 2.386–407 (pp. 75–76, 1712 ed.). Mather corresponded with Blackmore and greatly admired his philosophical poem. See Mather, *Diary,* 2:105, 141.

22:83–23:17 Mather takes this long paragraph, except for the two biblical quotations, from Alsted, pp. 234–40, 244–45. Alsted repeats part of this information in his *Encyclopaedia scientiarum omnium,* 4 vols. in 2 (Leiden, 1649), 3:327. Alsted cites sources for only Hugh of St. Victor and Stigelius, but his reference to Hugh's *De arca,* 2.3, is inaccurate.

22:86 Mather had earlier written that Hugh tells us there is a threefold voice of all the creatures unto mankind, "*Accipe, Redde, Fuge.*" See Mather's *Christianus per Ignem: Or, a Disciple Warming of Himself and Owning of His Lord* (Boston, 1702), pp. 13–14.

23:91 Adrianus Heereboort (1613–61), Dutch professor of philosophy, published *Meletemata philosophica* (Leiden, 1654). The book was used as a text in ethics at Harvard College in the seventeenth century.

23:99–4 This concept appears in *Plutarch's De Iside et Osiride* (Of Isis and Osiris), ed. J. Gwyn Griffiths ([Cardiff]: University of Wales Press, 1970), pp. 205–7.

23:5 The ultimate source is Cicero, *Tusculan Disputations,* 1.28, 70. Mather alters the original.

23:14–17 Stigelius, *Epist. ad Eberum,* as quoted in Alsted, p. 244. Johann Stigel (1515–62), professor of German and Latin at Wittenberg, was an evangelical humanist who wrote Latin verse. Alsted quotes eight lines of Stigel. Mather takes lines 1, 2, 7, and 8, changing "terris" to "Rebus" in line 3 (7).

2. OF THE STARS

26:82–83 Huygens advanced this thesis in *Celestial Worlds Discover'd,* pp. 149–50.

26:88–93 Mather quotes George Buchanan, *De sphaera* (1584), 1.650–52, 678, in *Opera omnia* (Edinburgh, 1715), pp. 126–27. See James R. Naiden, trans., *The "Sphera" of George Buchanan (1506–1582): A Literary Opponent of Copernicus and Tycho Brahe* (n.p., 1952), p. 108.

26:94–4 Mather takes this information from Joshua Childrey, *Britannia Baconica: Or, the Natural Rarities of England, Scotland, and Wales. . . . Historically Related, According to the Precepts of the Lord Bacon* (London, 1661), pp. 183–84. Mather discusses the same point in a letter of 24 November 1712 to Richard Waller, then secretary of the Royal Society, calling the phenomenon first noticed by Childrey "the *Evening Glade.*" This letter, the eighth in the first series of Mather's "Curiosa Americana," is described in "An Extract of Several Letters from Cotton Mather, D.D. to John Woodward, M.D. and Richard Waller, Esq; S.R. Secr.," *Phil. Trans.,* 29 (Apr.–June 1714), 65–66. See also Kittredge, p. 25.

324 *Endnotes*

26:5–27:10 The Jewish "fancy" goes as follows: "In the east, the west, and the south, heaven and earth touch each other, but the north God left unfinished, that any man who announced himself a god might be set the task of supplying the deficiency, and stand convicted as a pretender." Quoted from Louis Ginzberg, *The Legends of the Jews*, trans. Henrietta Szold, 6 vols. (Philadelphia: Jewish Publication Society of America, 1914), 1:12.

27:14–16 The source is Walker, p. 1.

27:23–38 The source is Walker, pp. 2–3.

27:39–41 Mather could have taken this quotation (my translation) from either Arndt, bk. 4, pt. 2, chap. 4 (2:327) or from Alsted, p. 265. Alsted does not identify the ancient author on p. 265, but on p. 244 he attributes the idea to Plutarch, *De Iside*. In treating natural theology in his *Encyclopaedia scientiarum omnium*, 3:327, however, Alsted names Plato as the source of the concept. He writes (in my translation): "Plato correctly said in *Timaeus*: 'The world is an epistle of God the Father to the human race. And gentile philosophers and Christian savants rightly advise that the creatures are mirrors, in which we observe the singular art of him who created the world.'" Plato actually wrote in *Timaeus* 37C: "When the father who had begotten it [the physical creation] saw it set in motion and alive, a shrine brought into being for the everlasting gods, he rejoiced and being well pleased he thought to make it yet more like its pattern." See Francis M. Cornford, *Plato's Cosmology: The Timaeus of Plato Translated with a Running Commentary* (London: Kegan Paul, Trench, Trubner, and Co., 1937), p. 97. I am indebted to Professor Richard Mohr of the University of Illinois at Urbana-Champaign for help on Plato and Plutarch.

27:42–28:48 Mather probably takes the quotation from Ray, pp. 76–77. Derham 2, p. 104, also quotes the passage, but only in English translation. The ultimate source is Cicero, *De natura deorum*, 2.21, 56.

28:49–56 The source is Harris, *LT*, 2:s.v. "Fixed *Stars*."

28:59–61 Mather quotes Arndt, bk. 4, pt. 1, chap. 4, 9 (2:357). Boehm's translation.

3. OF THE FIXED STARS

29:85–87 Mather's source is Walker, pp. 1, 5.

29:87–94 Bayer reformed the method of naming the stars visible to the naked eye which had prevailed since antiquity, but Mather's account of the number of stars and constellations known by earlier astronomers is not entirely accurate. The earliest astronomers defined star positions in terms of constellations imagined as legendary creatures and objects. Hipparchus (fl. second century B.C.) was the first to number and name the fixed stars, and his star catalog contained a large number of stars. According to David Gregory, Hipparchus counted the number at 1,026 (Gregory, *The Elements of Astronomy, Physical and Geometrical. To which is Annex'd Dr. Halley's Synopsis of the Astronomy of Comets*, 2 vols. [London, 1715], 1:303), though F. Boll, writing in 1900 (as cited by G. J. Toomer, "Hipparchus," in the *DSB*, 15:222), states that Hipparchus put the number at 850.

Ptolemy, who compiled a star catalog around A.D. 150 partly based on that of Hipparchus, organized the stars by constellations and described each by its position within the constellation figure. His catalog covered 1,028 (1,025 plus 3 duplicates) visible to the naked eye. The stars are grouped into 48 constellations, 12 in the zodiac, 21 to the north, and 15 to the south.

The Ptolemaic star catalog remained the standard through the late sixteenth century, with all new catalogs being simply revisions of Ptolemy. Bayer introduced a new departure. His novel method assigned to each star in a constellation one of the twenty-four letters of the Greek alphabet or, where a larger number of stars required it, a letter of the Latin alphabet. Bayer placed the Greek and Latin letters on his star charts; he also reproduced the traditional numeration of the stars in the constellations as well as the names used by Ptolemy and his successors. Bayer devised a stellar nomenclature still used by astronomers for most stars visible to the naked eye. In the explanation to his maps, Bayer suggests that there are 1,706 stars of magnitudes 1 to 6, plus about 325 scattered or unformed stars. The popularity of Bayer's *Uranometria* was further enhanced by a plate which displayed twelve new southern constellations which had recently been defined by the Dutch navigator Pietr Dirksz Keyser. See Deborah J. Warner, *The Sky Explored: Celestial Cartography, 1500–1800* (New York: Alan R. Liss, 1979), pp. ix, 18–19.

Kepler listed more than 1,440 stars in his *Tabulae Rudolphinae* (Ulm, 1627), which included Tycho Brahe's catalog of 1,000 fixed stars. The latest star catalog known when Mather wrote was Johannes Hevelius (1611–87), *Prodoromus astronomiae* (Danzig, 1690), which contained 1,888 stars in all (Gregory, *Elements of Astronomy*, 1:308). The *DSB* account of Hevelius gives 1,564 as the number.

29:94–99 Mather bases these lines on Walker, pp. 5, 6.
29:3–6 Mather's source is Derham 2, pp. xlvi–xlvii.
29:7–9 Mather's source is Walker, p. 6.
30:10–15 Here Mather quotes Cheyne, chap. 3, p. 109.
30:17–33 Mather bases this material on Cheyne, chap. 3, pp. 114–15, 116–17.
30:36–32:82 Mather relies on Walker, pp. 23–30, for this information. But the dates on the appearance of new stars are incorrect. The nova

in Cassiopeia occurred in 1572, the nova in Cygnus (Swan's Breast) in 1600, and another nova appeared in 1670 rather than 1671. The correct dates are given in Gregory, *Elements of Astronomy,* 1:311, 312, 313.

32:83–87 Here Mather follows Derham 2, p. 46.

32:88–5 Mather quotes Cheyne, chap. 3, pp. 137, 138, slightly modifying the quotation and adding italics.

32:6–7 Here Mather follows Derham 2, pp. 22–23.

32:8–10 Here Mather draws on Grew, p. 6.

32:11–34:21 The Reformation and Counter-Reformation spurred attempts to depaganize the heavens from about 1580 to 1630, and Schiller's atlas, which was essentially a revision of Bayer's *Uranometria,* redefined all of the constellations, substituting biblical figures for ancient mythological figures. Schiller, a Catholic and like Bayer a resident of Augsburg, was a cartographer rather than astronomer. Bayer undertook the astronomical revisions, which were based on the latest astronomical information, while Schiller, after corresponding with Jesuit priests, converted the Greco-Roman constellations into Judeo-Christian constellations. In Schiller's atlas, New Testament figures represent the constellations in the northern hemisphere, the twelve apostles the signs of the zodiac, and Old Testament figures represent signs in the southern hemisphere. Schiller depicted the Milky Way as All Saints' Way and identified the sun with Christ, the moon with the Virgin Mary, etc. Schiller's effort to Judeo-Christianize the heavens was the most radical of several proposed, and it did not gain sufficient popularity to take hold. But there is no good reason why celestial names cannot be changed, and some biblically named constellations which date from this period—for example, Crux and Columba (Noah's Dove)—are recognized today by the International Astronomical Union. See Warner, *Sky Explored,* pp. xi–xii, 229–32.

34:22–32 This paragraph is based on Jacques Gaffarel, *Unheard-of Curiosities: Concerning the Talismanical Sculpture of the Persians, the Horoscope of the Patriarkes, and the Reading of the Stars* (London, 1650), pp. 386–87.

4. OF THE SUN

35:64–65 Plato relates the episode, *Apology,* 26D, as does Diogenes Laertius, *Lives of Eminent Philosophers,* 2.8, 12.

35:65–69 Mather takes this material from Walker, p. 8. Kircher's comments on the sun are in *Iter exstaticum coeleste* (Würzburg, 1671), p. 186, and *Mundus subterraneus in XII libros digestus,* 2 vols. in 1 (Amsterdam, 1664–65), bk. 2, pp. 57–62.

35:70–75 Mather bases this material on Hooke, *Posthumous Works*, pp. 91, 89, 94.

35:75–37:86 This material is taken from Harris, *LT*, 2:s.v. "Sun." Harris is paraphrasing Newton's *Opticks* (1704).

37:87–9 This material, except for the clause ending the first paragraph (37:91–92) is taken from Walker, pp. 7–9. Walker asserts that sunspots revolve in twenty-six or twenty-seven days, whereas Mather writes twenty-five days. Here he may follow Harris, *LT*, 1:s.v. "Sun," or Derham 2, p. 74.

37:7–9 The reference is to Vergil, *Georgics*, 1.463–68, and Ovid, *Metamorphoses*, 15.785–86. This darkness is said to have taken place in 44 B.C., the year of Caesar's assassination.

37:10–38:17 This information is a close paraphrase of Harris, *LT*, 1:s.v. "Sun." Derham 2, pp. 90–91 n. 2, repeats the last sentence.

38:22–25 This paragraph is based on Harris, *LT*, 1:s.v. "Sun."

38:26–27 These lines are probably from Derham 2, p. 19 n. 2, which gives 172,102,795 miles.

38:28–37 This paragraph is based on Derham 2, pp. 12–13.

38:38–39:54 Mather's source is Grew, pp. 7–8.

39:55–68 Mather paraphrases Harris, *LT*, 1:s.v. "Heat."

39:69–79 The source is Cheyne, chap. 3, pp. 142–43.

39:80–40:88 Mather paraphrases Robert Wittie, Ουρανοσκοπια: *Or, A Survey of the Heavens. A Plain Description of the Admirable Fabrick and Motions of the Heavenly Bodies as They are Discovered to the Eye by the Telescope* (London, 1681), p. 51.

40:89–91 The quotation is from Arndt, bk. 4, pt. 1, chap. 1, 20 (2:323). My translation.

40:92–9 This material is based on Cheyne, chap. 1, pp. 95–98.

41:22–23 The quotation is from Alsted, p. 286. My translation.

41:29 The first clause in the paragraph is from Derham 2, p. 10.

41:31–36 These lines are from Harris, *LT*, 1:s.v. "Orbis *Magnus*."

41:45–52 Mather slightly condenses a passage from *Theologia ruris*, pp. 201–2. This anonymous twenty-three-page homily appears at the end of George Ashwell's English translation of Edward Pococke's Latin translation of Abu Bakr Ibn Tufayl, *The History of Hai Eb'n Yockdan, An Indian Prince: Or, the Self-Taught Philosopher* (London, 1686). On Ibn Tufayl's book, see the Endnotes (11:7–12:39). *Theologia ruris* has no connection with the story of Hai Eb'n Yockdan, and neither the author of this work nor its source (if it is a translation) are known. George Ashwell may have written it himself, though the nature of his acknowledged writings does not suggest his authorship. The Augustan Reprint Society published *Theologia ruris* with an excellent introduction by H. S. V. Ogden (Los Angeles: William Andrews Clark Memorial Library, 1956).

41:53–42:57 Mather most likely borrows the quotation from Derham 2, p. 91, where the English translation appears in the text and the Latin in a footnote. I have altered the capitalization. The ultimate source is Hugh of St. Victor, *Didascalicon,* 7.8 (PL, 176:818).

5. OF SATURN

42:61–69 The source is Grew, pp. 6–7.

42:72–43:83 The source is Walker, pp. 16–17, but Mather changes "Arms" into "*Ansae.*"

43:84–85 Mather takes these lines from Harris, *LT,* 1:s.v. "Ring of *Saturn.*"

43:85–88 This material is probably drawn from Harris, *LT,* 1:s.v. "Saturn." Much of the information is available in fragmentary form in Derham 2, p. 188 n. 3, and in Huygens, *Celestial Worlds Discover'd,* p. 124.

43:89–90 The source is Harris, *LT,* 1:s.v. "Saturn."

43:2–44:20 This material is probably based on Harris, *LT,* 1:s.v. "Satellites of *Saturn.*" But Mather appears to have supplemented his account with Huygens, *Celestial Worlds Discover'd,* pp. 116, 113, 114. Similar information is found in Cheyne, chap. 2, p. 101.

44:21–32 The quotation is from Blackmore, *Creation,* 2.458–66 (p. 80, 1712 ed.).

44:33–40 Mather takes these quotations from Derham 2, pp. 61, 63. The ultimate source of the first one is Lactantius, *The Divine Institutes,* 2.5; of the second, Plato, *Epinomis,* 983A-B.

6. OF JUPITER

44:42–45 Mather bases this information on Harris, *LT,* 1:s.v. "Jupiter."

44:46–45:57 The source is Walker, pp. 18–19.

45:58–70 These paragraphs are based on Cheyne, chap. 2, pp. 100–101, but similar information is found in Huygens, *Celestial Worlds Discover'd,* pp. 115–16.

45:71–72 The source is Walker, p. 20.

45:73–82 The source is Harris, *LT,* 1:s.v. "Jupiter."

45:83–46:88 The source is probably Walker, p. 20, but it could be Harris, *LT,* 1:s.v. "Jupiter."

46:88–91 The source is Harris, *LT,* 1:s.v. "Jupiter."

46:92–98 These data are based on Derham 2, pp. 92–93.

46:99–12 Mather undoubtedly takes the quotation from Derham 2, pp. 96–97, omitting the word "plainly" before "shews" in the last sentence. The ultimate source is Molyneux, *Dioptrica nova,* p. 273.

7. OF MARS

46:14–47:38 The entire essay paraphrases Harris, *LT*, 1:s.v. "Mars."

8. OF VENUS

47:40–47 These paragraphs obviously paraphrase Walker, p. 21. But Walker writes that the planet moves from south to north. Mather may have drawn also upon "An Extract of a Letter Written by Signor Cassini . . . in [Bologna] to Monsieur Petit at Paris . . . Concerning Severall Spots . . . in the Planet Venus," *Phil. Trans.*, 3 (Feb. 10, 1668), 615–17. Cassini writes that the motion is from south to north in the inferior part of the disc, but from north to south in the superior part.

47:47–54 This material is based on Harris, *LT*, 1:s.v. "Venus."

47:54–55 The source is Cheyne, chap. 2, p. 102.

9. OF MERCURY

48:57–59 This paragraph is based on Walker, p. 22.

48:60–73 This material is taken from Harris, *LT*, 1:s.v. "Mercury."

48:74–49:99 These tables are probably based on Cheyne, chap. 3, pp. 98–99. But Cheyne gives periodical times in years, days, and hours, and distances and diameters in statute (English) miles.

49:1–7 This information is taken from Derham 2, pp. 20–21.

49:10–11 The story is told in Diogenes Laertius, *Lives of Eminent Philosophers*, 1.1, 33–34.

49:17–50:34 Mather took this material from Roberts, "Concerning the Distance of the Fixed Stars," in *Misc. Cur.*, 1:267–69. The original was in the *Phil. Trans.*, 18 (Mar. Apr. 1694), 101–3. Harris also reprinted Roberts's article in *LT*, 1:s.v. "Fixed *Stars*."

10. OF COMETS

50:41–52:16 These paragraphs, over half of the essay, are taken nearly verbatim from key portions of Harris, *LT*, 1:s.v. "Comets," with exceptions noted in the following two entries.

51:84–85 The source is Cheyne, chap. 2, p. 119. Cheyne also drew on Harris, using the 1704 ed. of *LT*, s.v. "Comets."

52:1–2 Mather concludes this paragraph with his own observations.

52:17–53:46 Mather's historical account is taken from Edmond Halley, "A Synopsis of the Astronomy of Comets," in *Misc. Cur.*, 2:App., pp. 2–4. The article had appeared earlier as "Astronomiae cometicae synopsis," in *Phil. Trans.*, 24 (Mar. 1705), 1882–99.

52:17–20 Surely Mather takes note of Seneca's prediction from Halley's "Synopsis." The ultimate source, with which he was undoubtedly familiar, is Seneca, *Natural Questions*, 7.25, 2–3.

53:47–62 The source of the first quotation, which Mather slightly alters, is Cheyne, chap. 2, pp. 121–22; of the second, p. 151.

54:75–76 The source is Horace, "The Apotheosis of Romulus," in *Odes*, 3.3, 1–2, 7–8.

50:36–54:76 This entire essay was published anonymously and posthumously as *An Essay on Comets: Their Nature, the Laws of their Motions, the Cause and Magnitude of their Atmosphere, and Tails; With a Conjecture of their Use and Design* (Boston, 1744). This rare eight-page pamphlet, published by Rogers and Fowle, is identical to Mather's essay "Of Comets" with three exceptions. The pamphlet omits the paragraph at 51:84–85. The pamphlet considerably expands the quotations from Cheyne at 53:49–62, adding material from Cheyne, pp. 118–23, 150–51. The pamphlet reproduces the table at 48:77–49:7 under the title, "A Synopsis of Certain Matters Relating to the Planets, As They are Determined by the Latest and Most Accurate Astronomers."

Appendix. Of Heat

54:81–55:99 This material is taken almost verbatim from Harris, *LT*, 1:s.v. "Heat." In an experiment performed in 1694, Frederick Slare had poured spirit of nitre (nitric acid) over the oil of caraway seeds in a vacuum, with the result that a violent explosion blew up the glass container. Slare was impressed with the great quantity of air produced from small amounts of these liquids. He reported his findings in "An Account of Some Experiments Relating to the Production of Fire and Flame, together with an Explosion: Made by the Mixture of Two Liquors Actually Cold," *Phil. Trans.*, 18 (July–Aug. 1694), 212–13.

55:1–17 This material is taken almost verbatim from Harris, *LT*, 2:s.v. "*Burning Glasses.*" Harris takes his description of the concave glass in Germany from *Phil. Trans.*, 16 (July–Aug. 1687), 352–54, which had taken its account from the January 1687 *Acta Eruditorum* of Leipzig. His account of the work of Tschirnhaus is from *Histoire de l'Acadmie des Sciences, Anne 1699.*

56:34–35 Mather began *Christianus per Ignem*, a work of nearly two hundred pages, in 1683 and finished it in 1701. With John 18:18 as a text, Mather uses fire as a figure and symbol of thought and writes to encourage religious meditation and reflection. This book discusses a number of ancient and early modern authors treated in *The Christian Philosopher*. Mather's purpose was to draw "profitable instructions" from all the "Creatures of God"—i.e., from the whole created physical universe. See Mather, *Diary*, 1:381.

11. Of the Moon

56:44–47 These lines are based on Walker, p. 10.

56:48–58:30 All of this material, with two exceptions to be noted, is drawn, nearly verbatim, from Harris, *LT,* 1:s.v. "Moon."

57:85–58:95 One of the exceptions is based on Harris, *LT,* 2:s.v. "Moon."

58:96–2 The other exception is from Hooke, *Posthumous Works,* pp. 80–81. Mather, however, writes 288 where Hooke writes 228.

58:31–59:37 The source is Cheyne, chap. 2, pp. 147–48.

59:38–40 Mather's source is almost certainly Harris, *LT,* 1:s.v. "Moon," which reprints Newton's "Theory of the Moon."

59:41–48 The source is Huygens, *Celestial Worlds Discover'd,* pp. 130–32.

59:50–53 The source is Derham 2, pp. li–liii, lvi, 121–22 n. 1.

60:67–70 Grew, pp. 88–89. Grew's enumeration comes in a chapter entitled "Of the Nature of God's Government, or of Divine Providence."

60:71–82 Mather draws here on an earlier account by himself, "Surprising Influences of the Moon," which he communicated either to Richard Waller, then secretary of the Royal Society, or John Woodward, in 1714. This account is the "Curiosa Americana," second series, no. 2. He omits from *The Christian Philosopher* another curiosity regarding the moon which he notes in his letter: "We very much observe it in o[u]r Countrey, and govern o[u]r affaires by the Observation, That, *as the Winds are in the Last Quarter of the* Moon, *so they generally govern in ye next Three Quarters.*" See Kittredge, pp. 29–30.

60:82–85 These lines appear again essentially the same in *Bethesda,* p. 205.

12. Of the Rain

62:97–1 These lines (through "*Drops.*") are taken from Ray, p. 101.

62:8–28 This material is a paraphrase of Ray, pp. 101–3.

63:46–52 Mather quotes Fénelon, *A Demonstration of the Existence, Wisdom, and Omnipotence of God,* trans. A. Boyer (London, 1713), p. 29. Fénelon wrote three philosophical works, including *Traité de l'existence de Dieu.* The two parts of this essay are really distinct works. The first part, *Démonstration de l'existence de Dieu, tirée de la connoissance de la nature, et proportionnée a la foible intelligence des plus simples* (Paris, 1713), is of an essentially popular character, except at the end. Published during Fénelon's lifetime, but without his knowledge, it went through four French editions by 1715 (two were published at Amsterdam). This part is better known than the second part, which is a metaphysical treatise. The first part was translated into English as indicated at the head of this note. The book expounds the argument from final causes and the argument based on the marvels of nature. In 1718 the two parts of the treatise were bound together in a single work.

13. OF THE RAINBOW

63:54–65:10 This long section is taken almost verbatim from Cotton Mather, *Thoughts for the Day of Rain. In Two Essays: I. The Gospel of the Rainbow . . . II. The Saviour with His Rainbow . . .* (Boston, 1712), pp. 3–5, ii–iv. The material in these lines from Mather's *Thoughts* is apparently based on two sources. One is Edmond Halley, "De Iride, sive de arcu coelesti, dissertatio geometrica," in *Phil. Trans.*, 22 (Nov.–Dec. 1700), 714–25, which appeared in English translation as "A Geometrical Dissertation Concerning the Rainbow," in *Misc. Cur.*, 2:App., 25–41. The other is Harris, *LT*, 1:s.v. "Rain-bow." Mather simplified Halley's analysis with a view to reaching a wider audience. On his motive for writing on the rainbow as a remembrance of God's covenant that He will preserve His church in the world, see Mather, *Diary*, 2:82, 87, 89–90, 165.

63:54–57 This allegedly "common definition" of the rainbow has left little trace; at least I cannot find it in ancient or medieval writers on natural phenomena. The closest approximation is in Plutarch, *On Isis and Osiris*, 358E, which reads as follows in translation: "The rainbow is an image of the sun made brilliant by the reflection of its appearance into a cloud." See Plutarch's *De Iside et Osiride*, p. 149.

65:3 Mather omits the following additional description of Newton (in his *Thoughts for the Day of Rain*, pp. iii–iv): "The most Victorious Assertor of an Infinite God, that hath appeared in the bright Army of them that have driven the baffled Herd of *Atheists* away from the Tents of Humanity."

65:6 Mather adds the parenthetical phrase, which is not in *Thoughts for the Day of Rain*.

65:11–14 Mather's source is Halley, "A Geometrical Dissertation concerning the Rainbow," in *Misc. Cur.*, 2:App., 40–41.

65:15–16 Mather took these lines from *Thoughts for the Day of Rain*, p. 6.

65:17–66:38 Mather had previously given the "Gospel of the Rainbow" by Frytsche in both Latin and English in *Thoughts for the Day of Rain*, p. vi. Almost certainly he borrowed the poem from Alsted, p. 376. The ultimate source is Frytsche, *Meteororum* (Wittenberg, 1581), p. A [1].

66:39–49 The source is Harris, *LT*, 1:s.v. "Halo," and 2:s.v. "Halo."

66:50–67:78 This material is taken largely verbatim from *Thoughts for the Day of Rain*, pp. 6–7, 15–16, 23, 25, 29. Mather had quoted Mercier on p. 7. His source was apparently comments on Gen. 9:13 in Jean Mercier, *In Genesin primum Mosis librum, sic a Graecis apellatum, commentarius* ([Geneva]), 1598), p. 202. Mercier writes, "Sane Iudaei mire sunt religiosi ubi arcum vident, egrediuntur, et peccata confitentur, se diluvio dignos fuisse, sed clementia sua usum Dominum ne orbem ultra perderet."

67:76–78 The ultimate source is Ovid, *Metamorphoses,* 1.256–58.

14. OF THE SNOW

67:82–68:24 This material, nearly the entire essay, is taken almost verbatim from Nehemiah Grew, "Some Observations Touching the Nature of Snow," *Phil. Trans.,* 8 (Mar. 25, 1673), 5193–94, 5196. Grew mentions Descartes and Hooke. Hooke discussed snow in *Micrographia: Or Some Physiological Descriptions of Minute Bodies Made by Magnifying Glasses. With Observations and Inquiries Thereupon* (London, 1665), pp. 91–92. Grew draws on his earlier account for a brief discussion of snow in *Cosmologia Sacra,* p. 16.

68:25–69:30 This account of "A Woolen Snow" is based on a letter of 1 December 1713 to Richard Waller, secretary of the Royal Society in London. The letter is the first in the second series of Mather's "Curiosa Americana." Mather notes that he has "unhappily mislaid the large and well-attested account of what follows, yet, however, my Memory sufficiently serves me to assert so much as may afford you a tolerable Satisfaction." He then describes events at Fairfield and encloses a specimen of the wool. An extract of the letter is in W[illiam] Derham, ed., *Philosophical Experiments and Observations of the Late Eminent Dr. Robert Hooke . . . and Other Eminent Virtuoso's in His Time* (London, 1726), p. 386. See also Kittredge, p. 29.

69:31–32 This quotation is from Gale, *Philosophia generalis* (London, 1676), p. 249.

15. OF HAIL

69:45 The quotation is taken from Gale, *Philosophia generalis,* p. 249.

69:46–70:60 This material is taken almost verbatim from John Wallis, "A Letter of Dr. Wallis to Dr. Sloane, Concerning the Generation of Hail, and of Thunder and Lightning, and the Effects Thereof," *Phil. Trans.,* 19 (Aug. 1697), 657–58; reprinted in *Misc. Cur.,* 2:319–20. The letter is dated 26 July 1697.

70:68–76 Mather most likely took these four Latin lines from Alsted, pp. 376–77. Alsted also quotes the first two lines in his *Encyclopaedia,* 3:328. The original source is Marcus Frytsche, *Meteororum* (Wittenberg, 1581), pp. A, A2. Mather also places the first two lines on the title page of his *Thoughts for the Day of Rain.* There he translates the second line as: "The wondrous works of the Eternal spies." My translation of the last two lines.

16. Of Thunder and Lightning

70:80–71:18 This material is taken nearly verbatim from Harris, *LT,* 2:s.v. "Thunder and Lightning." Harris took the first part (70:80–71:97) from Robert Hooke, *Posthumous Works,* pp. 169–70; the second part (71:98–18) from Wallis, "A Letter . . . to Dr. Sloane," in *Phil. Trans.,* 19 (Aug. 1697), 655–57; reprinted in *Misc. Cur.,* 2:317–18.

71:19–73:70 These paragraphs draw on previous publications by Mather. Except for two rhapsodies (72:38–42 and 73:58–60) they reorder and paraphrase material found in Mather's *Magnalia Christi Americana: Or, the Ecclesiastical History of New England from . . . 1620 unto . . . 1698* (London, 1702), bk. 6, chap. 3 (pp. 14–20). This chapter in the *Magnalia* reprints with a new introduction a work Mather had published anonymously, *Brontologia Sacra: The Voice of the Glorious God in the Thunder* (London, 1695). The ultimate sources of the Latin quotations and of the Plutarch item, all of which are in the earlier material, are indicated in the following five entries.

72:23–25 The source is Vergil, *Georgics,* 1.329–31.

72:34 The source is Horace, *Odes,* 2.10, 11–12.

72:46 The source is Ovid, *Tristia,* 2.1, 33. I am indebted to Dr. Dorothy S. Adelmann for the location of this quotation and the Juvenal quotation at 73:63–64.

73:53–58 The source is Plutarch, *Quaestiones convivales* ("Table Talk") 4.665B-C, 664D; 5. 684C. My translation.

73:63–64 The source is Juvenal, Satire 13, ll. 223–24. Mather may have been reminded of these lines by Alsted, p. 198.

73:72–74 Mather almost certainly knew Herlicius from Alsted, p. 342. The work cited was Herlitz, *Tractatus de fulmine* (Stettin, 1600).

17. Of the Air

73:76–74:14 This material is taken almost verbatim from Harris, *LT,* 1:s.v. "Air."

74:15–75:26 These paragraphs are based closely on Harris, *LT,* 1:s.v. "Air Weight."

74:17 The concept of vacuum is central in the history of spatial theories since classical antiquity. The Greek atomists used the doctrine of a separate, empty space to explain physical phenomena. They identified a real, though for the most part only hypothetically real, separate nothing between the discrete particles of which, they held, all matter is composed. Ontologically, they conceived of the nothingness of space as something. Aristotle disagreed, holding that void space could not possibly exist. He defined space as the extension of an object filling it. Two bodies cannot occupy the same space simultaneously. The vacuum is therefore a contradiction in logic. The Scholastics built on Aristotle in the Middle Ages. Expressions such as *fuga vacui* and *horror vacui* began to emerge in the thirteenth century along with the declaration *natura abhorret vacuum* (nature abhors a vacuum). The argument over vacuum and plenum flared up again during the Renaissance. Galileo opposed Aristotle on the vacuum, and in about the year 1613 he studied the weight of air experimentally. In Galileo's time Florentine well diggers had discovered that water could not be drawn by a pump or syphon higher than about thirty-four feet. Galileo inquired whether it was not from some cause other than *fuga vacui* that it could be drawn only a determinate rather than an infinite height and came up with the hypothesis of the countergravitation of the incumbent air. See Edward Grant, *Much Ado about Nothing: Theories of Space and Vacuum from the Middle Ages to the Scientific Revolution* (Cambridge: Cambridge University Press, 1981), pp. 3–8, 67.

74:21 The Torricellian experiment was intended not simply to produce a vacuum but to make an instrument which might show the weight of air. Torricelli instructed Vincencio Viviani to perform the experiment, which was conducted in Florence, most probably in 1644. The experiment involved a glass tube about one meter long with one end sealed and the other open. The tube was filled with quicksilver (mercury) and inverted, with the open end stopped with the finger while it was immersed in a bowl containing mercury, and then opened. The mercury in the tube ran into the bowl, leaving in the tube a column of mercury about twenty-nine inches high (the height depends on the temperature of the air) and an empty space (which came to be called the Torricellian vacuum) at the top of the tube. Torricelli concluded that an external force stopped the mercury in the tube from falling below a certain level. This external force was atmospheric pressure, the weight of the air on the mercury in the bowl. The instrument used in the experiment became the barometer. See W. E. Knowles Middleton, *The History of the Barometer* (Baltimore: The Johns Hopkins Press, 1964), pp. 3–32.

75:27–38 This material paraphrases Jeremiah Wainewright, *A Mechanical Account of the Non-Naturals . . .* (London, 1707), pp. 58–59, and quotes p. 61 in condensed form. Wainewright's book is based on Galen's medical thought. According to Galen, medicine involves both the theoretical and the practical. The former includes things both natural and non-natural. There are seven natural things (including, e.g., the parts of the body, the elements, temperaments, humors, etc.), and six non-naturals that maintain good health, prevent diseases, or restore a sick body to health. The six non-naturals are air, meat and drink, excretion and retention, sleep and wakefulness, motion and rest, and the passions of the soul (or affections of the mind). The non-naturals were so-called because they are not naturally causes of disease. The idea of the non-naturals is one of the most enduring contributions of Galen to medical thought, and Wainewright's book demonstrates the persistence of this doctrine into the early eighteenth century.

75:41–76:62 This material closely paraphrases Harris, *LT*, 2:s.v. "Air."

76:63–93 This paragraph paraphrases Ray, pp. 81–84, 86, 88. The ultimate source of the Pliny quotation is *Natural History*, 11.21, 66.

76:93–77:1 Mather quotes these lines from "A Canto on Isa. 53 Being Turned into Verse by Mrs. Wharton," in Waller, *Poems... Written Upon Several Occasions*, 8th ed. (London, 1711), pp. 369–70.

18. OF THE WIND

77:15–78:60 Mather probably takes this material from Harris, *LT*, 1:s.v. "Wind." But he adds the definition, which is in Plato, *Definitions*, 411C. Scholars no longer believe that Plato wrote this work. Harris takes his account primarily from Edmond Halley, "An Historical Account of the Trade Winds, and Monsoons, Observable in the Seas between and near the Tropicks, with an Attempt to Assign the Physical Cause of the Said Winds," in *Phil. Trans.*, 16 (July–Sept. 1686), 153–68, reprinted in *Misc. Cur.*, 1:61–80.

78:62–66 The source is Harris, *LT*, 2:s.v. "Wind."

79:67–71 The source is Grew, p. 9.

79:72–83 This paragraph is closely based on Ray, pp. 103–5. My translation.

79:84–92 These lines are partly and loosely based on Ray, pp. 104–5. They also draw on general information from traveler's accounts and other sources.

19. OF THE COLD

80:6–21 Mather's source is Robert Boyle, *New Experiments and Observations Touching Cold, or an Experimental History of Cold* (London, 1665), pp. 412–63 passim. Naturalists had long disputed the concept of a *Primum Frigidum*. According to Boyle (p. 412), Plutarch had argued for the earth, Aristotle and the Schools for water, the Stoics and some moderns for air, and Gassendi for water as that body "that is of its own nature supremely Cold, and by participation of which, all other cold Bodies obtain that quality." Boyle rejected the whole notion of a *"Primum Frigidum."*

80:22–26 This paragraph is taken nearly verbatim from Harris, *LT*, 2:s.v. "Freezing."

80:27–81:42 Mather draws on Boyle, *Cold*, pp. 526, 532–37, 539–44, 284–

85, 383–84. But he embroiders his source. Boyle makes no mention of Herberstein or Lucretius. According to Boyle, Baffin puts the highest "icy islands" at 240 feet, which Mather hikes to 300 feet. The Lucretius quotation is from *De rerum natura,* 6.530.

81:43–48 The source is Harris, *LT,* 2:s.v. "Freezing."

81:49–50 I do not find the phrase Mather quotes in Pliny's *Natural History,* but the mistaken idea mentioned is in that work, 2.61, 152 and 31.31, 33.

81:64 I do not find this phrase in Livy. My translation.

81:65–82:73 This passage shows indebtedness to Harris, *LT,* 1:s.v. "Cold." Mather must have taken the material on Martinus from Boyle, *Cold,* pp. 16–17, but Harris also mentions Martinus. The original is in Martini Martinus, *Novus atlas Sinensis* (Amsterdam, 1654), p. 27. My translation.

82:73–81 These lines as well as the reference to mitigations of the New England cold (81:61) reflect contemporary views on climate. The English people who settled New England had come with the notion that climate was uniform in any given latitude around the world, but they discovered that New England, which is nearer the equator than England, was not only hotter in the summer but also colder in the winter than England. The physical challenge of harsh weather was made more severe by the fact that North America was at the time in the grip of the Little Ice Age, and its coldest period was from 1550 to 1700.

The colonists had to adjust to changing patterns of weather in the seventeenth century. During the 1630s and 1640s the weather was very cold. The settlers, believing that human beings are responsible for their environment, took active measures to improve it. They experimented with different agricultural ways and cut down the woods to let in the sun to warm the country and lengthen the growing season. The weather became more temperate from the 1650s to the 1670s, leading the settlers to conclude that their efforts were responsible for transforming the climate. But the weather became extremely harsh during the 1680s and 1690s, with the worst winters of the century coming in 1696, 1697, and 1698. Mather's account of the wood in his fireplace (82:74–77) is based on a personal experience reported on 23 February 1697 in his *Diary,* 1:216. On this subject, see Karen Ordahl Kupperman, "Climate and Mastery of the Wilderness in Seventeenth-Century New England," in *Seventeenth-Century New England,* ed. David D. Hall and David G. Allen (Boston: Colonial Society of Massachusetts, 1984), pp. 3–37.

82:81–86 The quotation is from Petronius Arbiter, *Satyricon,* 99. Heseltine's translation.

20. OF THE TERRAQUEOUS GLOBE

82:91–96 Mather's source is Derham, p. 43 nn. Mather adds the cubic foot measurement, most likely taking it from Harris, *LT*, 1:s.v. "Earth."

82:96–84:1 Mather's source is Derham 2, pp. 18–19.

84:2–5 The source is Derham, pp. 43 n. 1, 34 n. 3. New England Puritans readily accepted the Copernican theory. It was first discussed in a New England almanac in Zechariah Brigden's *An Almanack of the Coelestial Motions* (Cambridge, 1659). John Foster diagrammed the planets as described by modern science in *An Almanac of Coelestial Motions* (Cambridge, 1675), which was reprinted in 1681. See Donald Fleming, "The Judgment upon Copernicus in Puritan New England," in *Mélanges Alexandre Koyré*, vol. 2: *L'Aventure de l'Esprit* (Paris: Hermann, 1964), pp. 160–75; and David D. Hall, "Literacy, Religion, and the Plain Style," in Jonathan L. Fairbanks and Robert F. Trent, eds., *New England Begins: The Seventeenth Century*, 3 vols. (Boston: Museum of Fine Arts, 1982), 2:110, 147.

84:6–17 This passage is taken almost verbatim from Cheyne, chap. 3, pp. 140–41.

84:18–27 This material is taken from Derham, pp. 44–45 and n. 3. Cicero quotes Cleanthes in *De natura deorum*, 2.5, 15. Derham gives an English translation of Plutarch in *Astro-Theology*, pp. 3–4. The ultimate source is *De placita philosophorum* ("Of Those Sentiments Concerning Nature with Which Philosophers Were Delighted"), in Plutarch's *Moralia*, 1.6, 880A-B. The Loeb edition of the *Moralia* omits *De placita* on the grounds that it is by Aetius, not Plutarch. The Teubner edition includes it in Greek in Plutarch, *Moralia* (Leipzig, 1971), vol. 5, fasc. 2, pt. 1. An English translation of the essay is in William W. Goodwin, ed., *Plutarch's Morals*, 5 vols. (Boston, 1871), vol. 3.

84:31–33 Mather's source is Derham, p. 45 n. 6. The ultimate source of the quotation is Cicero, *De natura deorum*, 2.53, 132. The entire passage reads (in translation): "Again the alternation of day and night contributes to the preservation of living creatures by affording one time for activity and another for repose."

85:39–63 Mather paraphrases and quotes Ray, pp. 227–36 passim. Ray took much of his passage verbatim from Henry More, *An Antidote against Atheism* (London, 1655), pp. 70–73, and from John Keill, *An Examination of Dr. Burnet's Theory of the Earth* [1698], 2d ed. (London, 1734), chap. 4.

85:49 Rochefort's book has been attributed, perhaps erroneously, to Louis de Poincy, whose initials close the dedicatory epistle of the first edition. Recent bibliographies distinguish between Cesar de Rochefort, the French jurisconsult, lexicographer, and putative author of the *Histoire . . . des Antilles,* and Charles de Rochefort of Rotterdam, assigning the *Histoire . . . des Antilles* to the latter.

85:50 John Keill spent most of his career at Oxford. He believed that Newton's discoveries should be employed in combating "atheistic" Cartesianism and mechanical philosophy, but rejected the idea that this should be accomplished by means of natural theology. As a Newtonian with high church patronage, Keill sought to counter the low church influence of men like Richard Bentley and William Whiston. For Keill, natural theology should be subordinated to Scripture, while natural philosophy should acknowledge Providence and even miracles. He advanced this argument in his book examining Dr. Burnet's theory of the earth. Newton himself accepted Keill's criticism of cosmogonical theories with which he had sympathized.

86:72–78 The source is Gregory, *Elements of Astronomy,* 1:129–32.

86:79–93 Mather takes this material from Harris, *LT,* 1:s.v. "Earth."

87:4–7 Mather adapts this quotation from Arndt, bk. 4, pt. 2, chap. 3, 2 (2:414–15). My translation.

87:17–19 The quotation is taken verbatim from Arndt, bk. 2, chap. 42, 12 (2:89). My translation.

87:23–25 The quotation is from Arndt, bk. 4, pt. 2, chap. 21, 3 (2:441). My translation.

87:26–88:30 The idea expressed in the first part of this quotation is found in Arndt, bk. 2, chap. 42, 12 (2:88–89). Mather may have adapted these Latin lines from that source. My translation.

88:32–37 The source of both quotations is Paul Egard(us) or Eggers, Γνωθι Σεαυτον: *sive Tractatus utilissimus de vera microcosmi cognitione tum naturali tum supernaturali,* p. 8. My translation.

88:41–63 Mather paraphrases Cheyne, chap. 1, pp. 101–4.

21. Of Gravity

89:92–90:31 Mather takes these two paragraphs almost verbatim from Derham, pp. 31–35, adding the quotation from Prov. 26:10.

90:95 This material is drawn primarily from Harris, *LT,* 1:s.v. "Gravity." Harris in turn draws heavily on Edmond Halley, "A Discourse Concerning Gravity, and Its Properties, Wherein the Descent of Heavy Bodies, and the Motion of Projects [Projectiles] is Briefly, But Fully Handled," *Phil. Trans.,* 16 (Jan.–Feb. 1686), 3–21, reprinted in *Misc. Cur.,* 1:304–28. Mather follows Harris closely, but he elaborates. His elaborations are indicated in the five entries that follow.

90:95 Both Halley and Harris mention Descartes, *Principia Philosophiae,* sects. 20–23, as the source of erroneous ideas. But Mather introduces the "*Materia Striata*" of Descartes. According to Reese P. Miller, coeditor with Valentine R. Miller of Descartes, *Principles of Philosophy* (Dordrecht: D. Reidel Publishing Co., 1983), Mather may not be talking about Descartes's views of gravity, though that seems to be the case. Descartes does not use "*Materia Striata*" in his *Principles of Philosophy*. He uses "*striata*," which is for him a technical term, in pt. 3, art. 90 of that work, entitled "What the shape of these particles, which from now on we shall call *striatae*, is." The Millers have translated *striata* as "grooved" throughout. *Particulae striatae* occurs frequently in the *Principles*. These "grooved particles" form the basis of Descartes's exposition of magnetism and have nothing to do with gravity. Reese P. Miller to author, n.d. (received 21 March 1985).

90:97-5 On Halley's explanation of the cause of gravity, Mather quotes Harris nearly verbatim. But Halley himself had written "Tho' the efficient Cause of *Gravity* be so obscure, yet the final Cause thereof is clear enough; for it is by this single *Principle,* that the *Earth* and all the *Celestial Bodies* are kept from *Dissolution;* the least of their *Particles* not being suffer'd to recede far from their *Surfaces,* without being immediately brought down again by virtue of this *Natural Tendency;* which, for their Preservation, the Infinite Wisdom of their *Creator* has ordained to be towards each of their *Centers;* nor can the *Globes* of the *Sun* and *Planets* otherwise be destroy'd, but by taking from them this Power of keeping their Parts united." See *Misc. Cur.,* 1:307.

90:20-21 This detail is incorporated from Derham, p. 32 n. 1.

90:23-24 Here Mather shows familiarity with Harris, *LT,* 1:s.v. "Specifick Gravity."

90:25-27 Besides Harris, these lines draw on Derham, p. 32 n. 1. It was Galileo who first determined the acceleration of falling bodies. Halley, the source on which Harris drew and thus Mather's ultimate source at this point, wrote that the "Properties of *Gravity,* and its manner of acting upon *Bodies falling,* have been in a great measure discovered, and most of them made out by *Mathematical Demonstration* in this our *Century,* by the accurate diligence of *Galilaeus, Torricellius, Hugenius,* and others, and now lately by our worthy Countryman, Mr. *Isaac Newton.*" See Halley, "A Discourse concerning Gravity, and Its Properties," *Misc. Cur.,* 1:307-8.

91:32-39 The conclusion of the paragraph appears to be Mather's improvisation on the previous material.

91:50-59 Mather takes this material from Derham, pp. 35-36 nn. 6-7.

91:60-63 Mather takes these lines from Harris, *LT,* 1:s.v. "Gravity." Harris cites as his source Edmond Halley, "An Account of the Circulation of the Watry Vapours of the Sea, and of the Cause of

Springs," *Phil. Trans.,* 17 (Jan.–Feb. 1691), 468–73. That article, along with an earlier one by Halley entitled "An Estimate of the Quantity of Vapour Raised out of the Sea by the Warmth of the Sun: Derived from an Experiment," *Phil. Trans.,* 16 (Sept.–Oct. 1687), 366–70, was published as "An Estimate of the Quantity of the Vapours Raised out of the Sea Derived from Experiment: Together with an Account of the Circulation of the Watry Vapours of the Sea, and of the Cause of Springs," in *Misc. Cur.,* 1:1–12. Mather's reference is at p. 7.

91:64–67 Mather bases this on Gregory, *Elements of Astronomy,* 1:iv–xii.

91:68–92:76 Mather draws here on J[ohn] Keill, *An Examination of Dr. Burnet's Theory of the Earth* (Oxford, 1698). In the second edition (London, 1734), the reference is at pp. 96–97.

92:76–78 Mather most likely took the quotation from Derham, p. 34 n. 4. The original is in Cicero, *De natura deorum,* 2.45, 115. My translation. Cicero, discussing the stability and coherence of the world, says that all its parts gravitate with a uniform pressure toward the center. Some encompassing bond binds them together; "this function is fulfilled by that rational and intelligent substance which pervades the whole world as the efficient cause of all things and which draws and collects the outermost particles toward the center." Loeb translation.

92:79–88 In 1697 Clarke published a new Latin translation of Jacques Rohault's *Traité de physique,* and his notes made that Cartesian treatise a means of disseminating the ideas of Newton. Clarke's Boyle lectures in 1704–5, a theological application of Newtonian science, were published as *A Demonstration of the Being and Attributes of God* (London, 1705) and *A Discourse Concerning the Unchangeable Obligations of Natural Religion, and the Truth and Certainty of the Christian Revelation* (London, 1706). Mather's material in these lines is paraphrased from *A Demonstration,* pp. 129–30 (in the 1706 edition).

92:88–95 The passage is quoted from Samuel Clarke, *A Discourse Concerning the Being and Attributes of God, the Obligations of Natural Religion, and the Truth and Certainty of the Christian Revelation* [1706], 9th ed. (London, 1738), p. 161. Theodore Hornberger in "Cotton Mather's Annotations on the First Chapter of Genesis," *University of Texas Publications,* no. 3826 (1938), 119, says that Mather made use of Richard Bentley's *Folly and Unreasonableness of Atheism* (London, 1693), at pp. 84–85 in *The Christian Philosopher* (presumably 92:79–95 in the present edition). Mather knew Bentley's book, but here he draws on Clarke.

92:96–93:31 This long passage is a skillfully abbreviated paraphrase of Cheyne, chap. 1, pp. 47–53.

94:67–95:90 This paragraph closely follows Cheyne, chap. 1, pp. 3–6. Mather substitutes "Great God" for "Author of Nature," and makes no reference to Cheyne's discussion of the universe as a clock.

94:73-74 Greek science and philosophy used many polar expressions (e.g., hot and cold) for want of or in place of such abstract terms as temperature, density, size, and weight, and the concept of innate heat was fundamental in ancient physiology. Both Hippocrates and Aristotle associated heat with life and cold with death. The body of the living fetus is warm, deriving its heat from the mother. When a child is born it begins to breathe, inhaling air that is cooler than its body. In respiration, the innate heat and the cold air inhaled alternately retreat and pursue each other. The body's innate heat is at its most intense in the heart. Galen reflected the teaching of Hippocrates and Aristotle, and these beliefs persisted until Borelli disposed of them in the seventeenth century by using a thermometer. See William A. Heidel, *Hippocratic Medicine: Its Spirit and Method* (New York: Columbia University Press, 1941), pp. 50–87, 93–94.

95:91-5 This paragraph is based on Cheyne, chap. 2, pp. 7–11. Cheyne merely states Newton's scientific principle of *vis inertiae*. Mather adds the reference to God as the cause of the preservation of a body in rest or in motion after the first instant.

22. OF THE WATER

96:9-19 This material paraphrases Harris, *LT,* 2:s.v. "Water." Harris bases that part of his account which Mather borrows on Newton's description of water.

96:20-42 These lines are taken almost verbatim from Wainewright, *Mechanical Account,* pp. 135–37.

97:45-48 This section paraphrases Cheyne, chap. 1, p. 66.

97:49-52 This observation is from Mariotte, *Mouvement des eaux* [1686] (Paris, 1700), pp. 182, 184. Mariotte writes "according to the Duplicate Ratio." In the English translation by J. T. Desaguliers, *The Motion of Water, and Other Fluids. Being a Treatise of Hydrostatics* (London, 1718), the reference is at pp. 193, 195.

97:53-98:83 This passage paraphrases and quotes Ray, pp. 89–92, 97–99. Ray includes the Psalms which Mather quotes. Ray merely mentions an objection raised against the wisdom of God in dividing sea and land; Mather attributes the objection to "A fanciful and presumptuous Gentleman." His reference is to Thomas Burnet, author of *The Theory of the Earth* (Latin, 1681; English, 1684), whose theory Mather knew as early as 1689.

98:84-9 This paragraph paraphrases Halley, "An Estimate of the Quantity of Vapour Raised out of the Sea by the Warmth of the Sun: Derived from an Experiment," *Phil. Trans.,* 16 (Sept.–Oct. 1687), 368–69; reprinted along with another article by Halley on a related subject in *Misc. Cur.*, 1:3–5. Mather adds the reference to "our vast *Canadian Lakes.*"

98:10–99:19 This paragraph paraphrases Cheyne, chap. 3, pp. 145–47. Mather adds the two concluding lines.

99:22–100:69 This material is taken selectively but largely verbatim from Harris, *LT,* 1:s.v. "Tides." Harris based his account on Edmond Halley, "The True Theory of the Tides, Extracted from that Admired Treatise of Mr. Isaac Newton, Intituled, *Philosophiae Naturalis Principia Mathematica,*" *Phil. Trans.,* 19 (Mar. 1697), 445–57.

100:70–76 This paragraph closely paraphrases Cheyne, chap. 3, p. 147.

100:79–101:90 These lines are based on Derham, pp. 50–53 and nn. 7–11. Mather adds the reference to Antigua, and he takes the Latin quotation from Arndt, bk. 4, pt. 1, chap. 3, 26 (2:342). My translation. Mather gives the length of the Danube and the Ganges in English miles, of the Nile and the Niger in Italian miles.

101:93–97 The quotation is from Arndt, bk. 2, chap. 42, 6 (2:85). My translation.

101:98–4 The quotation is from Alsted, p. 313. My translation.

101:12–102:37 This long passage closely paraphrases Cheyne, chap. 3, pp. 187, 188–89, 191.

102:42–44 The quotation is from Grew, p. 13.

23. OF THE EARTH

102:46–103:52 Mather bases these two paragraphs on Ray, pp. 100–101, which includes the biblical and classical quotations. The Latin is from Cicero, *De natura deorum,* 2.39, 98. Cicero is also quoted by Derham, p. 37 n. 1.

103:53–57 This material, including the quotations, is from Derham, pp. 61–62 and n. 3. The Latin quotation is from Vergil, *Georgics,* 2.109.

103:58–62 Mather takes these lines almost verbatim from John Evelyn, *A Philosophical Discourse of Earth* (London, 1676), pp. 11–12. But Evelyn adds that no more than eight or nine kinds of earth were eminently useful to our purposes. Alsted had discussed the ideal of a "combinatorial art" for exploring the possibilities of human knowledge through combinations of all possible categories of knowledge in *Clavis artis Lullianae* (Key to the art of Lull) (Strassburg, 1609). Alsted's commentary on the *ars magna* of Lull influenced Leibniz, who published *Dissertatio de arte combinatoria* (Leipzig, 1666). Evelyn may have known the works of Leibniz and Alsted. See Leroy E. Loemker, "Leibniz and the Herborn Encyclopedists," *Journal of the History of Ideas,* 22 (July–Sept. 1961), 323–28, and idem, *Struggle for Synthesis: The Seventeenth-Century Background of Leibniz's Synthesis of Order and Freedom* (Cambridge, Mass.: Harvard University Press, 1972), pp. 109, 268

n. Mather may have seen the review of Kircher's *Ars magna sciendi sive combinatoria* in *Phil. Trans.*, 4 (Nov. 15, 1669), 1093. But Evelyn's reference on the "combinatorial art" was to Kircher in the *Mundus Subterraneus* (1665), as revealed by a marginal note in his *Philosophical Discourse of Earth* printed in the fourth edition of *Sylva* (London, 1706), p. 2.

103:63–73 Here Mather follows Derham, pp. 61 n. 1, 63–67. Derham quotes Boyle, *The Skeptical Chemyst* (1661). Mather drops Derham's statement that "it is greatly probable" that minerals, metals, and stones "have a power of *growing*" in the beds in which they have lain "ever since *Noah's* Flood, if not from the Creation." Influenced by Paracelsus, Helmont was the first great theoretical chemist of the seventeenth century, devoting the greater part of his effort to making chemistry an independent theoretical science. He pioneered in the attempt to apply chemical theory to physiological functions and explained digestion in terms of chemical processes. Helmont opposed the traditional doctrine of the elements and ostensibly accepted only one element—water. His one-element theory was rendered attractive by a pseudoquantitative experiment in which he planted a willow tree weighing 5 pounds in 200 pounds of earth; five years later the tree weighed 169 pounds while the earth had lost no weight. Helmont thereupon concluded that plants consisted largely of water. But he neglected the air, and his results had no meaning.

103:74–104:78 This paragraph draws primarily on Ray, p. 249, but it is perhaps also informed by Derham, pp. 70–71 and n. 1, which quotes Pliny.

104:79–84 Here Mather follows almost verbatim Ray, p. 250, which cites Halley, "An Account of the Circulation of the Watry Vapours of the Sea, and of the Cause of Springs," *Phil. Trans.*, 17 (Jan.–Feb. 1691), 468–73; reprinted along with another article by Halley on a related subject in *Misc. Cur.*, 1:1–12.

104:84–88 Here Mather follows Derham, p. 75 n. 1, which gives the Latin quotation from Johann Jakob Scheuchzer, *Helveticus, sive itinera Alpina* (London, 1708), 2d description, p. 20. Derham's translations.

104:91–10 This material is taken almost verbatim from Cheyne, chap. 3, pp. 183–84.

104:11–105:14 Mather takes this passage from Ray, pp. 250–51, omitting Ray's reference to the generation of minerals in the earth.

105:14–15 The quotation is from Olaus Magnus, *Historia de gentibus septentrionalibus* (Antwerp, 1562), p. 70 recto. The translation is from *A Compendious History of the Goths, Swedes, and Vandals, and Other Northern Nations* (London, 1658).

105:16–18 The quotation is from Arndt, bk. 4, pt. 1, chap. 3, 21 (2:340). My translation, a modification of Boehm's.

105:19–24 This paragraph is based on Ray, pp. 252–54, but the biblical passage which Mather quotes (105:22–23) is in Derham, p. 71 n. 1. Derham cites as his source John Wilkins, *The Discovery of a World in the Moone: Or, a Discourse Tending, to Prove, That 'Tis Probable there May be Another Habitable World in that Planet* (London, 1638), p. 114. The proper reference is p. 118.

105:25–26 These lines are based on Arndt, bk. 4, pt. 1, chap. 3, 22 (2:340).

105:27–31 Mather draws here on Ray, p. 255, and Derham, p. 71 n. 1, which quotes Wilkins, *World in the Moon*, p. 119. Mather's "One" is "a good Author" in Derham and Wilkins.

105:32–41 Mather's source is Derham, pp. 71–72, 80 n. 12. The ultimate source of the Latin quotation is Galen, *De usu partium*, 10, 9. For a translation of the passage, see May, *Galen*, 2:486.

106:44–58 The quotation is from Grew, pp. 98–99.

106:59–65 This paragraph paraphrases John Woodward, *An Essay toward a Natural History of the Earth, and Terrestrial Bodies, Especially Minerals, As Also of the Sea, Rivers, and Springs, with an Account of the Universal Deluge, and of the Effects that It Had upon the Earth* (London, 1695), pp. 139–40. But the word "ignivomous" is not in Woodward. Woodward's notion that volcanoes act as safety valves was a very common idea at the time. The Swiss naturalist Johann J. Scheuchzer translated the *Essay* into Latin under the title *Specimen geographiae physicae quo agitur de terra, et corporibus terrestribus speciatim mineralibus: nec non mari, fluminibus, et fontibus accedit. Deluvii universalibus effectuumque ejus in terra descriptio* (Zurich, 1704), and it was widely read both in Britain and on the Continent. The *Essay* aroused controversy, with Dr. John Arbuthnot, John Ray, and others criticizing it and John Harris answering the critics with *Remarks on Some Late Papers Relating to the Universal Deluge* (London, 1697). Woodward defended and enlarged upon his *Essay* in *Naturalis historia telluris* (London, 1714), which was translated into English as *The Natural History of the Earth* (London, 1726). He devised a classification of minerals, dividing them into six classes. John Harris reported this scheme in his *Lexicon Technicum* (London, 1704), s.v. "Fossils" (also in vol. 1 of the 1708 edition). Mather corresponded with Woodward, sending many of his American *curiosa* letters to his friend, who sent Mather a gift copy of *Naturalis historia telluris*.

106:66–108:77 Mather almost certainly based this account on *The Vulcano's*, pp. 30, 45–53, 6–7, 15–16. This little work is drawn from Kircher, *Mundus subterraneus in XII libros digestas*, 2d ed., 2 vols. (Amsterdam, 1678), 1:205–9. Kircher writes about "Bartholomaeus" and "Nicolaus Zenetus," identifying them as of Venice (1:194–95). The English translation referred to Bartholomew Zenet and Nicholas

Zenet (pp. 15–16). Only one individual, properly named Niccolò Zeno, is involved.

109:3–8 Mather may have been inspired here by John Ray, *Three Physico-Theological Discourses: Concerning I. The Primitive Chaos, and Creation of the World. II. The General Deluge, Its Causes and Effects. III. The Dissolution of the World and Future Conflagrations* (London, 1693), pp. 164, 183, 163. In all likelihood Mather draws on his own knowledge of classical antiquity to embellish this description. Ray merely quotes Ovid, *Metamorphoses*, 15.293–95. The story about the twelve cities is in Pliny, *Natural History*, 2.86, 200; the account of Atlantis is in Plato, *Timaeus*, 24E–25D.

109:8–12 Mather probably took this point from Varenius, *Geographia generalis*, chap. 18, prop. 17, par. 1 (p. 217, 1681 Latin ed.; 1:407–8, 1733 English trans.) Mather may have seen the first English translation of Varen by Richard Blome, *Cosmography and Geography* (London, 1683).

109:13–16 Mather could have obtained this information from Vincentius Bonajutus, "An Account of the Earthquakes in Sicilia, on the Ninth and Eleventh of January, 1693," *Phil. Trans.*, 18 (Jan. 1694), 2–10, or from John Shower, *Practical Reflections on the Late Earthquakes in Jamaica, England, Sicily, Malta, etc. Anno 1692 [1693]* (London, 1693), pp. 9–29. The Bonajutus account, which Malpighi communicated to the Royal Society, put the number of dead at 59,963. Shower gave the figure of 120,000. Robert Mallet and John W. Mallet, "The Earthquake Catalogue of the British Association, with the Discussion, Curves, and Maps," *Transactions of the British Association for the Advancement of Science, 1852–1858* (London, 1858), Report for 1852, p. 101, put the number at 93,000.

109:17–18 Mather is quoting from Ray, *Three Physico-Theological Discourses*, p. 13. Ray does not mention the year 365, but he writes that the earthquake occurred in the time of the emperor Valentinian I. The original source is Ammianus Marcellinus, *Rerum gestarum libri qui supersunt*, 26.10, 15.

110:47–48 Here Mather draws on *The Homilies of S. John Chrysostom, Archbishop of Constantinople, on the Statutes, Or, To the People of Antioch* (Oxford, 1856), Homily 7 (p. 138). He also alludes to a work by Athenaeus (fl. c. A.D. 200) that describes a learned banquet at which the guests discussed several weighty topics as well as the dishes before them.

110:50–52 Mather bases these lines on Alsted, p. 326. He condenses the Latin quotation and changes its syntax to take the statement out of the mouth of God.

111:69–71 These lines are based ultimately on the Koran. Surah 37 ("The Rangers"): 6–10 reads: "We have adorned the lower heaven with the adornment of the stars / and to preserve against every rebel Satan; / they listen not to the High Council. / For they are pelted from every side, / rejected, and theirs is an everlasting chastisement, / except such as snatches a fragment, / and he is pursued by a piercing flame." Arthur J. Arberry, *The Koran Interpreted*, 2 vols. (London: George Allen and Unwin, 1955), 2:150. The angels in the High Council, which is in the highest heaven, have some knowledge of God's plan and will. Though evil is foreign to this exalted assembly, evil tries by stealth to overhear the divine secrets. But it is repulsed by a flaming fire which resembles the trail of a shooting star. Surah 72 ("The Jinn"): 8-9 reads: "And we stretched towards heaven, but we / found it filled with terrible guards and meteors. / We would sit there on seats to hear; but / any listening now finds a meteor in wait for him." Arberry, *Koran Interpreted*, 2:305.

A variant of the notion reported by Mather is perpetuated in modern Arab superstition, according to Edward W. Lane, who wrote that Egyptians believe a falling star to be "a dart thrown by God at an evil ginnee." *The Manners and Customs of the Modern Egyptians* (1836; New York: Everyman's Library, 1966), p. 203. The novelist Lawrence Durrell repeats this popular belief as follows: "Ali says that shooting stars are stones thrown by the angels in heaven to drive off evil djinns when they try to eavesdrop on the conversations in Paradise and learn the secrets of the future." *Mountolive* (New York: E. P. Dutton, 1959), pp. 31-32. I owe these references to Professor Joel Gordon.

111:72-74 The Koran contains several references to thunder or thunderbolts; they usually refer to acts of divine anger. One of these references (Surah 18:40) has a positive connotation related to agricultural fertility.

111:82-88 The Koran makes several references to earthquakes: Surah 2 ("The Cow"): 214; Surah 33 ("The Clans"), and, most notably, Surah 99 ("The Earthquake"): 1-8. These references usually relate to acts of divine anger and make no mention of angels, pulleys, or any such mechanical operation. Surah 99:1-8 reads: "When earth is shaken with a mighty shaking / and earth brings forth her burdens, / and Man says, 'What ails her?' / upon that day she shall tell her tidings / for that her Lord has inspired her. / Upon that day men shall issue in scatterings to see their works, / and whoso has done an atom's weight of evil shall see it." Arberry, *Koran Intepreted*, 2:348. For help on this point I thank Professor Joel Gordon.

24. OF MAGNETISM

111:97-114:84 This extensive description is taken practically verbatim, except for omissions, from Harris, *LT*, 1:s.v. "Magnet." The sentence at 111:92 is also from Harris. Mather elaborates on Harris as indicated in the three notes that follow.

111:98–112:1 Mather adds the references to Pliny and Aristotle, which he probably takes from Derham, p. 315 n. 21, who actually names Plato and Aristotle. There Derham describes "Dr. Gilbert" as "the most learned and accurate Writer on the *Magnet*." Harris mentions Gilbert along with others who discovered the properties of the magnet. Mather misses the significance of William Gilbert. Aristotle's statement is in *On the Soul*, 1.2.405a 21.

112:6–7 The reference to Gioia is from Derham, p. 315 n. 21.

112:12–19 These lines are added from Derham, pp. 315–16 n. 21.

112:26–27 These lines are not taken from Harris.

114:85–90 These lines draw partly on Harris, *LT*, 1:s.v. "Magnet," but they are based primarily on Harris, *LT*, 2:s.v. "Magnetism." Derham reported his experiment in "An Account of Some Magnetical Experiments and Observations," and "Farther Observations and Remarks on the Same Subject," *Phil. Trans.*, 24 (Sept.–Oct. 1705), 2136–44.

114:91–7 This description is based on Martin Lister, *A Journey to Paris in the Year 1698*, ed. Raymond P. Stearns (Urbana: University of Illinois Press, 1967), pp. 83–96.

114:98–1 The mechanistic path of the lines of force indicate the lines of the magnetic field.

115:8–116:39 These paragraphs closely paraphrase Edmond Halley, "A Theory of the Variation of the Magnetical Compass," *Phil. Trans.*, 13 (June 10, 1683), 215–16, reprinted in *Misc. Cur.*, 1:35–37.

116:40–117:79 This long passage closely paraphrases Halley, "An Account of the Cause of the Change of the Variation of the Magnetical Needle, with an Hypothesis of the Structure of the Internal Parts of the Earth," *Phil. Trans.*, 17 (Oct. 19, 1692), 568, 574–77, reprinted in *Misc. Cur.*, 1:48, 55–58. This essay explains the phenomena described by Halley in the article cited immediately above. Halley quotes the lines from both Vergil (*Aeneid*, 6.640–41) and Claudian, ("Rape of Proserpine," 2.282–84) which Mather incorporates.

117:82–87 Mather probably takes this account from Harris, *LT*, 2:s.v. "Magnetism." The original is in Derham, "Farther Observations and Remarks on the Same Subject [Magnetism]," *Phil. Trans.*, 24 (Sept.–Oct. 1705), 2139.

117:98–118:14 The source is Robert Boyle. The quotation is from *Some Considerations about the Reconcileableness of Reason and Religion. By T. E. a layman. To Which is Annex'd a Discourse of Mr. Boyle about the Possibility of the Resurrection* (London, 1675), pt. 1, sect. 2. T. E. are the final letters of Boyle's names. The quoted material is in *The Works of the Honourable Robert Boyle,* ed. Thomas Birch, 6 vols. (London, 1772), 4:160–61. Mather quotes the passage often nearly verbatim, but he condenses, changes a few words, and reorders the sequence of the material. He adds the introductory clause about the *"natural Imbecillity of the Human Intellect."*

118:19–50 Mather paraphrases and condenses Robert Jenkin, *The Reasonableness and Certainty of the Christian Religion,* 2 vols. (London, 1698–1700), 2:7, 4–6. Mather changes Jenkin's "ten thousand" and "many thousand" to "million" or "millions."

118:51–119:66 These paragraphs quote Grew, p. 55, often nearly verbatim.

119:81–83 Mather again quotes Grew, p. 55.

120:91–92 The source of the quotation is Samuel Ward, *Magnetis reductorium theologicum tropologicum* (London, 1637), p. 21. My translation. Ward takes the line from an unnamed Greek source, but his Latin translation cannot express the play on words (involving "foolish" and "rock") which the original Greek contains. The 1640 translation of Ward's book translated the phrase as follows: "I am not such a one, / As thinks there is a God-head in a stone."

120:9 "Elephantine Book" was used in 1695 to mean Book of Nature. "Elephantine" was also applied to books composed of ivory leaves or tablets on which the ancient Romans recorded official transactions, but this usage dates from 1751, according to the *OED,* and thus Mather would not have known it.

120:10–20 The quotation is from Henry Scudder, *The Christians Daily Walke in Holy Securitie and Peace* (London, 1631), p. 352. For Cotton Mather's youthful reliance on Scudder, see Samuel Mather, *The Life of the Very Reverend and Learned Cotton Mather* (Boston, 1729), p. 8, and Cotton Mather, *Paterna* (Delmar, N.Y.: Scholars' Facsimiles and Reprints, 1976), p. 9.

120:21–121:33 This passage is based on Ward, *Magnetis;* the quotation is at p. 13. Mather changes *"potisimus"* to *"potentissimus."* My translation, based on the 1640 translation of Ward's book.

121:57–60 Mather quotes from Ward, *Magnetis,* p. *2 [preface, p. 5], with slight omission. My translation.

25. OF MINERALS

122:65–69 Mather's source is Arndt, bk. 4, pt. 1, chap. 5, 9 (2:380–81). My translation.

122:70–124:88 This account is based on Harris, *LT,* 1:s.v. "Fossils." Harris gives a table "extracted out of" John Woodward, *An Essay toward a Natural History of the Earth* (London, 1695). Mather organizes his essay around the headings in Woodward: earths, stones, salts, bitumens, metallic minerals, and metals. He adds the *"Nimok"* (Nichols?) reference (117:93); lines 117:4 and 11 are drawn from Ray, pp. 106–7. Oddly, in dealing with fossils, Mather does not mention the teeth and leg presumed to belong to an antediluvian giant man found near Albany, New York, in 1705. He had reported this discovery to John Woodward in the first letter of the first series of his "Curiosa Americana." See "An Extract of Several Letters from Cotton Mather, D.D. to John Woodward, M.D. and Richard Waller, Esq; S.R. Secr.," *Phil. Trans.,* 29 (Apr.–June 1714), 62–63, and Kittredge, pp. 22–23. This letter may be "the earliest printed account of an American vertebrate fossil." See R. M. Hazen, ed., *North American Geology: Early Writings.* Benchmark Papers in Geology, vol. 51 (Stroudsburg, Pa.: Dowden, Hutchinson and Ross, 1979), p. 68. Professor Albert V. Carozzi of the University of Illinois at Urbana-Champaign called this reference to my attention.

124:90–94 The quotation is from Alsted, p. 378. My translation.

124:95–97 This quotation is from Alsted, p. 381. My translation.

124:99–3 Here Mather draws on Ray, pp. 107–9, which quotes from a manuscript "Itinerary of Italy" by Tancred Robinson.

124:3–4 The source is Harris, *LT,* 2:s.v. "Colaptice."

124:5–125:11 This paragraph apparently draws on diverse sources. Kircher included drawings of rocks which resemble such figures as John the Baptist, Jesus Christ, and Jerome, in *Mundus subterraneus,* 8.1, 9, table 4 (p. 36, 1665 ed.; p. 39, 1678 ed.). Ole Worm, *Museum Wormianum: Seu historia rerum rariorum* (Amsterdam, 1655), 1.2.13 (pp. 81–92) may be the source of the Wormius reference. Monconys discusses many rocks that resemble human figures in his *Voyages,* but none of his references precisely fits Mather's description.

125:13–39 The source is Harris, *LT,* 1:s.v. "Fossils."

126:40–43 Here Mather paraphrases and quotes Alsted, p. 380. My translation.

126:50 The source is Harris, *LT,* 1:s.v. "Fossils."

126:51–57 The source is Johann Bicker, *Hermes redivivus, declarans hygieinam, de sanitate vel bono valetudine hominis conservanda* (Giessen, 1612), *Epistola Dedicatoria* [p. 4].

126:58–62 Mather's sophisticated use of classical learning is evident here. The Latin phrase "Animam habent pro Sale" is found in various forms in Varro, Cicero, Pliny, Plutarch, Clement of Alexandria, and Porphyry. Mather's wording differs from that of the authors named, which suggests that he is rephrasing a familiar idea. The wordplay is clever. The Latin *sal* means both "salt" and "wit" or "intelligence." The English "insipid" means "without savor or taste" or "lacking salt," and more generally "dull" or "uninteresting." In the classical writers, the point of the Latin clause, to have a soul instead of salt, is that the pig, being good for nothing but banquets, has a soul for no other reason than to serve as a preservative (salt) for the flesh. Mather employs the classical tag not only to characterize the brute beast but also to suggest a vital function of the mineral salt.

126:63–70 The source is Harris, *LT,* 1:s.v. "Fossils," except for the description of mercury.

126:71–72 Cardan argues that rocks have a soul in *De subtilitate* (1550), bk. 6; in *De rerum varietate* (1557), bk. 4, chap. 16; and in *De natura,* bk. 1, chap. 3. These references are in Cardan, *Opera omnia,* 10 vols. (Leyden, 1663), 3:478, 47, and 2:295–98 respectively. It seems unlikely that Mather took this information directly from Cardan.

126:73–127:81 Mather loosely paraphrases and then quotes Alsted, p. 397. My translation. Alsted's quotation parallels Cicero: "Quotus enim quisque philosophorum invenitur qui sit ita moratus, ita animo ac vita constitutus, ut ratio postulat?" ("How few philosophers are found to be so constituted and to have principles and a rule of life so firmly settled as reason requires!") *Tusculan Disputations,* 2.4, 11.

127:90–91 This statement is quoted nearly verbatim from Grew, p. 100, who is affirming that we are capable of commanding and enjoying the world as it is furnished.

127:92–2 These lines are from Ray, pp. 111–13, who is quoting Cockburn, *An Enquiry into the Nature, Necessity, and Evidence of Christian Faith in Several Essays,* 2 pts. (London, 1696–97), pp. 87–91. Ray does not quote the Latin phrase. Mather probably borrowed it from or was reminded of it by Alsted, p. 399, which quotes a line from Ovid, *Metamorphoses,* 1,138–40, without attribution. My translation. Ovid wrote that not only did men demand crops and sustenance of the bounteous fields "sed itum est in viscera terrae, / quasque recondiderat Stygiisque admoverat umbris, / effodiuntur opes, inritamenta malorum." ("But they delved as well into the very bowels of the earth; and the wealth which the creator had hidden away and buried deep amidst the very Stygian shades, was brought to light, wealth that pricks men on to crime.")

127:3–11 This material is from Harris, *LT,* 2:s.v. "Ductility." Harris summarizes Edmond Halley, "An Account of the Measure of the Thickness of Gold upon Gilt-Wire, Together with a Demonstration of the Exceeding Minuteness of the Atoms or Constituent Particles of Gold," *Phil. Trans.,* 17 (July–Sept. 1691), 540–42; reprinted in *Misc. Cur.,* 1:245–47.

128:12–18 This passage is paraphrased from Grew, pp. 12–13.

128:18–20 These lines are quoted verbatim from William Dampier, *A New Voyage Round the World* (London, 1697), p. 153.

128:25 The Latin is from Vergil, *Aeneid,* 3.57.

128:29–32 The source is Alsted, p. 402, which Mather reorders and condenses. My translation.

128:33–129:43 This entire passage is borrowed from Barker, *Natural Theology*, pp. 62–63. Mather changes Barker's "Tsabii" to "Zabians." Barker quotes "the Poet": "Jupiter esse pium statuit quodcunque juvaret." ("Jupiter fixed that virtue was to be in whatever brought us pleasure.") The quotation is from Ovid, *The Heroides*, 4.133. Barker also quotes "O Sanctas, Gentes, quibus haec nascuntur in hortis, / Numina, &c.," and provides this translation: "An holy people ye I trow, / Who have your Godds in Gardens grow." The source is Juvenal's satire on the Egyptians, Satire 15, ll. 10–11. Barker gives the source of the Pliny quotation as 1.7; the correct reference is 2.5, 18. Barker quotes Philo in Greek, citing *De mundi opificio* as his source. Mather silently "improves" Barker's Greek.

26. Of the Vegetables

129:47–49 These lines echo Derham, pp. 444–45, especially the reference to vegetables as "the Creator's Contrivance."

129:50–131:16 Mather takes this account from Harris, *LT*, 1:s.v. "Vegetation," skillfully paraphrasing and condensing the long entry. Harris reports in detail on John Woodward, "Some Thoughts and Experiments Concerning Vegetation," in *Phil. Trans.*, 21 (June 1699), 193–227; reprinted in *Misc. Cur.*, 1:205–44.

131:17–35 This account of two "Curiosa Botanica" was sent to James Petiver, a fellow of the Royal Society, in a letter dated 24 September 1716 ("Curiosa Americana," third series, no. 12). The letter demonstrates Mather's zeal in supplying descriptions of "Curiosa Americana" to the Royal Society, for it was sent *after* Mather had dispatched the manuscript of *The Christian Philosopher* to London. See Kittredge, pp. 42–43. The dark colors of corn are dominant to yellow, and wind-pollination would yield colored kernels in the first generation. Gourds and squashes cross freely, but the vegetative first generation would not be affected by the bitter gourd contribution. Seeds planted in subsequent years from cross-pollinated plants would have problems. Mather may have missed a season in his account. I thank Professor David Nanney of the University of Illinois at Urbana-Champaign for advice on this point.

131:36–39 This paragraph is based on Harris, *LT*, 1:s.v. "Plants."

131:40–42 This paragraph is based on Harris, *LT*, 1:s.v. "Trees."

131:43–133:84 This material comes from Ray, pp. 116–18, 119–24, 129–30. Ray's translation. Plastic nature possessed many attributes of the Platonic *anima mundi*. The theory held that God endowed matter with a plastic nature at the creation; this spiritual power ordered matter during the six days of creation, and thereafter it continued to form matter into various organisms. Plastic nature operates blindly to achieve divine ends of which it is unconscious, acting internally and immediately upon passive matter, and thus regulating the growth and magnitude of vegetables. The doctrine dominated the best English thought for half a century, but it was flawed from the beginning. Plastic nature was an occult power, explaining the unknown by another hypothetical unknown, and it could not be justified by observation. Belief in plastic nature waned in the late seventeenth century, and the concept was outmoded by the time Mather wrote. See William B. Hunter, Jr., "The Seventeenth Century Doctrine of Plastic Nature," *Harvard Theological Review*, 43 (July 1950), 197–213.

133:84–134:88 This sentence is taken from Arndt, bk. 4, pt. 1, chap. 3, 15 (2:337). My translation, based largely on that of Boehm.

134:93–135:18 This material draws upon Derham, pp. 445–48 and nn. 4–8.

135:24–34 Mather again borrows from Derham, pp. 448–49 n. 9. Derham's translation.

135:35–42 These lines are quoted from Cheyne, chap. 3, p. 231.

135:43–136:54 This paragraph draws on Derham, pp. 449–50 n. 9. Derham's translation of the first Latin quotation, my translation of the second one.

136:55–56 The ultimate source is Thomas Browne, *Hydriotaphia* and *The Garden of Cyrus* (London, 1658), p. 136.

136:56–58 The ultimate source is Ottonis de Guericke, *Experimenta nova (ut vocantur) Magdeburgica de vacuo spatio* ([Amsterdam, 1672]; rpt., Aalen: Otto Zeller, 1962), p. 67. This author gives the number of fifty-three ciphers (or zeros).

136:61–68 This paragraph is taken nearly verbatim from Grew, pp. 11–12.

136:69–71 The source is Derham, pp. 444 n. 1.

136:77–137:85 This story comes from Joseph Pitts, *A True and Faithful Account of the . . . Mohammetans* (Exeter, 1704), p. 168.

137:86–88 The sentence comes from Grew, p. 102.

137:88–92 The source is Derham, p. 445 n. 3.

137:93–95 These lines are based almost verbatim on John Speed, *The History of Great Britaine under the Conquests of Ye Romans, Saxons, Danes and Normans* (London, 1611), p. 167.

137:96–138:28 These lines are based on Derham, pp. 450–55 and nn. 10–16.

138:30–140:69 The source is Derham, pp. 456–58 and nn. 18–20, but the ultimate source of the Latin quotation is John Ray, *Historia plantarum*, 3 vols. (London, 1686–1704), 1:46. Derham's translation.

140:76–78 The source is Plutarch, *Quaestiones convivales,* 7.724E in his *Moralia.* Mather had made this statement earlier in his *Nets of Salvation* (Boston, 1704), p. 4.

140:79–141:91 This material is based on Ray, pp. 240–41.

141:91–93 The Hernandez quotation is from Derham, p. 459 n. 22.

141:94–142:22 The source is Ray, pp. 241–44 passim.

141:6 The reference to *Bandura cingalensium* is from Ray, p. 241, who wrote: "The *Bandura Cingalensium,* called by some the *Priapus Vegetabilis,* at the end of whose Leaves hang long Sacks or Bags, containing a pure limpid Water of great use to the *Natives,* when they want Rain for eight or ten Months together." Ray also describes this plant under the heading *Bandura Cingalensium Gentianae Indicae species* P. Amman" in his *Historia plantarum,* 3 vols. (London, 1686–1704), 1:721–22. Ray notes that the plant grows near Colombo (in Ceylon, now Sri Lanka) in humid and shady woods. This plant is *Nepenthes* (Pitcherplant) of the family Nepenthaceae, classically any plant which is capable of producing euphoria. The plant is highly ornamental, and its root has a vegetable bitter which is valued for medicinal uses.

142:23–28 Mather paraphrases Grew, p. 103.

142:35–36 In Christian fancy the Garden of Eden was also called the earthly paradise to distinguish it from the heavenly paradise, the seat of God and his angels and the final abode of the righteous. In Islam the ideal garden, Paradise, is portrayed in the Koran, which paints a detailed picture of the state of blessedness reserved exclusively for believers that might have served as a model for creators of gardens in East and West. These gardens incorporate lawns interspersed with streams of running water, trees bowed with fruit (but no flowers), seats, pavilions, and summer houses. The Muslim world looked to Iran for the art of landscape gardening. Persian horticulture flourished long before the birth of Islam and was associated with princely life. Xenophon refers to a beautiful park planned at Sardis by Cyrus the Younger (407 B.C.). The Muslim world adopted two styles of garden—gardens in an architectural framework, such as courtyards planted with trees, and spacious parks outside the towns, embellished by pavilions—and these spread across the world and centuries. The first style was preserved in Persia, and it inspired the Greco-Roman garden. Roman gardens were rather formal in their plan, with regular walks lined by closely clipped hedges of box, yew, and cypress; and diversified with statues, pyramids, and summer houses. The trees and shrubs were often cut into figures of animals, ships, letters, and grotesque forms. The principal flowers known to the ancients were the rose, violet, crocus, narcissus, lily, iris, poppy, amaranth, and gladiolus. The Roman garden inspired the Italian garden and the

"French-style" of garden, of which Versailles is the most notable example.

142:47–143:52 The source is Derham, p. 460 n. 23. My translation.

143:53–56 These lines are used again in *Bethesda,* p. 263.

143:57–60 The probable source is Alsted, pp. 412, 425.

143:61–62 This Latin apothegm was probably well known in the early modern period. It has a corresponding English proverb: "He that would live for aye, / Must eat Sage in May." See M. Grieve, *A Modern Herbal,* 2 vols. ([1931]; New York: Hafner Publishing Co., 1959), 2:701. Mather cites the Latin quotation again in *Bethesda,* p. 263.

143:63–144:67 The basis for this paragraph is "Of Health and Long Life," in *The Works of Sir William Temple,* 2 vols. (London, 1720), 1:284–86. Mather repeats these lines in *Bethesda,* p. 263.

145:99–4 The source is Derham, p. 461 n. 25, quoting Ray, *Historia plantarum,* 1:833.

145:5–8 The source is Ray, p. 131.

145:9–11 This passage is lifted from Derham, p. 461 n. 25.

146:23–24 The reference is to Alsted, pp. 444–50.

146:43–49 The lines are quoted from Nahum Tate, *Panacea: A Poem upon Tea* (London, 1700), title page and p. [35].

146:50–147:60 The source is Thomas Robinson, *An Essay towards a Natural History of Westmoreland and Cumberland . . . To Which Is Annexed, A Vindication of the Philosophical and Theological Paraphrase of the Mosaick System of the Creation* (London, 1709), p. 97.

147:61–69 The source is Robert Beverley, *The History and Present State of Virginia* [1705], ed. Louis B. Wright (Chapel Hill: University of North Carolina Press, 1947), p. 139.

147:75–77 Mather's source is probably Derham, p. 448 n. 7. The quotation is from Pliny, *Natural History,* 21.1, 2.

148:86–87 The source is Alsted, pp. 452–54.

148:96–98 Mather's source is Alsted, p. 245, quoting Bernard of Clairvaux.

148:3–13 The quotation is from Arndt, bk. 4, pt. 1, chap. 3, 10 (2:335). Boehm's translations.

149:14–17 This paragraph is based on Alsted, p. 423.

149:18–25 This paragraph is based on Arndt, bk. 2, chap. 37, 5 (2:48). Boehm's translations. I have not located the quotation from Augustine.

149:26–37 This paragraph paraphrases and quotes Arndt, bk. 4, preface (2:314–15). Boehm's translations.

150:46–55 The quotations are from Arndt, bk. 4, pt. 2, chap. 19, 1 (2:437) and bk. 2, chap. 42, 12 (2:89). Boehm's translations.

27. OF INSECTS

150:57–151:64 Mather's source, which he greatly condenses, is Harris, *LT,* 2:s.v. "Insects."

151:64–65 Mather was probably reminded of the quotation from Ray, p. 196. The original is in Pliny, *Natural History,* 11.1, 2.

151:71–152:22 This material is condensed from Harris, *LT,* 2:s.v. "Insects." Swammerdam created a system in which he classified insects as ametabolic ("no" metamorphosis), hemimetabolic (simple metamorphosis), and holometabolic (complete metamorphosis), although he did not use these terms. Swammerdam's manuscripts were posthumously published by Herman Boerhaave in two volumes with Latin and Dutch on facing pages. The former was entitled *Biblia naturae, sive historia insectorum* (Leiden, 1737–38). This work was translated into English as *The Book of Nature: Or, the History of Insects,* trans. Thomas Flloyd, rev. by John Hill, 2 pts. (London, 1758).

152:23–153:35 The source is Barker, *Natural Theology,* p. 26.

153:36–154:64 Mather follows Harris, *LT,* 2:s.v. "Insects." On Garden, Harris cites an article entitled "A Letter from Dr. Geo. Garden . . . Concerning Caterpillars that Destroy Fruit," in *Phil. Trans.,* 20 (Feb. 1698), 54–55.

154:65–77 Mather takes this material from Ray, pp. 347–58. My translation of the French phrase, Ray's translation of the Latin phrase.

154:77–94 Here Mather draws upon Cheyne, chap. 2, pp. 23–24.

155:98–156:63 Mather takes this material from Ray, pp. 360–73, 356–57. The Latin at 155:10–11 is apparently his own, based on information in Ray.

156:64–157:74 This material is taken from Derham, pp. 400–401 n. 1.

157:75–76 The source is "Part of a Letter from Mr. Anthony van Leeuwenhoek, F.R.S. Concerning the Eyes of Beetles," *Phil. Trans.,* 20 (May 1698), 169–75.

157:77–160:61 This material is based on Derham, pp. 401–8 and notes. The ultimate source of the Galen quotations is *De usu partium* 17.1. The translation is from May, *Galen,* 2:731–32. I thank Mr. Matthew P. Berg, Rare Books Assistant in the Joseph Regenstein Library, University of Chicago, for providing information on Oswald Nerlinger.

160:61–161:64 These lines are from Harris, *LT,* 1:s.v. "Animalcula."

161:66–75 The first sentence of this paragraph is based upon Jewish legend, with the Latin quotation from a translation of Solomon ben Isaac's commentary on Exod. 8:14 (in the A.V. 8:14 deals with the plague of frogs; the proper reference would be 8:16–29). My translation. See Ginzberg, *Legend of the Jews,* 2:347–52, and Solomon ben Isaac, *Commentarius hebraicus in Pentateuchum Mosis . . . Latine ver-*

sus . . . a Joh. Friderico Breithaupto (Gotha, 1710), p. 444. The quotation there reads: "Quia daemon nullam potestatem habet in creaturam, quae minor est grando hordaceo." I thank Mr. James Green, Rare Book Bibliographer, Special Collections, The Joseph Regenstein Library, University of Chicago, for locating the reference. As Mr. Green writes, "The fact that the text differs from Mather's, while the sense is the same, indicates that Mather was using a different translation from Breithaupt's. Since there seems to be no earlier separately published Latin translation of Solomon ben Isaac's commentary on the Pentateuch, Mather must have been using an edition of the Bible with commentary in Latin. It looks as if the field is all too wide." The remaining lines build on this Jewish legend, with the last two sentences in the paragraph drawing primarily upon New Testament accounts (Matt. 12:24–28, Mark 3:22–26, and Luke 11:15–20).

161:76–162:96 These paragraphs derive from Derham, pp. 409–10.

162:97–98 This line is from Thomas Robinson, *Vindication of the . . . Mosaick System*, p. 91.

162:99–165:60 This material is paraphrased from Moses Rusden, *A Further Discovery of Bees: Treating of the Nature, Government, Generation and Preservation of the Bee* (London, 1685), pp. 3–9, 16–20, 25–27, 31–33, 42, 50, 57, 53. The ultimate source of the Pliny quotation is *Natural History*, 11.16, 51. The lines of verse are from Vergil, *Georgics*, 4.212–13, 215–18.

165:61–66 The source is Samuel Purchas (the Younger), *A Theatre of Politicall Flying-Insects* (London, 1657), pp. 257–387.

165:66–68 The source is Pliny, *Natural History*, 11.4, 14.

165:69–171:56 This account is taken primarily from Derham, pp. 411–28 and notes. But Mather interpolates other material into Derham's description as indicated by the following nine notes.

165:79–166:90 These two paragraphs are from Thomas Robinson, *Vindication of the . . . Mosaick System*, pp. 90–91.

166:13–18 These lines closely follow John Jonston, *Thaumatographia naturalis, in decem classes distincta* (Amsterdam, 1632), p. 359. In the *History of the Wonderful Things of Nature: Set Forth in Ten Severall Classes* (London, 1657), the reference is at p. 253.

166:18–19 The reference is to *The Geography of Strabo*, 2.1, 9.

167:36–41 This paragraph is not from Derham.

168:64–70 These lines are based on Mather's personal knowledge, but he may have reported two different cases of intestinal worms. The event described here, involving a worm that measured 150 feet long, must have occurred before Mather sent his manuscript of *The Christian Philosopher* to London in 1715. He subsequently described "A Prodigious Worm" in a letter to John Woodward of London dated 12

December 1717. This communication, one of his "Curiosa Americana" for the Royal Society, included a copy of a letter from John Perkins of Boston, Mather's family physician. At Mather's request, Perkins had examined the worm vomited up by "a lusty tall new Negro" and found 128 feet remaining after neighbors had taken away many pieces as souvenirs. *The Boston News-Letter* of 7–14 January 1717 reported this event as occurring the previous day (6 January). If that date is accurate, Kittredge errs in concluding that the 1717 episode is the one described in *The Christian Philosopher*. See Kittredge, pp. 44–45. Perhaps Mather's letter to Woodward was sent on 12 January 1717 (old style) rather than 12 December 1717 (and thus *after* the episode reported in the *Boston News-Letter*). But the point cannot be checked now because there is a gap between 1714 and 1723 in the Cotton Mather correspondence to Woodward in possession of the Royal Society. N. H. Robinson, Librarian of the Royal Society, to author, 24 September 1986. Mather tells this story again in essentially the same terms in *Bethesda,* p. 203.

169:81–83 These lines, not in Derham, are based on Athanasius Kircher, *Mundus subterraneus,* 12.2, 7 (p. 370, 1665 ed.; p. 389, 1678 ed.).

169:99–2 These lines are taken from Ray, p. 196.

170:10–11 Mather introduces these lines into Derham's account. I am unable to locate the Latin quotation in the person who seems to be the most likely author—Guilielmus Arvernus (or William of Auvergne), "De fide et legibus" and "De universo" in his *Opera omnia,* 2 vols. ([Paris, 1674]; rpt., Frankfurt: Minerva, 1963).

170:21–22 Mather most likely took this information from Jonston, *Thaumatographia naturalis,* pp. 379–406. A translation is in Jonston, *History of the Wonderful Things of Nature,* pp. 268–87.

171:56–60 This quotation is from Ray, p. 351.

171:61–172:71 These lines, except for the rhapsody (161:73–77), are from Grew, p. 42.

172:72–78 These lines are based upon Thomas Robinson, *Vindication of the . . . Mosaick System,* p. 73.

172:86–94 This description is from Dampier, *New Voyage,* p. 430.

172:94–173:96 The story is related in *Diodorus of Sicily,* 3.29 (2:160–65), and in *The Geography of Strabo,* 16.4, 12 (7:326–27).

173:18–20 The source is Dampier, *New Voyage,* p. 464.

174:32–37 The source is Edward Terry, *A Voyage to East-India* (London, 1655), pp. 357–58.

175:76–176:27 Except for Mather's rhapsody (176:99–3) the source is Cheyne, chap. 2, pp. 17–22.

28. Of Reptiles

177:36–179:63 These paragraphs are closely based on Derham, pp. 433–36 and nn. 1–5, 164. Derham's translation.

179:64–67 These lines are from Samuel Bochart, *Hierozoicon, sive bipertitum opus de animalibus Sacrae Scripturae* (London, 1663), reprinted in his *Opera omnia*, 3 vols. (Leiden, 1692), 2:430–31.

179:67–69 The reference is to Konrad Gesner, *Historia animalia*, 4 vols. (Zurich, 1551–58). (But it is unlikely that Mather had seen this work.)

179:70–80 This paragraph is from Bochart, *Hierozoicon*, 2:431.

179:83–88 This verse is in *Poems by the Earl of Roscommon* (London, 1717), p. 55. There the third line reads: "Change your fierce Hissing into joyful Song."

180:91–3 These lines are based on Derham, pp. 436–39 nn. 8–9. Mather's discourse, *A Treacle Fetch'd Out of a Viper* (Boston, 1707), described falls into sin and how to recover from them. See Mather, *Diary*, 1:580.

180:3–12 These lines draw on Mather's American experience. He reported on rattlesnakes three times in his "Curiosa Americana," once before and twice after completing the manuscript of *The Christian Philosopher*. These lines are a condensed version of the first series, no. 11 (27 November 1712). See Kittredge, pp. 26, 39, 51.

180:13–16 This material comes from Derham, pp. 437–38 n. 9.

180:20–181:25 This paragraph is from Derham, p. 438.

181:26–33 The source is Frantze, *Historia animalium sacra* (Amsterdam, 1643), 3.1 (p. 508). My translation.

181:34–37 The source is Théodore Turquet de Mayerne, "A Discourse of the Viper," *Phil. Trans.*, 18 (June 1694), 164.

181:38–49 The source is Thomas Robinson, *Vindication of the ... Mosaick System*, pp. 71, 69.

182:52–183:98 This account is from Georgius Baglivi, *The Practice of Physick Reduc'd to the Ancient Way of Observations... Together with Several New and Curious Dissertations; Particularly of the Tarantula, and the Nature of Its Poison* (London, 1704), pp. 363–65, 380–82, 359.

183:13–184:24 These two paragraphs are based on Alsted, pp. 531–32, except for the Latin statement. My translation. Mather had used that same phrase in a different context in his *Christianus per Ignem*, p. 194.

29. Of the Fishes

184:27–33 These lines are from Derham, p. 440 and n. 3.

184:33–37 Mather's source is Ray, p. 22.

184:38–40 Mather has probably taken the lines from Derham, but the ultimate source is Pliny, *Natural History*, 9.1, 2.

184:41–45 This material is from Harris, *LT*, 2:s.v. "Fishes." Rondelet had classified fishes according to the places they were found—i.e., sea fish, river fish, and lake or pond fish.

184:46–186:56 This paragraph is based on Ray, p. 175.

186:57–61 The original source is Athanasius Kircher, *Musurgia universalis: sive ars magna consoni et dissoni in X libros digesta*, 2 vols. (Rome, 1650), 1:13. My translation.

186:62–70 These lines are based on Ray, pp. 175–76. But the reference to Borelli comes from Derham, p. 442 n. 10.

186:70–75 The source is Derham, pp. 442 n. 12, 441 n. 6.

186:76–82 This paragraph is borrowed from Ray, p. 31.

187:86–99 This material is from Derham, pp. 185–86 and n. 13.

187:16–188:24 These lines are from Derham, pp. 440–41 nn. 5, 4.

188:25–50 This material is derived from Ray, pp. 176–77, 28.

188:50–189:58 Mather's "Curiosa Americana" communications of 1720 included a piece on "The Whale," and Kittredge concluded that this is "doubtless preserved in part" on pp. 176–78 of *The Christian Philosopher*. See Kittredge, pp. 45–46. But that is not the case. I am unable to identify the author of these lines of verse on the whale.

189:59–60 The quotation is from Alsted, p. 496. My translation.

189:61–75 This paragraph is from George Hakewill, *An Apologie of the Power and Providence of God in the Government of the World* (Oxford, 1627), pp. 126–27.

189:76–192:40 Most of these paragraphs are Mather's own rhapsody based on material that cannot be identified, except for the following two items.

191:94–96 The source of this story is Aelian, *On the Characteristics of Animals*, 8.4.

191:1–3 John Peckham, archbishop of Canterbury (d. 1292) wrote a life of Anthony. The story Mather relates is reported in S. Baring-Gould, *The Lives of the Saints* [1872], new and rev. ed., 16 vols. (Edinburgh: John Grant, 1914), 6:188.

191:18–20 Mather had used this Latin phrase without attribution in an anonymously published lecture sermon, *The Nets of Salvation*, p. 48.

30. Of the Feathered

192:44–47 The probable source is Alsted, p. 493.

192:48–193:69 This material is a skillful condensation of Harris, *LT*, 2:s.v. "Birds."

193:70–196:64 This long account, except for rhapsodical interpolations by Mather at 194:18–23 and 195:43, closely paraphrases Derham, pp. 372–78 and notes. My translation.

196:65–73 The source of this paragraph is William Dampier, *Voyages and Descriptions* (London, 1699), bk. 2, pt. 2, pp. 67–68. Dampier, traveling

Endnotes 361

at the time in the Gulf of Campeche, reported seeing some carrion crows that are "all over white." The snowy cotinga (*Carpodectes nitidus*) found from Honduras southward is entirely white, but it is much smaller than a crow.

196:84–198:2 Mather paraphrases Ray, pp. 147, 181–82. The ultimate source of the Pliny quotation is *Natural History*, 10.76, 155.

198:3–5 These lines are from Derham, p. 380.

198:11–199:25 These paragraphs are taken from Derham, pp. 380–83 and notes. My translation, based on Derham's.

199:29–34 These lines are based on Dampier, *Voyages and Descriptions*, bk. 2, pt. 2, p. 25.

199:38–46 This paragraph draws upon Derham, pp. 383–84 and n. 5.

200:55–63 Here Mather borrows from Grew, p. 24.

200:64–73 Mather relies on Derham, p. 384 and n. 6.

200:74–79 This paragraph is taken verbatim from Ray, pp. 29–30.

200:80–201:99 This material, except for 201:94–95, is taken directly from Derham, pp. 385–86 and notes. My translation.

201:1–5 These lines paraphrase Ray, pp. 138–39.

201:5–15 Mather communicated this same account, "The Nidification of Pigeons," to John Woodward in a letter dated 4 July 1716 (that is, *after* he had dispatched the manuscript which became *The Christian Philosopher*). The letter was the third series, no. 3, of Mather's "Curiosa Americana." See Kittredge, p. 38.

201:16–202:22 These lines paraphrase Ray, p. 139. "Subtle Jacks are Birds as big as Pigeons," wrote William Dampier, who went on to describe their nest-building. "They are called by the English *Subtle Jacks,* because of this uncommon way of building." Dampier described this bird, almost certainly the weaver, in an account of his travels along the Mexican coast in the Gulf of Campeche. See Dampier, *Voyages and Descriptions*, bk. 2, pt. 2, pp. 68–69. Dampier's "Subtle Jacks" are probably Montezuma's Oropendolas (*Psaracolius montezuma*).

202:38–203:61 These paragraphs are based upon Derham, pp. 390–91. The ultimate source of the Pliny quotation is *Natural History*, 10.74, 148.

203:68–78 Mather again follows Ray, pp. 134–35.

203:79–204:94 This long paragraph paraphrases and quotes Cheyne, chap. 3, pp. 231–33.

204:95–4 This paragraph is based on Derham, pp. 391–92.

204:5–21 Here Mather follows Ray, pp. 135–36.

204:22–207:41 This account, probably to John Woodward, is reproduced from Mather's "Curiosa Americana," second series, no. 5 (June 1714). Mather had previously (19 November 1712) written to Woodward about the pigeons in America ("Curiosa Americana," first series, no. 3), and Woodward had requested further information. Mather obtained the details about cocks and hens alternating the care of their young from a letter sent him by his friend Captain Billings. See Kittredge, p. 31.

207:42–43 These lines are taken from Alsted, p. 466, who quotes them without attribution. My translation.

207:44–46 These lines are taken verbatim from Derham, p. 391 n. 2, which paraphrases Ray, p. 137.

207:46–55 Mather paraphrases Ray, p. 137.

207:56–208:63 Mather borrows from Derham, pp. 392–93.

208:64–70 This material paraphrases Ray, p. 138.

208:71–78 Again Mather turns to Derham, p. 394.

208:79–209:7 Mather closely paraphrases Ray, pp. 136, 139–40.

209:7–12 These lines are based upon Thomas Robinson, *Vindication of the . . . Mosaick System*, p. 65.

209:12–15 This material follows Ray, p. 146.

209:16–28 This paragraph is taken from Derham, p. 317 n. 25, but Mather adds the reference to Du Bartas. He probably knew Du Bartas, but may have taken this addition from either Nath[aniel] Wanley, *The Wonders of the Little World: Or, a General History of Man in Six Books* (London, 1678), p. 224, or from one of Wanley's sources, namely, Hakewill, *Apologie of the Power and Providence of God*, p. 255.

209:29–210:36 This paragraph is based on Derham, pp. 395–96. The ultimate source of the quotation is Augustine, *De civitate Dei*, 5.11. Oates's translation.

210:37–43 The Mishnah is the oral Law developed from the first half of the second century B.C. to the close of the second century A.D. for the purpose of applying "the Law" (*Torah*) of the Pentateuch to the daily life of the Jews. In Judaism the accepted belief was that the oral Law was delivered to Moses on Mount Sinai at the same time as the written Law and was preserved by transmission to successive generations until it found expression in the Mishnah. Thus the oral Law was of the same divine origin and possessed the same authority as the written Law. Rabbi Judah the Patriarch brought together this mass of traditional laws (halakoth) and compiled the oral Law as it was taught in many schools at the end of the second century A.D. The Gemara is the comments and discussion on the text of the Mishnah by several generations of scholars and ecclesiastical lawyers in the rabbinical schools of Palestine and Babylonia. The Mishnah and the Gemara commentary together constitute the Talmud. Mather shows awareness of material found in *The Babylonian Talmud: Seder Kodashim. Hullin*, trans. Eli Cashdan, 2 vols. (London: Soncino Press, 1948), 1:343.

210:44–211:64 These two paragraphs closely paraphrase Grew, p. 99.

211:72–79 This material is from Alexander Clogie, *Vox Corvi: Or, the Voice of a Raven* (London, 1694), [preface].

212:84–87 This paragraph probably uses as a point of departure Alsted, p. 472, but Mather must have been familiar with Epictetus in a context other than Alsted as well. The ultimate source is Epictetus, *The Discourses as Reported by Arrian*, 1.16. My translations.

212:4–5 The source is Alsted, p. 487. My translation.

212:6–9 This rhapsody is probably based on Alsted, p. 474, who cited Simone Maiolo, *Dies caniculares* (Rome, 1597), pp. 107–8, as one of his sources for a discussion that Mather condenses.

31. OF THE FOUR-FOOTED

212:11–215:80 This classification, except for 215:66–73, is a précis of Harris, *LT*, 2:s.v. "Quadrupeds."

215:81–216:88 This paragraph is based on Fénelon, *Demonstration of the Existence, Wisdom, and Omnipotence of God*, pp. 57–58.

216:89–217:32 These paragraphs are based on Derham, pp. 354–57 and notes.

217:33–218:58 These lines are based on Ray, pp. 378, 387–89.

218:59–66 These lines follow Derham, pp. 357–58 and nn. 1–2.

218:66–74 The remainder of the paragraph is based on Ray, pp. 161–65.

218:75–219:1 These two paragraphs are derived from Derham, pp. 358–60.

219:2–7 This material follows Ray, p. 385.

219:8–20 This paragraph is based on Grew, p. 24.

219:21–220:34 Here Mather follows Ray, pp. 183–84.

220:35–48 These paragraphs draw upon Derham, pp. 361–62 and n. 1. The Cicero quotation is ultimately from *De natura deorum*, 2.47, 123. The Galen quotations are ultimately from *De usu partium*, 11.8. Derham has apparently modified the original of the first Galen quotation; at least Derham's text differs from that given in Karl G. Kühn, *Medicorum Graecorum opera quae exstant*, 26 vols. in 28 (Leipzig, 1821–33), 3:876. The translations of Galen are from Derham, *Physico-Theology* (1798 ed.), but the first translation is also based partly on May, *Galen*, 2:520.

220:49–221:62 This paragraph draws upon Ray, pp. 382–84.

221:63–90 Mather bases this material on Derham, pp. 363 and n. 1, 199–200 n. 49, 198 n. 45.

222:98–8 This paragraph is from Derham, pp. 197–98 n. 44. Derham's translation.

222:9–21 Mather paraphrases Ray, pp. 390–91.

222:22–223:31 This material is taken from Derham, pp. 364 and n. 3, 158 and n. 14.

223:32–34 This material restates Ray, p. 382.

223:35–224:83 This long description paraphrases Derham, pp. 158 n. 14, 365–69 and n. 4. Derham's translation.

224:87–225:94 The quotation is condensed from John Smith, *Christian Religion's Appeal from the Groundless Prejudices of the Skeptick, to the Bar of Common Reason* (London, 1675), 3.1, 3 (bk. 3, p. 7).

225:95–1 Here Mather paraphrases and quotes from Derham, p. 371 and n. The ultimate source is Lactantius, *Divinarum institutionem*, 7.9. The library of the Mathers possessed a copy of bk. 7 of the *Divine Institutes* (Basel, 1521).

225:2–16 This paragraph condenses an account in Ray, pp. 402–6.

225:20–226:28 Mather's source is Girolamo Rorario, *Quod animalia bruta ratione utantur melius homine* (Paris, 1648), pp. 12–13. My translation.

226:33–53 Mather is quoting Richard Blackmore, "An Essay upon the Immortality of the Soul," in *Essays upon Several Subjects*, 2 vols. (London, 1716), 1:346, 348, 349–50. Mather sent his manuscript to London in 1715, which raises the question as to how Mather saw this essay. Perhaps Blackmore's book actually appeared in 1715 and Mather used it before sending off his manuscript (Mather's own book appeared in 1720 though it bears a 1721 publication date.) Mather admired and corresponded with Blackmore, and may have seen this essay in manuscript.

226:54–227:73 These two paragraphs paraphrase Ray, pp. 61–62.

227:74–79 Camerarius was counselor of the city of Nuremberg and the first pro-chancellor of the academy established by the city council at Altdorf. One of his annual functions in this position was to deliver an informal oration on a light topic following a serious address by the dean when degrees were conferred. These orations, which represent reflections gathered in spare time from reading and experience, are contained in the one work left by Camerarius, *Operae horarum subcisivarum*. The most complete of the many editions of that work was the one published in three volumes at Frankfurt in 1624. The first book was translated into English by John Molle as *The Living Librarie, or Meditations and Observations Historical, Natural, Moral, Political, and Poetical* (London, 1621); a second English edition appeared in 1625. The library of the Mathers contained a three-volume Latin edition published at Frankfurt from 1644 to 1650. Mather's source for these lines is Camerarius, *Operae horarum subcisivarum* (1591), pp. 102–8. The same material is in Camerarius, *Living Librarie* (1621), pp. 82–86.

227:80–84 Mather's source is probably Camerarius, *Operae horarum subcisivarum*, pp. 107–8; also in *Living Librarie*, p. 86. The ultimate source is Jacobus Micyllus, *Sylvarum libri quinque* ([Frankfurt], 1564), pp. 436-38.

227:85-2 This material is a paraphrase of Ray, pp. 386-87.

228:3-5 The reference is to Plutarch, *"De sollertia animalium,"* in *Moralia*, 959-85. Mather is obviously familiar with other classical writers on the cleverness of animals.

228:6-7 The story of Reynard pitting his force and wit against other animals appeared on the Continent in several versions of beast-epics starting in the Middle Ages. A French one was probably composed beginning in 1175, a German one was written in 1180, and a Middle Dutch one, the most important and influential version, was almost certainly written in the thirteenth century. A prose version of the latter printed in 1479 at Gouda was the basis for the first English edition published by Caxton in 1481. Perhaps Mather knew *The History of Reynard the Fox* in one of the versions which had appeared since that time.

228:8 The probable source on the cleverness of the fox is Derham, p. 204 n. 57.

228:13-15 These lines are based on Valeriano Bolzani, *Hieroglyphica* (Lyons, 1611), pp. 20-21. According to the story Valeriano related, Manuel I decided to send Hanno to Rome as a present to Pope Leo X (1475-1521), but when the ship was ready to depart, the elephant would not embark. The king, much annoyed at this obstinacy, promised great reward to whoever would transport Hanno, but no one came forward. The king then learned that the impasse was owing to Hanno's master, who did not want to leave his girl friend and therefore persuaded Hanno not to depart, informing the beast that Italy was a disagreeable country where the elephant would be ridiculed, deprived of necessities, and made miserable. The king, discovering this malice, told the servant to depart with Hanno within three days or he would be executed as an example to others who might have the temerity to disobey royal commands. The master then convinced Hanno that Rome was one of the best cities in the world and that he would be fed with delicacies there. Thus Hanno voluntarily embarked on the long voyage.

228:18-23 These lines are based on Alsted, pp. 518-20, 523-24. My translation.

228:24-25 The source is Solinus, *Collectanea rerum memorabilium*, translated by Arthur Golding as *The Excellent and Pleasant Worke of Julius Solinus Polyhistory* (London, 1587), 45.5-15.

228:26-30 These lines are taken nearly verbatim from Grew, p. 99.

230:33-34 The source of these lines is Dio Cassius, *Roman History*, 69.10, and Edward Topsell, *The History of Four-Footed Beasts and Serpents*

366 Endnotes

(London, 1658), pp. 222, 336. The inscription to Borysthenes is in Otto Hirschfeld, *Corpus Inscriptionum Latinarum* (Berlin, 1888), vol. 12, no. 1122.

230:44–49 The source is Cheyne, chap. 2, p. 89.

230:52–231:64 Mather's source is Carlo Emanuel Filiberto Giacinto de Simiana, *The Truth of the Christian Religion* (London, 1703), p. 74.

231:65–84 The probable source of the first of these two paragraphs is Thomas Robinson, *Vindication of the . . . Mosaick System*, pp. 77–78, but Mather rephrases his source. The second paragraph is based on Robinson, p. 77.

231:93–232:97 Mather shows knowledge of Frantze, *Historia animalium sacra*, and of Archibald Simson, *Hieroglyphica animalium . . . cum eorum interpretationibus* (Edinburgh, 1622–24).

232:2–7 The quotation is from Seneca, *On Anger*, 3.30, 1–2.

232:8–233:32 This paragraph paraphrases François Leguat, *A New Voyage to the East Indies* (London, 1708), pp. 71–74. Leguat and nine other French Protestant refugees who had fled Holland were put off the expedition's frigate and lived for two years on Rodriguez, an uninhabited island in the Mascarene Islands in the Indian Ocean east of Madagascar, after which they built a boat and worked their way back to Flushing, with sojourns at Mauritius and Batavia, via the Cape of Good Hope and the island of St. Helena. Only three of the original ten returned home after an absence of eight years. Geoffroy Atkinson contends that Leguat's *New Voyage* is a desert-island novel of ideas written almost exclusively from French travel accounts with the personal observations on natural history traceable to nine earlier authors. François Maximilien Misson wrote the "Author's Preface" to the *New Voyage*, and "it is easily possible that Misson was assisted in the writing of the story itself by one or more collaborators." See Geoffroy Atkinson, *The Extraordinary Voyage in French Literature from 1700 to 1720* (Paris: Édouard Champion, 1922), pp. 35–65 (quotation at p. 64).

233:33–234:68 Mather paraphrases and quotes Grew, pp. 100, 102–3.

234:75–235:93 All of this material is from Alsted, pp. 459, 510, 514–17. Professor David Sansone of the University of Illinois at Urbana-Champaign kindly provided the translation of Gaza. My translation of the second Latin quotation.

235:3–4 These lines are based on Alsted, pp. 512–13.

235:13–16 The source is *Dris Martini Lutheri Colloquia Mensalia: Or, Dr Martin Luther's Divine Discourses at his Table, Collected by Dr. Antonius Lauterbach, and . . . disposed into Common-places by John Aurifaber*, trans. Captain Henrie Bell (London, 1652), p. 43. I thank Professor H. G.

Haile of the University of Illinois at Urbana-Champaign for this reference.

235:20–24 Mather's source is Paul Egard, Γνωθι Σεαυτον: sive Tractatus utilissimus de vera microscosmi cognitione tum naturali tum supernaturali, pp. 38–39. My translation.

236:25–43 The conclusion is closely paraphrased from Grew, p. 30. But the phrase "whatever is natural is delightful" is Mather's.

236:46–47 These lines are quoted from Alsted, p. 510. My translation.

32. OF MAN

236:53–237:55 Mather quotes Raymond of Sebonde, *Theologia naturalis, sive liber creaturarum* [c. 1484] (Frankfurt, 1635), Prologue, p. [9]. This book is the first to use the words "natural theology" as a title. Raymond apparently wrote to refute the fideism of the Nominalists, who opposed reason to faith. God gives us both the Book of Nature and the Holy Scriptures, he wrote, and one is able to learn everything necessary, and especially may understand the Scriptures and have an infallible certainty of the truth of Christianity, by human reason, through the study of the creation. *Theologia naturalis* went through ten editions by the mid-seventeenth century. The Roman Catholic church placed it on its 1558–59 Index of Prohibited Books, though the text was removed and only the Prologue retained on a revised list in 1564. Montaigne translated *Theologia naturalis* into French; his longest, most influential essay was entitled an "Apology for Raymond Sebond." The essay emphasizes the vanity of human reason and says little about Raymond of Sebonde.

237:56–59 These lines are based on Augustine, *Confessions*, 10.8. My translation. I thank Mr. J. van Sint Feijth of the *Thesaurus Linguae Augustinianae*, Eindhoven, The Netherlands, for identifying the passage.

237:62–75 Mather quotes from *Theologia ruris*, pp. 200–201.

237:76–78 The lines from Stigelius are taken from Alsted, p. 591; also in Alsted, *Encyclopaedia omnium scientiarum*, 3:330. My translation.

237:83–238:89 These lines are based on Vitruvius Pollio, *On Architecture*, bk. 3, chap. 1, sects. 1–4.

238:99–12 The source of this paragraph is Derham, pp. 323–25 and nn. 2–5. *Os sublime* is from Ovid, *Metamorphoses*, 1.84. The ultimate source of the Cicero quotation is *De natura deorum*, 2.56, 140; of the Galen, *De usu partium*, 1.3 (translation from May, *Galen*, 1:69).

238:12–239:17 These lines are based on Ray, pp. 257–60.

239:17–31 These lines, except for 239:21–22, closely paraphrase Derham, pp. 326–28 and nn. 8–9.

239:39–43 The quotation is from Arndt, bk. 4, preface (2:315). My translation.
239:44–47 This paragraph is based on Alsted, pp. 242–44.
239:51–240:69 These lines are based on Ray, pp. 262–64. My translation.
240:69–72 Mather's source is Alsted, p. 596. My translation. Alsted is quoting Camerarius.
240:73–241:1 These two paragraphs are derived from Derham, pp. 334–38 and notes, except for the phrase "the *Sol Microcosmi*" (241:94), which is taken from Ray, p. 267. The ultimate sources of the Latin quotations are Galen, *De usu partium*, 1.18 and 3.10 (translations from May, *Galen,* 1:100, 178) and Cicero, *De natura deorum,* 2.56, 140–41. Galen discusses the four senses lodged in the head in 8.6, 7 (May, *Galen,* 1:400–408).
241:2–3 The source is Ray, p. 266.
241:4–244:70 This material is mainly from Derham, pp. 339–46 and notes, but with exceptions indicated in the following two entries.
243:28–29 Mather interpolates this phrase into Derham's account. My translation.
243:39–41 Mather adds this quotation from Grew, p. 98.
244:66 Derham, p. 346n., cites "Fletus aerumnas levat" from Seneca's *Troades,* 762. The phrase is from Ulysses, who told Andromache, "Weep thy fill. Weeping lightens woes."
244:80–81 These lines are from Derham, p. 347.
244:82–245:99 This paragraph is mainly from Ray, pp. 283–84, but lines 39–41 are from Derham, p. 347 n. 1. The ultimate source of the Pliny quotation is *Natural History,* 7.1, 8.
245:1–5 The source is Derham, pp. 347–48 n. 2.
245:10–246:27 This paragraph quotes from Ray, pp. 284–86.
246:31–247:53 This material is drawn primarily from Cheyne, chap. 3, pp. 221–28, 235–40. But Mather supplements Cheyne from his general reading in anatomical literature.
247:56–58 Mather could have obtained this information from Derham, p. 219 n. 5, though the figure given there is forty ounces a day.
247:63–250:30 This long section, except for two interpolations in which Mather embroiders what he borrows (247:82–249:85 and 250:24–27), closely paraphrases Ray, pp. 277–80, 281–83, 272–75, 254–55.
250:31–33 These lines are from Cheyne, chap. 3, pp. 254–55.
250:34–42 This paragraph relies on Ray, pp. 286–87.
250:43–251:55 This paragraph draws upon Cheyne, chap. 3, p. 255.
251:56–74 This material is based upon Grew, pp. 21, 58.
251:79–80 This sentence is from Ray, p. 288. Mather repeats it in *Bethesda,* p. 155.
252:83–87 Mather's source is undoubtedly "An Extract of a Letter to

Dr. Edward Tyson from the Reverend Mr. Charles Ellis, Giving an Account of . . . the Friesland Boy with Letters in His Eye," *Phil. Trans.*, 23 (July–Aug. 1703), 1418. Ellis described "this Cheat" in scientific terms and with complete incredulity.

252:91–99 Mather's source is Derham, p. 90 and nn. 4–6. The ultimate source of the Pliny quotations is *Natural History*, 5.8, 46 and 7.2, 23–24.

252:96 Mandeville's book is based mainly upon travel accounts by others, with some details from the author's own experience and the addition of fanciful particulars. The work is usually identified with Jean de Bourgogne à la Barbe, a Liège physician who wrote under the pseudonym Jean de Mandeville. The *Travels,* written in French and translated into many other languages, enjoyed an enormous success. In English the earliest dated edition is 1499.

252:1 This line paraphrases Derham, p. 89.

252:1–253:23 This material is based on Ray, pp. 290, 288–89, 291.

253:23–35 These lines draw upon Derham, pp. 101–2 and nn. 25–26. The Pliny quotation is ultimately from *Natural History*, 11.54, 143.

253:36–43 The source here is Ray, p. 292.

253:43–45 Briggs is quoted from Derham, p. 97 n. 16. Derham's translation.

253:46–49 The source is Grew, p. 18.

253:50–254:58 Mather relies on Ray, pp. 292–93.

254:58–73 This material is taken from Derham, pp. 103 n. 27, 106 n. 29.

254:74–85 This paragraph is based on Ray, pp. 293–94.

255:86–92 These lines derive from Derham, p. 107.

255:92–7 This material paraphrases Ray, pp. 294–95.

255:7–14 These lines rely on Derham, p. 113 n. 38. Olscamp's translation.

255:15–256:26 This material borrows from Ray, p. 296.

256:26–29 These lines are based on Derham, p. 108 n. 33.

256:30–257:57 This material is a close paraphrase of Ray, pp. 297–98. My translation of Scheiner. The "Orator" (Cicero) is quoted from *De natura deorum*, 2.57, 143.

257:57–60 This information is taken from Derham, p. 110 n. 34.

257:60–62 These lines draw upon Ray, pp. 298–99.

257:62–66 These lines are based on Derham, p. 109 n. 34.

257:66–76 This material closely paraphrases Ray, p. 299. The ultimate source of the Cicero quotation is *De natura deorum*, 2.57, 142.

257:77–79 These lines derive from Derham, p. 110 n. 34.

258:81–91 The source is [Richard Saunders], *A View of the Soul* (London, 1682), p. 68.

258:92–94 My translations. Mather uses these lines again verbatim in *Bethesda,* p. 156.

Endnotes

258:95–1 Mather relies here on Ray, p. 300.

258:2–6 Mather's source is probably James Keill, *The Anatomy of the Humane Body Abridg'd*, 2d ed. (London, 1703), p. 160, but he could have take the same material, except for the reference to the French Academists, from Cheyne, chap. 3, p. 264.

258:7–259:24 This material is taken from Cheyne, chap. 3, pp. 268–69.

259:25–41 These lines are based on Derham, pp. 95–96 n. 15.

259:41–260:57 The quotations and related material are from Daniel Tauvry, *A New Rational Anatomy... According to the Rules of Mechanicks* (London, 1701), pp. 212, 210–11.

260:58–63 These lines are based on Derham, pp. 111–12 and n. 37. Mather uses them again in *Bethesda*, p. 155.

260:63–67 These lines are based on Cheyne, chap. 3, pp. 271–72. Mather repeats them in *Bethesda*, p. 155.

260:77–80 The source is Matthaeus Tympe, *Mensae Theolophilosophicae* (Münster, 1645), pt. 2, p. 192. I am indebted to Professor John Kevin Newman of the University of Illinois at Urbana-Champaign for this translation. Mather quotes Tympe again in *Bethesda*, p. 156.

260:81–268:37 This long description of the ear, hearing, and sound closely paraphrases the text and notes in Derham, pp. 114–16, 118–25, 127–37. The ultimate source of the Pliny quotation (262:35–37) is *Natural History*, 28.8, 40. Derham's translation of the Schelhammer quotation. John Clarke's translation of Rohault. My translation of Willis. The Olaus Magnus translation is from *A Compendious History*. Mather quotes Pliny again in *Bethesda*, p. 161. Mather's three interpolations in this account are indicated in entries that follow.

261:86–88 Mather adds a comment to Derham's account.

266:48–50 Pietro Martire d' Anghiera's *Decades*, first published in 1516, reported the exploits of voyagers in the service of Spain. Richard Eden included his own translation of Anghiera's *De rebus oceanicis et orbe novo decades tres* (Basel, 1533), in *The Decades of the Newe World or West Indies* (London, 1555). An enlarged edition of Anghiera's history was published in English translation as *De orbe novo, or the Historie of the West Indies, etc. Comprised in Eight Decades. Whereof Three Have Beene Formerly Translated into English by R. Eden, Whereunto the Other Five Are Newly Added by the Industries and Painfull Travails of M. Lok* (London, 1612).

266:64–70 Mather may have seen Hennepin's description of Niagara Falls in *Nouvelle découverte d'un très grand pays situé dans l'Amerique* (Utrecht, 1697) or in the translation, *A New Discovery of a Vast Country in America* (London, 1698), pp. 29–32.

268:25–28 Mather adds comments which are not in Derham.

269:50–70 This material is paraphrased from Derham, pp. 138–39. The ultimate source of the Cicero quotation is *De natura deorum*, 2.56, 141.

270:74–76 Mather indicates familiarity with Grew, "A Discourse of the Diversities and Causes of Tast[e]s Chiefly in Plants," in *The Anatomy of Plants* ([London], 1682), pp. 279–96.

270:77–271:23 These paragraphs paraphrase Derham pp. 141–45 and notes. My translation of Willis; Derham's translation of Malpighi. Both Cicero quotations are from *De natura deorum*, 2.56, 141. The Pliny quotation is from *Natural History*, 10.90, 195.

271:24–31 Here Mather draws on Cheyne, chap. 3, p. 238.

271:32–273:77 This description is taken nearly verbatim from Ray, pp. 305–9.

273:81–9 This material paraphrases, alternately loosely and closely, Ray, pp. 309–11. Mather interpolates the reference to Andreas Vesalius (273:83–85), which shows familiarity with *De humani corporis fabrica* (Basel, 1543), p. 253. I thank Dr. Dorothy S. Adelmann for this reference.

274:23–25 Mather discussed stammering speech in general and the alleged infirmity of the apostle Paul in *Bethesda*, pp. 226–32. Robert Boyle described how he became a stutterer by imitating other children in "An Account of Philaretus [Robert Boyle] during His Minority," in *The Works of the Honourable Robert Boyle*, ed. Thomas Birch, 6 vols. (London, 1772), 1:xiii–xiv.

274:25–31 These lines are based on Derham, p. 150 n. 4.

274:32–37 Mather's wonderful tale is based on historical events. The Roman empire included parts of Africa, but in 429 the Vandals under Gaiseric invaded and captured the fairest provinces, establishing their capital at Carthage. The Vandals treated the Libyans harshly during Gaiseric's reign of nearly forty years. The Arian Vandals persecuted orthodox Christians and imposed their faith upon them, and this cruel policy continued during the reign of Honoric (477–84). Victor, bishop of Vita in Africa, left an eyewitness account of Vandal persecutions in these years. In Carthage, Victor reported, the Vandal king ordered the tongues and right hands of orthodox Christians "cut off at the very roote and stumpe: yet through the assistance of the holy Ghost, they so spake and speake still, as they did neuer before. If any man be incredulous, let him goe now to *Constantinople*, and there shal he find *Reparatus* a Sub-deacon, one of that company, speaking (and that eloquently) without any impediment. For which cause he is greatly reuerenced in the palace of the Emperor *Zeno*." Victor, Bishop of Vita, *The Memorable and Tragical History of the Persecution in Africke, 1605* in *English Recusant Literature, 1558–1640*, ed. D. M. Rogers [Menston, Yorkshire: Scolar Press, 1969], p. 102.

The Vandal kings assured the peace of the Christians beginning in 496, but the emperor Justinian (527–65) valued religious uniformity and decided to recover Africa nevertheless. The army was sent in 533, and Procopius of Caesarea, the historian of the reign of Justinian, went along as secretary of the commanding general. Procopius reports that Honoric forced Christians in Libya to change over to the Arian faith or suffer death, "and he also cut off the tongues of many from the very throat, who even up to my time were going about in Byzantium having their speech uninjured, and perceiving not the least effect from this punishment." (Procopius, *History of the Wars*, 3.8, 4.) Justinian or his lawyers knew these events well; Mather's Latin quotation is from the Code of Justinian (1.27, 1). The relevant portion of the code says (in translation): "We saw venerable men who with difficulty related their sufferings, whose tongues had been cut out by the roots." See S. P. Scott, *The Civil Law: Including . . . the Enactments of Justinian*, 17 vols. (Cincinnati: Central Trust Co., 1931), 12:130. Mather discusses the same episode more fully and from a different perspective in *Bethesda*, pp. 229–30.

275:45–276:82 These paragraphs paraphrase Ray, pp. 311–13.

276:83–4 This material is taken from Derham, pp. 150, 149, and notes.

276:11–277:43 These paragraphs rehearse material in Derham, pp. 146–49 and notes.

277:44–278:54 This material is taken nearly verbatim from Wainewright, *Mechanical Account*, pp. 55–56, 57.

278:61–70 Mather relies here on Derham, pp. 151–53 and notes. The Latin quotation is from Galen's *De usu partium*, 5.15.

278:71–279:89 This information is drawn from James Keill, *Anatomy of the Humane Body*, pp. 124–25.

279:90–280:33 This material is mainly from Ray, pp. 313–18. One sentence (279:97–1) is from Cheyne, chap. 3. p. 256.

280:34–36 These lines are based on Tauvry, *New Rational Anatomy*, pp. 46–57.

280:36–49 These lines are based on James Keill, *Anatomy of the Humane Body*, p. 111, 117–18.

280:55–59 This quotation is probably taken from Derham, p. 197 n. 43. But Mather could have borrowed it directly from Grew, p. 18.

280:59–281:64 These lines borrow from Ray, p. 333.

281:64–282:28 This long passage closely paraphrases James Drake, *Anthropologia nova: Or, A New System of Anatomy*, 2 vols. (London, 1707), 1:180–86.

282:32–283:39 This paragraph paraphrases Ray, p. 333.

283:49–64 This material is based on Ray, pp. 333–34.

283:65–68 Mather's source is Wainewright, *Mechanical Account*, p. 31. Wainewright gives the figure of 79,200.

284:74–2 These paragraphs, except for 284:80–81, are taken nearly verbatim from Ray, pp. 334, 326–27.

284:3–285:6 Mather's source is Grew, p. 18.

285:8–286:73 Mather closely paraphrases Ray, pp. 327–32, 335, except for the rhapsody at 286:60–62. Mather changes Ray's *"Mechanicks"* (286:66) to *"Mathematicks."*

286:74–288:95 This material is drawn from Grew, pp. 28–29, 18, 27. Mather glosses the text at 288:76–77 and 83–85.

288:96–289:35 With the exception of 289:14–17, this material paraphrases Ray, pp. 335–39.

289:36–47 This paragraph paraphrases Cheyne, chap. 3, pp. 242–43.

289:48–290:54 Mather's main source here is apparently Harris, *LT*, 1:s.v. "Blood," but he probably draws also on Wainewright, *Mechanical Account*, pp. 38–50 (a chapter describing the effect of acute distempers, particularly fever, on the blood). Mather errs in reporting Harris, which admittedly is not entirely clear. Allen Moulin, one of Harris's sources, had found by experimentation that a man's blood bears the same proportion to its weight as that of an animal, namely, one-twentieth. See Allen Moulin, "A Conjecture at the Quantity of Blood in Men, together with an Estimate of the Celerity of Its Circulation," *Phil. Trans.*, 16 (Dec. 1687), 433–34.

290:75–83 This paragraph is based on Cheyne, chap. 3, pp. 276–78.

291:1–292:37 Mather's description paraphrases Ray, pp. 321–25, but Mather adds the material noted in the following three entries.

291:3–5 This quotation is no doubt borrowed from Derham, p. 337 n. 4. The original source is Cicero, *De natura deorum*, 2.50, 150.

291:21–292:25 The reference is to Camillo Baldi, *De naturali ex unguium inspectione praesagio comment. Ab eodem Hyppolito Scaffiliono medicinae doctore. Ex eiusdem Camilli Baldi Bonon. philosophi sermonibus collectus, ac typis mandatus* (Bologna, 1629).

292:27–29 The quotation is most likely taken from Derham, p. 323 n. 3. The original is in Galen, *De usu partium*, 1.5 (translation from May, *Galen*, 1:100).

292:37–40 Mather draws here on Grew, pp. 26–27.

292:41–42 These lines draw on Derham, p. 335 n. 2. The ultimate source is Galen, *De usu partium*, 1.18.

292:51–293:58 This material paraphrases Cheyne, chap. 2, p. 35.

293:67–294:16 These two paragraphs are taken primarily from Wanley, *Wonders of the Little World*, pp. 37–39. But Mather supplements Wanley. The description of bones and nerves in the first paragraph probably draws on Cheyne, chap. 3, p. 249. Samson is mentioned along with other strong men in Derham, p. 332 n. 4. The authors mentioned at the end of the first paragraph, with the exception of Valerius Maximus, are all cited by Wanley. Mather quotes all this material and the rhapsody that follows (294:17–295:22) in a discussion of Judges 15, which deals with Samson, in his "Biblia Americana." Professor Edward H. Davidson of the University of Illinois at Urbana-Champaign called my attention to this entry in the "Biblia Americana."

295:23–25 This quotation is from Alsted, p. 510. My translation.

295:26–296:54 Some of this material is taken nearly verbatim from and the rest is loosely based on Derham, pp. 302–3 and notes. The original sources are Cicero, *De natura deorum*, 2.59, 147; Juvenal, Satire 15, ll. 146–47; and Claudian, "The Fourth Consulship of Honorius," ll. 234–35. Galen is quoted from *De usu partium*, 17.1 (translation from May, *Galen*, 2:731).

296:65–297:77 Wishing to revive primitive Christianity, Mather gave three sermons on the subject and published them as *The Good Old Way: Or, Christianity Described, from the Glorious Lustre of It, Appearing in the Lives of the Primitive Christians* (Boston, 1706). These lines are not found in *The Good Old Way*. In 1711 he thought about writing a book on the "Christian Asceticks." See his *Diary*, 1:559, 561, 588–89; 2:87–88, 93, 235, 249. Elsewhere Mather wrote, "It is thro' the Neglect of the Christian Asceticks, which were much maintained in the days of *Primitive Christianity*, that the *Power of Godliness* is now almost lost in the World." See *Bethesda*, p. 42.

297:93–298:19 This account of learned persons is taken primarily from Wanley, *Wonders of the Little World*, pp. 219–21 (my translations of the two Latin quotations.) But there are exceptions, as indicated by the following three entries.

297:97–99 Mather's reference is to Jean J. Bouchard, ed., *Monumentum romanum Nicolao Claudio Fabricio Perescio . . . factum* (Rome, 1638). He may have seen a biography by Pierre Gassendi translated by William Rand and published as *The Mirrour of True Nobility and Gentility: Being the Life of the Renowned Nicolaus Claudius Fabricius Lord of Peiresk* (London, 1657).

298:10–11 Wanley makes no reference to Samuel Bochart.

298:15–16 Both Witsius and Baxter died after publication of Wanley's book.

298:17–19 Mather's source is Wanley, *Wonders of the Little World*, p. 219. My translation.

298:34–36 The reference is to Adrien Baillet, *Des enfans devenus célèbres par leurs études ou par leurs écrits* (Paris, 1688).

299:55–57 There were two major Jewish academies (yeshivas) which possessed central religious authority in the Babylonian Diaspora—

Sura and Pumbedita. The academy at Sura was founded in 219 in southern Babylonia, where the Euphrates divides into two rivers, and it flourished for almost eight hundred years. The academy at Pumbedita, to the north on the bank of the Euphrates, was founded in 259. Joseph headed it for two and a half years. Pumbedita academy was overshadowed by the academy at Sura until the beginning of the ninth century; Mather has probably confused the two. The academy at Pumbedita was transferred to Baghdad in the ninth century, and it appears that the Baghdad academy continued in existence until the thirteenth century.

300:68–75 On Mather's relations with William Jameson and the degree of doctor of divinity conferred upon Mather in 1710 by the University of Glasgow, see *The Diary of Cotton Mather . . . for the Year 1712*, ed. William R. Manierre II (Charlottesville: University Press of Virginia, 1964), p. 71, 73n, and Mather, *Diary*, 2:247, 40.

300:76–88 Mather borrows this material from Derham, p. 303 n. 4. The original source of the Latin quotation is Pliny, *Natural History*, 7.24, 88–89, which refers to Charmades.

300:88–301:3 These stories of the memory, save for that regarding Suarez, are taken from Wanley, *Wonders of the Little World*, pp. 97–98.

301:3–7 Mather may have used an intermediate source; the original of two of these items is in Theodore Zwinger, *Theatrum vitae humanae* (Basel, 1565), pp. 18, 1166, and Johannes Schenck von Grafenberg, *Observationes medicae de capito humano* (Basel, 1584), pp. 150–56. Pieter van Foreest wrote *Observationum et curationum medicinalium* (Leiden, 1596).

301:7–10 Mather adapts these lines from Derham, p. 304 n. 5, taking one of five lines of Latin verse quoted by Derham. The ultimate source of the Latin quotation is Marcello Palingenio, *Zodiacus vitae: hoc est, De hominis vita, studio, ac moribus optime instituendis, libri XII*, bk. 8 (Scorpio). The author, Pietro Angelo Manzolli of Stellata, near Ferrara (fl. 1528), wrote under the pseudonym of Marcellus Palingenius Stellatus or Marcello Palingeneo. His Renaissance Latin poem was first published in the 1530s; it went through more than sixty editions, ten of them in English. A translation by Barnabe Googe was published as *The Zodiake of Life* in five editions from 1560 to 1588. In the 1576 edition, a translation of the Latin phrase is at p. 141. Palingenio's astronomical poem enjoyed enormous popularity in England. The late Professor Luitpold Wallach of the University of Illinois at Urbana-Champaign provided the translation.

301:11–14 The source is Horace, *Odes*, 1.1, 2–4.

301:15–19 The source is Derham, p. 305 n. 6. The ultimate source of the Cicero quotation is *De finibus bonorum et malorum*, 3.2, 7.

301:20–302:35 This material is most likely derived from a work by John Edwards. Mather corresponded with Edwards and praised his works, eight of which were in the library of the Mathers (see Mather, *Diary*, 1:550; 2:81, 243.) Despite a search in many of Edwards's works, I have been unable to locate Mather's source in those volumes. The statement about "dung and urine" and this entire paragraph are based ultimately upon a provision in Berakoth, the first tractate in the Mishnah (and thus in the Talmud). Early Judaism drew no sharp line of demarcation between the secular and the religious, and the rabbis formed rules, usually but not necessarily derived from the Pentateuch, to apply the Mosaic law to even the most trivial situations in daily life. These rules, called halakhah, form the bulk of the Mishnah. The theme of Berakoth is the daily prayer which the Jew has a duty to offer. Since sanitary arrangements in ancient times were very defective, the presence of defiling matter had to be considered in connection with the saying of prayers. Thus the rabbis in their halakhic decisions instructed people how to perform religious acts when their surroundings were unclean. The tractate Berakoth specifies (3.5) that one remove oneself four cubits from urine and excrement if one wishes to read the Shema, and in comments (Gemara) on the passage, rabbis explored whether the Shema may be recited when it is uncertain whether dung and urine are present in the house and when such pollutants may be considered to have lost their defiling power. See Herbert Danby, *The Mishnah* (Oxford: Clarendon Press, 1933), p. 4, and A. Cohen, *The Babylonian Talmud: Tractate Berakoth* (Cambridge: University Press, 1921), pp. xxxii, 163–66.

302:42–43 These lines show knowledge of Bernardino Ramazzini, *De morbis artificum diatriba* (Modena, 1700). Ramazzini published a revised and enlarged edition of his treatise at Padua in 1713. A translation of this Latin text by Wilmer C. Wright was published as *De morbis artificum Bernardini Ramazzini, diatriba: Diseases of Workers* (Chicago: University of Chicago Press, 1940). Mather also mentions Ramazzini's treatise in *Bethesda*, p. 290.

302:46–49 John Wallis discusses the subject in "A Letter . . . to Robert Boyle Esq. Concerning . . . Teaching a Person Dumb and Deaf to Speak, and to Understand a Language," *Phil. Trans.*, 5 (July 18, 1670), 1087–97, and "A Letter . . . to Mr. Thomas Beverly, Concerning His Method for Instructing Persons Deaf and Dumb," *Phil. Trans.*, 20 (Oct. 1698), 353–60.

303:58–75 This material is condensed from Derham, pp. 305–17 and notes.

303:75–80 Mather's source is Baglivi, *Practice of Physick*, p. 56.

303:80–84 Mather usually quotes Pliny from an intermediate source, but here he may be relying here directly upon *Natural History*, 7.60, 213–14.

304:87–99 These lines are taken nearly verbatim from Grew, p. 69.

304:1–305:23 This description is derived mainly from William Derham, *The Artificial Clock-maker: A Treatise of Watch and Clock-work* (London, 1714), pp. 84–88. But Mather interpolates in this account, as indicated by the following entry.

304:13–16 These lines are probably based on Gaffarel, *Unheard-of Curiosities*, p. 236 (also in *Curiositez inouyes*, p. 173). Gaffarel does not mention Haarlem. The same material is in Wanley, *Wonders of the Little World*, p. 227.

305:30–34 These lines are loosely based on Ray, p. 201. The original source of the quotation is Seneca, *Natural Questions*, 7.30, 5.

305:35–41 This paragraph paraphrases Tauvry, *New Rational Anatomy*, pp. 183–84.

306:84–307:94 These lines paraphrase and quote Grew, pp. 79–80.

307:10–15 This paragraph is based on Barker, *Natural Theology*, p. 64.

310:17 The Latin phrase is from Alsted, p. 246.

311:58–59 My translations.

312:84–85 The late Professor John Heller of the University of Illinois at Urbana-Champaign kindly provided the translation.

313:95–99 The source is Raymond of Sebonde, *Theologia naturalis*, p. 63 (heading to Title 46). My translation.

313:7–19 This quotation is from *The Works of Thomas Goodwin*, 12 vols. (Edinburgh, 1861), 4:541.

313:20–314:24 Origen, a Platonist who attempted to amalgamate Greek thought with Christianity, was a prolific writer. His biblical exegesis and theological speculation were inseparable. Viewing the visible world as pervaded with symbols of the invisible world, he sought to determine the deeper meaning of Scripture through allegorical interpretation. In his masterly *Commentary on the Gospel of John* he refers to Jesus as "having tasted of death" (1.40). In *De principiis*, his most important theological work, Origen advances the doctrine of the preexistence of souls, rebirth, and universal salvation. His eschatology was based on belief in the justice and goodness of an omnipotent Creator and in the absolute free will of every rational being, including animated stars. Ultimately all creatures, even the Devil, will be saved. Origen exerted great influence, particularly on the Eastern church, but he came under repeated ecclesiastical condemnation. His doctrine of universal salvation was condemned as heretical by the Council of Alexandria in 400 and by the Fifth Ecumenical Council at Constantinople in 553. The Renaissance revived interest in Origen and stimulated fresh controversy over the ultimate salvation of all. Origen's reference to Jesus' having "tasted of death" comes in an allusion to Heb. 2:9. See *The Ante-Nicene Fathers,* ed. Alexander Roberts and James Donaldson, 10 vols. (New York: Charles Scribner's Sons, 1905–17), 9:318–19. Vol. 9 (1906) was edited by Allan Menzies. George Rust, a member of the Cambridge Platonist school and bishop of Dromore, defended Origen in *A Letter of Resolution Concerning Origen and the Chief of His Opinions* (London, 1661). Mather may have known this book.

314:29–40 The source of this paragraph is Goodwin, *Works,* 4:539.

317:28–35 This material is based on *Theologia ruris,* pp. 216–17.

317:36–318:72 This long passage is paraphrased from George Cheyne, *Philosophical Principles of Religion: Natural and Revealed. In Two Parts* (London, 1715), pt. 2, pp. 78–83. This material is not in the first edition of Cheyne, entitled *Philosophical Principles of Natural Religion* (London, 1705).

BIOGRAPHICAL REGISTER

[A LAPIDE]. *See* LAPIDE

[ABU BAKR]. *See* IBN TUFAYL

ABU MA'SHAR, usually known as Albumasar (787–886), an Arab astrologer, drew on a complex cultural inheritance and the diverse intellectual trends current in Baghdad, where he made his career, to expound and justify astrology. His fame as an astrologer was immense both among contemporaries and later, and Latin translations of some of his works are said to have influenced European philosophy in the twelfth century.

[AEGARDUS]. *See* EGARDUS

AELIANUS, CLAUDIUS (170–235), derived his philosophical notion of universal reason manifested in the animal creation from Stoicism, and he bitterly opposed the Epicureans. His major works are *Varia historia,* a miscellany dealing with human life and history, and *De natura animalium,* a similar collection on the characteristics of animals. Aelian's works enjoyed great popularity and were much drawn upon by Christian writers. Gesner translated both titles into Latin, and was the first to edit all of Aelian's writings (1556). He patterned his own history of animals on that of Aelian.

[ALBUMAZER HALY]. *See* ABU MA'SHAR

ALDROVANDI, ULISSE (1522–1605), a representative Renaissance polymath, was born in Bologna of a noble family. He studied mathematics, humanities, law, and medicine, and took a medical degree. Though he practiced medicine and taught at the University of Bologna, he became interested in natural history and devoted himself to intensive research in its various fields. He published four folio volumes during his lifetime, including *De animalibus insectis libri septem* (Bologna, 1602; another ed., Frankfurt, 1618), which was important in establishing entomology as a science. Much of his *Storia naturale* was published posthumously.

ALSTED, JOHANN HEINRICH (1588–1638). *See* the Introduction, pp. lxiii–lxiv.

AMBROSE (339–97), bishop of Milan, is one of the four traditional Doctors of the Latin church. His *Hexaemeron,* based on that of Basil the Great and the first written in Latin, treats the work of the six days in nine homilies.

AMERBACH, BASILIUS (1535–91), was a professor of jurisprudence at Basel, grandson of the celebrated fifteenth-century printer Jean Amerbach of Basel, and son of Bonifacius, a famous professor of jurisprudence at Basel and a friend of Erasmus.

AMMIANUS MARCELLINUS (c. 330–95), a Roman soldier and historian, was the author of *Rerum gestarum libri qui supersunt.*

ANDRY, NICOLAS (1658–1742), took the surname de Boisregard after a brief career as an ecclesiastic and turned to the study of medicine. He became professor of medicine at the College of France in 1701, an editor of the *Journal des Sçavans,* and dean of the faculty of medicine at the college. Andry wrote *De la génération des vers dans le corps de l'homme* (Paris, 1700), a study of the new science of parasitology. It appeared in English translation as *An Account of the Breeding of Worms in Human Bodies* (London, 1701). The book was reprinted at Amsterdam in 1701 and at Paris in 1715 and 1741.

ANGHIERA, PIETRO MARTIRE D' (1457–1526), an Italian-born scholar and torchbearer of the Renaissance in Spain, served there as soldier, priest, writer, diplomat, and member of the Council of the Indies. He was the first historian of the New World.

ANTHONY OF PADUA (1195–1231), a native of Lisbon, became the most celebrated follower of Francis of Assisi. Learned and eloquent, he gained a considerable reputation as a preacher. Like Francis he was a lover of nature, and it is said that when men were deaf to his preaching, Anthony preached to the fishes of the sea, who came in shoals to hear the saint discourse.

ARISTOTLE (384–322 B.C.) was born at Stagira, and from the age of seventeen to thirty was associated with Plato's school at Athens, first as a pupil and then as a largely independent investigator. He then spent a dozen years at Assos and Mytilene before returning to Athens, where he founded his own school, the Lyceum, whose covered porch gave its name to his followers—the Peripatetics. He left Athens after twelve years when a charge of impiety was brought against him; he died a year later.

Aristotle wrote in many fields, and while many of his works are now lost, his philosophical and scientific writings are still extant. The main lines of his thought were largely determined by his association with Plato and the Academy, yet he exhibited many differences with the Platonic school. One of his early works, now lost, asserted the eternal preexistence of the world and opposed the Platonic doctrine of Ideas. Aristotle had a passion for research and was curious about all sorts of natural phenomena. He wrote about the heavens, physics, meteorology, and zoology; the science in which he was most at home was biology. He contributed significantly to the organization of knowledge by his systematic survey and classification of the sciences. During the Middle Ages, a series of commentators harmonized Aristotle's writings with Christianity. He was accepted as the final authority on natural science down to early modern times, when investigators emancipated themselves from his iron grip.

ARNDT, JOHANN (1555–1621). *See* the Introduction, pp. lxiv-lxv.

AUGUSTINE (354–430), born in North Africa (present-day Algeria) of a pagan father and a Christian mother, was reared a Christian. As a young man he was given to sensual pleasure and for fifteen years lived with a concubine, by whom he had a son. A serious student, Augustine took up philosophy at nineteen and became a Manichean, remaining attached to that religion for nine years. Leaving Carthage, he taught briefly in Rome and then became professor of rhetoric in Milan. Here he came under the influence of Bishop Ambrose, who gave him a spiritual interpretation of Scripture and introduced him to a Christian Platonism that depended primarily on Origen. Neoplatonism brought him closer to Christianity, but for a time he could not break away from his strong sensual desires. Finally, in 386, after hearing a story of how monastic life had inspired conquest over self and after reading Paul, he returned to Christianity. He went back to Africa in 388, became a priest living a monastic life, and began his polemical activity. He was consecrated bishop of Hippo in 395, an office he held until his death.

A fertile author, Augustine formulated his own theology mainly in refuting three heresies during his episcopate. In attacking Manicheanism, he defended the essential goodness of all creation against the Manichean notion of an evil agency eternally opposed to the good God. Augustine maintained that God was the sole creator of all things and alone sustained them in being. In his struggle against Donatism, a schismatic movement in the African church, Augustine defined the doctrine of the church, the sacraments, and sacramental grace with new precision. The Pelagian controversy evoked his teaching upon the Fall, original sin and grace, and freedom and predestination. Augustine often rested his arguments on Neoplatonist foundations. He also wrote *On the Trinity*, an important treatise on triune nature of God; the *Confessions*, an autobiographical account of his first thirty-three years; and *The City of God*, a reply to pagans who attributed the fall of Rome in 410 to the abolition of heathen worship. Augustine contrasted Christianity and the world, and his work is the supreme exposition of the Christian philosophy of history.

The most important Father of the early church, Augustine shaped the theology of the Middle Ages, and even the reaction against him arising from the rediscovery of Aristotle in the thirteenth century was never complete. The Reformers appealed to elements of Augustine's teaching, and the Reformed churches especially, including Puritanism, were founded on the Augustinian tradition.

AUZOUT, ADRIEN (1622–91), was a French astronomer who contributed significantly to the development of astronomical instruments and made systematic astronomical observations with micrometers and telescopic sights.

BACON, FRANCIS (1561–1626), English statesman, philosopher, and author. He is best remembered for his *Advancement of Learning* (1605) and *Novum organum* (1620), which aimed at replacing the deductive logic of Aristotle with inductive method in interpreting nature.

BACON, ROGER (c. 1219–c. 1292), a member of the Franciscan order, was devoted to the reform of scientific and philosophical learning and the welfare of the Church. He wrote on languages, mathematics, logic, astronomy, astrology, alchemy, experimental science, and especially optics. His *Opus minus* describes original experiments, possibly his own, with a lodestone held above and below a floating magnet. Bacon argued that it was not the north part of the heavens that causes its orientation but all four parts equally.

BAFFIN, WILLIAM (d. 1622), navigator, made many voyages of exploration and is best remembered as pilot of the *Discovery* on its attempt to locate a northwest passage in 1615. He made remarkably accurate observations of latitude and of tides and was one of the first navigators who endeavored to determine longitude at sea by astronomical observations. Several of his accounts of his voyages have been published.

BAGLIVI, GEORGIUS (1668–1707), settled in Bologna as a pupil and assistant of Marcello Malpighi soon after receiving his medical degree. He conducted biological research while serving as a physician, and his careful observations made him suspicious of medical systems and hypotheses. Baglivi removed to Rome with Malpighi and found favor at the papal court after his master's death two years later. He became a fellow of the Royal Society and professor of theoretical medicine at the Sapienza in Rome. A master clinician, Baglivi enjoyed a high reputation throughout Europe for his lectures, anatomical demonstrations, and medical consultations. Among his writings are *De praxi medica ad priscam observandi rationem revocanda* (Rome, 1696), translated as *The Practice of Physick Reduc'd to the Ancient Way of Observations . . . Together with Several New and Curious Dissertations; Particularly of the Tarantula, and the Nature of Its Poison* (London, 1704).

BAILLET, ADRIEN (1649–1706), French scholar, priest, and librarian, wrote prolifically in the fields of theology, philosophy, and science. His works include a life of Descartes which was translated into English and *Des enfants devenus célèbres par leurs études ou par leurs écrits* (Paris, 1688).

BALDUS, CAMILLUS, or CAMILLO BALDI (1550–1637), professor of logic, philosophy, and humane letters at the University of Bologna, initiated graphology, the study of handwriting, especially for the purposes of character analysis. He published two works which tell how to discern the character of the writer of a letter from the handwriting as well as from the literary style and the thoughts expressed. Baldi's master work, *In physiognomica Aristotelis commentarius* (Bologna, 1621), was a commentary on the physiognomy attributed to Aristotle. Physiognomy was popular at the time because it was regarded as having a natural basis, whereas other forms of divination were condemned as superstitious. Baldi also published a treatise upon how to tell human propensities from a person's temperament as well as a treatise upon natural prediction from inspection of human nails—*De naturali ex unguium inspectione praesagiis commentarius* (Bologna, 1629). I am indebted to Dr. Dorothy S. Adelmann for the identification of "Camillus."

BARKER, MATTHEW (1619–98), a graduate of Trinity College, Cambridge, taught school until the outbreak of the civil war and then became a nonconformist divine. He was the author of a variety of works, including *Natural Theology: Or, the Knowledge of God from the Works of Creation, Accommodated, and Improved, to the Service of Christianity* (London, 1674).

BARLETIUS, MARINUS, or MARINO BARLEZIO or BARLESIO (b. c. middle fifteenth century), an Albanian, was the author of *De vita moribus ac rebus praecipue adversus Turcas gestis, Georgii Castrioti, clarissimi Epirotarum principis* (Strassburg, 1537). The work was translated into several other languages. The English translation was entitled *The Historie of George Castriot, surnamed Scanderbeg, King of Albanie* (London, 1596).

BARTHOLIN, CASPAR (1585–1629), a celebrated Danish polymath, was familiar with leading savants in Germany, France, Italy, and England. He lectured in various countries and published in many fields, including anatomy, medicine, physics, and theology. His anatomical studies formed the basis for his *Anatomicae institutiones corporis humani* (Wittenberg, 1611). This anatomical manual made him famous and was reprinted five times; it became even more famous when his son Thomas brought out an enlarged and illustrated edition in 1641.

BARTHOLIN, THOMAS (1616–80), the Elder, a member of a famous Danish family, obtained a medical degree at Basel and became professor of anatomy at the University of Copenhagen. In 1641 he published the first revised edition of his father's *Institutiones anatomicae* (many other revised editions followed, all published at Leiden). The third edition reflected Bartholin's extensive anatomical studies and was much superior to the second. Bartholin discovered the thoracic duct in humans and contributed significantly to physiology by his discovery that the lymphatic system is a separate system. His *Historarium anatomicarum rariorum centuria I–IV* (Copenhagen, 1654–61) deals with numerous problems of human and comparative anatomy. Bartholin always retained a strong interest in the humanities, and he published on a variety of topics. The most distinguished physician in Denmark in his day, he was responsible for the royal decree of 1672 that decided the organization of Danish medicine for the next century. In 1673 he began publication of the first Danish scientific journal, *Acta medica et philosophica hafniensa*.

BASIL THE GREAT (c. 330–79), one of the three Cappadocian fathers, became bishop of Caesarea in 370. He was the most distinguished hexameral authority after Philo. His *Hexaemeron* introduced animal descriptions and anecdotes.

BAUHIN, GASPARD (1560–1624), son of a French Protestant physician who fled to Switzerland to avoid religious persecution, was an anatomist and professor at Basel. His Προδρομοσ *theatri botanici* (Frankfurt, 1620), and Πιναξ *theatri botanici* (Basel, 1623) were the culmination of Renaissance botanical science. Bauhin laid the basis for a binomial system of nomenclature by giving each plant a generic and specific name, and in classification he abandoned the old alphabetical arrangement and grouped plants according to natural affinities. Bauhin is an important precursor of Linnaeus.

BAXTER, RICHARD (1615–91), was a prominent English Puritan divine and the author of a number of theological works, including *The Saints Everlasting Rest* (London, 1650), *A Call to the Unconverted* (London, 1657), and the autobiographical *Reliquiae Baxterianae* (London, 1696).

BAYER, JOHANN (1572–1625), German lawyer and amateur astronomer whose *Uranometria, omnium asterismorum continens schemata, nova methodo delineata, aereis laminis expressa* (Augsburg, 1603) established the modern nomenclature of the visible stars.

BEAUPLAN, GUILLAUME LE VASSEUR, sieur de (c. 1600–1673), architect and engineer, was stationed in the Ukraine from 1630 to 1647 as captain of artillery in the service of the Polish king. His *Description d'Ukraine* (Rouen, 1651), went through several editions in various languages, with the first English edition in 1704. The fourth English and first American edition was published as *A Description of Ukraine* (New York: Organization for the Defense of Four Freedoms of Ukraine, 1959).

BECHER, JOHANN JOACHIM (1635–82), was a curious polymath, interested in alchemy, medicine, theology, politics, economics, and even the formation of a universal language. He tried to adapt alchemy to the growing chemical knowledge of the seventeenth century, and is important mainly for his influence on George Ernst Stahl, who refined Becher's ideas into the phlogiston theory a generation later. While in England Becher unsuccessfully sought membership in the Royal Society, and in 1680 dedicated a short book on clock design to the society. It was considered to be of little value, and he was not elected.

BELLARMINE, ROBERT (1542–1621), Italian theologian, entered the Society of Jesus in 1560 and was created a cardinal in 1599. He was an outstanding opponent of Protestant doctrines of the Reformation and aided in preparation of a revision of the Vulgate.

BELLINI, LORENZO (1643–1704), became professor of theoretical medicine at Pisa at the age of twenty and acquired a chair in anatomy five years later. A founder of Italian iatromechanism, which he learned under the influence of Borelli, he pioneered in applying mechanical philosophy to explaining the functions of the body. His *Exercitatio anatomica . . . de structura et usu renum* (Florence, 1662) rejected Galenic theory on the function of the kidneys for a mechanical explanation of urinary secretion. Bellini's mechanical manner of physiological explanation won the admiration of contemporary scientists, and he remained a major scientific authority until a reaction to iatromechanism began in the mid-eighteenth century.

[BENEROVINUS]. *See* BEVERWYCK

BERENGARIO DA CARPI, JACOPO (c. 1460–1530?), was a member of the medical faculty at the University of Bologna and physician to members of the Medici family. A popular lecturer in surgery, he also taught anatomy and gradually made that his major interest. After editing the medieval dissection manual of Mondino de' Luzzi, he published his own book on the subject entitled *Commentaria cum amplissimis additionibus super anatomia Mundini cum textu eiusdem in pristinum et verum nitorem redacto* (Bologna, 1521). This work, based on considerable personal observation, was the most important forerunner of Vesalius's *Fabrica*. Berengario's *Isagogae breves* (Bologna, 1522) was a compendium of the *Commentaria* intended as a manual for students. Translated into English by Henry Jackson, it was published as Μικροκοσμογραφια: *Or, A Description of the Little World or Body of Man* (London, 1660; another ed., 1664).

BERNARD (1090–1153), abbot of Clairvaux, one of the chief centers of the Cistercian Order, and, owing primarily to his saintliness and personality, an influential religious force in medieval Europe.

BERNOULLI, JAKOB (1654–1705), elder brother of Johann Bernoulli, was professor of mathematics at Basel. He improved differential calculus as invented by Newton and Leibniz and solved many problems in celestial mechanics. The library of the Mathers had a copy of Bernoulli's *Conamen novi systematis cometarum* (Amsterdam, 1682).

BERNOULLI, JOHANN (1667–1748), Swiss mathematician and professor at Groningen and Basel, was the leading mathematical preceptor to all of Europe after Newton's death. He published several papers on experimental physics.

BEVERWYCK, JOHAN or JAN VAN (Johannes Beverovicius) (fl. 1634–44), a physician of Dort whose works included Αυταρχεια *Bataviae, sive, introductio ad medicinam indigenam* (Leiden, 1644). The library of the Mathers contained a copy of Beverwyck's *Epistolica quaestio de vitae termino, fatali, an mobili? Cum doctorum responsis* (Rotterdam, 1644). This work had first been published at Dort in 1634.

BICKERUS, JOHANN (fl. early seventeenth century), was a north German writer on astrological medicine whose *Hermes redivivus, declarans hygieinam, de sanitate vel bona valetudine hominis conservanda* (Giessen, 1612) attempts to reconcile Galenic with Hermetic medicine.

BIDLOO, GOVARD (1649–1713), professor of anatomy at The Hague and later professor of medicine and surgery at Leiden, became physician in ordinary to William III in England and was elected a fellow of the Royal Society. Bidloo's chief work was *Anatomia humani corporis* (Amsterdam, 1685); it was translated into Dutch and published (Amsterdam, 1690) with 105 copperplate figures drawn from life. This book was the first large-scale anatomical atlas since Vesalius's *De humani corporis fabrica* in 1543.

BILS, LODEWIJK DE (Ludovicus Bilsius) (1624–71), a Dutch nobleman, was a self-taught and unqualified anatomist. He stirred considerable controversy by publications in which he claimed to have discovered methods for preserving corpses for years and dissecting living animals without spilling blood. He also published fantastic theories on the lymph vessels and the chylus which aroused strong opposition.

BLACKMORE, RICHARD (1655–1729), was an English physician and indefatigable writer. His *Creation: A Philosophical Poem* (London, 1712) was much admired by contemporaries and went through many editions. Blackmore also published *Essays upon Several Subjects,* 2 vols. (London, 1716).

[BLANCARDIUS]. *See* BLANKAART

BLANKAART, or BLANCKAERT, STEVEN (1650–1702), Dutch physician and surgeon, was a prolific author of medical works. His *Anatomia reformata, sive concinna corporis humani dissectio* (Amsterdam, 1687) was often reprinted. He also published on the subject of entomology.

BLASIUS, GERARDUS, or GERARD (often GERHARD) BLÄES (d. 1692?), a Belgian-born professor of medicine at the University of Amsterdam, edited the works of many prominent researchers, especially in comparative anatomy, and published a number of his own writings on medicine and anatomy.

BOCCONE, PAOLO (1633–1704), spent much of his life traveling throughout his native Sicily as well as Italy and Europe studying and collecting natural history, especially plants. His many works, published in Italian, Latin, and French, constitute an encyclopedia of the natural history of his time. In his later years he joined a Cistercian order, assuming the name Silvis, and retired to a monastery near Palermo.

BOCHART, SAMUEL (1599–1667), an erudite theologian and orientalist, was for many years pastor of the French Reformed Church in Caen. His *Geographia sacra* (Caen, 1646), an account of the places where the descendants of Noah settled after the dispersion and of the colonies and language of the Phoenicians, made his reputation. His *Hierozoicon, sive bipertitum opus de animalibus Sacrae Scripturae* (London, 1663), a learned treatise on all the animals mentioned in Scripture, draws heavily on classical and oriental sources. These and other writings are collected in his *Opera omnia*, 2 vols. (Leiden, 1675).

BOETHIUS, ANICIUS MANILIUS SEVERINUS (c. 480–524), Roman aristocrat who wrote *De consolatione philosophiae* while in prison under suspicion of treason. Boethius built a water clock and a sundial.

BOHUN, EDMUND (1645–99 [1702?]), a tract writer and compiler, served briefly as chief justice of Carolina. His publications included *A Geographical Dictionary* (London, 1693).

BORELLI, GIOVANNI ALFONSO (1608–79), a multifaceted Italian scientist, was important in extending the new mathematical-experimental physics along Cartesian lines. His reputation is based upon his mechanics, including celestial mechanics, and his physiology, which was of the iatrophysical school. He was a leader of the iatromechanical school. His last and most important scientific writing was *De moto animalium*, 2 pts. (Rome, 1680–81), the first comprehensive textbook of physiology and a work which explained the functioning of the muscles and limbs mechanically in strict conformity to Cartesian principles. It treats the external motions, or the motions produced by the muscles, and the internal motions, such as the movements of the muscles themselves, along with circulation, respiration, the secretion of the glands, and nervous activity.

BORELLUS, PETRUS, or PIERRE BOREL (c. 1620–71), a Parisian medical man who became physician to the king, was an active collector of objects of natural history, antiquities, and rarities, on which he published a catalogue. His works included *De vero telescopii inventore, cum brevi omnium conspiciliorum historia . . . accessit etiam centuria observationum microcospicarum* [sic], 2 vols. (The Hague, 1655–56), a book on the telescope which included an account of a hundred microscopical observations, including some of insects. He also published *Historiarum, et observationum medico-physicarum, centuriae IV*, 3 pts. (Paris, 1656).

BORRICHIUS, or BORCH, OLAUS (1626–90), professor and twice rector of the University of Copenhagen, was famous as a physician and defender of Hermeticism as well as a prolific writer on chemical, botanical, and philological topics.

BOYLE, ROBERT (1627–91), an aristocrat who early demonstrated a passion for natural philosophy, settled in Oxford about 1656. One of a group of eminent Oxford scientists, his first scientific interest was chemistry. He conducted skillful chemical experiments at a time when experimentation was still novel, separating chemistry from utilitarian medicine on the one hand and mystical dogma on the other. Stimulated by the achievements of Torricelli and von Guericke, he established his fame in the field of pneumatics, publishing his results in *New Experiments*

Physico-Mechanicall, Touching the Spring of the Air, and Its Effects (Oxford, 1660). He was one of the few chemists to interest himself in the related problems of the discovery of new gases and combustion and calcination. Boyle's reputation helped establish the Royal Society, of which he was an original fellow and the most influential figure during his lifetime.

A strong opponent of Aristotelian and Scholastic physics, with its supposed substantial forms, real qualities, and occult forces, Boyle strove to establish an empirically based mechanistic theory of matter and by means of such a theory to establish a scientific and rational chemistry. He pioneered in promoting the mechanical philosophy in England, with its assumption that nature is governed by immutable laws concerned with matter and motion. Boyle was influenced by the original Epicurean atomism and its revival by contemporaries such as Pierre Gassendi and Walter Charleton, but his Christianity and his familiarity with current science prevented any literal Epicureanism. An eclectic, he was influenced by the ideas of Bacon and Descartes in his search for rational, mechanical explanations. He rejected both the "atoms" of Epicurus and the "hierarchy of particles" of Descartes, preferring the neutral word *corpuscle*. Boyle conducted experiments and published a number of books to advance his corpuscularian or mechanical philosophy, with its basic principles of matter and motion and its repudiation of the occult, and he converted members of the Royal Society to his views.

Boyle promoted the idea of chemistry as a theoretical physical science and interested himself in theoretical problems, including the nature of chemical composition, the identification and classification of substances in terms of real chemical entities, and the determination of the course of chemical reactions. He used his corpuscularian philosophy to reject the old definition of an element and to provide a clearer one, though his was not the same as the modern definition of an element. Boyle insisted on experimentation as an essential ingredient of proof in an age when Rationalists doubted its value, and he contributed significantly to the establishment of scientific method in the seventeenth century. Newton was greatly influenced by Boyle.

A devout Christian, Boyle was from an early age a prolific writer on natural theology, the point of intersection between science and religion. The God who could create a mechanical universe which obeyed laws was for Boyle more to be admired than a God who created a universe without scientific laws. Always eager to show that the new science was an incentive rather than a threat to established religion, Boyle proclaimed this view in *The Christian Virtuoso* (London, 1690) and other writings. His will provided for a series of lectures to prove the Christian religion against infidels of various kinds. Eminent divines gave these sermons to large London audiences, and the lectures were published.

In the first and most famous series in 1692, Richard Bentley confuted atheism drawn from scientific advances of the day. Subsequent Boyle lecturers, including John Harris (1698), Samuel Clarke (1704 and 1705), William Whiston (1707), John Woodward (1710), and William Derham (1711–12), followed this pattern to produce a form of natural religion or physicotheology which is characteristic of the early eighteenth century. Here as elsewhere Boyle set the tone and inspired methods of thought which were widely accepted by the next two generations.

BRAHE, TYCHO (1546–1601), a Danish astronomer, made accurate observations of the sun, moon, stars, and planets over some three decades. His observations of the nova of 1572 from 11 November 1572 to the end of March 1574 and of several comets necessitated the alteration of prevalent cosmological theories. Yet Tycho did not accept the Copernican system. In his own geoheliocentric system the earth is the center of the universe, the planets circle the sun, while the sun circles the earth. Kepler used records of Tycho's accurate observations, especially of Mars, to discover the laws of planetary motion, and Tycho's compromise theory of the universe helped win acceptance of the Copernican system.

BRERE WOOD, EDWARD (1565?–1613), an Oxford graduate and a zealous student, was the first professor of astronomy in Gresham College (1596). He was an antiquary who lived in retirement; his posthumous publications include *Enquiries touching the Diversity of Languages, and Religions through the Cheife Parts of the World* (London, 1614, and later eds.). The work was translated into French (1640) and into Latin by John Jonston (1650).

BRIGGS, WILLIAM (1642–1704), was a physician with a successful practice, especially in diseases of the eye. His *Opthalmo-graphia, sive oculi ejusque partium descriptio anatomica* (Cambridge, 1676; 2d ed., 1686) was an anatomical description of the eye. His "New Theory of Vision" was first published in English ("A New Theory of Vision," in Robert Hooke, ed., *Philosophical Collections* [Mar. 1682], 167–77; and "A Continuation of a Discourse about Vision," *Phil. Trans.*, 13 [May 10, 1683], 171–82) and then translated into Latin at the request of Isaac Newton, who wrote a commendatory preface for it, and published as *Nova visionis theoria* (London, 1685).

BROTHERTON, THOMAS (c. 1657–1702), was a barrister of Gray's Inn and sometime justice of the peace in his native county of Lancaster. He published "An Account of Several Curious Observations and Experiments, concerning the Growth of Trees," *Phil. Trans.*, 16 (Apr.–June, 1687), 307–13.

BROWNE, THOMAS (1605–82), physician of Norwich and a scholar of prodigious knowledge, was the author of *Religio medici* (published surreptitiously in 1642 and in an authorized edition the following year), *Pseudodoxia epidemica* (1646), and *Hydriotaphia, Urne-Buriall, Or, A Discourse of the Sepulchrall Urnes Lately Found in Norfolk, together with The Garden of Cyrus, Or the Quincunciall, Lozenge, or Net-work Plantations of the Ancients, Artificially, Naturally, Mystically Considered* (London, 1658). *The Garden of Cyrus* traces the history of horticulture from the Garden of Eden to the time of the Persian Cyrus. Browne's posthumous publications included two essays on plants.

BUCHANAN, GEORGE (1506–82), was a Renaissance Scottish humanist and author of much widely acclaimed Latin verse. His *De sphaera* (1584) defended the older Ptolemaic cosmology and the central position of man in the universe.

BUDÉ, GUILLAUME (1467–1540), laid the foundations of the French national library, the Bibliothèque Nationale. His *Commentarii linguae Graecae* (Paris, 1529) and his establishment of the College of France (1530) were instrumental in bringing about the revival of classical studies in France.

BUTLER, CHARLES (1560–1647), vicar of Wotton, wrote on philology, music, and affinity as a bar to marriage as well as on bees. He published *The Feminine Monarchie. Or a Treatise Concerning Bees, and the Due Ordering of Them* (Oxford, 1609). New editions of this classic appeared in 1623 and 1634; the book was translated into Latin in 1673.

CABOT, SEBASTIAN (c. 1474–1557), English-born son of the Venetian John Cabot, was an explorer who won considerable recognition as a cartographer.

[CAESIUS]. *See* CESI

CAMERARIUS, PHILIPP (1537–1624), member of a distinguished German family and a celebrated jurisconsult, left one work, *Operae horarum subcisivarum, sive meditationes historicae*. The first "century" (of which three ultimately appeared) was published at Nuremberg in 1591 and 1599 and at Frankfurt in 1602. The work was often reprinted; it included a chapter entitled *De canum fidelitate et sagacitate cum elogio memorabili*.

[CAMILLUS]. *See* BALDUS

CAMPANI, GIUSEPPE (1635–1715), an Italian astronomer, produced telescopic instruments of great focal length and made significant observations of the satellites of Jupiter and of the rings of Saturn. He published his results in *Ragguaglio di due nuove osservazioni una celeste in ordine alla stella di Saturno* (Rome, 1664).

CARDANO, GIROLAMO (1501–76), an Italian Renaissance physician and mathematician of brilliant intellect, wrote on various topics, including philosophy, religion, ethics, and music. A pioneer in psychiatry, he was preoccupied with divination and occult learning.

CAREW, RICHARD (1555–1620), a member of one of the leading families of Cornwall, devoted his leisure to study of the history and antiquities of his native country and to the mastery of foreign languages. He published *The Survey of Cornwall* (London, 1602).

CARNEADES (214/3–129/8 B.C.), founder of the New or Third Academy in Athens, was a skeptic philosopher. He argued that grasp of reality could only come through "presentations," mental events of which we are directly aware. But since we can never be certain that any presentation, whether arising from the senses or from the mind, is veracious, we must always reserve judgment. Carneades attacked the dogmatic philosophers, exposing their inconsistencies and improbabilities.

[CARPENSIS]. *See* BERENGARIO

CASAUBON, ISAAC (1559–1614), was born in Geneva of French-Huguenot parents. He was professor of Greek at Geneva and then Montpellier, and might have been a professor in Paris had he embraced Roman Catholicism. In 1610 he went to England, where he was made prebendary of Canterbury and Westminster. His works belong mainly to the field of classical scholarship, though he published some theological pamphlets. He was considered the most learned man in Europe after Joseph Scaliger.

CASSINI, GIAN-DOMENICO (1625–1712), was an astronomer who made a reputation with many important discoveries concerning the planets. He left Italy in 1669 to become a member of the recently founded Academy of Sciences in France, where he continued his work as a gifted observer and director of research at the Paris Observatory.

CASTRIOTA, GEORGE, known as SKANDERBEG, a form of the Turkish name Iskender Bey, meaning Prince Alexander (1405–68) was early given as a hostage to the Turkish sultan. He converted to Islam and was educated in Turkey, but he abandoned Turkish army service, returned to his own people, embraced Christianity, and became a national hero by leading the struggle for Albanian independence against the Turks.

CATO, MARCUS PORCIUS (234–149 B.C.), Roman statesman, known as Cato the Censor in allusion to the severity with which he discharged the office of a censor, endeavored to restore by legislation what he believed to be the high morals and simplicity of life characteristic of the early republic. In *De agri cultura,* written about 160 B.C. and his only extant work, he gives more attention to cabbage than to any other vegetable, taking several pages to explain the preparation of cures made from cabbage.

CESI, FEDERICO (1585–1630), duke of Acquasparta, dedicated himself to research at an early age. In 1603 he founded at Rome the Accademia dei Lincei (Academy of the Lynxes, or Lincean Academy), the earliest scientific society in Italy and the first truly modern scientific academy. A pioneer of modern scientific botany, Cesi formulated a rational system for the classification and nomenclature of plants using family names, and he made microscopic studies of plant anatomy during which he reportedly discovered the spores of cryptogams and the sexuality of plants. Cesi is an important precursor of Linnaeus.

CHARLES V (1500–1558), Holy Roman Emperor from 1519 to 1556 and king of Spain as Charles I from 1516 to 1556. He resigned both crowns and retired to a monastery in Spain.

CHESELDEN, WILLIAM (1688–1752), surgeon, anatomist, and fellow of the Royal Society, won fame for his lateral operation for the stone. He also made important contributions to anatomy, publishing, among other works, *The Anatomy of the Human Body* (London, 1713.)

CHEYNE, GEORGE (1671–1743), a physician, mathematician, and theologian, was born in Aberdeenshire and received a classical education with a view to entering the ministry. But he turned to medicine under the influence of Archibald Pitcairn, professor of medicine at Edinburgh and chief representative of the iatromathematical school of medical science. This school made medicine a branch of mathematics and drew close analogies between the human body and a machine. In 1702 Cheyne moved to London, was elected a fellow of the Royal Society, and established a medical practice. He soon became a member of a circle of "Newtonian" medical and scientific writers and won a reputation as a convivial companion and tavern-house wit. Later, probably by 1720, he moved to Bath and became a serious and dedicated medical practitioner while continuing to write medical books.

Cheyne was a prominent spokesman of "Newtonianism" during his London residence. His first book was an elaborate explanation of fevers which followed Pitcairn's supposedly "mathematical" and "Newtonian" variety of iatromechanism in viewing the body as a system of pipes and fluids. Cheyne demonstrated his continuing mathematical interest in his next book, a work on the calculus which claimed that the mathematical method was applicable to medicine as well as mechanical science.

Cheyne explored the theological significance of Newtonian science in *Philosophical Principles of Natural Religion* (London, 1705), which he revised and expanded as *Philosophical Principles of Religion: Natural and Revealed. In Two Parts* (London, 1715). Part 1 was a corrected and enlarged second edition of his earlier treatise; Part 2 mingled theology and mathematics in a fantastic manner. This work went through a fifth edition by 1736. Part 1 is based on Newton and other authorities. In discussing the physical laws of nature, Cheyne demonstrates that he is no materialist and he "proves" that the Epicurean hypothesis is both inconsistent and impossible. Along with other arguments for the existence and providence of God, Cheyne asserted that the observed phenomena of attraction in the universe argued for a Supreme Being. His argument proved very popular with contemporaries, and perhaps even inspired Newton, as Newton's discussion of attraction in his *Optics* demonstrated.

Cheyne never gave up his interest in Newtonian science or in theological and philosophical speculation, but after removing to Bath he wrote mainly on practical and popular medical subjects, emphasizing moderation in diet and drink and often repeating the same ideas. His writings usually went through several editions and were translated into other languages. He was a significant figure in eighteenth-century medicine.

CHILDREY, JOSHUA (1623-70), a Dorset parson, wrote on theology, meteorology, and natural history. He is best remembered for *Britannia Baconica: Or, the Natural Rarities of England, Scotland, and Wales. According as They are to be Found in Every Shire. Historically Related, According to the Precepts of the Lord Bacon* (London, 1661).

CHRYSOSTOM, JOHN (c. 347-407), a native of Antioch, lived for a time as a monk before being ordained deacon and later priest. His preaching at Antioch won him the name of Chrysostom ("golden-mouthed") and there he delivered a series of homilies and produced the bulk of the literary works, including treatises and scriptural commentaries, which established his title as the greatest of Christian expositors. Made bishop of Constantinople against his will in 398, he set out to implement much-needed reforms. But they alienated people in high places, and he was deposed and exiled.

CICERO, MARCUS TULLIUS (106-43 B.C.), a politician, orator, and philosopher familiarly known to early Americans as "Tully," received an excellent education in philosophy and rhetoric. He pursued a public career in Rome, and through his powers of eloquence and his merits as a civil magistrate attained the highest honors of the state. As consul he crushed the conspiracy of Catiline, winning admiration as the deliverer of Romans, and when Clodius drove him into voluntary exile, he was recalled to Rome by the people. Although devoted to the existing republican constitution, in politics Cicero was irresolute and timid, and his vacillation between members of the First Triumvirate—Pompey, Caesar, and Crassus—at the opening of the civil war dashed his hopes for political influence. Isolated, he devoted himself to literature and philosophy. After the assassination of Caesar he hoped to regain political influence, but he misjudged the young Caesar (Octavian), and when Octavian formed an alliance with Antony and Lepidus known as the Second Triumvirate, Cicero feared that liberty was at an end and attempted to escape by sea. Antony's soldiers caught Cicero, who courageously submitted to his own beheading.

A prolific author in many fields, Cicero's writings include a number of orations, many works on rhetoric, a large number of letters, and some verse. His best known political work, *De republica,* discussed the best form of government. Cicero favored a constitution containing elements of monarchy, oligarchy, and democracy. *De republica* is also valuable for its assertion of human rights and the brotherhood of man, notions which Stoic beliefs helped to foster. Philosophical subjects had always interested Cicero, and he turned to them in his political inactivity. His susceptibility to Stoic teaching on moral questions is revealed in *De finibus bonorum et malorum,* a discussion of theories of the highest good, and *Tusculan Disputations,* which deals with the problems of death, grief, passion, and what is essential for happiness. In theological speculation Cicero's major work was *De natura deorum,* which sets forth the views of Epicureans, Stoics, and Academics on the nature of the gods and the existence of a divine providence. Cicero abhorred Epicurus and concluded that the Stoics' argument is likely to be correct. In other theological works he reaffirmed his belief in the existence of a divine being. Less a profound thinker than a popular exponent of the leading principles of the ancient philosophical schools, Cicero's theological and philosophical writings attracted readers in the Middle Ages when the work of Cicero the politician and orator was almost forgotten, and they were rediscovered at the time of the Renaissance and cherished during the eighteenth century.

CLARKE, SAMUEL (1675–1729), mastered the chief parts of the Newtonian philosophy as a Cambridge University undergraduate and then turned to divinity, applying Newtonian principles to metaphysical and theological problems. His Boyle lectures of 1704 and 1705, which established his reputation, were first published separately and afterwards together as *A Discourse Concerning the Being and Attributes of God, the Obligations of Natural Religion, and the Truth and Certainty of the Christian Revelation* (London, 1706). As rector of St. James's Church, Westminster, Clarke continued to defend Newton's ideas in theological and scientific matters.

CLAUDIANUS, CLAUDIUS (c. 370–c. 404), an Alexandrian Greek who became court poet under Honorius. His *Panegyric on the Third Consulship of the Emperor Honorius* mentions the new star (ll. 172–74). Claudian was the last notable representative of the classical tradition in Latin poetry. His works include the unfinished epic *The Rape of Proserpine.*

CLAYTON, JOHN (1657–1725), rector of Crofton at Wakefield in Yorkshire, "made anatomies" together with Allen Moulen in London. He published accounts of his observations of the natural history of Virginia in *Phil. Trans.* 17 (June 1693), 781–95; (Nov. 1693), 941–48; (Dec. 1693), 978–98; and 18 (May 1694), 121–35. His account of the air, water, earth, soil, birds, and beasts was extracted in *Misc. Cur.,* 3:281–355. The rector of Crofton is not to be confused with John Clayton (1693–1773), the celebrated botanist who went to Virginia in 1705 and is remembered for his part in the publication of *Flora Virginica,* 2 vols. in 1 (Leiden, 1739–43).

CLEANTHES (331–232 B.C.), disciple of Zeno and his successor as head of the Stoic School. Famous for his poverty and iron will, Cleanthes imparted religious fervor to the sober philosophy of Zeno. He considered God as the soul of the living universe and the sun as its heart. His best-known work is his hymn to Zeus.

COCKBURN, JOHN (1652–1729), Scottish-born clergyman, served parishes in Scotland, the Netherlands, and England, and received the D.D. degree. His many publications included *An Enquiry into the Nature, Necessity, and Evidence of Christian Faith in Several Essays,* 2 pts. (London, 1696–97).

COCKBURN, WILLIAM (1669–1739), was physician to the fleet before settling in London. He made a fortune with a secret remedy for dysentery, became a member of the Royal Society, published four papers in its *Philosophical Transactions,* and wrote on various medical topics.

COITER, VOLCHER (1534–76), left the Netherlands to study for five years under eminent teachers in Italy and France. He obtained a doctorate in medicine at Bologna and after imprisonment in Rome and Bologna (probably for offending the Inquisition) became physician to the city of Nuremberg. An eminent physician, he contributed significantly to advances in knowledge of human and comparative anatomy. Coiter's investigation of avian anatomy was particularly significant. He discovered the tongue of the woodpecker and made epochal discoveries on the development of the chick.

COLES, or COLE, WILLIAM (1626–6?), graduated from New College, Oxford, and became the most famous herbalist of his time. He published *The Art of Simpling: An Introduction to the Knowledge and Gathering of Plants* (London, 1656), and *Adam in Eden: Or, Natures Paradise. The History of Plants, Fruits, Herbs and Flowers* (London, 1657).

COSTER, LAURENS JANSZOON (1405–84) of Haarlem has been credited with the invention of printing with movable types about 1440. But the "Coster legend" was fabricated by a Dutchman in 1568, and modern scholarship proves it to be a fable.

COWPER, WILLIAM (1666–1709), surgeon and anatomist, published *Myotomia Reformata: Or, a New Administration of the Muscles of the Humane Bodies* (London, 1694; new ed., 1724). His *Anatomy of Humane Bodies* (Oxford, 1698) was the best English anatomy when it appeared. Govard Bidloo, the Dutch anatomist, charged Cowper with plagiarizing the excellent plates from his anatomical atlas. A fellow of the Royal Society, Cowper also published several papers in the *Philosophical Transactions.*

CRASSUS, MARCUS LICINIUS (d. 53 B.C.), a contemporary of Cicero, served in various public offices and in various ways attained influence. Maintenance of a piscina was apparently an important Roman status symbol, and Crassus is best remembered for his wealth and greed.

CREITLOVIUS, JOACHIM FRIDERICUS, is cited by Johann Scheuchzer (1684–1738), brother of Johann Jakob, in his dissertation Οὑρεσιφοιτης helveticus, sive itineris Alpini descriptio physico-medica (Zurich, 1702).

CROONE, WILLIAM (1633–84), a physician, was a founder and one of the first fellows of the Royal Society and a fellow of the College of Physicians. He reported on various subjects to the Royal Society, especially embryology and muscular activity. His De ratione motus musculorum (London, 1664) set the terms for much of the subsequent discussion of the subject.

CUJAS, or CUJACIUS, JACQUES (1522–90), French jurist and scholar, was the most profound student of Roman law in the sixteenth century. Largely self-taught in the humanities, he was professor of civil law in many French and Italian cities, but mainly at Bourges. He studied Roman law, and above all the Justinian law, in historical context. At Bourges Cujas developed a school of jurisprudence equal in fame to that of Bologna.

DAMPIER, WILLIAM (1652–1715), spent most of his life at sea, where he joined a party of buccaneers, circumnavigated the globe, and became a captain in the English navy. He possessed a marvelous talent for observing and recording natural phenomena and was an excellent hydrographer. Dampier published A New Voyage Round the World (London, 1697). The book enjoyed great success. He followed it with a second volume entitled Voyages and Descriptions (London, 1699), and a third entitled A Voyage to New Holland (London, 1703).

DANHAWER, or DANNHAUER, JOHANN KONRAD (1603–66), a Lutheran minister and professor at Strassburg, became preacher of the cathedral church and dean of the chapter. A staunch upholder of the Augsburg Confession, he opposed reunion of Lutherans and Calvinists. He wrote a large number of works.

DEE, JOHN (1527–1608), English mathematician who wrote treatises on navigation and navigational instruments and was interested in the occult.

[DE GUERIC]. See GUERICKE

[DE LA VEGA]. See GARCILASO

DEMOCRITUS of Abdera (460/457–357 B.C.), Greek mathematician and physicist, allegedly assigned atoms of cold a cubical figure.

DERHAM, WILLIAM (1657–1735). See the Introduction, p. lx.

DESCARTES, RENÉ (1596–1650), the French philosopher, mathematician, and scientific thinker, was one of the founders of modern thought. Educated at the Jesuit college of La Flèche, he later settled in Holland, where he remained for over two decades working out his system. He was preoccupied with method as the clue to scientific advance and with establishing absolute certainty as the basis of human knowledge. By 1634 he had completed a scientific work, *Le Monde,* but at the news of the condemnation of Galileo for teaching the Copernican system he withheld it from publication. In 1637 he published a book containing treatises on geometry, dioptrics, and meteors prefaced by a *Discourse on Method.* This celebrated work, written in French rather than Latin so as to reach beyond a restricted audience to educated persons of good sense, expounded the foundations of the Cartesian system. Descartes rejected scholastic subtleties and based his speculation upon pure reason. Willing to take nothing for granted, he started with the self-evident, certain principle stated in his famous axiom "I think, therefore I am," and proceeded to proof of the existence of God and from that to a demonstration of the reality of the material world. In 1641 he published a metaphysical work, *Meditations on First Philosophy,* together with *Objections* from various learned authors, and his own *Replies.* A more formal treatise, *Principles of Philosophy* (1644), contained, besides philosophical matter, a cautious exposition of Descartes's views on cosmology. In 1649 he finally yielded to the requests of Queen Cristina to join the Swedish court and instruct her in philosophy. In this year also he published *The Passions of the Soul.* The following winter, owing to the Swedish climate and the rigorous schedule demanded by the queen, he caught pneumonia and died.

Descartes posited a radical division of created existence into matter (extended substance given motion at the creation) and mind (unextended thinking substance), a conclusion he held to be guaranteed by the perfection of God. These first principles led Descartes to view cosmogony and cosmology as products simply of matter in motion, making the laws of motion the ultimate "laws of nature." The Cartesian doctrine of gravity supposed the revolution of fine invisible particles in a vortex whirling in planes perpendicular to the earth's axis; his vortex theory was widely accepted. For Descartes, all scientific explanation was ultimately mechanistic. His system aimed to be as complete as Aristotle's, which it was designed to replace, and he dealt with many of the same phenomena (such as light and the rainbow) and with others more recently investigated (such as magnetism).

Descartes's physiology was an integral part of his philosophy, an expression of his dualist ontology. He portrayed animals and the human body in terms of mere mechanism. Animal and subrational human movement are controlled solely by unconscious mechanisms that operate without any aid from the reason. Mind was outside the body and independent of it but interacted with it through the pineal gland in the brain. The rational soul is most closely associated with this gland, located centrally in the substance of the cerebral marrow. Descartes believed that the pineal gland was not found in lower animals. Without it, these brute creatures lacked mind and soul and were merely living machines, automata.

A leading figure in the early modern scientific movement, Descartes exerted a strong influence on seventeenth-century English thought. While his general system of physics was challenged and eventually overthrown by Newton's theory of gravitation, the impact of Descartes's physiological doctrines was enormous. His notion of mind-body dualism and animal automatism was picked up by Henry More and many others in the seventeenth century, and the idea of the "beast-machine" had continuing ramifications in the eighteenth century. The Cartesian program enabled many prominent medical scientists to wed the Cartesian method of mechanical explanation to careful anatomical investigation and thus to circumvent traditional categories of physiological explanation.

DIO CASSIUS, or DION CASSIUS, also known as Cassius Dio, surnamed Cocceianus (c. 155–c. 230) of Bithnyia, was a Roman official and historian. His main work, written in Greek, is a history of Rome from the beginnings to 229.

DIONIS, PIERRE (d. 1718), was professor of anatomy and surgery at the Jardin des Plantes in Paris. His publications include *L'anatomie de l'homme suivant la circulation du sang et des derniers découvertes* (Paris, 1690, and many later eds.), a treatise on surgery, and works on other medical topics.

DOD, JOHN (1549?–1645), a Puritan divine often called "Decalogue Dod" for his authorship of a book on the Ten Commandments, was instituted to a church in Oxfordshire when he was probably past thirty. While there he married Anne Bownd or Bound, daughter of Nicholas Bownd, author of an influential treatise on the morality of the Sabbath.

DOMINIS, MARCO ANTONIO DE (1560–1626), ecclesiastic and physicist. Dominis was bishop of Senj and later archbishop of Split in Dalmatia. His main theological work, *De republica ecclesiastica* (London, 1617), which urged the unity of all churches, made it necessary for him to flee to England. After returning to Rome he was imprisoned by the Inquisition. Dominis wrote two works on physics. The greater part of *De radiis visus et lucis in vitris perspectivis et iride* (Venice, 1611) is concerned with the rainbow. His *Euripus seu de fluxu et refluxu* (Rome, 1624) is concerned with the tides.

DOUGLAS, JAMES (1675–1742), a London physician who made his career in obstetrics and anatomy, was a fellow of both the Royal Society and the Royal College of Physicians. He published various works on anatomy, including *Myographiae comparatae specimen, or a Comparative Description of All the Muscles in a Man and in a Quadruped* (London, 1707).

DRAKE, JAMES (1667–1707), best known as a Tory pamphleteer, was also a physician who became a fellow of both the Royal Society and the College of Physicians. He published "A Discourse Concerning Some Influence of Respiration on the Motion of the Heart," *Phil. Trans.*, 23 (Sept.–Oct. 1702), 1217–40 (reprinted in *Misc. Cur.*, 1:173–204), and *Anthropologia Nova: Or, A New System of Anatomy*, 2 vols. (London, 1707).

DU BARTAS, GUILLAUME DE SALUSTE (1544–90), a Protestant poet from Gascony, was in the service of Henry of Navarre. His most important work, *La semaine, ou création du monde* (Paris, 1578), assembled all the scientific knowledge of his day in the guise of a description of the seven days of the Creation. *La seconde semaine* (Paris, 1584) enlarged the subject but was never completed. Du Bartas won fame as a poet. His work appeared in Joshua Sylvester's English translation as *Bartas his Devine Weekes and Workes*, 2 pts. (London, 1605–6). It made a strong impression in England, where its Protestant teaching was more generally accepted than in France.

EDWARDS, JOHN (1637–1716), preached for a time in and around Cambridge, England. He devoted his later years to study and writing, publishing more than forty books and winning a high reputation as a Calvinist divine.

EGARD(US) or EGGERS, PAUL (1570–1655), a zealous Protestant preacher of Rendsburg and Nortorf in northwest Germany (Schleswig-Holstein), was an ardent admirer of Johann Arndt and his type of spirituality. Like Arndt, Egard urged a biblical rather than scholastic theology. The majority of his twenty-two publications in the German language and four in Latin had a practical, edifying purpose. The German pietist Philip Jacob Spener had some of Egard's works published in three volumes at Frankfurt and at Giessen from 1679 to 1683.

ETTMÜLLER, MICHAEL (1644–83), was professor of botany and of surgery and anatomy at the University of Leipzig. An iatrochemist, he wrote many works on medicine, with several of them appearing posthumously. An English translation of the last edition of his works appeared under the title *Etmullerus Abridg'd: Or, A Compleat System of the Theory and Practice of Physic* (London, 1699).

EVELYN, JOHN (1620–1706), an English virtuoso, was a staunch royalist and ardent Anglican, a founder of the Royal Society of London, and a public-spirited country gentleman whose publications touched on many subjects. He wrote an essay on the first book of Lucretius, *De rerum natura* (London, 1656). His *Diary*, which he kept throughout his life, has great historical value. Horticulture was Evelyn's enduring interest. His primary work on the subject was *Sylva: Or, a Discourse of Forest-Trees, and the Propagation of Timber* (London, 1664). He also published *A Philosophical Discourse of Earth* (London, 1676), which in later editions was prefaced with the title *Terra*.

FLAMSTEED, JOHN (1646–1719), was the first astronomer royal in England. His accurate telescopic observations made possible a substantial advance in positional astronomy. Apart from *The Doctrine of the Sphere* (London, 1680), his major publications, which included a 3,000-star "British Catalogue," appeared after Mather completed his manuscript.

FLETCHER, GILES (1549?–1611), Cambridge-educated poet, statesman, and ambassador. In 1588 Elizabeth I sent him on an embassy to Moscow. He was treated to great indignity on his brief and unhappy mission, though he secured concessions for English merchants. On returning home he published *Of the Russe Commonwealth* (London, 1591), which has been often reprinted. Fletcher's account of the Russian winter drew on Herberstein's book, described below.

FRACCUS, AMBROISE NOVIDIUS (fl. 1513–49), was a Latin poet during the pontificates of Leo X and Paul III. His *Sacrorum fastorum libri XII* (Rome, 1547), an extremely rare work written in imitation of Ovid's *Fasti*, attempted to do for Christian Rome what Ovid did for pagan Rome.

FRANTZE, WOLFGANG (1564–1628), German Lutheran theologian, was first appointed professor of history and later professor of theology at Wittenberg University. He published many theological works; the Mather library owned his *Tractatus theologicus* (Wittenberg, 1619). Frantze also wrote *Historia animalium sacra* (Wittenberg, 1612). This description of the animals mentioned in the Bible enjoyed extraordinary success, going through many editions. The library of the Mathers had a copy of the fifth edition (1653). The part which treats quadrupeds was translated into English by N. W. as *The History of Brutes: Or, a Description of Living Creatures* (London, 1670).

[FRANZIUS]. See FRANTZE

FREDERICK I, called FREDERICK BARBAROSSA (1123?–90), Holy Roman Emperor from 1152 to 1190, enlarged the empire, advanced learning, maintained internal peace, and encouraged the development of towns and cities.

FROISSART, JEAN (1333?–1400?), French poet and court historian, whose *Chroniques* of the Hundred Years' War are among the most important documents of feudal times. An English translation by John Bourchier was published in London, 1523–25.

FROMONDUS, LIBERTUS, or LIEBERT FROIDMENT (1587–1653), successor of Jansenius to the chair of philosophy and theology at the University of Louvain, wrote under both his own name and the pseudonym of Vincentius Lenis. He treated many subjects: the *Naturales quaestiones* of

Seneca, the comet of 1618, meteorology, commentaries on various books of the Bible, philosophy (including the Cartesian theory of animal automatism), and theology. Opposed to the heliocentric theory, he defended a Roman Catholic decree which denounced Copernican doctrine.

FRYTSCHE or FRITSCHE, MARCUS (fl. 1563), a German humanist, presided over a school at Lauban and served at the court of King Ferdinand I. His works included *Meteororum: hoc est, impressionum aerearum et mirabilium naturae operum, loci fere omnes* (Nuremberg, 1563), which includes the lines quoted by Mather. Other editions were published at Wittenberg in 1581, 1583, and 1598.

FULLER, THOMAS (1608–61), was an Anglican divine of moderate principles and an exceedingly popular preacher. He published many works, including *The Holy State, and the Profane State* (Cambridge, 1642), *The Church-History of Britain* (London, 1655), and *The History of the Worthies of England* (London, 1662).

FUST, JOHANNES (c. 1400–1466), a Mainz lawyer who provided financial backing and then became a partner with Gutenberg in the production of books, is the first printer next to Gutenberg.

GAFFAREL, JACQUES (1601–81), a French priest and doctor of canon law, gained distinction as a Hebraist and orientalist. He lectured on the works of the rabbis. Gaffarel visited Venice, Greece, and the coast of Asia, where he acquired many precious objects. His *Curiositez inouyes, sur la sculpture talismanique, des Persans. Horoscope des Patriarches. Et lecture des estoilles* (Paris, 1629) was translated into English by Edmund Chilmead and published as *Unheard-of Curiosities: Concerning the Talismanical Sculpture of the Persians, the Horoscope of the Patriarkes, and the Reading of the Stars* (London, 1650).

GALE, THEOPHILUS (1628–78), English nonconformist divine and tutor, was best known for *The Court of the Gentiles: Or, A Discourse Touching the Original of Human Literature, Both Philologie and Philosophie, from the Scriptures and Jewish Church*, 2 vols. (Oxford and London, 1669–72). This erudite work, which was in the library of the Mathers, traces the origin of all human arts and sciences, including the languages and learning of pagan antiquity, to Hebrew sources with a view to confirming the authority of Scripture and the Christian religion. Most of Gale's writings, including *Philosophia generalis, in duas partes disterminata* (London, 1676), were restatements of his system in Latin. A favorite of the New England Puritans, Gale bequeathed most of his library to Harvard College.

GALEN (c. 130–c. 200) was born in Pergamum, a Hellenistic city in Asia Minor famous for its temple of the healing god Asclepius. His society was predominantly pagan and hostile to Christianity. Galen believed in divination, especially in dreams. He was educated at home by his father and by eminent teachers of philosophy and medicine, next in Smyrna and Corinth, and finally in Alexandria. He returned to Pergamum as physician to the gladiators and then went to Rome, where he established a medical practice, held public anatomical lectures, and composed his first major works. He interrupted his first stay in Rome to return to Pergamum and to travel for scientific purposes, but was back in Rome by 169, summoned by Marcus Aurelius.

Galen was a man of great knowledge. In medicine he was most influenced by Hippocrates. A skillful anatomist (he dissected many kinds of animals but no humans) and physiologist, he was also a physician, surgeon, and pharmacologist. His commentaries on Hippocrates show much philological knowledge. In philosophy Galen was most influenced by Plato. He held a Platonic vision of truth and was zealous in the search for basic principles. He wrote *De placitis Hippocratis et Platonis* (On the opinions of Hippocrates and Plato) as well as a work on logic.

Galen was also a prolific author. *De usu partium corporis humani* (On the usefulness of the parts of the body) was one of the greatest as well as the most influential of all his writings. Begun about 165, it was completed between 169 and 175. A treatise of anatomy and physiology in seventeen books, *De usu partium* is primarily an elaborate exposition of the teleological connection between the two. The work is a splendid example of the argument from design. This doctrine was already familiar when Galen wrote, but he was the first to marshal the facts to support the design argument in a scientific way. His starting point was Aristotle's dictum, "Nature does nothing in vain" (*De partibus animalium* 691b 4.11). His thesis is that an intelligence is at work in every part of the heavens and on earth, and that the human body as a whole as well as the adaptation of its individual parts to their several functions and to each other is the result of conscious design. Galen usually called this formative agency Nature, but he also called it the Creator. Galen's Nature was like Plato's Demiurge; and Galen called *De usu partium* a "sacred discourse," composed "as a true hymn to our Creator" (3.10; Margaret T. May, *Galen: On the Usefulness of the Parts of the Body,* 2 vols. [Ithaca: Cornell University Press, 1968], 1:189). He endowed the Creator with the virtues of power, wisdom, and benevolence. For Galen, Nature decides everything from the height of a superior wisdom and has the power and skill to accomplish her decisions for the benefit of any creature. A pagan, Galen believed that Nature does not create matter but arranges it in a way which we cannot improve. Behind Nature looms necessity, and Nature never attempts the impossible. The Creator chooses from the possible what is best to be done.

Galen made brilliant use of the argument from design to prove the existence of God. He bitterly opposed the atomists and mechanists and repeatedly denounced as stupid those who attributed the perfection of the animal body to blind chance. He was ingenious in finding excuses for Nature, and at times distorted facts to support his theory. Yet his work as a whole bears witness to his competence and his devotion to scientific truth. His scientific authority made Galen highly agreeable to theologians and helps explain his enduring fame.

De usu partium was read and revered by the ancients. After knowledge of the book perished in the West during the Dark Ages it still persisted in the Eastern Empire. The treatise was translated from the Greek into Arabic, and then from the original Greek text into Latin in the fourteenth century. Printed Latin editions were published during the Renaissance, the first at Paris in 1528. At least seven editions appeared in the sixteenth century, and the book remained influential well into the nineteenth century. It expounded a system of medicine and reinforced the foundations of natural theology. Philosophers and theologians bowed to Galen's scientific authority because it bolstered their own dogma. The only complete translations of Galen's famous treatise into modern languages appear to be two in French (1566, 1854–56) and one in English (May, 1968).

[GALILAEUS]. *See* GALILEI

GALILEI, GALILEO (1564–1642), Italian mathematician, astronomer, and physicist, was a founder of the experimental method. He suggested the use of the pendulum for clocks and proposed the law of uniform acceleration of falling bodies. Galileo devised the simple open-air thermometer and demonstrated that the path of projectiles is a parabola. He improved the refracting telescope for astronomical use, and with it discovered craters on the moon, sunspots, the phases of Venus, and the satellites of Jupiter. His support of the Copernican thesis brought him into conflict with the Church.

GARCILASO DE LA VEGA (1539–1616) was the American-born son of an aristocratic Spanish conquistador and an Incan princess. Called "El Inca" (to distinguish him from the Spanish poet of the same name), he was consciously a *Latin* American. He wrote on Inca history and the Spanish conquest. His *Commentarios reales, que tratan del origen de los Yncas* (Lisbon, 1609; reprinted in 1617 and often thereafter) was translated into English as *The Royal Commentaries of Peru* (London, 1688). It was intended to demonstrate that Peruvians could claim two classical heritages, one represented by Rome, the other by Cuzco, the old Inca capital.

GARDEN, GEORGE (1649–1733), of Aberdeen, Scotland, contributed seven articles to the *Philosophical Transactions* between 1677 and 1700, one of which was reprinted in *Misc. Cur.*, 1:142–52. According to the *DNB*, this is probably the George Garden who was a Scottish episcopal divine. He sympathized with mystical theology and refused to take the oath to William and Mary, though he never approved the arbitrary policy of James II.

GARENCIÈRES, THÉOPHILE DE (1610–80) went to England as the physician of the French ambassador. In addition to a book on the prophecies of Nostradamus, he wrote on pulmonary phthisis, preservatives from the plague, and *The Admirable Virtues and Wonderful Effects of the True and Genuine Tincture of Coral in Physic* (London, 1676).

GASSENDI, PIERRE (1592–1655), was a French theologian, philosopher, and student of the exact sciences. An anti-Aristotelian, he opposed the Cartesian emphasis on pure reason and the Baconian emphasis on mere collection of facts. He revived the philosophy of Epicurus while accommodating it to Christianity and rejecting the doctrine of the supremacy of chance. Gassendi adopted the atomic physics of Epicurus but ascribed the creation of atoms to God, and within the Epicurean system he explained natural phenomena by natural causes. A partisan of Copernicus and Galileo, he opposed judicial astrology and made several contributions in astronomy. He vigorously opposed the Cartesian notion that animals lack minds and souls and are merely living machines.

GAZA, THEODORE (1398–1475), Byzantine humanist, left his home in Thessalonica to become professor of Greek at Ferrara and a leader of the revival of learning in Italy. His Latin translations of Aristotle and Theophrastus became famous.

GELLIBRAND, HENRY (1597–1636), Puritan minister and professor of astronomy at Gresham College, wrote on navigation and mathematics. His best-known scientific discovery, based on observations by John Marr as well as himself, was that of secular change in the magnetic variation (declination). That is, the magnetic needle in its horizontal position does not retain the same declination, or variation from the true north, in the same place at all times, but varies that declination from time to time. He announced this discovery in *A Discourse Mathematical on the Variation of the Magneticall Needle. Together with Its Admirable Diminution Lately Discovered* (London, 1635).

GELLIUS, AULUS (c. 130–c.180), author of *Noctes Atticae*, a collection of chapters dealing with philosophy, history, literary questions, and other topics. He began collecting his material during the winter nights of Attica and assembled it later to entertain and instruct his own children.

GERARD, JOHN (1545–1612), barber-surgeon and garden superintendent to Lord Burghley, minister to Queen Elizabeth I, grew many exotic plants in his Holborn garden. He gained a reputation with publication of *The Herball: Or Generall Historie of Plantes* (London, 1597). A much-amended edition was published in 1633 and reprinted in 1636. Gerard's *Herball*, one of the first botanical books published in English rather than Latin, describes Pimpernell (*Anagallis arvensis*) as the "weather-wiser." The plant is also known as the poor man's weatherglass and the shepherd's clock because it opens only in fine weather.

[GERHARD]. *See* GERARD

GESNER, KONRAD (1516–65), chief physician at Zurich, was typically Renaissance in his wide interests and immense erudition. His two dominant passions were letters and natural science, especially natural history, and he wrote prolifically in both areas. He collected many plants which were unknown to the ancients, but never finished his *Opera botanica,* which was published in two volumes from 1551 to 1571. He wrote a massive *Historia animalia,* 4 vols. (Zurich, 1551–58), to replace the earlier and less complete work of Aristotle, and it won immediate acclaim.

GIBSON, THOMAS (1647–1722), obtained a medical degree at Leiden. A practicing physician, he became an honorary fellow of the Royal College of Physicians. Gibson published *The Anatomy of Humane Bodies Epitomized* (London, 1682), long a popular work.

GIOIA, FLAVIO (fl. early fourteenth century), of Amalfi was long credited with first discovering the use of the mariner's compass in navigation in 1302, but the attribution is incorrect.

GLISSON, FRANCIS (1597?–1677), was regius professor of physic at Cambridge University and a member of both the College of Physicians and the Royal Society. His *Anatomia hepatis* (London, 1654) and *Tractatus de ventriculo et intestinis* (London, 1677) together, according to Owsei Temkin in the *DSB,* "constitute a monumental work on general anatomy and on anatomy and physiology of the digestive organs."

GOEDAERT, JOHANNES (1617–68), was a Dutch painter whose subjects were birds and insects. His knowledge of entomology was based on firsthand observation rather than authority. He published two volumes of his only work, *Metamorphosis naturalis,* in Dutch during his lifetime. A Latin translation in three volumes, edited and with commentary by Johannes de Mey, appeared as *Metamorphosis et historia naturalis insectorum* (Middleburg, 1662–69). An English translation with notes by Martin Lister was entitled *Of Insects* (York, 1682).

GOODWIN, THOMAS (1600–1680), a leading Independent divine, was a member of the Westminster Assembly, president of Magdalene College, Oxford, during the 1650s, and a member of the Savoy Assembly. Deprived of his office as president of Magdalen in 1660, he founded an Independent congregation in London and published his sermons in many volumes.

[GORDEN]. *See* Garden

GREGORAS, NICEPHORUS (c. 1295–c. 1359), a Byzantine scholar, was the author of a history of Byzantium covering the period from 1204 to

1359. Gregoras also wrote on astronomy, theology, mathematics, and other subjects.

GREGORY of Nazianzus (329–89), Greek theologian and bishop of Constantinople, retired to Cappadocia. His writings put Christian content into traditional forms. His *Poemata dogmatica* and *Orationes* discuss both the work of God in creation and divine providence.

GREGORY, DAVID (1659–1708), a Scottish astronomer, purveyed the "new" science of Newton as a professor of mathematics at the University of Edinburgh starting in 1683. In 1691 he was appointed Savilian professor of astronomy at Oxford, and the following year became a fellow of the Royal Society. In 1695 he published in Latin a work on the science of Descartes, James Wallis, and Newton which was translated into English as *Elements of Catoptrics and Dioptrics* (London, 1715). Gregory's principal work, *Astronomiae, physicae et geometricae elementa* (Oxford, 1702), was the first textbook attempt to graft the gravitational principles propounded by Newton's *Principia* onto traditional astronomy. An English translation was entitled *The Elements of Astronomy, Physical and Geometrical*, 2 vols. (London, 1715).

GREW, NEHEMIAH (1641–1712), a plant anatomist and plant morphologist, was born in Warwickshire and educated at Pembroke Hall, Cambridge. He earned a medical doctorate at the University of Leiden in 1671, and upon returning to England relied primarily upon the practice of medicine as a means of livelihood for the rest of his life. He was admitted as an honorary fellow of the College of Physicians in 1680.

As a medical man, Grew was interested in the structure of both animals and plants. His religious and philosophical beliefs led him to regard both as "the Contrivances of the same Wisdom." He turned to the study of plant anatomy as early as 1664 and wrote an essay on the subject which led to his election as a fellow of the Royal Society in 1671. He moved to London the following year, and the Royal Society, which made its compound microscope available to him, became the center of his activities. In 1678 he succeeded Henry Oldenberg as secretary of the society upon the latter's death. Grew's scientific distinction rests on his contribution to plant anatomy. His first important book was *The Anatomy of Vegetables Begun: With a General Account of Vegetation Founded Thereon* (London, 1672). His researches on plant anatomy and those of Marcello Malpighi moved in the same direction, but Grew's results were independent of Malpighi's. Grew applied the mechanistic science of his day to the study of botany and attempted to find a mechanistic cause for each phenomenon. His medical training led him to study the chemistry of plants to determine their medical uses. He was interested in other topics also, and published "Some Observations touching the Nature of Snow" (*Phil. Trans.*, 8 [Mar. 25, 1673], 5193–96). Grew's outstanding accomplishment is *The Anatomy of Plants. With an Idea of a Philosophical History of Plants. And Several Other Lectures* ([London], 1682). He probably deserves credit for first observing the existence of sex in plants.

Grew's last work was *Cosmologia Sacra: Or, A Discourse of the Universe As It is the Creature and Kingdom of God* (London, 1701). This treatise sets forth the religious and philosophical beliefs which serve as a background to his scientific work. Grew directed his argument especially against Spinoza. He deduces the nature of God both a priori, from the necessity of His being, and a posteriori, from His handiwork, the universe. As in similar works of the period, the argument begins with much borrowed astronomical learning, though Grew also exhibits a good acquaintance with the whole body of divinity. Grew saw all of nature glorifying God in its fine design. "The culpa communis of Grew and his contemporaries," Coleridge later charged, "was to assume as the measure of every truth its reduction to Geometric Imaginability."

GRIMALDI, FRANCISCO MARIA (1618–63), Italian Jesuit active in astronomy and optics, whose primary contribution to science was the discovery of optical diffraction. He published *Physico-mathesis de lumine, coloribus, et iride* (Bologna, 1665), which was reviewed in *Phil. Trans.*, 6 (Jan. 22, 1672), 3068–70.

GROTIUS, HUGO (1583–1645), established his reputation by editing various works and writing dramas in Latin. A leader of the Remonstrants, he was imprisoned for his religious involvement but escaped to France. A jurist and scholar, his *De jure belli ac pacis* (Paris, 1625) is one of the great contributions to modern international law.

GUERICKE, or GERICKE, OTTO VON (1602–86), made important experiments on space and matter and dealt with astronomy. He invented an instrument, the air pump, by means of which it was possible to produce a vacuum in a given vessel. His major work was *Experimenta nova (ut vocantur) Magdeburgica de vacuo spatio* (Amsterdam, 1672). It gives the number in Mather's text as fifty-three ciphers (i.e., naughts or zeros).

HAKEWILL, GEORGE (1578–1649), Anglican divine and rector of Exeter College from 1642, was an excellent prose stylist and author of several works. His book entitled *An Apologie of the Power and Providence of God in the Government of the World* (London, 1627, and later eds.), argued against the notion put forward by Bishop Godfrey Goodman that nature and man were decaying.

HALLEY, EDMOND (1656?–1743), demonstrated a wide interest in science and published in many fields, but his most notable achievements were in astronomy. He left Oxford in 1676 without taking a degree and went to St. Helena, an island off the western coast of Africa, where he observed the positions of the fixed stars in the Southern Hemisphere in order to make astronomical calculations more accurate. On returning to England he dedicated a planisphere of the southern constellations to Charles II and was rewarded with a royal mandamus to Oxford for the M.A. degree. He was elected a fellow of the Royal Society in 1678, and with the blessing of that body went to visit Johannes Hevelius in Danzig.

Halley contributed significantly in determining the distance of the sun from the earth by observing and timing a transit of Mercury in 1677. He also worked out plans for obtaining greater precision in the result by observing a transit of Venus, and his method, slightly modified and used in 1671 after his death, produced a more accurate result. His outstanding achievement in stellar astronomy was the discovery that the stars which had long been regarded as fixed possessed individual motion. He was able to detect proper motion only in the case of three bright stars, but he correctly deduced that dimmer stars possessed motion too small to be detected, and with more accurate instruments the study of the proper motions of the stars was later extended.

Keenly aware of the importance of the problem of gravity, Halley suggested that Newton write down his own views of the subject. Halley gave an account of Newton's findings to the Royal Society, and in 1686 paid for publication of the *Principia*. After it appeared, Halley began an intensive study of the motion of comets. He hypothesized that the path of comets must be an ellipse and made computations that enabled him to conclude that the bright comets of 1531, 1607, and 1682 were the same object and to predict its return in 1758. He published his views in 1705. When the comet reappeared in 1758 it was named "Halley's Comet." The accuracy of Halley's prediction offered compelling proof of the Newtonian cosmology. The return of the comet not only reinforced Newton's demonstration in the *Principia* that the notion of heavenly spheres was a fallacy but also dealt a blow to the proponents of Cartesian vortices.

Halley was also the founder of geophysics. His main contributions in this field, all published in the *Philosophical Transactions,* were on terrestrial magnetism and terrestrial physics, including the rate of evaporation of water, the salinity of lakes and oceans, and the age of the earth.

Halley edited the *Philosophical Transactions* of the Royal Society from 1685 to 1693 and published eighty-four papers in that journal. He also edited *Miscellanea Curiosa,* 3 vols. (1705–7; 2d ed., 1708), a collection of papers, including many of his own, previously published in the *Philosophical Transactions.* He became secretary of the Royal Society in 1713 and was appointed royal astronomer in 1720.

HARRIS, JOHN (1666–1719), Oxford graduate and clergyman, whose most famous work, *Lexicon Technicum: Or, An Universal English Dictionary of Arts and Sciences* (London, 1704; 2d ed., 2 vols., London, 1708–10), was the first general scientific encyclopedia.

HARVEY, WILLIAM (1578–1657), English physician and anatomist, published his discovery of the circulation of the blood in Frankfurt in 1628. The first edition in English was published in London in 1653. Harvey observed animal life in all its forms, and he published *Excercitationes de generatione animalium* (London and Amsterdam, 1651).

HAVERS, CLOPTON (c. 1655–1702), a medical practitioner and fellow of the Royal Society, published *Osteologia nova: or Some New Observations of the Bones and the Parts Belonging to Them* (London, 1691). This book, the first complete and systematic study of the subject, was reviewed at length in *Phil. Trans.*, 17 (June–Sept. 1691), 544–54, and immediately translated into Latin.

HELMONT, FRANCISCUS MERCURIUS VAN (1618–99), son of Johannes Baptista van Helmont, wrote on medicine, chemistry, theosophy, and linguistics.

HELMONT, JOHANNES BAPTISTA VAN (1579–1644), a Belgian savant and pioneer in chemistry, combined both scientific and mystical elements in his writings, which were often obscure. His work was widely read in its collected form in either the Latin *Ortus medicinae* (Amsterdam, 1648) or an English translation *Oriatrike: Or, Physick Refined* (London, 1662).

HENNEPIN, LOUIS (c. 1640–c. 1701), a Recollet missionary in Canada, accompanied La Salle on his exploration of the Great Lakes and the Mississippi River in 1679. His *Nouvelle découverte d'un très grand pays situé dans l'Amerique* (Utrecht, 1697), describes Niagara Falls. Two different translations of the book appeared under the title *A New Discovery of a Vast Country in America* (London, 1698).

HERBERSTEIN, SIGISMUND VON (1486–1566), diplomat and historian, was sent by Hapsburg emperors on missions to Russia twice during the reign of Vasilii III (1517 and 1526). His *Rerum Moscoviticarum commentarii* (Vienna, 1549) was based on good knowledge of the written sources as well as intelligent personal observation. The book enjoyed great success, going through eighteen editions in various languages by 1589, and provided westerners their best knowledge of Russia in the sixteenth century.

HÉRIGONE, PIERRE (d. c. 1643), a little-known person apparently of Basque origin, spent most of his life in Paris as a teacher of mathematics. He published many works in the field, the most important of which is *Cursus mathematicus, nova, brevi et clara methodo demonstratus*, 6 vols. (Paris, 1634–42). John Ray attributes to this "learned mathematician" a work entitled *Optica*. Some authorities contend that the name Pierre Hérigone is a pseudonym for Clément Cyriaque de Mangin, and that he also used the pseudonym Denis (or Didier) Henrion, who published *Cosmographie, ou traicté general des choses tant célestes qu' élémentaires* (Paris, 1620).

HERLICIUS or HERLITZ, DAVID (1557–1636) was briefly professor of mathematics at the University of Gripswald. He practiced medicine in north German cities for most of his life and published works on medicine, history, philosophy, and other topics. Above all an astrologer, he cast over twelve hundred horoscopes over a period of more than five decades, mainly for money. His publications include *Tractatus de fulmine, et aliis impressionibus, prodigiis et miraculis ignitis; vom Blitze, Donner, und allerley Feur-Zeichen und Wunderwercken; über das Begrabniss* (Stettin, 1600).

HERNANDEZ, FRANCISCO (1517–87), physician to Philip II, was ordered to Mexico, where from 1570 to 1577 he studied its natural history and wrote many books on the subject. Two abridged versions were published in Mexico (1579 and 1615), and members of the Accademia dei Lincei edited from his manuscripts *Rerum medicarum novae Hispaniae thesaurus* (Rome, 1649) and *Nova plantarum, animalium et mineralium Mexicanorum historia* (Rome, 1628).

HEVELIUS, JOHANNES (1611–87), the leading mid-seventeenth-century astronomer, established an astronomical observatory atop his house in Danzig and corresponded with astronomers throughout Europe. His *Selenographia: sive, lunae descriptio* (Danzig, 1647) was an atlas which named the features on the moon's surface and discussed its movement of libration. His *Cometographia* (Danzig, 1668) discussed the comet of 1652 and the physical constitution of comets. Hevelius described the design, manufacture, and engraving of his own observational instruments in *Machina coelestis*, 2 pts. (Danzig, 1673–79). His best-known compendium of observations was the *Prodromus astronomiae* (Danzig, 1690), a catalogue of 1,564 stars, and an accompanying volume of plates illustrating the constellations, *Firmamentum Sobiescianum, sive Uranographia* (Danzig, 1690).

HIPPOCRATES (460–c. 370 B.C.) was the most important medical person of antiquity, and in later centuries he became regarded as the father of medicine, the embodiment of the ideal physician. The best-known physician of his time, he headed the flourishing medical school on the island of Cos. He is connected with the Hippocratic Corpus, some seventy treatises, several of which Hippocrates wrote, which were brought together in the library at Alexandria. Hippocrates represented the stage of medicine in which war was waged on magico-religious medical practice. His chief contribution was to establish the fact that disease was a natural process, that its symptoms were the reaction of the body to the disease, and that the chief function of physicians was to aid nature in its healing process.

HOOKE, ROBERT (1635–1702), a mechanical prodigy, was active in many fields of science. As a student at Oxford he took his place in a group of scientists—including John Wilkins, Thomas Willis, John Wallis, and Robert Boyle—which later developed into the Royal Society of London. His mechanical skill enabled him to develop the air pump, and he assisted Boyle in the experiment that concluded in Boyle's law. He became a member of the Royal Society in 1663, and from 1662 on was the society's curator of experiments.

His most important book, *Micrographia, or Some Physiological Descriptions of Minute Bodies* (London, 1665), identified Hooke with microscopical investigations of the mineral, vegetable, and animal kingdoms. His examination of the porous structure of cork led Hooke to coin the modern biological usage of the word *cell*. He inaugurated the study of insect anatomy. *Micrographia* is also filled with ingenious scientific ideas. It expounds the author's theory of light, a subject on which he later lectured before the Royal Society. He proposed a wave theory of light which contradicted Newton and anticipated Huygens. Hooke was difficult and argumentative as a person, and the theory of light was the occasion of his initial controversy with Newton. A rift opened between the two then that was never closed. Hooke also studied the nature of air, its function in combustion and respiration, at the time that Boyle, Richard Lower, and John Mayow were independently studying the same problem. He demonstrated that a continued supply of fresh air is as essential to life as it is to fire. With Mayow and the others, he identified the nitrous salt or spirit in the air as the ingredient essential to life.

His contribution to the development of the spring that made small and accurate clocks and watches possible is difficult to assess, but in 1678 he enunciated Hooke's law regarding the action of springs. Hooke also worked out a theory of universal gravitation which, although imperfect, contributed to Newton's later achievement. He stated the law of inverse squares. Venturing into astronomy, he was the first to infer the rotation of Jupiter from the movement of a spot noted in 1664. His drawings of Mars fixed that planet's exact rate of rotation.

Geology was one of his enduring interests, and with the possible exception of Nicholas Steno, according to Richard Westfall in the *DSB*, Hooke was easily the most important geologist of his day. He discussed the controversy over the origin of fossils, dividing "figured stones" into two categories—those with forms characteristic of the organism and those with forms characteristic of the substance. He dealt at length with earthquakes, identifying the upheavals on the surface of the earth with cataclysmic earthquakes. He was the first catastrophist. His *Posthumous Works* (London, 1705) contain several important papers, including discourses on light, gravity, and earthquakes.

HUGH of St. Victor (d. 1142), a monk of the house of Augustinian canons in Paris, was a wide-ranging intellectual whose writings covered theology, biblical commentaries, homilies, and works of spirituality. He was a most influential theologian.

HULSE, EDWARD (1631–1711), who studied medicine at Leiden and joined the College of Physicians in 1675. He apparently published no writings, but original letters with his scientific observations were available to contemporaries. His eldest son was Edward Hulse (1682–1759), also a physician.

HUYGENS, CHRISTIAAN (1629–95), a member of a prominent Dutch family, received an excellent education as a youth before studying law and medicine at the University of Leiden. He then devoted himself completely to the study of nature, at first concentrating on mathematics, and later branching out into astronomy, physics, and other sciences. On journeys to Paris he met members of the circle of scholars which later formed the Royal Academy of Sciences, and on a later visit to London in 1663 he met prominent scientists and became a member of the Royal Society. Lured to Paris when the Academy of Sciences was founded in 1666, he was its most prominent member until he returned to Holland in 1681. Much of his scientific work was accomplished while he enjoyed an ample stipend from the academy.

Huygens was one of the greatest mathematicians of his age. He established his reputation in this field by improving existing methods and applying them to problems in the natural sciences rather than by developing new mathematical theories.

Together with his brother, Huygens ground lenses and built microscopes and telescopes. In 1655, with new lenses that minimized the aberration of light, he discovered a sixth satellite of Saturn and recognized that Saturn was surrounded by a thin ring which nowhere touched the planet. He published his discovery, in code, in *De Saturni luna observatio nova* (The Hague, 1656), and fully in *Systema Saturnium* (The Hague, 1659).

Huygens made a great contribution in the measurement of time. Although Galileo had noted isochronal chandeliers and recognized the possibility of attaching a pendulum to the gears of a clock, Huygens brought the idea to fruition, and the first clock using a pendulum as a regulator for the works dates from 1657. He described his invention in *Horologium* (The Hague, 1658), and subsequently spent much time studying the application of the pendulum clock to the determination of longitudes at sea. He also studied the relation between period and length of the pendulum and developed the theory of the center of oscillation, embodying the results in his most important work, *Horologium oscillatorium* (Paris, 1673). He invented the spring balance as the regulator of clocks in 1674–75, and his invention gave portability and accuracy to clocks and watches. Huygens and Hooke engaged in a priority dispute over this discovery.

Huygens adhered to a mechanistic philosophy of nature. He followed

Descartes in viewing the motions of particles of matter and their interactions by direct contact as the only valid starting points for philosophizing about nature. According to Huygens, the particles of matter move in a vacuum. In working out a pattern of size relations between particles, Huygens concluded that four or five discrete classes of particles exist. Those of the first class are the components of ordinary bodies and of the air. Those of the second class form the "ether," and the phenomenon of light may be explained by shock waves in this medium. Particles of the third class are the carriers of magnetic phenomena, and those of the fourth class form the "subtle matter" which causes gravity.

Huygens realized the importance of Newton's *Principia,* but his mechanistic outlook led him to oppose Newton's use of attractive force as a fundamental explanatory principle, and it strongly influenced his studies on light and gravity. He expounded his own theory of the cause of gravity in 1669. Developing the ideas of Descartes, he presupposed a vortex of particles of subtle matter to be circling the earth with great velocity. Because of their circular movement these particles have a tendency to move away from the earth's center. They can follow this tendency if ordinary bodies move toward the center. The centrifugal tendency of the vortex particles thus causes a centripetal tendency in ordinary bodies, and this latter tendency is gravity. This explanation dealt with fundamental problems that Newton avoided and left unsolved.

Huygens completed his *Traité de la lumière* in 1678, and he published it in 1690 as his answer to Newton's *Principia.* Huygens viewed light as an irregular series of shock waves which proceed with very great but finite velocity through the ether. He was the founder of the wave theory of light.

Convinced that complete certainty cannot be achieved in the study of nature, Huygens nevertheless held that the philosopher must pursue the highest degree of probability in his theories. His *Cosmotheoris* (The Hague, 1698) assigned a very high degree of probability to his own conjectures about life on other planets. The wisdom of God is most manifest in the creation of life and living beings, Huygens wrote, and in the Copernican world system the earth holds no privileged position among other planets. In all probability there must be life and living beings endowed with reason and having a culture similar to humans on other planets. The second part of this work discussed the different movements of the heavenly bodies and how they must appear to the inhabitants of the planets. *Cosmotheoris* was translated into English and several other languages and attracted many readers. But Huygens's other studies exerted little influence.

IBN TUFAYL MUHAMMAD IBN 'ABD AL-MALIK (c. 1100/1110–1185/1186), also known as Abu Bakr ibn Tufayl, was born near Granada and belonged to a prominent Arab tribe. For a time secretary to a provincial governor, he became court physician to an Almohad sultan. A renowned philosopher, he was the author of *Risalat Haiy b. Yaqzan,* one of the most celebrated books of the Middle Ages. He was known to Muslim scholars as Ibn Tufayl and to Europeans as Abubacer (or Abubeker), a corruption of Abu Bakr.

INGRASSIA, GIOVANNI FILIPPO (c. 1510–80), professor of anatomy and medicine at the University of Naples and from 1563 at Palermo, was a distinguished osteologist. His anatomical studies were published posthumously under the title of *In Galeni librum de ossibus doctissima commentaria* (Palermo, 1603). According to Fallopio, Ingrassia orally described the third auditory ossicle or stapes to students in 1546, calling it *stapha* because of its resemblance to the shape of the stirrup commonly used in Sicily.

JAMES, THOMAS (1593?–1635), English explorer, commanded an expedition which sailed from Bristol in 1631, going to Greenland, Hudson's Bay, and James's Bay, and wintering at Charleton Island. An account of his adventure was published as *The Strange and Dangerous Voyage of Captain Thomas James in His Intended Discovery of the Northwest Passage into the South Sea* (London, 1633).

JAMESON, WILLIAM (fl. 1689–1720), lecturer on history at Glasgow and a Presbyterian controversialist, published various works. Mather corresponded with him, and the University of Glasgow conferred the degree of doctor of divinity upon Mather in 1710.

JARCHI. *See* RASHI

JENKIN, ROBERT (1656–1727), Anglican divine and master of St. John's College, Cambridge, was the author of several works on church history and theology. His book *The Reasonableness and Certainty of the Christian Religion,* 2 vols. (London, 1698–1700), went through a sixth edition by 1734.

JESSOP, FRANCIS (fl. 1670s–1680s), a country virtuoso of Broomhall in Yorkshire, published two articles in the *Philosophical Transactions,* one on uncommon mineral substances found in coal and iron mines in England, another on the damps in mines. He was also the author of *Propositiones hydrostaticae ad illustrandum Aristarchi Samii systema destinatae, et quaedam phaenomena naturae generalia* (London, 1687).

JEWEL, JOHN (1522–71), a Marian exile and excellent scholar, became bishop of Salisbury in 1560. His *Apologia ecclesiae Anglicanae* (London, 1562), the first methodical defense of the Church of England against the Church of Rome, made him the official champion of Anglicanism.

JONES, JOHN (1645–1709), a physician, became chancellor of the diocese of Llandaff. He authored a number of works, including *The Mysteries of Opium Reveal'd* (London, 1700).

JONSTONUS, JOANNES, or JOHN JONSTON or JOHNSTONE (1603-75), was born in Poland of an English father, attended the University of St. Andrews, studied botany and medicine at Cambridge, and later gained a reputation by the practice of medicine in Leiden. He produced many compilations which were greatly esteemed in seventeenth-century England. His *Thaumatographia naturalis, in decem classes distincta* (Amsterdam, 1632) was translated into English by "a Person of Quality" as *An History of the Wonderful Things of Nature: Set Forth in Ten Severall Classes* (London, 1657). The library of the Mathers contained a duodecimo copy of a Latin version of this title. Cotton Mather knew Jonston's *Historia naturalis, de quadrupedibus,* which was part of Jonston's *Historia naturalis,* 4 vols. (Frankfurt, 1650-53).

JUSTINIAN I, called the Great, in full FLAVIUS PETRUS SABBATIUS JUSTINIANUS (483-565), ruled the eastern Roman empire from 527. He did much to restore the greatness of the Roman empire by recovering the lost provinces of the West. He sought God's favor for the empire by suppressing heresy and paganism, and he appointed commissions to codify and rationalize the legal system. The *Corpus Juris Civilis* of Justinian still serves as a foundation of actual law in continental Europe.

KEILL, JAMES (1673-1719), the younger brother of John Keill, the distinguished Newtonian mathematician, was born in Scotland. Educated partly at home and partly on the Continent, he unofficially lectured on anatomy at Oxford and Cambridge and received an honorary M.D. degree from the latter in 1705. He combined research with a successful practice as a physician in Northampton, and was elected to the Royal Society in 1712.

His *Anatomy of the Humane Body Abridg'd* (London, 1698), largely copied from a French compendium, went through many editions. The most popular anatomical epitome of the time, it provided sound basic knowledge to generations of students. Keill actively supported the mechanical or iatromathematical school of medicine, and his chief work was *An Account of Animal Secretion, the Quantity of Blood in the Humane Body, and Muscular Motion* (London, 1708). He used measurement and mathematics to examine the matters named in the title. Inspired by Newtonian theories, Keill posited an attractive force between particles of matter, proposing that glandular secretions consisted of cohesions of particles in the blood, and that these particles had united through forces of attraction and were mechanically filtered by various glands according to size. He also studied the rate of blood flow, calculating the absolute velocity at which blood travels through the aorta and smaller vessels and recognizing that the blood's velocity must decrease as the number of arterial branches increases. Keill's physiology was a rational attempt at quantification, but his reputation declined in the eighteenth century as vitalistic trends overshadowed the quantitative approach.

KEILL, JOHN (1671–1721), an enthusiastic disciple of Newton, published *An Examination of Dr. Burnet's Theory of the Earth: Together With Some Remarks on Mr. Whiston's New Theory of the Earth* (Oxford, 1698). His most important work as a disseminator of Newtonianism was *Introductio ad veram physicam* (Oxford, 1702), based on a series of experimental lectures given at Oxford since 1694. It was translated as *An Introduction to Natural Philosophy* (London, 1720) and later into French. In 1700 Keill was elected fellow of the Royal Society, and in 1712 he became Savilian professor of astronomy at Oxford.

KEPLER, JOHANNES (1571–1630), German astronomer and ardent Copernican, whose notions about the harmony of the world were intimately linked with his theological view of God the Creator. Kepler is best remembered for his three laws of planetary motion, upon which Newton built his celestial physics.

KERCKRING, THEODOR (1639–93), a Dutch-born physician, conducted research and wrote on anatomy. His *Spicilegium anatomicum continens observationum anatomicarum rariorum centuriam unam, necnon osteogeniam foetum* (Amsterdam, 1670), discusses the intestines. Kerckring was elected a fellow of the Royal Society of London.

KIMCHI [or Kimhi], Rabbi David (1160–1235), a scholar and teacher known by the acronym Radak (*Rabbi David Kimchi*), a medieval Jewish way of rewarding intellectual distinction, wrote largely on philology and biblical exegesis. He prepared the *Mikhlol*, a Hebrew grammatical work in two parts, to bring order to a complicated field, and wrote commentaries on Chronicles, Psalms, Former Prophets, Later Prophets, and Genesis. His editions of the Bible contained annotations in which he scattered much of the learning of the rabbis, and his influence was widespread.

KING, EDMUND (1629–1709), converted from a surgeon to a physician and was physician to Charles II. He dissected animals and human subjects, conducted many scientific investigations, and was interested in insects. A fellow of the Royal Society, he published several papers in the *Philosophical Transactions*, including "Observations Concerning Emmets or Ants, Their Eggs, Production, Progress, Coming to Maturity, Life, Etc.," *Phil. Trans.*, 2 (Mar. 11, 1666), 425–28.

KIRCHER, ATHANASIUS (1601 or 1602–80), was born at Geisa in Germany and entered the Society of Jesus in 1616. Ordained in 1628, he became a professor at Würzburg. In 1631, during the Thirty Years' War, he fled, and after lecturing for a time in Avignon, he was appointed professor of mathematics at the College of Rome. He resigned after eight years and devoted the rest of his life to independent investigations. He benefited from living in the center of a worldwide network of Jesuit missionaries and others who reported on their travels.

Kircher studied nearly all fields of both humanities and sciences, including acoustics, archaeology, arithmetic, astronomy, chemistry, geography, geology, geometry, magnetism, medicine, music theory, optics, philology, philosophy, physics, and theology. He published some forty-four books that demonstrate the variety of his interests (he also left over two thousand letters and manuscripts). An early microscopist of note, he used the microscope in medical investigations. His inclination to deal with curious questions led him to study orientology and to attempt the interpretation of hieroglyphics found in ancient monuments and ruins. He wrote on China; Martin Martini was his disciple.

His printed works disseminated the knowledge at his disposal and became very popular. He published five books on magnetism, including *Ars magna lucis et umbrae in mundo* (Rome, 1646). Kircher's *Musurgia universalis, sive ars magna consoni et dissoni* (Rome, 1650; rpt., Hildesheim, 1970) dealt with sounds and echoes and recorded the first description of a speaking trumpet. His *Itinerarium exstaticum II* (Rome, 1656) treated natural phenomena both on and below the earth, whereas his *Itinerarium exstaticum coeleste* (Rome, 1671) contained dialogues on journeys to the heavenly bodies and on the providence of God. His *Mundus subterraneus* (2 vols. in 1, Amsterdam, 1664–65; 2d ed., Amsterdam, 1678), a mixture of odd but partly accurate speculation that comprised many branches of science, assumed the existence of vast underground reservoirs and described a hydrologic circle of water. Kircher, who had witnessed the eruptions in 1638 of Stromboli, Etna, and Vesuvius, regarded hot springs and volcanic eruptions as the consequence of subterranean regions of fire. A brief, popular translation of this work was published as *The Vulcano's* (London, 1669). Kircher's *Phonurgia nova* (Kempten, 1673; rpt., New York, 1966) treated sound and music.

A brilliant but peculiar genius, Kircher was both an ardent Aristotelian and an avid experimenter, an important link between medieval and modern science who exerted vast influence on the scientists of his day. His extensive writings afford a picture of the scientific understanding of the seventeenth century. He made particular contributions to specific scientific fields, but most of what he described was already known.

[KOSTER]. *See* COSTER

LA BRUYÈRE, JEAN DE (1645–96), French satiric moralist known for *Les caractères de Théophraste . . . avec les caractères ou les moeurs de ce siècle* (Paris, 1688), published in English translation as *The Characters, Or the Manners of the Age* (London, 1699).

LACTANTIUS FIRMIANUS, LUCIUS CAECILIUS (c. 240–c. 320), convert to and apologist for Christianity. His writings include *Divinae institutiones* and *De opificio Dei*, an attempt to prove the existence of God from the marvels of the human body.

LAMBECIUS, PETRUS, or PETER LAMBECK (1628–80), was one of the most learned men of his age. Born in Hamburg, he studied in foreign countries for two years at the expense of his uncle and published his first book at the age of nineteen. He was appointed professor of history in Hamburg in 1652 and rector of the local college in 1660. He married a rich old woman but quickly abandoned her, going to Vienna in 1662, where he became the emperor's librarian and historiographer. He published many books, but some, planned on a grandiose scale, were never completed. His works include *Origines Hamburgenses*, 2 vols. (Hamburg, 1652–61), a history of the city; *Prodromus historiae litterariae* (Hamburg, 1659), a world history; and *Commentarii de augustissimam bibliotheca caesaria Vindobonensi*, 8 vols. (Vienna, 1665–79), a catalogue of the imperial library in Vienna.

LANE, RALPH (d. 1603), joined an expedition which established a colony off North Carolina in 1585. Lane was the first governor of the colony, which was called Virginia. Tobacco and potatoes may have been first brought into England at this time by Lane and his companions, but there is no direct evidence of it.

LANGE, JOHANNES (1485–1565) of Löwenberg in Lower Silesia, who served for forty years as chief physician to the electors of the Palatinate. He was a leading physician of his time, and his writings include *Medicinalium epistolarum miscellanea* (Basel, [1554]).

LAPIDE, CORNELIUS À, or CORNELIS VAN DEN STEEN (1566–1637), a Dutch Jesuit scholar, teacher, and prolific exegete, published commentaries on the major and minor prophets in 1625 and 1628 respectively. According to Lapide, Flemings proved whether a man was a Frenchman or not by bidding him to pronounce *Acht en tachtentich*.

LAURENTIUS, or LAURENS, ANDRÉ DU (1558–1609), professor of medicine at the University of Montpellier, became first physician to King Henry IV of France. He published many medical works; his *Historia anatomica humani corporis et singularum ejus partium, multis controversiis et observationibus novis illustrata* (Paris, 1600) was one of the most widely used anatomical textbooks of the first half of the seventeenth century.

LE BLANC, VINCENT (1554–1640), French voyager and author of *Les voyages fameux du sieur Vincent Leblanc* (Paris, 1648), translated as *The World Surveyed: or the Famous Voyages and Travailes of Vincent Le Blanc* (London, 1660).

LE CLERC, JEAN (1657–1736), became a Protestant minister in his native Geneva, but a less dogmatic type of Christianity attracted him briefly to Saumur, Paris, and London, and he finally settled in Amsterdam. Here he became pastor of a Remonstrant church and professor of Hebrew, belles lettres, and ecclesiastical history in an Arminian college. Le Clerc edited and wrote numerous books and often engaged in religious and literary controversy. His works included *Bibliothèque universelle et historique*, 26 vols. (Amsterdam, 1686–93), which he edited with others, *Bibliothèque choisie*, 28 vols. (Amsterdam, 1703–13), and *Bibliothèque ancienne et moderne*, 29 vols. (Amsterdam, 1714–27).

LEEUWENHOEK, ANTON VAN (1632–1723), began his scientific activities about 1671 while employed as a civil servant in Delft. He ground lenses and made simple microscopes which were unsurpassed in quality until the nineteenth century. Although he lacked a university education and worked in relative scientific isolation, he acquired scientific knowledge from Dutch books and translations and from correspondence with the Royal Society of London.

Leeuwenhoek made microscopical investigations of organic and inorganic structures, and his most important contributions were in general biology. He presupposed that inorganic and organic nature are generally similar and that all living creatures are similar in form and function, and he generalized from observations of many types of any given group of organisms.

Leeuwenhoek's most important discovery, made in 1674, was the recognition of the true nature of the microorganism. Starting from the assumption that life and motility are identical, he concluded that the moving objects he saw through his microscope were little animals. Two years later he communicated his findings to the Royal Society, causing a sensation. In subsequent letters to the Royal Society he described many specific forms of microorganisms.

Leeuwenhoek put the microscope at the service of his lifelong study of sexual reproduction, which aimed at refuting the theory of spontaneous generation. Observing the spermatozoa of many species of animals, he postulated that fertilization occurs when the spermatozoa penetrates the egg. He denied any generative role to the motionless (and thus lifeless) egg, holding to an animalculist theory of reproduction. That fertilization represents the fusion of the nuclei of the spermatozoon and the egg was not demonstrated until 1875. Leeuwenhoek also denied the Aristotelian theory of generation, which held that certain tiny animals (such as insects and intestinal worms) were primitive in structure and originated from the putrefaction of organic matter.

Leeuwenhoek also put microscopy at the service of his lifelong study of the transport system of nutrients. He drew an analogy between the animal system and the plant system, minutely analyzing the transport canals, the transport media, and the nutritive matter to be transported. He paid special attention to the blood vessels and the blood. Unaware of Malpighi's pioneer work, Leeuwenhoek rediscovered the blood corpuscles (1674) and the blood capillaries (1683), but he did not fully understand the function of the blood. Leeuwenhoek's work in plant anatomy coincided with that of Malpighi and Grew, and the three are considered the founders of that study.

Leeuwenhoek made his work known through letters to scientific correspondents in and beyond Holland, and he was indebted primarily to the Royal Society for the publication of his discoveries. He addressed 190 of his letters to the society, which prepared English translations or summaries for publication in the *Philosophical Transactions* between 1673 and 1724. In 1680 he was elected a fellow of the Royal Society.

LEGUAT, FRANÇOIS (1638–1735), fled to Holland to avoid persecution after revocation of the Edict of Nantes, and in 1691 sailed on a voyage to the East Indies. He eventually went to England and published accounts of his adventures in both French and English in the same year. The English translation was *A New Voyage to the East Indies* (London, 1708).

LEUSDEN, JEAN (1624–99), a celebrated Dutch philologist, obtained the chair of Hebrew at Utrecht in 1649 and occupied it with distinction until his death. He published a large number of works on biblical, philological, and related subjects.

LIBAU, ANDREAS (1560–1616), a Saxon, wrote prolifically on alchemy and medicine. He was for a time professor of history and poetry at the University of Jena, and later was chief physician of Rothenburg-on-the-Tauber. His fame rests chiefly upon his *Alchymia* (1595), a survey of the chemical knowledge of the time. His *Singularium,* 4 pts. (Frankfurt, 1599–1601) describes an experiment in silkworm culture made in 1599. This account was published as an appendix to John Jonston, *An History of the Wonderful Things of Nature* (London, 1657).

LIPPERSHEY, or LIPPERSHEIM, HANS (c. 1570–c. 1619), a Dutch spectacle maker, formally offered his invention (the optical telescope) to the Estates General in 1608. But this instrument was probably invented independently and accidentally many times at the beginning of the seventeenth century, and the oft-repeated statement that the telescope was invented by Lippershey is incorrect. Galileo heard of Lippershey's invention, reinvented the telescope, and turned it on the heavens in 1609.

LIPSIUS, JUSTUS, or JOEST LIPS (1547–1606), a distinguished scholar at times Catholic and at times Protestant, was a professor at Jena, Leiden, and Louvain at different periods. His specialties were Latin literature and Roman history, and his editions of Tacitus and Seneca were long regarded as models of their kind. His knew Tacitus so well that he offered to repeat anything in that author word for word on penalty of forfeiting his life if he faltered. Lipsius, Scaliger, and Casaubon constituted a triumvirate of outstanding sixteenth-century humanists.

LISTER, MARTIN (1639-1712), a medical doctor by profession, gained prominence in the fields of zoology and geology. After graduating from Cambridge, he studied medicine and related subjects in Montpellier, and while still in France formed a friendship with John Ray. In 1670 he settled in York, where he practiced medicine and pursued natural science and antiquities for the next thirteen years. He then removed to London, became a fellow of the College of Physicians, and rose to a high place in his profession. He published a number of books and articles on medical topics.

Lister's interest in the new science was stimulated by his correspondence with Ray and other English natural philosophers. He carried out pioneer studies in many fields of natural history and published fifty-one papers in the *Philosophical Transactions,* especially on insects, spiders, parasites, and mollusks, but also on birds, plants, physiology, medicine, geology, meteorology, and antiquities. He was elected a fellow of the Royal Society in 1671. His first book, a collection of earlier articles, was *Historiae animalium Angliae tres tractatus* (London, 1678). This work offered the first systematic descriptions and figures of spiders, snails, and fossils in England. He published a translation of Johannes Goedartius's *De insectis* entitled *Johannes Godartius of Insects. Done into English, and Methodized with the Addition of Notes* (York, 1682).

Lister's concern with mollusk classification brought him into the controversy over the nature of fossils. This controversy flared in the late seventeenth century when the rejection of the idea of spontaneous generation caused a distinction to be made between the living and the nonliving, and it was centered in England. John Ray argued that fossils were the organic remains of creatures formerly living, both vegetable and animal, and he had support from other scientists. Lister contended that petrified stones found in the earth were "formed stones," and he was convinced that there could be no direct link between extant and fossil shells. He noted that the distribution of fossil shells is correlated with the distribution of rocks, and he believed that this was an argument for their geological origin. He explained the growth of fossils in rocks as a complex crystallization from lapidifying juice found naturally in the earth. In 1684 he proposed the basic principle of constructing a geological map, an idea which was first realized in England in 1815.

Lister's greatest and best-known works related to conchology, a subject popular at the time he wrote. He published parts starting in 1685, and the parts were drawn together in the *Historia sive synopsis methodica conchyliorium* published in two volumes in London (1692).

Lister accompanied a diplomatic mission to France as physician in

1698, and he left an account entitled *A Journey to Paris in the Year 1698* (London, 1698). His work was a travel book with an emphasis upon the scientific and technical scene.

LIVY (TITUS LIVIUS) (59 B.C.–A.D. 17 or 64 B.C.–A.D. 12), author of an immense history of Rome, of which only a part is extant. He sought to give Rome a history that in conception and style should be worthy of its imperial greatness, and to challenge his generation to resume the responsibilities of their position. He had a capacity for vivid historical reconstruction and enjoyed an immediate as well as lasting success.

LONGOLIUS, CHRISTOPHERUS, or CHRISTOPHE DE LONGUEIL (1488–1522), one of the most celebrated humanists of his time, studied law and became counselor to the Parlement of Paris. Then he turned to the study of the classics and fixed his residence in Padua. Noted for his excellent Ciceronian style, he combined a remarkable erudition with a prodigious memory.

LOWER, RICHARD (1631–91), physician and fellow of the Royal Society, was an outstanding seventeenth-century English physiologist. He made pioneering experiments on blood transfusion and on cardiopulmonary function, discovering that dark venous blood was converted into bright arterial blood on contact with air. His *Tractatus de corde* (London, 1669), which held that the heart is merely a muscle and does not have an innate ferment, was quickly recognized as a major work on physiology.

LUCRETIUS CARUS, TITUS (probably 94–55 B.C.), poet and philosopher, whose *De rerum natura,* a didactic poem in six books, expounds the materialist and atomist physical theory of the universe of Epicurus with a view to relieving mankind of the fear of death, the intervention of the gods in the world, and the punishment of the soul after death.

LYSER, MICHAEL (1626–59), went from his native Germany to Copenhagen, where he assisted Thomas Bartholin in research which led to discovery of the lymph system. He later took a medical degree at Padua and left a number of works, including an excellent manual of anatomy entitled *Culter anatomicus: hoc est methodus brevis facilis ac perspicua artificiose et compendiose humana incidendi cadavera* (Copenhagen, 1653).

MAGNUS, OLAUS, or OLAF MANSSON (1490–1557), Swedish prelate, remained a loyal Roman Catholic after the Lutheran Reformation erupted. Ordered to Rome in 1523, he never returned to Sweden, though he was appointed archbishop of Uppsala in 1544. His scientific works make him a pioneer in geographic research on Scandinavia. Both his map of the Scandinavian countries and his *Historia de gentibus septentrionalibus* (Rome, 1555) had great influence. His history, published in many editions and many languages, was translated as *A Compendious History of the Goths, Swedes, and Vandals, and Other Northern Nations* (London, 1658). For generations it informed educated Europeans about Scandinavians.

MALPIGHI, MARCELLO (1628–94), was a distinguished representative of the new experimental science whose main contributions were in medicine, anatomy (microscopic, comparative, and plant), and embryology. He graduated from the University of Bologna as a doctor of medicine and of philosophy, and, while practicing medicine and conducting research, taught at the universities of Bologna, Pisa, and Messina for a decade before returning to Bologna in 1666 to lecture in medicine. Having turned from Peripateticism to the new science of the school of Galileo, Malpighi became a correspondent of the Royal Society of London in 1668 and was elected to membership the following year. The *Philosophical Transactions* carried many accounts of his research, and the Royal Society supervised the printing of all of his later works. Malpighi wrote important treatises on many subjects, including the lungs, neurology, adenology, hematology, the silkworm, the development of the chick, and plant anatomy, and aroused fierce opposition from those who trusted the authority of ancient philosophers. In 1691 he was called to Rome as chief physician of Pope Innocent XI.

MARCHETTI, DOMENICO DE (1626–88), anatomist and physiologist, was appointed professor of anatomy at Padua in 1649. He published *Anatomia* (Padua, 1652), a response to works by Jean Riolan and Johann Vesling. It is no longer believed that baldness comes from dryness of the brain, but it is now recognized that shrinking creates an empty space between the brain and the skull.

MARGGRAF, GEORG (1610–44), physician, accompanied Willem Piso to Brazil and spent six years along the coast collecting observations on geography, astronomy, and natural history. Johann de Laet edited his writings on natural history under the title *Historiae rerum naturalium Brasiliae libri octi*, and they were included in Willem Piso, *Historia naturalis Brasiliae* (Leiden and Amsterdam, 1648). Marggraf also left a *Tractatus topographicus et meteorologicus Brasiliae* (Amsterdam, 1658).

MARIOTTE, EDME (d. 1684), a scientific eclectic, introduced experimental physics into France and was active in various fields of science. He published *De la végétation des plantes* (Paris, 1679), and as a plant physiologist first attracted the attention of the Royal Academy of Sciences in Paris. His *Traité du mouvement des eaux et des autres corps fluides* (Paris, 1686) was reprinted in 1700.

MARIUS, GAIUS (c. 157–86 B.C.), Roman general and statesman, was typical of the "new man" of the Roman republic. He rose from mean origins by his own merits, creating a client army, becoming a popular leader, and making alliances with demagogues and a noble faction. By these means he won some degree of recognition by the nobles.

MARTYR, PETER. *See* ANGHIERA

MAXIMINUS, GAIUS JULIUS VERUS (fl. 235–38), a Thracian peasant of gigantic stature, was made Roman emperor in 235 at Mainz in the mutiny which overthrew Alexander Severus. But his oppressive government and sanguinary excess caused a revolt in 238, and he was slain by his own soldiers.

MAYERNE, THÉODORE TURQUET DE (1573–1655), a great physician in his day, was born of French Protestant stock near Geneva. He obtained a medical degree at Montpellier and practiced for some time in Paris, where his defense of chemical remedies aroused strong professional opposition. Removing to England permanently in 1611, he treated many notables, was elected to the College of Physicians, and became physician to three Stuart kings. He wrote "A Discourse of the Viper, and Some Other Poysons," *Phil. Trans.,* 18 (June 1694), 162–66, after discussions with Mr. Pontaeus, a "Chymical Mountebank."

MAYOW, JOHN (1641–79), is best known for studies on respiration and the interrelated problems of atmospheric composition, aerial nitre (or oxygen), and combustion. His first publication, *Tractatus duo* (Oxford, 1668), includes "De respiratione." He clarified and modified his views on respiration in *Tractatus quinque medico-physici* (Oxford, 1674), a great contribution to physiological embryology. Mayow was elected a fellow of the Royal Society in 1678.

MEAD, RICHARD (1673–1754), studied medicine at Leiden under Alexander Pitcairne, chief of the iatromechanical school, and obtained a medical degree at Padua. The leading London practitioner of his day, he was elected a fellow of the Royal Society and of the College of Physicians. He dissected vipers and published *A Mechanical Account of Poisons* (London, 1702), which was quickly abstracted in *Phil. Trans.,* 23 (Jan.–Feb. 1703), 1320–28.

[MERCER]. *See* MERCERUS

MERCERUS, JOANNES, or JEAN MERCIER (d. 1570), professor of Hebrew at the Collège Royal in Paris. A prolific writer, Mercier's works included many Bible commentaries. The library of the Mathers contained a copy of his commentary on Job and other books of the Old Testament (Leiden, 1651). Mercier's commentary on Genesis was edited with a preface by Theodore Beza (Geneva, 1598).

MERRETT, CHRISTOPHER (1614–95), a physician of London and founding member of the Royal Society, translated Antonio Neri's *L'arte vetraria* (1612), a pioneer work, as *The Art of Glass* (London, 1662), adding his own observations. This work had considerable influence on glassmaking in England and in Europe. Merrett also published *Pinax rerum naturalium Britannicarum* (London, 1666) to replace William How's *Phytologia Britannica* (London, 1650).

MERSENNE, MARIN (1588–1648), was a French Minorite who used natural science to defend theological truth and the rationality of nature against skeptics and occultists. He was the central figure in a circle of scientists which became institutionalized in the French Academy of Sciences. Mersenne's large correspondence served as a means of disseminating the results of experimentation and speculation over much of Europe.

MERSENNUS. *See* MERSENNE

MICYLLUS, JACOBUS (1503-58), one of the best poets of his time in Germany, taught Latin and Greek at Frankfurt and was professor of Greek at Heidelberg. He wrote and edited various works. His verse included "De canum fidelitate," forty-nine lines in praise of dogs.

MINUCIUS FELIX, MARCUS (fl. 200-240), Roman lawyer and author of *Octavius,* a dialogue between a Christian and a pagan which defends Christianity from a philosophical point of view. The apology is modeled on Cicero's *De natura deorum,* while it draws on Stoic ideas and is probably dependent on Tertullian's *Apology.* An English translation by R. James was published as *Minucius Felix His Dialogue Called Octavius: Containing a Defence of Christian Religion* (Oxford, 1636). Other English translations followed in 1682, 1695, and 1708.

MITHRIDATES VI, EUPATOR DIONYSUS or "the Great" (120-63 B.C.), king of Pontus and in many ways Rome's stoutest oriental antagonist, occupied most of Asia Minor, the islands of the Aegean except Rhodes, and most of Greece.

MOFFETT (or MOUFET or MUFFET), THOMAS (1553-1604), an English physician, was the leading entomologist between Aldrovandi and Swammerdam. Drawing on unpublished notes of earlier observers, he wrote a valuable book on insects which was posthumously published as *Insectorum: sive minimorum animalium theatrum* (London, 1634). An English translation by John Rowland, *The Theater of Insects,* appeared as volume 3 of Edward Topsell, *The History of Four-Footed Beasts and Serpents* (London, 1658). Moffett's daughter was the "little Miss Muffett" of nursery rhyme fame who was frightened by a spider who sat down beside her.

MOLYNEUX, WILLIAM (1656-98), first secretary of the Dublin Philosophical Society, established in 1684, and author of *Dioptrica Nova: A Treatise of Dioptricks* (London, 1692), the first treatise on optics published in English.

MONCONYS, BALTHASAR (1611-65), a learned French traveler who cultivated the occult sciences, made an extensive voyage to Provence and the Mediterranean area from 1645 to 1649 to meet savants in countries with remaining traces of the philosophy of Hermes Trismegistus and Zoroaster. He made another voyage to England, the United Provinces, Germany, and Italy from 1663 to 1664. His son published the *Journal des voyages de Monsieur de Monconys,* 3 vols. in 4 (Lyon, 1665-66).

Other editions and a German translation followed. The journal contains much insignificant detail along with many bizarre recipes as well as medical and chemical formulas.

MORE, HENRY (1614–87), was reared by strong Calvinist parents but became an Anglican and took holy orders after graduating B.A. from Christ's College, Cambridge. Elected to a fellowship at Cambridge, he won an excellent reputation for saintliness and intellectual power but declined all preferment out of love of contemplation and a desire to serve the church and God. At Christ's College More became affiliated with the band of Christian thinkers known as the Cambridge Platonists. In close touch with leading scientists and philosophers of his time, More gradually abandoned his earlier admiration for the thought of René Descartes. He came to see mechanical naturalism and atheism as the inevitable result of Cartesian philosophy. More's religious views centered on his belief that there is no real clash between any genuine point of Christianity and what true philosophy and right reason allow. His *Antidote for Atheism* (1652) discusses the phenomena of nature as an argument for the existence of God, and John Ray's *Wisdom of God Manifested in the Works of the Creation* (1691) was probably an exposition and commentary on More's treatise. Thus More's *Antidote* may appropriately be viewed as the "grandparent" of Mather's *Christian Philosopher*.

MORHOF, DANIEL GEORG (1639–91), was professor of belles lettres and later professor of history and librarian at the University of Kiel. He combined vast erudition with a remarkable taste for poetry, and was a leading German poet of his age. He had seen an Amsterdam wine merchant break drinking glasses by raising his voice an octave above its natural tone, and described the experience in *Epistola de scypho vitreo per certum humanae vocis sonum rupto* (Kiel, 1672), published in another edition under the title *Stentor Ὑαλοκλάστης, sive de scypho vitreo per certum humanae vocis sonum fracto* (Kiel, 1682).

MORLAND, JOSEPH (fl. 1697–1713), was a physician and author of "Part of a Letter to Dr. Mead, Concerning Secretions in an Animal Body," *Phil. Trans.*, 23 (Jan.–Feb. 1703), 1292–96. He edited Samuel Morland's *Hydrostaticks: or Instructions Concerning Waterworks* (London, 1697) and published two works on medical topics.

MORLAND, SAMUEL (1625–95), an English virtuoso, served Oliver Cromwell as a diplomat but later promoted the restoration of the monarchy. His most important scientific work was in the field of hydrostatics. His mathematical studies and inventiveness enabled him to make hand calculators for pedagogic use, and he invented a speaking trumpet described in *Tuba Stentoro-phonica. An Instrument of Excellent Use, as Well at Sea as at Land* (London, 1671).

MORTON, RICHARD (1637–98), turned to medicine after being ejected as a minister because he could not comply with the requirements of the Act of Uniformity. He became a fellow of the College of Physicians and published two important medical works, one on phthisis, the other on fevers, entitled Πυρετολογια: *seu excercitationes de morbis universalibus acutis*, 2 pts. (London, 1692–94). A manuscript said to be by Morton deals with methods of preparing Peruvian bark.

MOULEN, or MOULIN, ALLEN (fl. 1687–91), a medical doctor and fellow of the Royal Society, made experiments on the human blood and on the presence of iron in black Virginia sand, and he made anatomical observations on the heads of fowls. He reported his findings in *Phil. Trans.*, 16 (Dec. 1687), 433–34; 17 (Jan.–Feb. 1691), 486–88; (Feb. 1693), 624–26; (Apr. 1693), 711–16.

MUNCHIUS, JANUS, or JENS MUNCK (1579–1628), Danish navigator, sent by his king to explore and give an account of Greenland.

MUNDINUS (RAIMONDO or MONDINO) DEI LIUCCI or LIUZZI (c. 1275–1326), Italian physician and anatomist, whose *Anatomia Mundini*, completed in 1316 and first printed at Padua in 1476, was the first book on anatomy based on human dissections. No real improvements on Mondino were made until 1521, when Berengario da Carpi wrote his commentary on it. Mondino's book remained the standard text until the time of Vesalius.

MÜNSTER, SEBASTIAN (1489–1552), a German humanist and professor at Basel, was a leading scholar in Hebrew studies in his day. His many publications in this field included *Hebraica Biblia*, 2 vols. (Basel: 1534–46). Called the "German Strabo," his *Cosmographia* (Basel, 1544), a geographical-historical encyclopedia which described the whole world and everything in it, was one of the most popular treatises of the sixteenth and early seventeenth centuries.

NEWTON, ISAAC (1642–1727), took his bachelor's degree at Trinity College, Cambridge, in 1665. During the next two years while the university was closed because of plague he returned home to Woolsthorpe, near Grantham, in Lincolnshire, where he formulated most of his major discoveries. He became a fellow of Trinity College in 1667 and served as Lucasian professor of mathematics at Cambridge from 1669 to 1701.

Newton's achievement in mathematics and natural philosophy was the culmination of the scientific revolution beginning with Copernicus in 1543. His contributions in natural philosophy were of prime importance to Mather. Newton studied light and colors and extended the understanding of those subjects when he read his *New Theory of Light and Colours* to the Royal Society in 1672. He developed an emission, or corpuscular, theory of light. He also formulated the three fundamental laws of the mechanics of planetary motion, and from Kepler's third law he derived the inverse square law that the force between the earth and the moon must be inversely proportional to the square of the distance between them. This discovery was crucial to his theory of universal gravitation, set forth in his *Philosophiae Naturalis Principia Mathematica* (1687), the seminal work for modern science. It was translated into English as "Mathematical Principles of Natural Philosophy" in 1729. Newton summed up his optical researches in *Opticks* (1704). He served as president of the Royal Society from 1703 to 1727, and was knighted in 1705. In addition to his mathematical and scientific work, Newton wrote much on the subject of biblical prophecies.

NORMAN, ROBERT (fl. 1590), established a reputation for constructing superior instruments for navigation and surveying. He related his discovery of the dip in the magnetic needle from the horizontal in *The New Attractive* (London, 1581).

NORWOOD, RICHARD (1590–1665), was active in mathematics, surveying, and navigation. Norwood's book, *The Seaman's Practice: Contayning a Fundamentall Probleme in Navigation, Experimentally Verified* (London, 1637), was especially concerned with the length of the degree of the meridian and improvements in the log line.

NOVIDIUS. *See* FRACCUS, AMBROISE NOVIDIUS

NUCK, ANTON (1650–92), a famous German anatomist of the seventeenth century, practiced medicine and surgery in The Hague before his appointment in 1687 as professor of anatomy and surgery at Leiden. He conducted research in anatomy, especially on the lymphatic glands. His many publications include *De vasis aquosis oculi* (Leiden, 1685) and *De ductu salivali novo, saliva, ductibus oculorum aquosis, et humore oculi aqueo* (Leiden, 1685), later published under the title *Sialographia et ductuum aquosorum anatome nova* (Leiden, 1690, and later eds.).

OLEARIUS, ADAMUS, or ADAM OELSCHLÄGER (c. 1600–1671), astronomer and mathematician, legate of Holstein in Russia, and secretary and counselor to a great Russian prince in the 1630s.

PAPIN, DENIS (1647–c. 1712), a French physician and a Calvinist with a scientific bent, went to England, where he worked with Robert Boyle on pneumatic experiments and became a fellow of the Royal Society. He invented a "steam digester," a pressure cooker for which he devised a safety valve that was technologically important in the development of steam power, and wrote about it in *A New Digester, or Engine for Softening Bones* (London, 1681) and in *A Continuation of the New Digester of Bones . . . Together with Some Improvements and New Uses of the Air Pump* (London, 1687).

PASCAL, BLAISE (1623–62), French mathematician, physicist, religious philosopher, and writer who in 1655 entered Port-Royal, where he wrote the *Provincial Letters,* a defense of Jansenism.

PECHLIN, JOHANN NICHOLAS (1644–1706), studied medicine in Leiden and in Italy. In 1673 he became professor of medicine and later dean of the medical faculty at the academy of Duke Christian Albert in Kiel. He published various works on medicine, including *Theophilus bibaculus, sive de potu theae dialogus* (Frankfurt, 1684).

[PECKER]. *See* PECQUET

PECQUET, JEAN (1622–74), a French physician and anatomist, made important advances in the study of the lymphatic system. His major publication was *Experimenta nova anatomica, quibus incognitum hactenus chyli receptaculum et ab eo per thoracem in ramos usque subclavios vasa lactea deteguntur* (Paris, 1651), which confirmed Harvey's theories on circulation of the blood. It was translated into English under the title *New Anatomical Experiments ... by Which the Hitherto Unknown Receptacle of the Chyle, and the Transmission from Thence to the Subclavial Veins by the Now Discovered Lacteal Chanels of the Thorax, is Plainly Made Appear in Brutes* (London, 1653).

PERRAULT, CLAUDE (1613–88), was a practicing physician and founding member of the Academy of Sciences in 1666. His group of Parisian anatomists made many dissections, and Perrault's anatomical descriptions contributed toward a natural history of animals. He expounded theories on the circulation of sap in plants and on embryonic growth from preformed germs which were highly influential, although subsequently shown to be erroneous. He translated Vitruvius from Latin and is celebrated as an architect whose remarkable monuments in Paris include the colonnade of the Louvre.

PETER MARTYR. *See* ANGHIERA

PETRONIUS ARBITER (first century A.D.), writer. His *Satyricon*, fragments of which survive, recounts the adventures of the disreputable pair Encolpius (the narrator) and his boyfriend Giton.

PEYER, JOHANN CONRAD (1653–1712), pursued collaborative research on physiology for a decade in Schaffhausen, Switzerland, and after quarreling with his colleagues became professor of logic, rhetoric, and medicine at the local gymnasium. He worked on the lymphatic system and intestinal tract and also studied the stomach of ruminants. His *Merycologia* was published at Basel in 1685.

PHILO (c. 20 B.C.–A.D. 50) was the most important Hellenistic Jew of his age. His theological ideas combined Greek and Jewish influences. His development of the allegorical interpretation of Scripture enabled him to discover much of Greek philosophy in the Old Testament, and it had a lasting influence on biblical exegesis in the Christian church.

PHILOPONUS, JOHN (500–565), of Alexandria, was an eccentric Christian philosopher, philologist, and theologian. His commentary on the Mosaic account of the creation of the world, published at Vienna in

1630, was based on the older hexameral literature but enriched with the author's own theories of nature.

PICARD, JEAN (1620–82), determined the length of a degree of meridian between 1668 and 1670 and published the results in *Mesure de la terre* (Paris, 1671). The work was translated by Richard Waller and published as *The Measure of the Earth* (London, 1688).

[PICKART]. *See* PICARD

[PIERIUS]. *See* VALERIANO

PISO, WILLEM (c. 1611–78), was a Netherlands physician of the Dutch settlement in Brazil from 1631 to 1644. A pioneer in tropical medicine, he related his findings in *Historia naturalis Brasiliae* (Leiden and Amsterdam, 1648).

PITTS, JOSEPH (1663–1735?), apprentice seaman, was captured off the Spanish coast by an Algerian pirate, taken to Algiers, and sold to a merchant. A second master tortured Pitts until he made formal submission to Mahomet. His conversion was taken to be genuine, and Pitts subsequently attended his master upon the pilgrimage to Mecca. Pitts later escaped, returned to England in 1694, and published *A True and Faithful Account of the Religion and Manners of the Mohammetans. In Which Is a Particular Relation of their Pilgrimage to Mecca* (Exeter, 1704). The *DNB* calls this work "the first authentic record by an Englishman of the pilgrimage to Mecca."

PLATO (c. 429–347 B.C.), a disciple of Socrates, taught at the Academy. His theory of knowledge distinguished between particulars, which are diverse and undergo constant change, and the Form or Idea of the particulars, which are universal and permanent. The Ideas themselves derive their existence from the supreme Idea, the Good. Plato also distinguished between knowledge and opinion. Knowledge can be only of the Ideas and is known by intellect; opinion deals with particulars which are known to the senses. Plato stressed the need for instruction in the sciences, especially mathematics, astronomy, and musical theory. In Plato's cosmology, the world is a spherical, living thing, a product of beneficent rational design having both soul and body. The movement of the heavens must reflect abstract mathematical perfection, and the only satisfactory explanation of physical facts is a teleological one. Plato exerted a strong influence on Christian thought throughout the early Middle Ages.

PLINY the Elder (GAIUS PLINIUS SECUNDUS) (A.D. 23/24–79), a Roman who served as a cavalry commander in Africa and Germany and as procurator in Spain, was also a scholar who wrote in the fields of history, rhetoric, military tactics, and natural science. Of his many writings, only the *Historia Naturalis* in thirty-seven books survives. Energetic and endlessly curious, Pliny draws upon 474 authors for this encyclopedia of natural science as it touches human life. The *Natural History* treats the universe, geography, man, various animals, botany, botany and zoology in medicine, and metals and stones. His massive compendium contains many mistakes and its contents are of unequal value, and yet it is a mine of information that portrays the science of the ancient world. Pliny's *Natural History* won a prominent place in the tradition of Western culture.

PLUTARCH (L. [?] MESTRIUS PLUTARCHUS) (before A.D. 50–after 120), a Greek philosopher and biographer, received philosophical training in Athens, visited Egypt, and lived for fifteen years in Rome before returning to Greece. For his last thirty years he was a priest at Delphi. A prolific author, he is famous for the *Parallel Lives,* which pair Greek and Roman worthies, and for the *Moralia*, "ethical essays" which touch on a wide variety of topics, including religion (*De Iside et Osiride*), the moral superiority of animals over man (*De sollertia animalium*), and table talk (*Quaestiones convivales*). Plutarch's essays defend Platonism while opposing Epicureanism and Stoicism.

POUPART, FRANÇOIS (1661–1708), anatomist, surgeon, and naturalist, was a member of the French Academy of Sciences from 1669 to his death. He published many works on his special fields, especially entomology. Three of these appeared in the *Philosophical Transactions*. He is credited with a *Mémoire sur les insectes hermaphrodites,* but it was apparently not issued as a separate title.

PYRARD, FRANÇOIS (c. 1570–1621), French voyager, sailed in 1601 with an expedition to discover a route to the East Indies. He was shipwrecked, enslaved, and forced into the Portuguese army, though he visited many countries and was a keen observer. He finally secured his liberty, returned home in 1611, and published the tale of his adventures under the title *Discours du voyage des François aux Indes Orientales* together with *Traité et description des animaux, arbres, et fruicts des Indes Orientales* (Paris, 1611). This work was reedited several times; the Hakluyt Society published an English translation (London, 1887–90).

RAINOLDS, or REYNOLDS, JOHN (1549–1607), president of Corpus Christi College, Oxford, and dean of Lincoln, was widely read and a voluminous author. A doctrinal Puritan, he occupied a prominent position at the Hampton Court Conference, where he urged the need for an authorized version of the Bible and shared in translation of the prophets.

RALEIGH or RALEGH, WALTER (1554?–1618), English courtier, adventurer, and historian, was a court favorite during the latter half of the reign of Elizabeth I. An early colonizer of America, he tried to establish a colony near Roanoke Island (present North Carolina) which he named Virginia. Accused of treason against James I, he was imprisoned in the

Tower of London from 1603 to 1616, where he wrote *The History of the World* (London, 1614).

RAMAZZINI, BERNARDINO (1633–1714), was a medical practitioner, a professor of medicine at Modena from 1682, and at Padua from 1700 until his death. He engaged in a number of medical, barometrical, and hydrostatic controversies and published a treatise which established his reputation as an epidemiologist of the first rank. His *De morbis artificum diatriba* (Modena, 1700) was the first systematic treatise on occupational disease. This work was anonymously translated into English as *A Treatise of the Diseases of Tradesmen* (London, 1705).

RAMUS, PETER (1515–72), a noted French pedagogue whose name became largely identified with logic and method, was an anti-Aristotelian who aimed his criticisms of Scholasticism primarily at contemporary Aristotelians. He embraced Calvinism and was murdered during the St. Bartholomew's Day Massacre.

RASHI (Rabbi SOLOMON BEN ISAAC, also known as SOLOMON BEN YITZHAK) (1040–1105), of Troyes, France, so called from an abbreviation of the initials of his name. He was sometimes erroneously called Yarchi (Jarchi) because of confusion with another Solomon. He was a popular expounder of the Bible and a famous commentator on the Babylonian Talmud, whose writings, first published between 1475 and 1525, circulated widely. His reputation was equally great in the fields of biblical and rabbinical commentary. His commentaries on the Old Testament were translated into Latin by Christian scholars and published at Gotha from 1710 to 1713.

RAY, JOHN (1627–1705). See the Introduction, pp. l–lx.

RAYMUNDUS DE SEBONDE (Sebonde has many variant spellings) or RAYMOND OF SEBONDE (d. 1436?), a native of Barcelona who taught philosophy, theology, and medicine at Toulouse. He is best remembered for *Theologia naturalis: sive liber creaturarum,* a treatise of 330 titles (chapters) first published at Deventer in 1484. The library of the Mathers possessed a copy published at Frankfurt in 1635.

REDI, FRANCESCO (1626–97 or 1698), a poet, physician, and scientific investigator, was a member of the Florentine Accademia del Cimento. He made model experiments in entomology, toxicology, and parasitology. Redi disproved abiogenesis, the doctrine of spontaneous generation inherited from Aristotle and still prevalent in the seventeenth century. He demonstrated by systematic experiments that insects do not come from decay but develop without exception from eggs deposited by the female. His most important works include *Esperienze intorno alla generazione degli insetti* (Florence, 1668). Yet Redi continued to believe that spontaneous generation occurred in such cases as gall flies and intestinal worms.

REEDE, or RHEEDE, TOT DRAAKESTEIN, HENDRIK ADRIANN VAN (1636 or 1637–1691 or 1692), was born in Utrecht and left home at an early age to make his career by helping to establish his native country in the Orient. He became governor-general of Malabar during the Dutch supremacy in India and won lasting fame for his *Hortus Indicus Malabaricus, continens regni Malabarici apud Indos . . . omnis generis plantas rariores, Latinis, Malabaricis, Arabicis, et Bramanum, characteribus nominibusque expressas,* 12 vols. (Amsterdam, 1678–1703), which first made known to Europe the luxuriant vegetation of India.

REGIOMONTANUS, JOHANNES, also known as JOHANN MÜLLER (1436–76), coauthored an *Epitome* of the astronomical thought of Ptolemy and wrote mathematical treatises for the purpose of facilitating astronomical calculations. He installed a printing press in his own house in Nuremberg and was the first publisher of astronomical and mathematical literature. He was noted for constructing two automata, a wooden eagle and an iron fly.

REGULUS, MARCUS ATILIUS (fl. 267–c. 249 B.C.), Roman militarist, defeated the Carthaginian fleet, invaded Africa, and captured Tunis (256) during the First Punic War. Defeated and captured by the Carthaginians (255), he was sent on an embassy to Rome, returned as promised, and died in Carthage in captivity. The story of his death by torture became a national epic.

[RHAETENSIS]. *See* RHAETICUS

RHAETICUS, GEORGE JOACHIM (1514–74), who adopted his surname (based on the geographical designation of the ancient Roman province of Rhaetia) because his father was beheaded for sorcery and the family surname (Iserin) could no longer be used, was a mathematician, astronomer, and professor at Wittenberg. He secured permission from Copernicus to publish a *Narratio prima* (1540) about the Copernican astronomical system. Its reception encouraged Copernicus to publish his famous book in 1543.

[RHODIGINUS]. *See* RICCHIERE

RICCHIERE, LODOVICO (1469–1525), better known as Lodovicus Caelius Rhodiginus from the latinized name of his birthplace, Rovigo, was professor of Greek and Latin in various Italian cities. He published the fruits of his observations on various classical texts in *Sicuti antiquarum lectionum commentarios concinnarat olim vindex Ceselius* (Venice, 1516). Other editions followed. The library of the Mathers owned a copy of this work published at Geneva in 1620.

RICCIOLI, GIAMBATTISTA (1598–1671), was a Jesuit astronomer who ardently opposed the Copernican theory (though he acknowledged it to be the best "mathematical hypothesis") and endeavored to disprove the ideas of Galileo. He made significant geographical and astronomical measurements, observed the topography of the moon, and introduced some of the nomenclature still used to describe lunar features.

ROBARTES, FRANCIS (1650?–1718), engaged in politics but devoted much of his interest to science. A fellow of the Royal Society, he published articles on astronomy, music, and mathematics in the *Philosophical Transactions* between 1692 and 1694 (with one exception under the name of Francis Roberts) and served as vice-president of the Royal Society in 1712.

ROBINSON, TANCRED (c. 1655/60–1748), naturalist and physician, was a fellow of the Royal Society and of the College of Physicians. He was instrumental in securing publication of John Ray's *Wisdom of God Manifested in the Works of the Creation,* and assisted Ray with others of his publications. Robinson's own writings dealt mainly with natural history generally, and especially plants. He contributed frequently to the *Philosophical Transactions,* and his seventeen letters to Ray were published in Ray's *Philosophical Letters* (London, 1718).

ROBINSON, THOMAS (d. 1719), rector of Ousby, Cumberland, and a Cambridge Platonist in his views, devoted his leisure to collecting facts about natural history. He disseminated them in *The Anatomy of the Earth* (London, 1694), *New Observations on the Natural History of this World of Matter, and this World of Life* (London, 1696; another ed. 1699), and *An Essay towards a Natural History of Westmoreland and Cumberland... To Which Is Annexed, A Vindication of the Philosophical and Theological Paraphrase of the Mosaick System of the Creation* (London, 1709).

ROCHEFORT, CÉSAR DE (1605?–90?) was the author of *Histoire naturelle et morale des iles Antilles de l'Amérique* (Rotterdam, 1658). Other editions followed, and the book was translated as *The History of the Caribby-Islands* (London, 1666).

ROHAULT, JACQUES (1620–75), was the leading exponent of Descartes's natural philosophy among the first generation of French Cartesians. His *Traité de physique* (Paris, 1671) was reviewed favorably and became a classic. A Latin translation (Geneva, 1674) enabled it to become the leading university textbook on physics at the time. But in 1697 Samuel Clarke provided a new Latin translation, and his notes, based on Newton's views, countered Rohault's Cartesian text. Clarke's 1710 edition of this work was the first to refute the text systematically in order to advance Newtonian principles.

ROLAND, JACQUES (fl. 1625–30), a surgeon of Saumur, published *Aglossostomographie, ou description d'une bouche sans langue, laquelle parle et faict naturellement toutes les autres fonctions* (Saumur, 1630). This "Account of a Tongue Cut Out, Or a Description of a Mouth without a Tongue, Which Speaks Perfectly and Performs Naturally Its Other Functions" was translated into Latin by Carolus Rayger under the title

Aglossostomographia, sive descriptio oris sine lingua, quod perfecte loquitur, et reliquas suas functiones naturaliter exercet (Leipzig, 1673).

RÖMER, OLE, or OLAUS ROEMER (1644–1710), Danish astronomer who first measured the speed of light in 1676. His greatest work dealt with the times of occultations of the satellites of Jupiter.

RONDELET, GUILLAUME (1507–66), was a physician before becoming regius professor of medicine (and later chancellor) at Montpellier University. His lectures in anatomy attracted scholars from all over Europe, but his reputation rests primarily on his *Libri de piscibus marinis in quibus verae piscium effigies expressae sunt: Universae aquatilium historiae pars altra* (Lyons, 1554–55), which actually treats all aquatic animals. A French translation appeared in 1558 as *L'histoire entiere des poissons* (Lyons, 1558). Rondelet offers experimental evidence for his argument that fish need air.

RORARIO, GIROLAMO (1485–1556), papal legate to Hungary and Poland, wrote *Quod animalia bruta ratione utantur melius homine,* apparently in 1544, to show that beasts are rational creatures and that they reason better than man. The book was first published in Paris in 1648; other editions followed.

ROSCOMMON, WENTWORTH DILLON, fourth earl of (1633?–85), was an Irish-born Protestant who was sent to study under Samuel Bochart at Caen. He returned to England after the Restoration, lived for a time in Ireland, and attempted the formation of a literary academy. He is best known for his poetical works, which include "A Paraphrase on the 148th Psalm," written at the age of twelve and published in 1709.

RUDBECK, OLOF (1630–1702), independently discovered the lymphatic system in 1650 and demonstrated his anatomical discoveries in 1652, but did not publish his results, *Nova excercitatio anatomica, exhibens ductus hepaticos aquosos, et vasa glandularum serosa,* until 1653 (in Västerås). This delay led to a priority dispute with Thomas Bartholin, who had conducted research on the lymphatic system about the same time as Rudbeck but was first to publish his findings. Rudbeck became professor of medicine at the University of Uppsala, established an important botanical garden there, and as a botanist laid solid foundations for natural history at Uppsala.

RUSDEN, MOSES (fl. 1679–85), who called himself an apothecary and bee master to the king, was the author of *A Further Discovery of Bees: Treating of the Nature, Government, Generation and Preservation of the Bee* (London, 1679). A second edition was entitled *A Full Discovery of Bees* (London, 1685).

RUYSCH, FREDERIK (1638–1731), had a youthful passion for anatomy, and although he was active in Amsterdam in medicine, obstetrics, and botany, he is best remembered for research in anatomy. He ended the dispute between Johannes van Horne and Louis de Bils over valves in the lymphatic vessels by discovering their presence, publishing the results of his research in *Dilucidatio valvularum in vasis lympathicis, et lacteis* (The Hague, 1665).

SALMASIUS, CLAUDIUS, or CLAUDE DE SAUMAISE (1588–1653) became professor at Leiden in 1631. Through his scholarship and his judgment he acquired great contemporary influence. His scholarly reputation rests upon his monumental commentary on Solinus's *Polyhistor* and upon his publication of Casaubon's notes on *Augustan History*.

SANCTORIUS (SANTORIO SANTORIO) (1561–1636), a professor at Padua and a leading representative of the iatrophysical school, opened a new line of medicine with his quantitative experiments. He showed variation in weight experienced by a human body as a result of ingestion and excretion, and with the aid of a chair scale demonstrated that a large part of excretion takes place invisibly through the skin and lungs. (This insensible loss may average 400–600 cc. per day in an adult.) Famous for static medicine, he published *De statica medicina sectionibus aphorismorum septem comprehensa* (Venice, 1614); a revision was published as *De medicina statica libri octo* (Rome, 1704). An English translation appeared in London in 1676, and another as *Medicina Statica: Being the Aphorisms of Sanctorius,* trans. J. Quincy (London, 1712).

SCALIGER, JOSEPH JUSTUS (1540–1609), son of Julius Caesar Scaliger, converted to Protestantism in 1562. After teaching in Geneva he returned to France, where he met violent opposition, and in 1593 accepted a call to the University of Leiden. A philologist and historian, he laid the foundations for modern textual criticism, established numismatics, and placed the study of classical chronology on a scientific basis. He became known as the most erudite scholar of his time—"a bottomless pit of erudition."

SCALIGER, JULIUS CAESAR (1484–1558) acquired a reputation as a writer on scientific and philosophical subjects. He opposed the new science, and his *Exotericarum exercitationum* (Paris, 1557) tore to shreds Girolamo Cardano's book *De subtilitate* (1550). Scaliger's frequently reprinted book, which Leibniz called the ablest exposition of Aristotelian physics by a modern, was used at Harvard College in the seventeenth century. The library of the Mathers possessed a copy published at Lyons in 1615.

SCHEINER, CHRISTOPH (1573–1650), a German Jesuit at the College of Rome, achieved distinction in astronomy. He made many discoveries, including sun spots, and engaged in a dispute with Galileo over priority of publication on the subject. Scheiner also experimented on the physiology of the eye. His *Oculus, hoc est: fundamentum opticum* (Innsbruck, 1619; 3d ed., London, 1652), showed that the retina is the seat of vision.

Biographical Register 437

SCHELHAMMER, GÜNTHER CHRISTOPH (1649–1716), pursued scientific studies on the Continent and in England for several years before becoming a physician. He was professor of botany at Helmstedt, professor of anatomy, surgery, and botany at Jena, and professor of medicine at Kiel from 1695. He published a number of works, including *De auditu liber unus* (Leiden, 1684).

SCHENCK von GRAFENBERG, JOHANNES (1530–98), the municipal physician of Freiburg, left a valuable treatise which combined his own observations of rare diseases with pathological reports from important works since antiquity. His *Observationes medicae de capito humano* (Basel, 1584) deals with *laesa memoria*.

SCHEUCHZER, JOHANN JAKOB (1672–1733), Swiss physician and scientist, made a systematic exploration of the Alps which provided the material for many communications with European scholars on science and medicine. He was elected a fellow of the Royal Society. Scheuchzer's *Helveticus, sive itinera alpina tria* (London, 1708) discusses three excursions to the Alps made from 1702 to 1704. Another edition, 4 vols. in 2 (Leiden, 1723), discusses six excursions made from 1705 to 1711.

SCHILLER, JULIUS (d. 1627), a German lawyer and monk, was author of *Coelum stellatum Christianum concavum* (Augsburg, 1627), an atlas which applied Judeo-Christian names to the constellations.

SCHÖFFER, PETER (c. 1425–1502), who was in Johannes Fust's service and later married his daughter, came into possession of the bulk of Gutenberg's presses and types when Fust the financier foreclosed upon Gutenberg the inventor. Schoeffer was superior to Gutenberg as a typographer and printer.

SCHUYL, FLORENTIUS (1619–69), was a Netherlands professor of philosophy and author of numerous medical and botanical studies. He also supervised the Botanical Garden of Leiden and published a catalogue of its plants. His Latin translation of Descartes's *Traité de l'homme* in 1662 constituted the first printed appearance of that influential work. Schuyl's preface was chiefly devoted to an endorsement of the Cartesian theory of animal automatism. Schuyl wished to put science on a purely mechanistic basis, and he sensed the significance of the theory of the beast-machine in winning universal acceptance for mechanical physiology.

SCUDDER, HENRY (d. 1659?), an English Puritan divine, was the author of *The Christians Daily Walke in Holy Securitie and Peace* (London, 1631). This devotional work was frequently reissued, the sixth edition appearing in 1635, and the fifteenth in 1813.

SEDILEAU (d. 1693), astronomer and natural philosopher, was a member of the Royal Academy of Sciences in Paris from 1681 to his death. A zealous observer, he made many meteorological observations and published the results of his studies on evaporation in *Observations de la quantité de l'eau de pluye tombée à Paris durant près trois années et de la quantité de l'évaporation* (Paris, 1692). Sedileau is not associated with any important discovery, and his given name seems lost to history.

SELDEN, JOHN (1584–1654), practiced law, but hard study was always his delight, and his writings—on legal, historical, theological, oriental, and other topics—established his reputation as a learned scholar. Despite his public duties he always maintained his literary work. His *Table-Talk* (London, 1680) reports his utterances on matters relating especially to religion and the state over the last twenty years of his life.

SENECA the Younger (LUCIUS ANNAEUS SENECA) (4 B.C./A.D. 1–65), was a Spanish-born orator, man of letters, statesman, and philosopher. After serving as tutor and political adviser to the emperor Nero, he retired and withdrew from public life to devote his remaining years to philosophy. His writings include ethical treatises, philosophical and poetical works, and prose works, including the *Naturales quaestiones*, which deals mainly with natural phenomena and is of great scientific interest.

[SIMPSON]. *See* SIMSON

SIMSON, ARCHIBALD (1564?–1628), Scottish poet and minister of a parish at Dalkeith, published *Hieroglyphica animalium terrestrium, volatilium, natatalium, reptilium, insectorum, vegetivorum, metallorum, lapidum, etc. quae in Scripturis Sacris inveniuntur, et plurimorum aliorum, cum eorum interpretationibus*, 2 vols. (Edinburgh, 1622–24). The library of the Mathers had a copy in which Cotton Mather inscribed his name while a student at Harvard College in 1674.

SLARE, FREDERICK (1647?–1727), physician and chemist, was a fellow of both the Royal Society and the College of Physicians.

SLOANE, HANS (1660–1753) made many natural history observations and collections during fifteen months in the West Indies as physician to the governor of Jamaica. He became secretary of the Royal Society in 1693. His publications include *Catalogus plantarum quae in insula Jamaica sponte proveniunt, aut vulgo coluntur* (London, 1696) and *A Voyage to the Islands Madera, Barbados, Nieves, S. Christophers and Jamaica, with the Natural History . . . of the Last of Those Islands*, 2 vols. (London, 1707–25). Sloane became president of the Royal Society upon Newton's death in 1717 and held the post until 1741.

SOLENANDER, REINERUS (1521–96), a physician of the duchy of Cleves, was skillful at discovering the cause of difficult illnesses. His works were published under the title *Consiliorum medicinalium* (Frankfurt, 1596).

SOLINUS, GAIUS JULIUS (fl. early third century), wrote, soon after 200, *Collectanea rerum memorabilium*, a geographical summary of the lands, peoples, and products of the world known to the Romans at the time. Solinus emphasizes the marvelous, taking almost all of his material from Pliny's *Natural History* without acknowledgment. An English translation by Arthur Golding was published as *The Excellent and Pleasant Worke of Julius Solinus Polyhistor* (London, 1587).

SPIEGEL, ADRIAAN VAN DEN (1578–1625), studied first at Louvain and Leiden and later at Padua. While a physician in Padua he studied botany and published *Isagoges in rem herbariam libri duo* (Padua, 1606). The Venetian senate appointed him professor of anatomy and surgery, and he performed anatomical demonstrations at Padua. He left important manuscripts, and *De humani corporis fabrica libri decem* (Venice, 1627), with many splendid engravings added by the editor, posthumously established his reputation as an anatomist.

[SPIEGELIUS]. *See* SPIEGEL

STENO, NICOLAUS, or NIELS STENSEN (1638–86), a learned Dane, obtained a medical degree from Leiden. A pioneer of the observational method of modern science, he made important discoveries regarding the anatomy of glands and of the lymphatic system, the structure of muscles, and geology. Reared a Lutheran, Stensen converted to Roman Catholicism, became a priest and bishop, and devoted the latter years of his life to God and the church. His published works include *Observationes anatomicae* (Leiden, 1662), *De musculis et glandulis* (Copenhagen, 1664), *Elementorum myologiae specimen* (Florence, 1667), and *De solido intra solidum naturaliter contento dissertationis prodromus* (Florence, 1669), a pioneering work in geology.

STEVIN, SIMON (1548–1620), a Dutch mathematician and engineer, wrote on diverse topics and published two works on mechanics, one on applied statics, the other a systematic treatise on hydrostatics, the first since Archimedes.

STIGEL, JOHANN (1515–62), an evangelical humanist and friend of Melancthon, was professor of Greek and Latin at Wittenberg and a prolific writer of neo-Latin poetry.

STRABO (64/63 B.C.–A.D. 21), a Greek of Pontus, spent several years in Rome and knew various parts of Asia Minor and Egypt. He wrote a history, which is lost, and a *Geography,* which survives. It is a storehouse of information.

STURMIUS, or STURM, JOHANN CHRISTOPHORUS (1635–1703), spent most of his career at Altdorf University near Nuremberg, where he was professor of mathematics and physics as well as dean of the philosophy faculty and rector on various occasions. He directed many dissertations and made his own contributions to physics. He wrote on the philosophy of nature, opposing a mechanical interpretation and stressing the idea of a divine power at work in the universe.

SUAREZ, FRANCISCO (1548–1617), taught philosophy and theology in various Spanish cities and in 1597 became professor of theology at the University of Coimbra. The most prolific of modern theologians, he was the foremost Jesuit theologian and an adherent of Thomas Aquinas.

SUETONIUS TRANQUILLUS, GAIUS (c. A.D. 69–c. 140), Roman biographer and historian, whose chief work is *De vita Caesarum,* twelve biographies from Julius Caesar to Domitian.

SWAMMERDAM, JAN (1637–80), earned a medical degree and then devoted his life to scientific research, focusing first on medicine and then on natural history, especially entomology. A religious man, he explained all biological development in terms of regular laws of nature which were instituted by God and operate like clockwork. He believed that the theories of spontaneous generation and metamorphosis allowed chance and accident to rule and thus led to atheism. His first comprehensive work was *Historia insectorum generalis* (Dutch ed., Utrecht, 1669; French ed., Utrecht, 1682; Latin ed., Leiden, 1685). His monograph on the mayfly, which became a hymn to the Creator, was written in Low Dutch and published as *Ephemeri vita* (Amsterdam, 1675). The biological portion, translated by Edward Tyson, was published as *Ephemeri Vita, Or The Natural History and Anatomy of the Ephemeron* (London, 1681).

SYLVIUS, FRANCISCUS DELE BOË (1614–72), also known as Frans de Le Boë, practiced as a physician and conducted scientific research before becoming professor of medicine at the University of Leiden, where his lectures drew students from all parts of Europe. The outstanding representative of the iatrochemical school of medicine, his main accomplishments were in anatomy and medical chemistry. His main work was *Praxeos medicae idea nova,* 4 vols. (Leiden, 1671–74).

TACITUS, CORNELIUS (c. A.D. 56–after 117), Roman orator, politician, official, and historian. He wrote on the Germanic tribes north of the Rhine and Danube and other short monographs as well as two larger works, usually called *Histories* and *Annals,* which narrate imperial history from A.D. 14 to 96.

TACQUET, ANDREAS (1612–60), a Jesuit scientist of Belgium, was important mainly for books which taught elementary mathematics to many generations. His *Opera mathematica,* published posthumously (Antwerp, 1669), included his *Astronomia,* which rejected the motion of the earth.

TATE, NAHUM (1652–1715), poetaster and dramatist, was poet laureate from 1692 to 1715. The poet and author Robert Southey pronounced him, except for his predecessor Thomas Shadwell, the lowest of the laureates. His only original poem worth naming, according to the *DNB,* is *Panacea: A Poem upon Tea* (London, 1700).

TAUVRY, DANIEL (1669–1701), earned a medical degree at Angers and went to Paris, where he published two works, one on anatomy and another on materia medica, which brought success and election as a fellow of the Royal Academy of Sciences. His publications include *Nouvelle anatomie raisonnée, ou les usages de la structure du corps de l'homme et de quelques autres animaux suivant les lois des mécaniques* (Paris, 1690 and later eds.). This work was translated into English as *A New Rational Anatomy, Containing an Explication of the Uses of the Structure of the Body of Man and Some Other Animals According to the Rules of Mechanicks* (London, 1701).

TAVERNIER, JEAN BAPTISTE (1605–86), one of the most celebrated travelers of the seventeenth century, recounted his adventures in *Les Six Voyages . . . en Turquie, en Perse et aux Indes*, 3 vols. (Paris, 1676–79).

TEMPLE, WILLIAM (1628–99), diplomat, statesman, and author, won notoriety for his letters, essays, and memoirs. In the debate of "Ancients and Moderns" in the 1690s, he defended the wisdom attained by the ancients. His works include the essays "Of Gardening" and "Of Health and Long Life." The latter discusses the plants of greatest virtue in England.

TERRY, EDWARD (1590–1660), graduated from Christ Church, Oxford, and went to India as chaplain with a fleet sent by the East India Company. He served for three years as chaplain to the English ambassador to the mogul's court, and after returning to England published *A Voyage to East-India* (London, 1655).

THEODORET (c. 393–c. 466), Greek theologian, exegete, church historian, and from 423 bishop of Cyrrhus in Syria; one of the greatest interpreters of Scripture in Christian antiquity. He discussed the creation as the handiwork of God in *De providentia*. An English translation appeared as *The Mirror of Divine Providence* (London, 1602).

THEOPHILUS (115–81 or 188), bishop of Antioch from 168 onward, wrote to counter prevailing heresies and to inform pagans about the Christian idea of God and the superiority of the doctrine of creation over their immoral myths. His "Apology" addressed to his friend Autolycus is the earliest Christian writing in the hexameral tradition.

THEOPHRASTUS (c. 371–c. 287 B.C.), pupil and collaborator of Aristotle and after Aristotle the most famous member of the Peripatetic school, wrote prolifically on scientific and philosophical topics. Only a small part of his output is preserved, and the text of extant works is defective. His writings described and classified plant life (nine books) and discussed the physiology of plant life (six books). The Latin translation by Theodore of Gaza in 1483 made the *Historia plantarum* of Theophrastus familiar during the Renaissance.

TORRICELLI, EVANGELISTA (1608–47), served as mathematician and philosopher for Grand Duke Ferdinand II of Tuscany from 1642 to his death. His *Opera geometrica* (Florence, 1644), the only work he published during his lifetime, won him fame. He made excellent telescope lenses which were widely admired, and gave lectures on physics. But his name is primarily identified with the barometric experiment named after him.

TOSTADO, ALONSO (1400?–1455), was professor of philosophy, theology, and law in Salamanca and bishop of Avila from 1449. Diligent and endowed with a prodigious memory, he wrote voluminously, mostly exegesis of Scripture and theological treatises.

TROUGHTON, JOHN (1637?–81), blinded from the effect of smallpox at an early age, graduated from St. John's College, Oxford, and ministered to a nonconformist congregation in Oxford. He published *Luther Redivivus: Or, the Protestant Doctrine of Justification by Faith Only, Vindicated*, 2 pts. (London, 1677–78).

TSCHIRNHAUS, EHRENFREID WALTHER (1651–1708), a German mathematician and physicist, constructed mirrors with which he obtained high temperatures by focusing sunlight. He also made burning glasses and published papers on the subject between 1687 and 1703.

TULP, NICOLAAS (1593–1674), enjoyed a lucrative medical practice in Amsterdam and as *praelector anatomiae* taught anatomy to the surgeons of the city. His main work, *Observationum medicarum libri tres* (Amsterdam, 1641), contains descriptions of 262 cases. He and his pupils are the subjects of a celebrated painting by Rembrandt.

TUSSER, TOM (1524?–80), poet and agricultural writer whose *A Hundreth Good Pointes of Husbandry* (London, 1557), first published during a time of quickened interest in agricultural subjects, was revised and frequently reprinted down to the early nineteenth century.

TYMPE, or TIMPE, MATTHAEUS (fl. c. 1588–1618), a Roman Catholic priest and professor who served churches and schools in several German cities, published many theological works, including *Mensae theolophilosophicae pars prima [et altera], seu conviviorum pulpamenta et condimenta suavissima*, 2 vols. (Münster, 1618 and later eds.). A copy of this work, a first edition of which may have been published in 1615, was in the library of the Mathers.

TYSON, EDWARD (1650/51–1708), physician, comparative anatomist, and man of considerable general learning, was a fellow of the Royal Society and of the College of Physicians. He translated Swammerdam's *Ephemeri vita* (1681), published a number of articles on medicine and natural history in the *Philosophical Transactions* and in Thomas Bartholin's *Acta medica et philosophica Hafniensa*, 5 vols. (Copenhagen, 1673–80), contributed to Ray's edition of Willughby's *Historia piscium* (1686), and published monographs on the anatomy of particular animals. Perhaps his best-known work was *Orang-Outang, sive Homo Sylvestris: Or, the Anatomy of a Pygmie Compared with That of a Monkey, an Ape, and a Man* (London, 1699). It included essays on the anatomy of other animals, including the round worm (*Ascaris lumbricoides*) bred in human bodies.

USSHER, JAMES (1581–1656), first exhibited his intellectual powers while a university student in Dublin. He was professor of divinity in Dublin, a leader in establishing the Protestant Church of Ireland, and from 1625 archbishop of Armagh. Contemporaries marveled at his erudition, and his friend John Selden called him "learned to a miracle."

VALERIANO BOLZANI, GIOVANNI PIERIO (1477–1560), added the latinized name PIERUS and was commonly known as PIERO VALERIANO. He studied letters and became acquainted with celebrated humanists, won the favor of two popes, and tutored two Medicis, thus forming close ties to the Medici family. Valeriano published on many subjects. His *Hieroglyphica, sive de sacris Aegyptiorum, aliarumque gentium literis commentarii* (Basel, 1556, and many later editions) attempts to explain most branches of science and art by commenting on the occult significance of Egyptian, Greek, and Roman symbols.

VALERIUS MAXIMUS (fl. c. A.D. 14–37), a Roman historian in the reign of the emperor Tiberius, who prepared a handbook of illustrative examples, mostly moral or philosophical, intended for rhetoricians. His main sources were Cicero and Livy. The convenience of the compilation, *Factorum et dictorum memorabilium,* assured it some success in antiquity and more in the Middle Ages and Renaissance.

VALSALVA, ANTON MARIA (1666–1723), the favorite pupil of Malpighi, devoted his life to teaching and research in anatomy as well as to the practice of medicine in Bologna. He published *De aure humana tractatus* (Bologna, 1704) with plates that illustrate the parts of the ear.

[VAN DRAAKENSTEIN]. *See* REEDE TOT DRAAKESTEIN

VARENIUS, BERNHARDUS, or BERNARD VAREN (1622–50), was the author of *Geographia generalis, in qua affectiones generales telluris explicantur* (Amsterdam, 1650), which ruled unchallenged as the standard geographic text for more than a century. It went through many editions—Isaac Newton edited two Latin editions—and was translated into English by Richard Blome under the title *Cosmography and Geography* (London, 1683).

VEER, GERRIT DE (c. 1570–after 1598), Dutch navigator, made a voyage in 1594 to the north of Norway, Muscovy, and legendary Tartary searching for a northeast passage to the kingdoms of Cathay and China. His account of this adventure, published at Amsterdam in 1598, was translated by William Phillips and published as *The True and Perfect Description of Three Voyages . . . by the Ships of Holland and Zealand* (London, 1609). In November 1596 de Veer set out with two ships on a further voyage of discovery to the northeast, and his party was forced to winter in Novaya Zemlya, where they remained until October 1597. This was apparently the first time that Western explorers had faced an Arctic winter. I am indebted to Professor Adriaan J. de Witte of the University of Illinois at Urbana-Champaign for help on de Veer.

VERZASCHA, BERNARD (1629-80), obtained a medical degree at Montpellier and became a successful practitioner in his native city of Basel. He wrote a book on plants (Basel, 1678) as well as several works on medicine, including *Observationum medicarum centuria* (Basel, 1677).

VESALIUS, ANDREAS (1514-64), a Flemish anatomist and physician, taught at Padua and other universities. A pioneer in dissecting the human body to study it, he contradicted the authority of Galen's anatomy, which was based upon dissections and observations of animals. Vesalius published *De humani corporis fabrica* (Basel, 1543), whose accurate and beautiful illustrations were superior to anything earlier. The teleological argument pervades the *Fabrica*, which provided the basis for a new anatomy and a new anatomical method. Mather errs in implying that the pericardium is attached by a ligament to the diaphragm.

VIEUSSENS, RAYMOND (c. 1635-1715), divided his time between medical practice as chief physician at the leading hospital of Montpellier and anatomical research. He contributed especially to the anatomy of the nervous system, the heart, and the circulation and composition of the blood. His publications include a work on the ear entitled *Traité nouveau de la structure de l'oreille* (Toulouse, 1714).

VIVES, JUAN LUIS (1492-1540), Spanish-born humanist and philosopher, was forced to flee Spain because of the Inquisition. He became professor of humanities at Louvain, went to England, where he lectured at Oxford, and later to Bruges. One of the great Catholic humanists of the sixteenth century, he was on intimate terms with the greatest humanists of the day.

VOETIUS, GISBERTUS, or GIJSBERT VOET (1589-1676), a Dutch pastor, was an ardent defender of Calvinist orthodoxy at the Synod of Dort. A man of vast learning, he became professor of theology and of oriental languages at Utrecht in 1637. His dispute with the more liberal Johannes Cocceius of Leiden long divided the Dutch Reformed Church into two warring camps.

VOLKAMER, or VOLCKAMER, JOHANN GEORG (1616-93), had a long and distinguished career as a practicing physician in Nuremberg. He also edited the journal of the Leopold-Caroline Academy of German Naturalists, interested himself in many branches of learning, and carried out experiments in physics. He constructed a sundial and made comparisons of its noon line with the axis direction of the magnetic needle suspended above the clock. These observations led to the discovery that magnetic declination is a variable quantity, and with his precise measurements the phenomenon became well known.

VOSSIUS, GERARDUS, or GERARD JAN VOSS (1577-1649), was a Dutch humanist and theologian whose main professorial positions were at Leiden and Amsterdam. His collected works were published in six volumes at Amsterdam, 1695–1701. Isaac Vossius or Voss (1618–89) was his son.

VOSSIUS, or VOSS, ISAAC (1618–89), a famous scholar who became a canon of Windsor, published many learned works. Perhaps the most original was *De poematum cantu et viribus rythmi* (Oxford, 1673), which treats the ancient alliance between poetry and music.

WAINEWRIGHT, JEREMIAH (fl. 1707), an English physician, is best known for *A Mechanical Account of the Non-Naturals: Being a Brief Explication of the Changes Made in Humane Bodies by Air, Diet, Etc. Together with an Enquiry into the Nature and Use of Baths upon the Same Principles* (London, 1707). This book went through a fifth edition in 1737 and was published in Latin translation at Avignon in 1748. Wainewright described a variety of things other than medicine, which he called "non-naturals," without which humans cannot live.

WALDSCHMIDT, or WALDSCHMIEDT, WILHELM ULRICH (1669–1731) gave up his post as physician to a Hessian regiment to become professor of anatomy and of botany at the academy in Kiel. A zealous scholar who wrote on many topics, his inaugural lecture was published as *De usu et abusu potus thee in genere, praecipue vero in hydrope* (Kiel, 1692).

WALLER, EDMUND (1606–87), a poet who also sat in the House of Commons. He was well recognized during his day, but his popularity derived more from his wealth, social position, and attractive personal qualities than from his poetry. The first edition of his poems appeared in 1645, and other editions followed.

WALLIS, JOHN (1616–1703), a Puritan divine, was appointed secretary of the Assembly of Divines at Westminster in 1644. He became Savilian professor of geometry at Oxford in 1649. His *Tractatus de loquela* (Oxford, 1653) laid theoretical foundations for his pioneer attempts to teach deaf-mutes how to speak. A founding member of the Royal Society, Wallis published on the apparent magnitude of the moon in the *Philosophical Transactions* in 1687.

WANLEY, NATHANIEL (1634–80), was educated at Trinity College, Cambridge, and became a clergyman. He published various works, the most important being *The Wonders of the Little World: Or, A General History of Man* (London, 1678). This book, which illustrates anecdotally the prodigies of human nature, shows "omnivorous reading and indiscriminate credence."

WARD, SAMUEL (1557–1640), a Puritan divine of Ipswich, England, was the older brother of Nathaniel Ward of Ipswich, Massachusetts. Ward published a moral treatise on magnetism entitled *Magnetis reductorium theologicum tropologicum* ([London], 1637; another ed. followed in 1639). The work has been erroneously attributed to Samuel Ward (d. 1643) of Cambridge. Ward's book was published in translation by

Harbottle Grimston as *The Wonders of the Load-stone: Or, the Load-Stone Newly Reduc't into a Divine and Morall Use* (London, 1640).

WEPFER, JOHANN JAKOB (1620–95), the municipal physician of Schaffhausen, Switzerland, conducted research in physiology and toxicology. His major research centered on the brain (he preceded Willis in publishing an anatomical description of the circle of anastomosed arteries at the base of the brain which became known as the circle of Willis), and he revealed the causative relation of cerebral hemorrhage to apoplexy. He made his greatest contribution in toxological analysis. In his study of poisons he experimented upon animals.

WHARTON, THOMAS (1614–73), practiced medicine in London and was elected a fellow of the Royal College of Physicians. His *Adenographia, sive glandularum totius corporis descriptio* (London, 1656), the first comprehensive survey of the glands of the human body, was widely acclaimed and often reprinted.

WHITAKER, TOBIAS (c. 1600/1601–66), physician in Norwich and later London, regarded wine as a universal remedy against disease. He published *The Tree of Humane Life, or the Bloud of the Grape* (London, 1638).

WIER, or WEYER, JOHANN (1515–88), studied under Cornelius Agrippa (a student of the occult), earned a medical degree in Paris, and became physician to the duke of Jülich and Cleves. An excellent observer of clinical conditions, he published *Medicarum observationum rararum liber I* (Basel, 1567) and described what appears to have been trichinosis. Wier's master work, *De praestigiis daemonum* (Basel, 1563), opposed the prevailing belief in and treatment of witches. Wier "was the first medical man to insist that normal and pathological mental processes differ in degree and form but not in substance," writes Gregory Zilboorg, "and that human will has nothing to do with mental sickness. It is this point of view, courageously stated and valiantly defended, that gives Weyer the right to be called the founder of modern psychiatry." Zilboorg, *The Medical Man and the Witch during the Renaissance*, Publications of the Institute of the History of Medicine, The Johns Hopkins University, 3d ser., vol. 2 (Baltimore: The Johns Hopkins Press, 1935), p. 205.

WILKINS, JOHN (1614–72), a clergyman who became bishop of Chester, was a promoter of experimental science and a founder of the Royal Society. A prolific author, he wrote on language, mathematics, philosophy, science, and theology. The work drawn on here is Wilkins's *Of the Principles and Duties of Natural Religion* (London, 1675).

WILLIS, THOMAS (1621–75), was an Oxford-educated Royalist and Anglican. An anatomist and physician, he combined scientific studies with a successful medical practice in Oxford for two decades after he was licensed to practice in 1646. He and other eminent Oxford scientists, including Boyle, formed a club that met weekly to conduct experiments, and Willis used the new corpuscular natural philosophy to link anatomical fact with acute clinical observations. His *Diatribae duae medico-philosophicae* (London, 1659) included works on fermentation and fevers. He described animal heat arising from a fermentation in the blood, fed by a nitreous aerial agent. His assistant John Mayow elaborated the notion into a concept of respiration and metabolism. Willis recorded the first reliable clinical description of typhoid fever, thus beginning the tradition of English epidemiology. The *Diatribae* also included a short piece on urine. Willis discovered the sugar content in urine among some people with diabetes, and thus distinguished diabetes mellitus, the most serious form, from other varieties.

After the Restoration, as Sedleian professor of natural philosophy at Oxford, Willis ignored the statutory injunction to lecture only from Aristotle and treated neurological topics. His *Cerebri anatome* (London, 1664), based on a series of dissections and a knowledge of the comparative anatomy of the nervous system of fishes, birds, and various mammals, was the most complete and accurate treatise on the brain and nervous system up to that time. It remained in use among neuroanatomists until the mid-nineteenth century. Willis described the location of voluntary and involuntary mental functions and, as a good Harveian, showed how various structures were bathed and nourished by circulating blood. He described the ring of communicating arteries at the base of the brain, known since as the circle of Willis. His *Pathologiae cerebri et nervosi generis specimen* (London, 1667) described nervous and convulsive diseases. In 1667 Willis removed to London, where he established a large and prosperous practice. A fellow of both the Royal Society and the College of Physicians, he found little time to participate in either body. His *De anima brutorum* (Oxford, 1672) argued that man has two souls, a corporeal, mortal soul that he shares with brutes, and a rational, immortal soul that is uniquely human.

WILLUGHBY, FRANCIS (1635-72), began his lifelong association with John Ray at Trinity College, Cambridge. An ardent student, Willughby devoted himself to scientific research upon graduation and became a founding member of the Royal Society. He and Ray together toured the British Isles and the Continent, collecting materials for an improved natural history. Ray took vegetables for his province, while Willughby concentrated on animals. After Willughby's early death, Ray published his friend's *Ornithologiae libri tres* (London, 1676), translated into English as *The Ornithology of Francis Willughby* (London, 1678). Ray supplemented Willughby's material with his own and that of others. Ray's *Historia insectorum* (London, 1710), which William Derham edited for the Royal Society, shows much indebtedness to Willughby.

WITSIUS, or WITS, HERMAN (1636–1708), a Protestant pastor, became professor of theology at Franeker, then Utrecht, and finally Leiden. A man of great erudition, he published many theological treatises, including one on federal or covenant theology.

WITTIE or WITTY, ROBERT (c. 1613–84), English medical doctor and author of various works, including a book on the spa at Scarborough, Yorkshire, and Ουρανοσκοπια: Or, A Survey of the Heavens. A Plain Description of the Admirable Fabrick and Motions of the Heavenly Bodies as They are Discovered to the Eye by the Telescope (London, 1681).

WOODWARD, JOHN (1665–1728), a man of great learning and dogmatic temper, was professor of physic at Gresham College, a leading member of the Royal Society, and a fellow of the College of Physicians. He is best remembered for his contributions to the earth sciences. In *An Essay towards a Natural History of the Earth* (London, 1695) he combined a clear statement of the organic origins of vegetable and animal fossils with a fantastic theory of the effects of the Flood. The *Essay* stirred up a lively controversy.

WORMIUS, OLAUS, or OLE WORM (1588–1654), practiced medicine briefly before his appointment as a professor—first in humanities, then in Greek, and finally in medicine—at the University of Copenhagen, where he was elected rector several times. He collected various kinds of objects, especially natural history objects and man-made artifacts, and established a museum of curiosities. His son published *Museum Wormianum, seu historia rerum rariorum* (Leiden, 1655). Worm also published accounts of Danish antiquities, was interested in runic inscriptions, and wrote on a gold horn found in Jutland.

ZEILLER, or ZEILER, MARTIN (1589–1661), traveled on the Continent for many years before settling in Ulm, where he was principal of a gymnasium and inspector of schools. He devoted his main energies to literary composition and published many works on politics, history, and belles lettres. His best writings deal with geography and travel, and his *Fidus Achates* (Ulm, 1651) has been called the first Baedeker in the German language.

[ZENET]. *See* ZENO

ZENO, NICCOLÒ (c. 1326–1402), was a Venetian navigator and voyager. In 1558 Niccolò Zeno the Younger published a book on a supposed voyage of Niccolò and his brother Antonio in the north seas, where they discovered many islands and lands. This account, allegedly based on letters of Niccolò and Antonio, was actually based on fantasy and on Niccolò the Younger's wide knowledge of literature, history, and geography. Mather confuses the two brothers and errs on the surname.

ZOROASTER or ZARATHUSTRA (fl. c. sixth century B.C.) was an Iranian religious reformer and founder of Zoroastrianism. This faith was dominant in western Asia for over a millennium ending about 650 and is still held by many thousands in Persia and India, where it is known as Parseeism. The theology is fundamentally dualistic, holding that the course of the universe is understood as a relentless struggle between the principle of light and goodness and the spirit of evil and darkness.

ZUCCHI, NICCOLÒ (1586–1670), a Jesuit mathematician and theologian, is remembered today for his research in optics. His writings include *Nova de machinis philosophica* (Rome, 1649) and his main work, *Optica philosophia experimentalis et ratione a fundamentis constituta*, 2 vols. (Leiden, 1652–56).

ZWINGER, THEODORE, the Elder (1553–88), a celebrated Swiss physician and humanist, was professor of Greek and of moral philosophy as well as of theoretical medicine at Basel. He published a number of works in these fields as well as *Theatrum vitae humanae* (Basel, 1565, and later eds.), a vast compilation of anecdotes on characteristic human traits based on materials left by his stepfather, Conrad Lycosthenes.

RECAPITULATION OF MATHER'S SOURCES

1. Essays on Astronomy

Essay 1. Of the Light

Lines of text		287
Mather's composition		107
Other authors		
Cheyne	89	
Alsted	41	
Blackmore	23	
Molyneux	12	
Harris	5	
Dee	4	
Arndt	3	
Cardan	2	
Derham 2	1	
Subtotal		180
Total		287

Essay 2. Of the Stars

Lines of text		128
Mather's composition		58
Other authors		
Walker	22+	
Childrey	13	
Harris	10	
Arndt	8+	
Ray	8	
Buchanan	6	
Huygens	2+	
Subtotal		70
Total		128

Essay 3. Of the Fixed Stars

Lines of text		214
Mather's composition		63+
Other authors		
Walker	67+	
Cheyne	50	
Derham 2	13+	
Gaffarel	14	
Grew	4	
Biblical	1	

Subtotal		150+
Total		214

Essay 4. Of the Sun

Lines of text		239
Mather's composition		25+
Other authors		
Harris	52+	
Cheyne	35	
Walker	32+	
Derham 2	23+	
Grew	20	
Unidentified	15+	
Wittie	10	
Theologia Ruris	10	
Hooke	7+	
Arndt	3	
Biblical	3	
Alsted	2	
Subtotal		213+
Total		239

Essay 5. Of Saturn

Lines of text		96
Mather's composition		3
Other authors		
Harris	46	
Walker	14	
Blackmore	12	
Grew	11	
Derham 2	10	
Subtotal		93
Total		96

Essay 6. Of Jupiter

Lines of text		85
Mather's composition		0
Other authors		
Walker	25+	
Derham 2	25	
Harris	20+	
Cheyne	14	
Subtotal		85
Total		85

Essay 7. Of Mars

Lines of text		31
Mather's composition		0
Other authors		

Essay 8. Of Venus

Harris	31	
Total		31
Lines of text		19
Mather's composition		0
Other authors		
Walker	10	
Harris	8	
Cheyne	1	
Subtotal		19
Total		19

Essay 9. Of Mercury

Lines of text		89
Mather's composition		11
Other authors		
Cheyne	27	
Misc. Cur.	22	
Harris	17	
Derham 2	8	
Walker	4	
Subtotal		78
Total		89

Essay 10. Of Comets

Lines of text		168
Mather's composition		22+
Other authors		
Harris	86+	
Halley	36	
Cheyne	21	
Horace	2	
Subtotal		145+
Total		168

Appendix. Of Heat

Lines of text		71
Mather's composition		28
Other authors		
Harris	43	
Subtotal		43
Total		71

Essay 11. Of the Moon

Lines of Text		183
Mather's composition		45
Other authors		
Harris	97	

454 *Recapitulation of Sources*

Cheyne	9
Huygens	9
Hooke	7
Derham 2	5
Walker	4
Unidentified	4
Grew	3
Subtotal	138
Total	183

Summary of Essays on Astronomy

Lines of text	1610
Mather's composition	362
(22.48 percent of the text)	
Other authors	
Harris	415+
Cheyne	246
Walker	178+
Derham 2	84
Alsted	43
Grew	38
Halley	36
Blackmore	35
Misc. Cur.	22
Arndt	14
Gaffarel	14
Hooke	14+
Childrey	13
Molyneux	12
Huygens	11
Lesser sources	46+
Subtotal	1222+
Total	1584+

2. Essays on Physics and Related Subjects

Essay 12. Of the Rain

Lines of text	72
Mather's composition	26
Other authors	
Ray	28+
Fénelon	9
Unidentified	8+
Subtotal	46
Total	72

Recapitulation of Sources 455

Essay 13. Of the Rainbow

Lines of text		148
Mather's composition		129
(all taken from his own *Thoughts for the Day of Rain*)		
Other authors		
Halley	5	
Harris	14	
Subtotal		19
Total		148

Essay 14. Of the Snow

Lines of text		76
Mather's composition		23
Other authors		
Grew	50	
Gale	3	
Subtotal		53
Total		76

Essay 15. Of the Hail

Lines of text		37
Mather's composition		8
Other authors		
Wallis	19	
Alsted	9	
Gale	1	
Subtotal		29
Total		37

Essay 16. Of the Thunder and Lightning

Lines of text		118
Mather's composition		61
Other authors		
Harris	46	
Alsted	4	
Vergil	3	
Juvenal	2	
Horace	1	
Ovid	1	
Subtotal		57
Total		118

Essay 17. Of the Air

Lines of text		165
Mather's composition		16
Other authors		
Harris	88	

Ray	38	
Wainewright	15	
Waller	8	
Subtotal		149
Total		165

Essay 18. Of the Wind

Lines of text		107
Mather's composition		14
Other authors		
Harris	62	
Ray	25	
Grew	6	
Subtotal		93
Total		107

Essay 19. Of the Cold

Lines of text		100
Mather's composition		30+
Other authors		
Boyle	49+	
Harris	14	
Petronius Arbiter	6+	
Subtotal		69+
Total		100

Essay 20. Of the Terraqueous Globe

Lines of text		222
Mather's composition		43+
Other authors		
Cheyne	44	
Ray	31	
Derham	27+	
Arndt	17+	
Harris	16+	
Unidentified	16	
Gregory	9	
Egard	7	
Derham 2	5+	
Biblical	4	
Subtotal		178+
Total		222

Essay 21. Of Gravity

Lines of text		281
Mather's composition		54
Other authors		
Cheyne	88	
Harris	47	

Recapitulation of Sources 457

Derham	43	
Clarke	21	
Unidentified	13	
Keill	11	
Gregory	4	
Subtotal		227
Total		281

Essay 22. Of the Water and Appendix

Lines of text		283
Mather's composition		24+
Other authors		
Harris	72	
Cheyne	59	
Ray	37	
Halley	30	
Wainewright	29	
Derham	11+	
Alsted	7	
Arndt	5	
Mariotte	5	
Grew	2+	
Subtotal		258+
Total		283

Essay 23. Of the Earth and Appendix

Lines of text		301
Mather's composition		60+
Other authors		
Unidentified	50+	
Ray	49	
Derham	37	
Cheyne	22+	
Grew	19	
Kircher	18+	
Woodward	8	
Arndt	7	
Evelyn	6	
Shower	5	
Varenius	4+	
The Koran	4	
Alsted	3+	
Chrysostom	3	
Olaus Magnus	2+	
Subtotal		240+
Total		301

Essay 24. Of Magnetism

Lines of text		456
Mather's composition		113
Other authors		
Harris	106+	
Halley	88	
Jenkin	39	
Grew	23+	
Lister	21	
Boyle	20+	
Ward	20+	
Derham	13+	
Scudder	11+	
Subtotal		343
Total		456

Essay 25. Of Minerals

Lines of text		208
Mather's composition		46+
Other authors		
Harris	71+	
Alsted	29+	
Ray	15	
Barker	13	
Grew	10+	
Bickerus	8	
Arndt	6	
Dampier	3+	
Cardan	2	
Vergil	2	
Subtotal		161+
Total		208

Summary of Essays on Physics and Related Subjects

Lines of text		2574
Mather's composition		646
(25.10 percent of the text)		
Other authors		
Harris	537+	
Ray	223+	
Cheyne	213+	
Derham	132+	
Halley	123	
Grew	111+	
Unidentified	88+	
Boyle	70	
Alsted	53	
Wainewright	44	

Recapitulation of Sources 459

Arndt	41+
Jenkin	39
Lister	21
Clarke	21
Ward	20+
Wallis	19
Kircher	18+
Barker	13
Gregory	13
Scudder	11+
Keill	11
Fénelon	9
Biblical	9
Bicker	8
Waller	8
Woodward	8
Egard	7
Petronius Arbiter	6+
Evelyn	6
Derham 2	5+
Mariotte	5
Shower	5
Vergil	5
Varenius	4+
Gale	4
The Koran	4
Dampier	3+
Chrysostom	3
Olaus Magnus	2+
Cardan	2
Juvenal	2
Horace	1
Ovid	1
Subtotal	1933
Total	2574

3. Essays on the Life Sciences and on Man
Essay 26. Of the Vegetables

Lines of text		725
Mather's composition		163+
Other authors		
Derham	177+	
Ray	103+	
Harris	91	
Arndt	54	
Unidentified	32+	

Grew	20+	
Alsted	18+	
Robinson	13	
Beverley	11	
Pitts	11	
Cheyne	9	
Tate	7	
Temple	6	
Otto von Guericke	3+	
Plutarch	2	
Browne	1+	
Subtotal		561+
Total		725

Essay 27. Of Insects

Lines of text		938
Mather's composition		117+
Other authors		
Derham	328+	
Harris	111+	
Ray	103+	
Cheyne	79+	
Rusden	76	
Robinson	24+	
Barker	15	
Unidentified	14	
Dampier	13+	
Rashi	12	
Grew	8+	
Jonston	8	
Terry	7	
Purchas	6+	
Pliny	3+	
Diodorus Siculus	3	
Kircher	2+	
Leeuwenhoek	2	
Strabo	1	
Subtotal		820+
Total		938

Essay 28. Of Reptils

Lines of text		233
Mather's composition		41+
Other authors		
Derham	63+	
Baglivi	57	
Bochart	17+	
Alsted	15	

Recapitulation of Sources

Robinson	15	
Frantze	10	
Roscommon	6	
Mayerne	5	
Gesner	2+	
Subtotal		191+
Total		233

Essay 29. Of the Fishes

Lines of text		259
Mather's composition		98+
Other authors		
Ray	69+	
Derham	46	
Hakewill	18	
Unidentified	10	
Kircher	6	
Harris	5	
Aelian	2+	
Alsted	2	
Peckham	1+	
Subtotal		160+
Total		259

Essay 30. Of the Feathered

Lines of text		684
Mather's composition		104
Other authors		
Derham	274	
Ray	135+	
Grew	37	
Unidentified	28+	
Harris	27	
Cheyne	19	
Dampier	17+	
Alsted	14	
Clogie	9+	
Babylonian Talmud	9	
Robinson	6	
Biblical	3	
Subtotal		580
Total		684

Essay 31. Of the Four-Footed

Lines of text		878
Mather's composition		83+
Other authors		
Derham	230+	

Ray	168+
Grew	88
Harris	77+
Alsted	32+
Leguat	31
Robinson	26
Blackmore	25
Unidentified	17+
Simiana	15
Rorario	11
Fénelon	10
Camerarius	9+
Smith	9
Seneca	8
Cheyne	7
Egard	5+
Frantze	5+
Luther	5
Plutarch	4
Piero Valeriano	3+
Reynard the Fox	2+
Dio Cassius/Topsell	1+
Solinus	1+
Subtotal	794+
Total	**878**

Essay 32. Of Man

Lines of text	3161
Mather's composition	705
Other authors	
Derham	854+
Ray	729+
Unidentified	170+
Cheyne	170
Wanley	103
Grew	92+
Drake	78
Keill	44+
Tauvry	31
Goodwin	29
Theologia Ruris	25+
Alsted	16+
Wainewright	18
Saunders	13
Vitruvius Pollio	9
Raymond of Sebonde	8+
Barker	8

Code of Justinian	8	
Hennepin	7	
Baglivi	6	
Arndt	5+	
Pliny	5	
Zwinger	4+	
Augustine	4+	
Tympe	4	
Baldi	3	
Baillet	3	
Ramazzini	3	
Bouchard	2+	
Subtotal		2456
Total		3161

Summary of Essays on the Life Sciences and Man

Lines of text		6878
Mather's composition		1314
(19.10 percent of the text)		
Other authors		
Derham	1974	
Ray	1309+	
Harris	311+	
Cheyne	284+	
Unidentified	273	
Grew	246+	
Wanley	103	
Alsted	98	
Robinson	84+	
Drake	78	
Rusden	76	
Baglivi	63	
Arndt	59+	
Keill	44+	
Dampier	31	
Leguat	31	
Tauvry	31	
Goodwin	29	
Blackmore	25	
Theologia Ruris	25+	
Barker	23	
Hakewill	18	
Bochart	17+	
Frantze	15+	
Simiana	15	
Wainewright	18	
Saunders	13	

Rashi	12
Beverley	11
Pitts	11
Rorario	11
Fénelon	10
Camerarius	9+
Clogie	9+
Babylonian Talmud	9
Smith	9
Vitrivius Pollio	9
Kircher	8+
Pliny	8+
Raymond of Sebonde	8+
Jonston	8
Code of Justinian	8
Seneca	8
Hennepin	7
Tate	7
Terry	7
Purchas	6+
Plutarch	6
Roscommon	6
Temple	6
Egard	5+
Luther	5
Mayerne	5
Augustine	4+
Zwinger	4+
Tympe	4
Otto von Guericke	3+
Piero Valerianio	3+
Baldi	3
Baillet	3
Biblical	3
Diodorus Siculus	3
Ramazzini	3
Aelian	2+
Bouchard	2+
Gesner	2+
Reynard the Fox	2+
Leeuwenhoek	2
Browne	1+
Dio Cassius/Topsell	1+
Peckham	1+
Solinus	1+
Strabo	1

Subtotal 5564
Total 6878

Summary of All Essays

Lines of text 11,062
Mather's composition 2322
    (20.99 percent of the text)
Ten major sources
    Derham 2104 (19.02 percent)
    Ray 1530 (13.83 percent)
    Harris 1263 (11.42 percent)
    Cheyne 743 (6.72 percent)
    Grew 394 (4.36 percent)
    Alsted 193 (1.74 percent)
    Walker 178 (1.61 percent)
    Halley 159 (1.44 percent)
    Arndt 114 (1.03 percent)
    Wanley 103 (0.93 percent)

INDEX OF BIBLICAL REFERENCES

OLD TESTAMENT

Genesis	page
1	312
1:1–31	174, 175
1:3	24
1:6	184
1:11, 24	103
1:26, 28	307
1:29–30	95
2:19	308
5:22, 24	93
9:8–17	67
13:14	314
15:5	34
22:17	297
39:9	94
41:45	99
46:1	135
46:3–4	110

Exodus	
4:10	274
7:19–25	161
7:20	101
8:19	102, 173, 300
9:18–35	70
31:2, 3	303
40:15	7

Leviticus	
26:4	62
26:19	127

Numbers	
23:11	283

Deuteronomy	
4:24	56
6:5	280
9:3	56
10:12	176
11:14	62
14:17	210
23:12–14	156
32:3	28–29, 30
32:22	108
33:19	187

Joshua	
11:4	297
23:14	191

Judges	
5:28	25
12:4–6	245
13:24–16:31	293
14:19	293

1 Samuel	
13:5	297

2 Samuel	
1:18	66
22:4	10
22:11	78

1 Kings	
1:52	297
2:2	191, 297
4:25	133
4:29	297
4:29–30	296
5:12	296

2 Kings	
1:3	161
5:20	11

Biblical References

1 Chronicles

6:32	283
16:25	41

2 Chronicles

17:4	135

Job

4:19	295
6:12	294
7:20	283
10:11	238, 247
11:7	307
11:12	103
12:7	211
12:8	184
14:14–15	238
22:2	87
22:15–16	66
23:8–9	94
23:10	94, 175
24:13	46
25:2–3	316
25:3	173
26:14	70, 72
28:15	128
31:26	56, 60
31:28	308
33:19	286
35:10	34, 60
35:10–11	224
36:24	17, 19
37:5	72
37:14–15	64
38:7	28
38:41	208
39:13–14	207–8
40:9	72
42:2	299

Psalms

5:9	276
8:1, 9	236
8:3–4	34
8:6	307
16:10	24
18:3	10
18:10	78
18:13	70, 72
18:14	72
19:1	34
19:3	22
21:9	53
25:15	260
29:1	9
29:3	71
29:9	7
31:15	303
33:1	11
35:5	106
35:10	286
36:6	226
36:9	24–25, 279
39:1	275
42:8	291
47:2	316
49:10	225
49:12	225
49:19	25
49:20	225
51:7	69
54:4	10
56:13	24
62:11	72, 230, 309
65:11	63
66:8–9	176, 247
72:6	63
76:7	79
77:18	72
78:12–43	284
78:48	72
84:3	202
86:17	292
89:5	299
89:6	249
89:8	295
89:9	89
89:47	151
90:11	108
94:8	86
95:3	316
96:4	41
103:20–21	306

104:3	78	Ecclesiastes	
104:7	72	9:12	192
104:8–9	89, 97	11:7	25
104:18	105	11:9	309
104:24	29, 106, 174	12:12	301
104:24–25	101	12:13	308
104:25, 24	97	Isaiah	
104:26	186		
104:27–28	186–87		
104:29	277	1:3	235
107:8, 15, 21, 31	27	1:18	69
119:68	236	7:14	315
119:91	42	8:8	315
135:7	79	12:5	135
136:1	24	13:21–22	309
136:4	230	23:8, 9	7
139:1	175	24:17	109
139:6	149	27:8	79
139:14	237, 247	28:2	70
145:3	41	33:14	55
145:21	275	40:15	161, 299
147:4	35	40:26	26, 27
147:9	208	40:28	311
147:16	68	45:7	24
147:17	81	45:18	151, 174
148:7	179	54:16	302
148:8	79	55:10	69
150:6	278	56:4	175
Proverbs		65:17	316
		66:2	176
3:19	102	66:22	316
4:23	306	Jeremiah	
5:11	232		
5:13	232		
6:6	167	8:8	151
8:10	109	9:21	260
8:19	128	10:13	79
10:20	275	14:22	63, 309
10:27	231	16:16	192
11:22	126	22:29	110
13:14	279	50:25	120
14:27	279	51:16	79
15:4	275	Lamentations	
23:26	280		
24:4	23, 299	4:3	191
26:10	19, 89	4:7	69
30:24–25	165		

Biblical References

Ezekiel		Nahum	
13:13	70	1:6	56, 108
18:4	299		
28:13	309	Habakkuk	
31:8, 9	309		
34:26	62	1:14	309
43:2	101	2:2	18
		2:14	120
Daniel		3:11	84
1:17	303		
2:21	302	Zephaniah	
4:30	313		
4:34	316	3:7	109, 311
4:35	316	3:9	274
5:23	278		
12:1	35	Haggai	
12:3	35	2:8	129
Hosea		Zechariah	
10:12	63		
		3:10	133
Amos		9:14	72
2:15	230	10:5	230
9:3	109	12:1	305
Micah		Malachi	
4:4	133		
6:8	93	1:14	122
7:8	25	3:17	126

New Testament

Matthew		12:41, 42	11
1:23	315	13:43	39, 42
4:19	192	14:19–20	192
5:8	156	14:30	124
5:13	126	15:27	221
5:45	63	22:37	280
5:47	66	28:18	316
6:26	211	Mark	
6:29	133		
7:9–10	187	1:17	192
7:12	297	1:24	76
8:12	25	3:22–26	161
11:23	49	6:41–42	192
12:24–28	161	7:28	221

7:32	274	1:25	128–29
8:24	147	1:31	191
12:30	280	2:11	121
16:15	237	10:17	269
16:18	177	11:33	17
		12:1	7
		12:4–5	309

Luke

2:14	7		
5:10	192	## 1 Corinthians	
6:31	297	1:15	312
9:16–17	192	1:20	306
11:11–12	187	2:1–5	274
11:15–20	161	2:14	25
14:34	126	3:16–17	7
17:10	87	4:7	10, 299
19:40	124	6:19	310
20:35	148	12:12–27	309
24:42–43	192	12:14–26	240, 244
		15:10	10
		15:40–41	40

John

1:3	314–15	## 2 Corinthians	
1:4	24	10:10	274
1:9	24, 25	## Galatians	
3:8	79, 80	3:1	260
3:27	10, 299	## Ephesians	
6:9–11	192	1:17–18	24
6:29	120	1:19	17
12:46	24	4:6	155
13:7	174, 316	4:21	24, 317
17:3	24	4:29	275
18:36	312	5:13	24
		5:16	305
		6:9	121

Acts

2:27	24	## Philippians	
3:21	316	1:11	148
4:2	148	2:5	310
7:50	314	2:9	316
14:17	62, 149	## Colossians	
17:27	93	1:12	25, 39
17:28	95, 176, 293, 307	1:15	312
25:27	25, 297	1:15–16	315
26:22	75, 314	1:16–17	313

Romans

1:4	148
1:16	311
1:18–32	310

472   *Biblical References*

1:19, 20	314	1 Peter	
3:8	275	1:12	315
3:16[15]	211		
4:5	305	2 Peter	
1 Timothy		1:19	25
3:16	9, 314	2:9	54
6:9	128	2:22	218
6:10	127	3:7	192
6:16	24	3:13	316
2 Timothy		1 John	
3:3	191	1:7	25
Titus		5:20	313
3:8	311	Jude	
3:14	302	13	49
Hebrews		25	174
1:3	315, 317	Revelation	
2:9	314	1:11	314
6:7–8	63	1:15	101
11:12	297	4:11	174–75
12:9	155, 295	11:6	187
12:22	306	12:1	39
12:29	56	13:18	208
James		14:2	101
		14:6–7	148
1:17	19, 34, 40, 299, 302	16:4	187
3:9	273	16:21	70
		19:6	101

Apocrypha

Ecclesiasticus		Song of Three Young Men	
43:11	65, 66	40	40
Wisdom of Solomon			
7:26	41		

INDEX

A Lapide. *See* Lapide
Abu Bakr. *See* Ibn Tufayl
Abu Ma'shar, 30, 379
Academy of Sciences. *See* Royal Academy of Sciences
Accademia dei Lincei, 391, 410
Accademia del Cimento, 305, 432
Account of the Origin and Formation of Fossil Shells, An, 139n
Acridophagi, 173
Adam, 308
Adrian. *See* Hadrian
AEgardus. *See* Egard
Aelian (Aelianus, Claudius), 189, 379
Aeschynomenae, 138
Aetna, Mount, 106
Aglossostomagraphia, 274
Air, 88, 102, 106; composition of, 73–74; description of, 73; elasticity of, 74; Hooke on, 411; necessary to animals, 76; Newton on, 75; specific gravity of, 75; weight of, 74, 335; worshipped by Syrians, 77
Albumazer Haly. *See* Abu Ma'shar
Aldrovandus, Ulisse, 166, 379, 425
Alexander the Great, 265, 268
Alexandrian library, 119
Alfred, king of England, 303
Alimentis, de, 243
Aloe muricata, 140
Alsted, Johann Heinrich: cited anonymously, 41, 101, 124, 126, 127, 128, 148, 149, 183, 189, 212, 234, 295; cited by name, 12, 146, 228, 235, 239, 298, 312; on a "combinatorial art," 343; his Encyclopedia at Harvard, xxv; life and writings of, lxiii–lxiv
Ambrose, 11, 379, 381
Amerbachius, Basilius, 298, 379
American Communications. *See* "Curiosa Americana"
Ammianus, Marcellinus, 109, 380
Anatomia Sambuci, 146
Anatomy of Plants. See Grew
Anaxagoras, 35n

Ancients, the, 11, 27, 29, 85, 111, 153, 159, 212, 224, 247n, 263
Andes Mountains, 108
Andry, Nicolas de Boisregard, 60, 151, 153, 380
Angermanland, 266
Anghiera, Pietro Martire d', 266, 370, 380
Animalibus, de, 234
Animal spirits. *See* Spirits
Anthony of Egypt, 18, 23, 321
Anthony of Padua, 191, 380
Antigua, 100n
Ants, 165–67
Aristotle: on animals, 223, 227, 234; biography of, 380; on birds, 196, 209; cited by Mather, cxv; cosmology of, lxxv, 397; on the hand, 292; on insects, 150, 159, 162, 167, 168, 419; on light, 19; and magnets, 112; maxim of, on Nature, 240, 402; and Peripatetic school, 441; physics at Harvard, xxv; rediscovery of, 381; on substance, 94; on Thales, 49n
Arndt, Johann: admired by Egard, 399; biography of, lxiv–lxv; cited anonymously, 101, 105, 133, 148, 150; on man, 239; on minerals, 122; on plants, 149; on the stars, 28; on the sun, 40; *Verus Christianismus* of, 19
Arte combinatoria, de, 103, 343–44
Artificial Clock-maker, The, 304
Asceticks, lxix, 296, 317, 374
Astrology, cxiii, 34, 379, 382, 385, 404, 410
Astronomer(s), 32, 35, 37, 41, 48, 49, 52, 86, 91, 381, 390, 391, 400, 403, 406, 410, 411, 412, 416, 433–34, 435, 437
Astronomy, 34, 49, 84, 382, 389, 400, 404, 406, 433, 434, 436
Astronomy's Advancement, lxiii, lxv, 32
Atheism, lxi, lxix, lxxii–lxxiii, lxxxiv, lxxxvi, lxxxviii, lxxxix, xci, cxvii, 203, 218, 239, 260, 308–9, 338, 339, 341, 389, 426. *See also* Mechanical philosophy

Atlantis, 109
Atlas chinensis, 81
Atmosphere, 78, 104, 129
Atoms: frigorific, 80, 81; primary, 88
Augermannia. *See* Angermanland
Augustine, lxix, lxx, cxv, 9, 149, 179, 210, 237, 300, 381
Augustissimam naturae scholam. See Theologia naturalis, exhibens
Augustus, 179
Ausonius, 321
Austin. *See* Augustine
Autolycus, 17
Auzout, Adrien, 48, 112, 381
Avicenna. *See* Ibn Sina

Bacon, Francis, xci, 313, 382, 388
Bacon, Roger, 112, 382
Baffin, William, 81, 382
Baglivi, Giorgio, 182, 303, 382
Baillet, Adrien, 298, 382
Baldi, Camillo. *See* Baldus
Baldus, Camillus, lxviii–lxix, 292, 382
Baliani, Giovanni Battista, lxxviii
Bandura cingalensium, 141, 354
Barbarossa. *See* Frederick I
Barker, Matthew, 153, 307, 321, 383
Barletius, Marinus, 294, 383
Bartholin, Caspar, 259, 276, 383
Bartholin, Thomas, 198, 240, 259, 270, 284, 290, 383, 422, 435
Basil the Great, 189, 192, 379, 384
Bath Kol, 72
Bauhin, Gaspard, 135, 270, 384
Bavarian poke, 240
Baxter, Richard, 298, 384
Bayer, Johann, lxxv, 29, 324–25, 326, 384
Bazaleel. *See* Bezaleel
Beall, Otho T., Jr., xxxvi, cviii–cix, 320
Beauplan, Guillaume Le Vasseur, sieur de, 81, 384
Becher, Johann Joachim, 305, 384
Beelzebub, 161
Bees, 162–65, 390, 435
Bellarmine, Robert, 297, 384
Bellini, Lorenzo, 283, 384–85
Benerovinus. *See* Beverwyck
Bentley, Richard, xxxvi, lxxii, 339, 389
Berakoth. *See* Jews
Berengario da Carpi, Jacopo, 263, 385, 427
Bernard of Clairvaux, 23, 385

Bernoulli, Jakob, 52, 385
Bernoulli, Johann, 40, 289, 385
Beverwyck, Johan van, 145, 385
Bezaleel, 303
"Biblia Americana." *See Christian Philosopher;* Mather, Cotton
Bickerus, Johann, lxviii, 126, 385–86
Bidloo, Govard, 247, 386, 395
Bils, Lodewijk de, 284, 386, 436
Biological concepts: animalculist theory, 419; epigenesis, lxxxv; preformationism, lxxxv, 135, 152; vitalism, lxxi, lxxxii, 416
Bird of paradise. *See* Bird(s)
Bird(s): bill of, 198, 199; crop of, 200, 207; eagle, 201; ear of, 198; eggs of, 203–4, 207–8; eye of, 198; feathers of, 193, 194; flamingos, 202; food of, 208; incubation of, 202, 203, 207; Indian, 201; man-of-war, 199; nidification of, 201, 361; ostrich, 207–8; of paradise, 198; passenger pigeon, 201, 204–7, 361–62; of prey, 199, 200, 209; rapacious, 195–96; scart, 198; solitary, 232–33; swan, 198; white crows, 196, 361; wings of, 194; woodpecker, 198–99, 200
Bishop of Paris, lxix, 170
Bitumens, 126
Blackmore, Richard, 22, 44, 226, 322–23, 364, 386
Blancardius. *See* Blankaart
Blankaart, Steven, 239, 386
Blasius, Gerard, 198, 222, 288, 386
Blochwitz, Martin, 146
Boccone, Paolo, 240, 386
Bochart, Samuel, lxviii, 179, 298, 386–87, 435
Boethius, Anicius Manilius Severinus, 304, 387
Bohun, Edmund, 100, 387
Book of Nature. *See Theologia ruris*
Borelli, Giovanni Alfonso, lxxvi, 91, 186, 195, 239, 277, 279, 286, 385, 387
Borellus, Petrus (Borel, Pierre), 60, 169, 387
Borrichius, Olaus, 217, 387
Borysthenes, 230, 366
Boston Philosophical Society, xxxiv–xxxv, xl, xli–xlii
Boyle lectures, lx, lxi, lxxi–lxxii, 341, 388–89, 394

Boyle, Robert: on air, 74, 75; biography of, 387–89; and Boyle lectures, 71–72; his *Christian Virtuoso*, 44; and cold, lxxviii, 336–37; corpuscularian (mechanical) philosophy of, xxv, xxvi; and human reason, lxxxi, 117–18, 348–49; on magnetism, 114; on plants, 103; and Royal Society, 411; as source for Mather, xxxvi, lxv; a stutterer, 274, 371; writings of, xxxvi, 344
Brahe, Tycho, lxxiv, 31, 52, 325, 389
Brerewood, Edward, 188, 389
Bridgewater Treatises, civ
Briggs, William, 253, 259, 389
Brotherton, Thomas, 133, 389
Browne, Thomas, 136, 389–90
Bucephalus, lxix, 229
Buchanan, George, lxxiv, 26, 228, 390
Bud, Guillaume, 297
Buris, 109
Burnet, Thomas, lxxxi, 97n, 103n, 338, 339, 341, 416
Butler, Charles, 162, 390

Cabot, Sebastian, 112, 390
Cadew-worm, 170
Caesius. *See* Cesi
Caligula, 72n, 228
Callus, 243
Cambray, Archbishop of. *See* Fénelon
Cambridge Platonists, l, lvi, lxix, lxxi, lxxxii, xcii, 426, 434
Camerarius, Philipp, lxviii, 227, 240, 293, 364, 390
Camerarius, Rudolf J., lxxxiv
Camillus. *See* Baldus
Campani, Giuseppe, 46, 390
Canadian lakes, 98
Cantharides, 174
Canum fidelitate et sagacitate, de, 227, 390, "Canum fidelitate, de," 425
Cardano, Girolamo (Jerome Cardan), lxxiv, 21, 126, 294, 303, 390, 436
Carew, Richard, 294, 390–91
Carneades, 300, 391
Carpensis. *See* Berengario
Cartesian: mechanism, lxxi, lxxii, 339, 387, 388, 397, 426, 434; physics, xxv–xxvi. *See also* Descartes; Rohault
Casaubon, Isaac, 297, 298, 391, 421
Cassini, Gian Domenico, 31, 38, 44, 45, 46, 47, 52, 391

Castriota, George, 294, 383, 391
Catania, 108
Caterpillars, 177
Cato, Marcus Porcius, 142, 391
Cause(s): first (final), lxxi, lxxiii, lxxx, 18, 44, 93, 331, 340; second(ary), lxxi, lxxiii, cxiv, cxvii, 18, 93, 95
Cesi, Federico, 135, 391
Chain of being, lxxxii, xci, 237, 306, 307
Chalazae, 202–3
Chance, lvi, lxxv–lxxvi, lxxxviii, cxvii, 46, 208, 240, 403
Charlemagne, 37
Charles V, 126, 225, 392
Chauncy, Charles, xciv
Cheesman, Mr. (a blind preacher), 298
Cheselden, William, 223, 239, 392
Cheyne, George: on astronomy, 18, 20–21, 30, 32, 39, 40, 47, 51, 53, 58–59, 84; biography of, lxii, 392–93; on eggs, 203; on man, 250, 259, 260, 271, 277, 289, 290, 292–93; his *Philosophical Principles of Natural Religion*, lxii, lxv, 94, 317, 392; on physics, 88, 92–93; on plants, 135; on a power superior to mechanism, 94–95, 175, 230; a source for Mather, l, lxv, lxvi; on spontaneous generation, 154; on vapors, 104; on water and fluids, 97, 100, 101–2
Childrey, Joshua, 26, 393
China, 81
Chiromancy, 292
Christian Philosopher, 17, 18, 22, 23, 231, 295, 303, 309, 312
Christian Philosopher, The:
—character of, xxi
—contents of: anatomy and physiology of man, lxxxviii–lxxxix; anti-atheism, xliii–xliv; astronomy, lxxiii–lxxvi; borrowings from other authors, xlix; design argument in, lxix–lxxii, cxiii, cxvi–cxx; fables, cxiii; natural history (life sciences), lxxii–lxxxviii; natural theology and social ideology, lxxii–lxxxiii; original composition in, xlix–l, lxvi–lxvii; physics and related sciences, lxxvi–lxxxii; promotion of piety, xci–xcii; religious significance of nature, xc–xcii, cvi–cvii, cix–cxx; soul of man, lxxxix–xc; superstitions in, cxiii, 60, 235, 344

—origins of: Mather's "Biblia Americana," xxxvi–xl, xlii, lxvii, lxviii; Mather's "Christian Virtuoso," xliii–xlvi; Mather's "Curiosa Americana," xl–xliii, lxvii; Mather's scientific interest and "do-goodism," xxxv–xxxvi

—publication of, xlvi

—reception of: in Boston and New England, xlvii, xciii–xcvi; Collier's "easy and familiar" edition (1815), civ–cvi; on the Continent, ci–ciii; in England, c–ci; Mather's solicitude for, xlvii–xlviii; in the middle colonies, xcviii–c; in the southern colonies, xcvi–xcviii

—scholarly reception in the twentieth century, cvi–cxi; Beall and Shryock, cviii–cix; Bercovitch, cx; Hornberger, cvii–cviii; Middlekauff, cix–cx; Miller, cviii; Murdock, cvii; Piercy, cix; Riley, cvi–cvii; Silverman, cx–cxi; Stearns, cix

—significance of: for American intellectual history, xxi; a contribution to science in America, civ–cv; a departure from New England Puritanism, cxii–cxiii, cxvii; a harbinger of the Enlightenment in America, cxiii–cxiv; as an interpretation of nature, cxvi–cxvii, cix–cxx; its literary artistry, cvi; in transmission of the humanistic legacy, cxv–cxvi

—sources of, xlviii–l; Alsted, lxiii–lxiv; Arndt, lxiv–lxv; Cheyne, lxii; Derham, lx; Grew, lxii–lxiii; Halley, lxiii; Harris, lx–lxii; Jewish and Muslim authors, lxvii–lxviii; miscellaneous, lxviii–lxix; Ray, l–lx; Scripture and the ancient classics, lxvii; Wanley, lxiii

—storytelling in, cxvi

—structure of, xlviii–xlix; astronomy, lxv; the life sciences and man, lxv–lxvi; physics and related phenomena, lxv. *See also* Ray, John; Reason, human

Christian Religion's Appeal, 224

Christian Virtuoso, The, xliv

"Christian Virtuoso, The." See *Christian Philosopher*, origins of

Chrysostom, John, 17, 110, 393

Cicatricle, 203

Cicero, Marcus Tullius, lv, cxv, 12, 23, 27, 28, 84, 92, 103, 220, 238, 241, 256, 257, 269, 291, 295, 301, 341, 393–94, 425

Circle, 118–19

Clarke, Samuel, lxxii, lxxix–lxxx, 92, 341, 389, 394, 434

Claudian. *See* Claudianus

Claudianus, Claudius, 30, 116, 295, 394

Clayton, John, 199, 394–95

Cleanthes, 84, 395

Clement of Alexandria, 23

Clogie, Alexander, 211

Clouds, 60, 62, 63, 64, 68, 69, 79

Cocceianus. *See* Dio Cassius

Cochineel, 167

Cockburn, John, 127, 245, 395

Cockburn, William, 241, 395

Coelo-Syria, 179

Cohesion (of bodies), 88

Coiter, Volcher, 198, 395

Colanders, 103

Colaptice, 124

Cold, 39, 54; animals and, 217; cause of, 80, 336; in China, 81–82; force of, 80–81; force of congelation, 81; in New England, lxxviii, 82, 337; varies in different climates, 81

Cole. *See* Coles

Colerus, 311

Coles, William, 135, 395

Collier, William: career of, civ; as evangelical publisher, civ–cv; his edition of *The Christian Philosopher*, cv–cvi. *See also Christian Philosopher*

Colors, 20, 427

Columella, 60

Combustio Animae, lxxvii, 72, 73

Comet(s), lxxvi, 50–54, 84, 410; atmosphere (vapors) of, 51–52; Cheyne's ideas on, 53; Cotton Mather's *Boston Ephemeris*, xxxiv; Cotton Mather's *Essay on Comets*, 330; as divine portents, 53; Halley's theory on, 53, 408; heat of, 50–51; Increase Mather's *Kometographia*, xxxiv; Kepler's hypothesis on, 52; in New England, xxxiv

Cometomania, lxxvi, 52

Compass: mariner's, 112; variation of the, 115

"Compendium Physicae," 26
Constantine, 62
Constantinople, 106
Constellations, 29, 34, 324–25; Andromeda, 31; Cassiopeia, 30, 31; Cepheus, 26; Cygnus (Swan's Breast), 26, 31; Eridanus, 31; the Hare, 31; Libra, 42; Orion, 29; Pisces, 31, 42; Pleiades (Seven Stars), 26, 29, 31, 32; renaming of, lxxv, 326; Swan's Bill, 31; Ursa Minor, 31; Whale's Neck, 31
Copernican hypothesis, lxxiii, lxxix, 84, 338, 401, 403, 433
Copernicus, Nicholas, lxx, lxxiv, lxxv, 404
Cortex peruvianus, 143
Cosmographia (Münster), 427
Cosmography (Heylin), 304
Cosmologia Sacra, 60, 407
Coster, Laurens Janszoon, 303, 395
Covenant, 67, 332
Cowper, William, 241, 270, 271, 277, 278, 395
Crassus, Marcus Licinius, 191, 395–96
Creation (poem), 44, 386
Creatures: emblems of, 233–34, 235
Creitlovius, Joachim Fridericus, 104, 396
Croone, William, 286, 396
Cudworth, Ralph, l, lxxxii. *See also* Cambridge Platonists
Cujas, Jacques, lxviii, 274, 396. *See also* Justinian I, the Great
"Curiosa Americana," lxvi, lxvii, 10–11, 11n, 320, 323, 331, 333, 350, 352, 357–58, 360, 361–62
Curtius Rufus, Quintus, 105
Cyprus, 62
Cyrus, 300

Dampier, William, 128, 141, 172n, 199, 396
Danhawer, Johann Konrad, 312, 396
Darwin, Charles, xx, cxvii–cxviii
Dee, John, lxxiv, 21, 396
de Gueric. *See* Guericke
Deipnosophist, 110
De la Vega. *See* Garcilaso
Democritus of Abdera, 80, 396
Demonstration of the Existence of God, A, 215
Derham, William: on astronomy, 29, 32, 38, 49, 59, 82, 84; biography and works of, lx, 304; on birds, 194, 198, 202, 204; Boyle lectures of, lxxii, 389; on the earth, 103; on fishes, 184, 187; on generation of frogs, 155; on insects, 157, 160, 166, 169, 171; on man, 241, 243–44, 254, 260, 262, 264, 267, 268, 269, 270, 276; on physics and related phenomena, 78, 89, 112, 114, 117; on quadrupeds, 220, 221, 224; a source for Mather, lxii, 10, 11; on vegetables, 177; on vipers, 180–81; on the wisdom of God, 160, 166, 171, 180–81, 184, 202, 204, 221, 224, 241, 243–44, 261
Descartes, René: and animal automatism (the "beast machine"), 226, 437; biography and works of, 382, 396–98, 437; on the eye, 255; on gravity and *"materia striata,"* 89, 340, 397; on the moon's distance from the earth, 58; on the rainbow, lxxvii, 64; the science of, 387, 406, 434; on snow, 67
Design and design argument, xlviii, lxix, lxxvi–lxxix, lxxxi, lxxxii, lxxxv, xc, cxiii, cxvi–cxx, 13, 46, 86, 95, 121, 223, 271n, 402, 407; origins of, xix–xx; relation to natural theology and natural religion, xx
Dio Cassius or Dion Cassius, 179, 189, 398
Diodorus Siculus, 173
Dionis, Pierre, 261, 263, 398
Disease, lxxxii, lxxxiv, lxxxv, 410
Dod, John, 212, 398
Dominis, Marco Antonio de, lxxvii, 64, 398
Douglas, James, 240, 398–99
Dove of Archytas, 209
Dowglass. *See* Douglas
Draakestein. *See* Reede
Drake, James, 262, 277, 281, 399
Du Bartas, Guillaume de, lxxxvii, 209, 309

Earth, 28, 30, 35, 38, 41, 43, 44, 48, 49, 51, 52, 56, 57, 58, 62, 79, 82, 84, 91, 116, 117, 129; atmosphere of, 70; axis of, 90, 99; diameter of, 28, 41, 90, 102–11; fuller's, 122; measure of the, 430; minerals in the, 122; *orbis magnus* of, 41, 50; potter's, 122; sorts of, 103, 122; strata of, 103, 116, 117, 129

Earthquakes, 106, 108, 109, 110, 111, 411
Eclipses: on satellites of Jupiter, 21; 111; solar, 58;
Ecliptic: inclination to, 86; plane of, 85
Edwards, John, lxix, 302, 376, 399
Egard(us) or Eggers, Paul, lxviii, 18, 88, 235, 321–22, 399
Egypt, 62
Elephant, 213, 216, 220–21, 228
Ellegård, Alvar, cix
Elogium canis, 227
Elysian Fields, 116
Ephemerides. *See* "German Ephemerides"
Ephemeron, 151
Ephraimites, 245
Epictetus, 212
Epicurean(s), lxxii, 85, 176, 303, 379, 388, 392, 394, 404, 431
Epicurism, 142
Epicurus, 85n, 404
Ericus, 268
Ettmüller, Michael, 277, 278, 399
Evelyn, John, 103, 343, 399–400
Experiments, 131, 352

Fables, lxxxiv, cxiii, 34, 154, 179, 181, 274
Fairfield (Conn.), 68
Faust. *See* Fust
Fénelon, Francis de Salignac de la Mothe, lxviii, 63, 215, 331–32
Fidentinus, 10
Fire, 55–56, 76, 106, 217
Fishes: anatomy of, 184, 186; food of, 186–87; shape of bodies of, 184; variety of, 184
Fixed star(s), 28, 29, 30, 32, 51, 111; distances of, 50; like the sun, 30; motion of, 408
Flamsteed, John, 45, 56, 86, 400
Fletcher, Giles, 80, 400
Fluids, 101–2
Fontenelle, Bernard Le Bouyer (or Bovier) de, lxxiv
Foreest, Pieter van, 301
Forestus. *See* Foreest
Fossils, lix, 122, 349–50, 411, 421
Foster, M. B., xix
Fraccus, Ambroise Novidius, 34, 400
Fraedlichius. *See* Froelich
Frantze, Wolfgang, 181, 400
Franzius. *See* Frantze

Frederick, John, 311, 400
Frederick I (Frederick Barbarossa), 225, 400
French Academists. *See* Royal Academy of Sciences
Froansberg, George of, 194
Froelich, David, 267
Froissart, Jean, 294, 400
Fromondus, Libertus, 155, 400–401
Frytsche, Marcus, 65, 70, 401
Fuga Vacui, lxxviii, 74. *See also* Vacuum
Fuller, Thomas, 245, 401
Fust, Johannes, 303, 401

Gaffarel, Jacques, lxviii, 34, 304, 401
Gale, Theophilus, 69, 401
Galen (Claudius Galenus): on insects, 160; life and writings of, 401–3; on naturals and non-naturals, 335–36; on the poison of a viper, 180; on quadrupeds, 217, 220, 225; rejection of Galenic theory, 385; on the skill of the Creator, 105, 160, 220, 225, 240–41, 292; a source for Mather, cxv; on the usefulness of the parts of the body, 238, 247, 252, 259, 271–72, 278, 280, 288, 289, 292, 295–96
Galilaeus. *See* Galilei
Galilei, Galileo, lxxvii, 74, 304, 335, 340, 397, 403, 404, 420, 423, 437
Garcilaso de la Vega, 140, 403
Garden, 142; of Eden, 354, 390
Garden, George, 153, 403–4
Garencières, Théophile de, 146, 404
Gassendi, Pierre, lxxii, 226, 388, 404
Gaza, Theodore, 234, 404, 441
Gellibrand, Henry, 112, 404
Gellius, Aulus, 209, 404
Generation: anomalous, 167; equivocal, 154, 159; spontaneous, lxxxiv–lxxxv, 154, 155–56, 419, 421, 432, 440
Génération des vers dans le corps de l'homme, de la, 153, 380
Georges, the, 60
Gerard, John, 138, 404–5
Gerhard. *See* Gerard
"German Ephemerides," xl, xciii, 146, 168, 243, 268, 312
German Ocean. *See* North Sea
Gesner, Konrad, lxxxvi, 179, 189, 379, 405
Gibraltar, Straits of, 98

Gibson, Thomas, 259, 261, 405
Gilbert, William, lxxxi, 348
Gileadites, 245
Gioia, Flavio, 112, 405
Glisson, Francis, 283, 405
Globe, terraqueous, 56, 82–89, 99; ambit of, 82; axis of, 85; distance from sun of, 82; inclination to the ecliptic, 85, 86; motion of, 82, 84, 86; perihelion of, 86; size of, 82; speed of, 89; spherical shape of, 84–85, 86; waters on the, 85
Goedaeret, Johannes, 170, 405, 421
Goodwin, Thomas, 314, 405
Gorden. *See* Garden
Granadilla, 312
Gravity, lxxix–lxxx, 88, 408, 410, 413; absolute, 90; affect of, on bodies, 90–91; Cartesian doctrine of, 397; cause of, 90, 92, 93, 94–95, 340, 397; comparative, 90; definition of, 89, 92; force of, 90, 91–92, 93, 99; Huygens's theory of, 413; mechanical accounts of, 93, 397
Greenwood, Isaac, xciv, xcv
Gregoras, Nicephorus, 52, 405–6
Gregory of Nazianzus, 11, 406
Gregory, David, 41, 47, 86, 91, 325, 406
Gresham College, 117, 404
Grew, Nehemiah: *Anatomy of Plants*, lxii, 134, 406–7; on astronomy, 32, 38–39, 60; on birds, 200; *Cosmologia Sacra*, lxii–lxiii, 60, 407; divine and human reason, 118–19, 171–72; on emblematic qualities of animals, 233–34; life and works of, lxii, 406–7; on lunar influences, 60; on man, 243, 251, 253, 254, 261, 270, 276, 284, 286, 288, 292; on minerals, 127, 128; on nature as a divine regulator, 102; on the origin of watches, 304; on physics and meteorology, 67, 79; on quadrupeds, 219, 221, 228; on the scale of being, 307; as a source for Mather, l, lxiii; teleology of, lv, lxii–lxiii; 236; on the usefulness of animals, 210–11, 236; on vegetables, lxxxiv, 134, 136, 137, 142
Grimaldi, Francisco Maria, 114, 407
Grotius, Hugo, 298, 407
Grotta degli Serpi, La, 180
Guericke, Otto von, lxviii, 136, 387, 407

Habakkuk, 84
Hadrian, 230
Hail: origins of, 69; -storms, 70
Hakewill, George, 293, 407
Halley, Edmond: on air, 75; on astronomy, 43, 53 (Halley's Comet), 76; on the earth, 86, 104; on gravity, 90, 91, 340; life and works of, 407–8; on magnetism, 115, 116; *Miscellanea Curiosa*, lxiii, 49; *Philosophical Transactions*, lxiii; on the rainbow, 64; as a source for Mather, lxiii, lxv; on water, 98; on wind, 77
Hanno, 228, 365
Harris, John, lxv, lxvi, 160, 345, 389, 408
Hartenius, 189
Harvey, William, lxxxvi, 200–202, 409, 429
Havers, Clopton, 285, 409
Hearne, Urban, 267
Heat: of bodies, 54; of burning glasses, 55; of comets, 50; innate, 342; of Mercury, 48; operations of, on our senses, 54; radical, 94; Slare's experiments on, 330; of the sun, 37, 38–39, 40; vital, 188
Helmont, Franciscus Mercurius van, 302, 409
Helmont, Johannes Baptista van, lxxx, 103, 344, 409
Hennepin, Louis, 266, 409
Herberstein, Sigismund von, 80, 409
Herbs, 131, 143–44, 167
Hercules (plant), 143
Hérigone, Pierre, 47, 254, 409
Herillus, 9
Herlicius or Herlitz, David, 73, 410
Hernandez, Francisco, 141, 410
Hevelius, Johannes, 48, 51, 52, 59, 112, 325, 408, 410
Hexaemeron, 189, 379, 384
Heylin, Peter, 304
Hipparchus, 324–25
Hippocrates, cxv, 94, 105, 243, 402, 410
Historia Singularis, 170, 420
History of Plants, 136
Hobbes, Thomas, lxxii
Holy Spirit, 317–18
Homer, 300, 301
Hooke, Robert, lxi; on astronomy, 19, 35, 45, 46, 58; on gravity, 90; life and

works of, 410–11; on meteorological phenomena, 67, 70, 74
Horace (Quintus Horatius Flaccus), 54, 209, 301
Horae subcesivae (*Operae horarum subcisivarum*), 227, 364, 390
Hornberger, Theodore, cvii–cviii, cxxiv
Hortus malabaricus, 140
Host of Heaven, 128
Hugh of St. Victor, 22, 41, 411
Hulse, Edward, 76, 159, 412
Hume, David, cxvii
Humors of the body, 247n
Hunter, John, lxxxvii
Huygenian satellite, 43
Huygens, Christaan, lxxiv, 21, 26, 28, 43, 59, 66, 90, 304, 412–13
Hybridity, lxxxiv
Hylarchic principle, 95

Ibn Sina, 155
Ibn Tufayl Muhammad ibn 'Abd al-Malik, lxxiii, 11, 320–21, 413–14
Immortality of the Soul, The, 226
Incitatus, 228
Indian corn, lxxxiv, 131, 134, 352
Ingrassia, Giovanni Filippo, 263, 414
Insects, 387; amphibious, 157; anatomy of, 153; antennae of, 157; care of young, 166–67; eyes of, 156–57; foresight of, 162, 166; generation of, 153, 154, 155; joints of, 160; legs of, 159; mutable, 152; nidification of, 169; nonmutable, 151; noxious, 173; numbers of, 172; sagacity of, 161–62, 164–65; webs of, 159
Invisible world, 306–18
Iron: and magnetism, 111–14, 121
Ismenias, 268

Jackson, Homer, 298
James, Thomas, 80, 414
James, William, cxviii
Jameson, William, 300, 375, 414
Jamestown-weed, 138, 147
Jansenists, 9, 10, 319–20, 428
Jarchi. *See* Rashi
Jenkin, Robert, 118, 414
Jerome, 306
Jessop, Francis, 168, 414
Jesus Christ, 311–16
Jewel, John, 300, 414

Jews, Jewish, 299, 301, 306, 401; academies, 374–75; Academy of Sura, 299; Berakoth, 301, 376; on the Egyptian plague, 161, 356–57; Gemara, 362; Mishna, 362, 376; Philo, 129, 384, 429; rabbins, 273; on the rainbow, 66; saying of the, 22; on the stars, 26, 324; Talmud(s), or Talmudists, 210, 244, 362, 376, 432
Jones, John, 250, 414–15
Jonstonus, Joannes, 166, 415
Joshua, 84
Judaism. *See* Jews, Jewish
Judicia Medica, 292
Julian, 9
Jupiter. *See* Planets
Justinian I, the Great, 274, 371–72, 415
Juvenal (Decimus Iunius Iuvenalis), 128n, 195

Keill, James, 241, 276, 278, 279, 280, 290, 415
Keill, John, 85, 90, 91, 97, 339, 416
Kepler, Johannes, lxxv, lxxvi, 29, 41, 47, 52, 325, 416
Kerckring, Theodor, 283, 416
Kimchi (or Kimhi), Rabbi David, 11, 416
King, Charles, 139, 416
King, Edmund, 165
Kircher, Athanasius: on a combinatorial art, 344; on fishes, 186; life and works of, 416–17; on rocks, 124; on serpents, 180; on sound, 266, 268; his speaking tube, 265; on the sun, 35; on volcanoes, 106, 108; on worms, 169
Klunder, Nicholas, 294
Koran, the, 346–47, 354
Koster. *See* Coster

La Bruyère, Jean de, 9, 417
Lactantius Firminiaus, Lucius, 44, 225, 417–18
Laesia Memoria, 301
Lambecius, Petrus, 311, 418
Lane, Ralph, 144, 418
Lange, Johannes, 169, 418
Lapide, Cornelius à, 245, 418
Laurentius, or Laurens, Andre du, 270, 418
Law(s): of attraction, 88; of gravity, 84, 90, 103; of mechanism, 102, 175; of nature, 18, 65; of optics, 254; of

planetary motion, lxxiii–lxxiv, 18-19, 293, 342, 416, 427–28; of statics, 77
Le Blanc, Vincent, 85
Le Clerc, Jean, 143
Leeuwenhoek, Anton van, 135, 137, 246, 254, 283, 419–20
Leeuwenhoekian examination, 160
Leguat, François, 232–33, 366, 420
Leibniz, Gottfried Wilhelm von, 343–44, 385
Leipsick ell, 55
Lethargus, lxix, 229
Leusden, Jean, 312, 420
Levi, Rabbi Joshua ben, 66
Levity, 91
Lexicon Technicum, lxi–lxii, xciii, 408. See also Harris, John
Libau, Andreas, 170, 420
Light, 102, 116; Aristotle's definition of, 19; bodies and, 20; Christ as the, of the world, 24–25; color of, 20; Hooke on, 19; Molyneux's description of, 19; rays of, 20; refraction and reflection of, 20; of the sun, 35, 37–41; swiftness of, 21, 50, 71
—theories of: corpuscular, 427; wave, 411, 413
Lightning, 72
Lippershey, or Lippersheim, Hans, 303, 420
Lipsius, Justus, 300, 420–21
Lister, Martin, 114, 154, 159, 168, 177, 207, 250, 405, 421–22
Livy (Titus Livius), 81, 422
Loadstone (or lodestone): described, 111, 112n; effluvia of, 114; like gravity, 120; at Gresham College, 117; poles of, 113; power of, 113, 114; Roger Bacon on, 382; Ward on, 445–46
Longolius, Christopherus, 301, 422
Lower, Richard, 223, 277, 411, 422
Lucretius Carus, Titus, lxxii, 81, 399, 422. See also Epicurus, Epicureans
Luther, Martin, 235
Lutherus redivivus, 299–300
Lyser, Michael, 247, 422

Magnet: attractive quality of the, 111, 112, 113; concave, 116; globe as a, 115, 116; orb of activity of a, 117; poles of a, 112; red-hot, 113; spherical, 112; verticity of a, 112; vertue of a, 113, 116

Magnetic(al): effluvia, 114; matter, 116; needle, 404; particles, 113; poles, 115, 116; power, 114; tendency, 112; vertue, 116
Magnetism, 120, 408, 417, 445–46
Magnetis reductorium theologicum tropologicum, 120
Magnus, Olaus, 105, 137, 266, 422
Mahomet, Mahometan, 11, 12, 110, 354, 430
Maiolo, Simon, 294
Maize. See Indian corn
Malapert, Charles, 37
Malpighi, Marcello, 382, 406, 420; on human physiology, 240, 270, 277, 278, 283; life and works of, 423; on plant anatomy, 76, 133, 135, 154, 171; on preformation, 135
Man: anatomy of, lxxxix; bladder of, 283; blood of, 278–80, 283, 284, 289–91, 373, 415–16, 419–20, 422, 426, 427, 444, 447; body of, 237–38, 239–40; bones of, 284–86; ear of, 260–65; emunctories of, 243; erect posture of, 238–39; eye of, 251–60; faces of, 244–45; glands of, 284; glottis of, 276; hand of, 291–93, 383; handwriting of, 245–46; heart of, 279–80; intestines of, 282–83; inventions of, 302–5; kidneys of, 283; liver of, 283; lodgment of the parts of the body of, lxxxviii–lxxxix, 241; lungs of, 278–79; lymph system, 256, 284, 289, 290–91, 383, 422, 428, 429, 435–36, 439; membranes of body of, 288; memory of, 300–301; mind of, 307; muscles of, 286, 288; nerves of body of, 243–44; parts of the body of, 246–47, 250; physiology of, lxxxix; a priest for the Creation, 236; prodigious learning of, 297–99; provision in body of, to ward off evils, 241; reason of, 297; respiration of, 276–78; the senses, 269, 270, 271, 273; soul of, 295–305; stomach of, 280–82; strength of, lxvii, 293–95; teeth of, 271–73; tongue of, 273–75; union of soul and body of, 305–8; uses of the parts of the body of, 247, 249; voices of, 245 See also Reason, human
Mandeville, Sir John, 252, 369
Manzolli, Pietro Angelo, of Stellata, 375

Marchetti, Domenico de, 250, 423
Marggraf, Georg, 140, 217, 423
Mariner's compass, 112, 405
Mariotte, Edme, 97, 133, 423
Marius, Gaius, 293, 423
Mars. See Planets
Martini, Martin, lxxviii, 81, 417
Martinius. See Martini
Martyr, Peter. See Anghiera
Materia striata, 90, 339–40. *See also* Descartes
Mathematical examples, 118
Mathematical Principles of Natural Philosophy. See *Philosophiae Naturalis Principia Mathematica*
Mather, Cotton: and Boston Philosophical Society, xxxiv–xxxv; contributions to science, lxvi, lxvii; early life and schooling, xxii, xxiv; education at Harvard, xxiv–xxvii; fear of atheism, lxiii–xliv; Fellow of the Royal Society, xlii, xlvii; hazing of, xxvi–xxvii, xcii–xciii; library of, xxvii; ministry and ordination, xxvii–xxviii; and New England's spiritual crisis, xxviii–xxxiii; and new science, xxxiii–xxxv; as prolific author, xlviii; stammer of, xxvi–xxvii, xxviii; transitional character of his mind, xli–xlii, xlix, lxix, cxiii; and transmission of the design argument to America, xx–xxi; and witchcraft, xxxiv
—writings: *The Angel of Bethesda*, lxvii; "Biblia Americana," xxxvi–xl, 11n; *Brontologia Sacra*, 334; *Christianus per Ignem*, 330, 359; *The Good Old Way*, 374; *Magnalia Christi Americana*, xxxiv, xxxix, xlii, lxvi, xcvii, 11n, 334; *Manuductio ad Ministerium*, cxii; *Nets of Salvation*, 354; *Thoughts for the Day of Rain*, 332; *A Treacle Fetch'd Out of a Viper*, 180, 359. *See also* *Christian Philosopher*
Mather, Increase: and Boston Philosophical Society, xxiv; on comets, xxix, xxxiv; life and career of, xxii; and new science, xxxiii–xxxv; obtains new charter, xxx, xxxi
—writings: *Essay for the Recording of Illustrious Providences*, xxxiv, xlii; *Heaven's Alarm to the World*, xxxiv; *Kometographia*, xxxiv
Maximinus, Gaius Julius Verus, 203, 423–24
Mayern. *See* Mayerne.
Mayerne, Theodore Turquet de, 181, 424
Mayolus. *See* Maiolo.
Mayow, John, 277, 411, 424, 447
Mead, Richard, 180, 424
Mechanical philosophy, lxii, lxxiii, lxxvi, civ, cvi, cviii, 18, 230; Bellini and, 384–85; Boyle and, 388; Descartes and, 397; Huygens and, 412; Mather resists, cxiv; More and, 426; Schuyl and, 437. *See also* Epicureans; Lucretius
Medicina Statica, 247
Medicine, schools of, lxxxix; of Hippocrates, 410; iatrochemical, 440; iatromathematical, lxii, 392, 415; iatromechanical, 385, 424; iatrophysical, 387
Meleteticks, 23, 323
Mercer. *See* Mercerus
Mercerus, Joannes, or Jean Mercier, lxviii, 66
Mercury (metal), 40, 75, 102
Mercury. *See* Planets
Merrett, Christopher, 137, 424
Mersenne, Marin, lxxiv, 21, 424-25
Mersennus. *See* Mersenne
Merycologia, sive de ruminantibus et ruminatione commentarius, 221
Methodus insectorum, 151
Microscope, 28, 290, 387, 411, 412, 417, 419
Micyllus, Jacobus, 227, 425
Milky Way, 26, 29
Miller, Perry, xxxvi
Milo of Crotona, 293
Mindelheim. *See* Froansberg
Minerals, 104, 122, 126–27
Minucius Felix, Marcus, 12, 425
Miscellanea Curiosa. *See* Halley
Mist, 62
Mithridates VI, Eupator Dionysius, or "the Great," 300, 425
Moffett, Thomas, 168, 425
Molyneux, William, 19, 46, 255, 425
Monconys, Balthasar, 125, 425–26
Money, 127
Moon, 27, 48, 91, 99, 100; distance from

earth of, 57; influence on sublunary bodies, lxxvi, 60, 331; librations of, 57; light of, 57–58; new and full, 100; periodical revolutions of, 56; as a satellite of Earth, 56; spherical figure of, 58; waters on, 59
Morbis Artificum, de, 302
More, Henry, l, lxxxvi, 85, 133, 256, 398; on atheism, lv, 203, 218; life and works of, 426; on plastic nature, lxxxii, 95. *See also* Cambridge Platonists
Morhoff, Daniel Georg, 268, 426
Morland, Joseph, 241, 426
Morland, Samuel, 265, 426
Morton, Charles, 26
Morton, Richard, 143, 426–27
Moses, 174
Motu animalium, de, 186, 286, 387
Moufet. *See* Moffett
Moulen or Moulin, Allen, 199, 264, 373, 394, 427
Mountains, 104, 105, 108
Movement des eaux, du. See Traité
Muffet. *See* Moffett
Munchius, Janus, or Jens Munk, 81, 427
Mundinus (Raimondo or Mondino) dei Liucci or Liuzzi, 303, 385, 427
Münster, Sebastian, 9, 427
Muslim. *See* Mahometan

Naturae Curiosi. See "German Ephemerides"
Natural theology, 339, 367, 403; origins of concept, xx; relation to design argument, xx, lxix, cxiii–cxiv. *See also* Design and design argument
Natural Theology, or the Knowledge of God from the Works of Creation, 307, 321, 383
Natural Theology (Paley), xx, xcix–c, ciii
Needle, 113, 120; common, 112; dipping, 112; magnetic, 404, 444; variations of, 116, 444
Nerlinger, Oswald, 160
Newton, Isaac, lxix, lxxii, lxxiv, 332, 339, 385, 388, 393, 406, 416; on air, 75; on astronomy, 20, 35, 39; on comets, lxxvi, 50, 51, 53; and design argument, lxxi; on the globe, 86; on gravity, 90; on heat, 48; and Hooke, 411; his laws of motion, 18–19; on optics, lxxvii, 65, 259; on the rainbow, lxxvii, 65, 66
New Voyage Round the World, A, 172, 396
Nieuwentijdt, Bernard, xcv, xcix, c, ciii
Nitre, 55, 69, 70–71, 80, 129, 330
Nitrous particles, 188, 217, 411
Non-naturals, 335–36. *See also* Galen; Wainewright
Norman, Robert, 112, 428
Norwood, Richard, 82, 428
Novidius. *See* Fraccus
Nuck, Anton, 256, 428
Nux vomica, 135

Ocean(s), 78, 115, 188
Ocher, 114
Olearius, Adamus, 80, 428
Opticks, 65
Origen (Origenes Adamantius), lxx, xci–xcii, 313, 377-78, 381
Osteology (Havers), 285, 409
Ovid (Publius Ovidius Naso), cxv, 37, 108, 351, 352

Pagan(s), lxix, 32, 72, 128, 133, 174
Paley, William, xx, lv, xcix–c, ciii. *See also Natural Theology*
Palingenio, Marcello. *See* Manzolli
Panglossie, 297
Papin, Denis, 282, 428
Paradise, 142
Parisian Anatomists. *See* Royal Academy of Sciences
Pascal, Blaise, 10, 428
Passenger pigeon, lxxxvi–lxxxvii, 201, 204–7, 361-62
Paul, xcii, 274, 314
Pechlin, Johann Nicolas, 146, 429
Pecker. *See* Pecquet
Pecquet, Jean, 290, 429
Peiresc, Nicolas Claude Fabri de, 297
Peireskius. *See* Peiresc
Perrault, Claude, 133, 155, 216n, 429
Peter Martyr. *See* Anghiera
Petiver, James, lxxxiv
Petronius Arbiter, 82, 429
Peyer, Johann Conrad, 221, 283, 429
Peyerus. *See* Peyer
Philo. *See* Jews
Philoponus, John, 80, 429–30
Philosophiae Naturalis Principia

Mathematica, xxvi, lxxi, lxxiii, 65, 406, 408, 428
Philosophical religion, 7, 9, 11
Philosophical Transactions, xxxiii, xlii, lxiii, xciii, xcv, 168, 420, 421, 423
Phonurgia nova, 266, 268
Pianezza. *See* Simiana
Picard, Jean, 82, 430
Picart. *See* Picard
Pierius. *See* Valeriano
Piety, xci, 10, 19, 22, 87, 298, 308, 309, 310, 313
Piso, Willem, 85
Pitts, Joseph, 137, 430
Plague, 161, 172, 181
Planets: Jupiter, 27, 38, 45, 58, 390, 411, 435; Mars, 27, 47, 411; master and lunar, 42; Mercury, 27, 48, 408; periodical motions of, 46; primary and secondary, 42, 46, 52; Saturn, 27, 38, 42–43, 52, 390, 412; synopsis of matters related to, 48–49; Venus, 27, 47, 48, 408
Plants: anatomy of, 134–35, 392, 406; capillary, 135; classification and nomenclature of, 391; dioecious, 131; growth of, 129; heat in, 130–31; Hercules, 143; matter in, 129–30; noxious, 137; parts of, 131, 133, 148; "shy," 138; spermatophorous, 135; usefulness of, 140–42; venomous, 147; water in, 129–30
Plastic: nature, lvi, lxxi, lxxxii, 131, 352–53; virtue, 94, 131
Plato: on Atlantis, 109; biography of, 430; cited and quoted, cxv, 9, 27, 44; and the design argument, xix–xx, lxix–lxx; on universal soul, 94; on the wind, 77, 336; on the world as a temple of God, lxxiv, 324
Platonic tradition, the, lxix–lxx, lxxxviii, 381, 402
Pliny the Elder (Gaius Plinius), cxv, 10, 184; biographical sketch of, 430–31; on birds, 196, 202; on the earth, 104; on insects, 151, 159, 162, 165, 169; on the magnet, 112; on man, 245, 252, 253, 262, 294, 300, 303; on meteorology, 76, 81; on minerals, 128; on plants, 133, 147 on quadrupeds, 216, 227; on whales, 133, 147, 151, 159, 162, 165, 169, 184, 188, 196, 202, 216, 227, 245, 252, 253, 262, 294, 300, 303, 430–31
Plotinus, lxx
Plutarch (L. Mestrius Plutarchus), cxv, 23, 73, 84, 140, 228, 332, 338, 431
Pocova. *See* Potocova
Poematum cantu et viribus rythmis, de, 268, 445
Pole(s), 26, 31, 81, 115
Porcius Latro, Marcus, 300
Port-Royal. *See* Jansenists
Potocova, 294
Potu theae, de, 146, 429
Poupart, François, 154
Primum Frigidum. See Cold
Providence (divine), xxxiii, lvii, lxxii, lxxvii, lxxxv, lxxxviii, xciv, 53, 62, 71, 72, 92, 97, 127, 130, 172, 179, 203, 209, 212, 226, 230, 231, 246, 272
Provost of Misna. *See* Klunder
Ptolemy, 325, 433
Purchas, Samuel, 165
Pyrard, François, 85
Python, Spirit of, 227

Quadrupeds: anomalous, 215; brain of, 218–19; clawed (digitate), 213; defensive armor of, 217; flyers, 215; food of, 221–22; head of, 218; heart of, 222–23; hoofed (ungulate), 212–13; neck of, 219–20; prone posture of, 216; sagacity of, 225–30; stomach of, 221
Queen Anne's War, 199n
Queen of the Night, 59

Racham, 210
Radiis et visus et lucis, de, 64
Rain: described, 62; dispersal over the earth, 62; reflection and refraction of sun's light on, 64
Rainbow: defined, lxix, 63, 332; Descartes's theory of, 64; Dominis on, 64, 338; gospel of, 65–66; halo of, 64, 66; iris of, 63–64; Newton's theory of, 65; as seal of a covenant, 67
Rainolds, or Reynolds, John, 300, 431
Rains, the, 78
Raleigh, Walter, 85, 431–32
Ramazzini, Bernardino, 302, 432
Ramus, Petrus, lxiv, lxxxvii, 209, 432
Rashi (Rabbi Solomon ben Isaac), 60, 161, 356, 432
Rational Anatomy, 259

Rational faculties, 12
Rational Thoughts on the Purposes of Natural Things (Wolff), cii–ciii
Rattlesnake, 180
Raven, Charles E., li, lvii, lviii, cxix
Ray, John, xx, lxxii, lxxxvi, 421, 434; biological vitalism, lxxv, lxxvii; botany, li–lii, 131, 133, 135, 139–40, 141n, 145; interpretation of nature, lv–lix; life and works of, l–lx; on man, 240, 245, 249, 250, 251, 252, 254, 275, 288; positive attitude toward nature, lvii, lviii, cxiii; as a source for Mather, lxv, 10, 11; and Willughby, li, lii, liii, 447; zoology, lii–liii, 151, 154, 155, 156, 184, 188, 196, 200, 204, 207, 208, 209, 217, 218, 225, 226
—writings of: *Three Physico-Theological Discourses*, lix; *Wisdom of God*, liii, lv–lix, 426
Raymundus de Sebonde (Raymond of Sebonde), lxviii, lxx, xci, cv, cxv, 237, 313, 367, 432
Reason: in brutes, lxxxvii–lxxxviii, 226–27, 228, 230, 397–98, 435; divine, lxxxviii, 153, 172; human, lxxxi–lxxxii, cxiii–cxiv, cxvii, 117, 119, 224–25, 256, 310, 348–49; light of, 25
Reasonableness of the Christian Religion, The, 118, 414
Redi, Francesco, 154, 167, 432
Reede, or Rheede, tot Draakestein, Hendrik Adriann van, 140, 433
Regiomontanus, Johannes, also known as Johann Müller, lxxxvii, 209, 304, 433
Regulus, Marcus Atilius, 179, 433
Religion of Nature Delineated, The (Wollaston), xcvi
Reptiles. *See* Serpents
Respirationis usu primario, diatriba, de, 277
Rete mirabile, 219
Reynard (Renard) the Fox, 228, 365
Rhaetensis. *See* Rheticus
Rheticus, George Joachim, 47, 433
Rhodiginus. *See* Ricchiere
Rhodologia, 145
Ricchiere, Lodovico, 293, 433
Riccioli, Giambattista, 59, 433–34
Rivers, 81, 98, 100, 101, 128, 187
Robartes, Francis, 50, 434

Robie, Thomas, xciv
Robinson, Tancred, 124, 154, 156, 434
Robinson, Thomas, 146, 181, 209, 231, 434
Rochefort, César de, 85, 434
Rohault, Jacques, xcvi–xcvii, xcviii, 264, 341, 434
Roland, Jacques, 274, 434–35
Römer, Ole, 21, 435
Romer glasses, 268
Rondelet, Guillaume, lxxviii, 76, 184, 189, 435
Rorario, Girolamo, lxviii, lxxxviii, 225, 435
Roscommon, Wentworth Dillon, fourth earl of, 179, 435
Rosenberg, Joannes Carolus, 145
Royal Academy of Sciences (Paris), 216, 219, 221, 227, 258, 268, 391, 412, 423, 424–25, 429
Royal Society (of London), lxiii, xcv; and Boston Philosophical Society, xxxiv; fellows of, lx, lxi, lxii, 384, 386, 398, 408, 409, 411, 412, 416, 419, 420, 421, 423, 424, 428, 434, 438; founding members of, 388, 410–11; 424, 445, 446, 447; Mather's letters to, xl–xliv, lxvi, lxvii; Ray and, li, lii, liii, lix. *See also Philosophical Transactions*
Rudbeck, Olof, 290, 435
Rusden, Moses, 162, 435
Rust, George, xcii, 378
Ruysch, Frederik, 290, 435–36

Sabeans, 128
Saggi Nehor, 299
Salmasius, Claudius, or Claude de Saumaise, 298, 436
Salt(s), 126, 129, 145, 180, 350–51
Salvius, Fufius, 294
Samson, 293
Sanctorius (Santorio Santorio), 247, 436
Saturn. *See* Planets
Saunders, Richard, 258
Scaliger, Joseph Justus, 297, 300, 436
Scaliger, Julius Caesar, xxv, 95, 436
Scheiner, Christoph, 37, 253, 256, 436
Schelhammer, Günther Christoph, 262, 437
Schenck von Grafenberg, Johannes, 301, 437
Scheuchzer, Johann Jacob, 104, 345, 437

Schiller, Julius, lxxv, 34, 326, 437
Schöffer, Peter, 303, 437
Schola et scala naturae. See *Theologia ruris*
Schuyl, Florentius, 187, 437
Scientific occupations of Mather's authorities: alchemy, 382, 384; anatomy, 382, 383, 385, 386, 392, 394, 395, 398, 399, 401–2, 405, 409, 411, 414, 415, 416, 418, 422, 423, 427, 428, 429, 431, 435, 437, 439, 440–41, 442, 444, 445, 446–47; astronomy, 381, 382, 384, 389, 390, 391, 400, 403, 404, 406, 407–8, 410, 412, 416, 422, 427, 428, 433, 435, 436, 437; biology/botany, 380, 384, 386, 391, 399, 406, 411, 419, 433, 437, 438, 439, 441, 444; chemistry, 387, 409, 438; embryology, 423; entomology, 379, 380, 386, 405, 411, 419, 421, 425, 431, 432, 435, 440, 447; geography, 387, 422, 443, 439, 448; geology, 400, 411, 421, 439, 448; hydrostatics, 426, 439; mathematics, 385, 392, 396, 403, 409, 412, 416, 427, 428, 433, 434, 441, 442, 445, 449; medicine, 379, 380, 382, 383, 384, 385, 386, 387, 389, 390, 392, 395, 396, 399, 401, 404, 405, 406, 409, 410, 412, 414, 415, 416, 418, 421, 422, 423, 424, 425, 426, 427, 429, 430, 432, 434, 436, 437, 438, 440, 442, 443, 444, 445, 446, 448, 449; optics, 382, 425, 428; parasitology, 380, 432; physics, 398, 403, 408, 412, 416, 423, 427, 428, 434, 436, 442; physiology, 387, 392, 397–98, 402, 415–16, 422, 423, 424, 427, 429, 446, 447; surgery, 392, 395, 398, 399, 428, 431, 434, 439; toxicology, 432, 446; zoology, 380, 447
Scudder, Henry, 120, 437
Sea(s), 98, 111, 115, 188, 189
Seals, 191
Sedileau, 91, 437–38
Seed, sowing of, 138, 147
Selden, John, 298, 438
Seneca the Elder, 300
Seneca the Younger (Lucius Annaeus Seneca), lviii, xc, cxiv, cxv, 232, 305, 438
Serpents: earthworms, 177; magnitude of, 179; motions of, 177; steam of, 180
Shrubs, 131

Shryock, Richard H., xxxvi, lxxxix,
Sicily, 109
Simiana, Carlo Emmanuel Filiberto Giancito de, 230
Simpson. *See* Simson
Simson, Archibald, 232, 438
Sirtori, Girolamo, 303
Six days, work of, 11, 320, 379
Slare, Frederick, 54, 330, 438
Sloane, Hans, 141, 241, 438
Smell, sense of, 269
Smith, John, 224
Snails, 177
Solar vortex, 26
Solenander, Reinerus, 145, 438
Solinus, Gaius Julius, 228, 438–39
Sollertia animalium, de, 228, 431
Solomon, 301, 307
Solomon ben Isaac, Rabbi. *See* Rashi
Son of Sirach, Jesus, 65
Soul, lxxxix–xci, 94, 255, 258, 364
Sound(s), 50, 71, 165–69, 268, 417
Spalato, clergyman of. *See* Dominis
Spiegel, Adriaan van den, lxxxviii, 133, 439
Spirits: animal, 224, 251n, 280, 289, 290, 293; vital (natural), 250, 250n, 293
Spontaneous generation. *See* Generation
Springs, 106
Star(s): new, 30–32, 325–26, 389; number of, 410; pole-, 32, 86
Steel, 113, 114
Steno, Nicolaus, or Niels Stensen, 198, 201, 286, 411, 439
Stephens, Edward, 296n
Stevin, Simon, lxxiv, 21, 322, 439
Stigel, Johann, 237, 323, 439
Stones: building, 124; sorts of, 124–25
Strabo, 166, 173, 179, 439
Sturmius, or Sturm, Johann Christoph, 65n, 111, 260, 439
Stutterers, 274, 371
Suarez, Francisco, 300, 439–40
Suetonius Tranquillus, 179, 440
Sun, 27, 29, 30, 38, 42; ancients on, 35; atmosphere of, 37, 38, 40; axis of, 37; diameter of, 37; distance from Earth, 37, 408; eclipse of, 37; heat of, 39–40; Kircher's description of, 35; Newton's description of, 35; spots, 37, 327
Superstitions, lxxvi, lxxxiv, lxxxvi, cxiii, 34, 60, 154, 155, 180, 203

Sura, Academy of. *See* Jews
Swammerdam, Jan, 152, 154, 166, 356, 425, 440
Sylvius, Franciscus dele Boë, 263, 440
Syrturus. *See* Sirtori
Systema Cometarum, 52

Tabella Hieroglyphica, 235
Tacitus, Cornelius, 300, 440
Tacquet, Andreas, 58, 440
Tailor, Francis, 298
Tarantula, 182–83, 268, 382
Taste, sense of, 270–71
Tate, Nahum, 146, 440
Tauvry, Daniel, xc, 259, 260, 280, 305, 440–41
Tavernier, Jean Baptiste, 138
Tea, 146
Telescope, 27, 28, 29, 30, 46, 57, 303, 387, 390, 403, 412, 420
Temple, William, 143, 441
Temple of God: the mind as, 310; the world as, 7, 23, 319, 324, 441
Terry, Edward, 174, 441
Thales of Miletus, 49, 112
Thaumantiadis Thaumasia, 65
Thaumatographia naturalis, 166, 415
Theodoret, 11
Theologia naturalis, exhibens augustissimam naturae scholam, 101, 148, 239
Theologia naturalis: sive liber creaturarum, 432
Theologia ruris, 41, 237, 317, 327–28
Theophilus, 17, 441
Theophrastus, 136, 138, 270, 441
Thermoscope, 48
Thruston, Malachi, 277
Thunder, 70, 71, 72, 111
Tiberius (Julius Caesar Augustus), 109, 253, 293
Tides, 99–100
Timothy the Musician, 268
Tobacco, 136, 144, 145
Tompion, Thomas, 38
Torricelli, Evangelista, 74, 387, 441–42
Torricellian experiment, 74, 335, 442
Tostado, Alonzo, 297, 442
Tractatus de Fulmine, 73
Traité du mouvement des eaux, 97, 423
Travis, Daniel, xciv
Trees, 131, 136–37, 140, 141, 147, 399–400

Troughton, John, 298, 442
Truth of the Christian Religion, The, 230n
Tschirnhaus, Ehrenfreid Walther, 55, 442
Tuba Stentoro-phonica, 265, 426
Tulipomania, 142
Tully. *See* Cicero
Tulp, Nicolaas, 288, 442
Tusser, Tom, 60, 442
Two books, the (Nature and Scripture), xix, xlvii, lxix, cxvi, 17, 321
Tympe, or Timpe, Matthaeus, lxviii, 260, 442
Tyson, Edward, 168, 177, 223, 442

Unheard-of Curiosities, 304, 401,
Universe: age of, lxxv; concept of infinite, lxxiv
Ursus Formicarius, 172
Ussher, James, 298

Vacuum, 334–35, 407, 413. *See also Fuga Vacuii;* Torricellian experiment
Valeriano Bolzani, Giovanni Pierio, lxviii, 228, 443
Valerius Maximus, 244, 293, 443
Valsalva, Anton Maria, 243, 262, 264, 443
Van Draakenstein. *See* Reede tot Draakestein
Vapors, 91, 97, 98, 104, 106
Varenius, Bernhardus, or Bernhard Varen, 100, 267, 443
Veer, Gerrit de, 80, 443
Vegetables. *See* Plants
Venetianello, 294
Venus. *See* Planets
Vergil (Publius Vergilius Maro), cxv, 37, 116
Vero Christianismo, de, lxiv–lxv, 19
Verus Christianismus. See *Vero Christianismo*
Verzascha, Bernard, 168, 256, 444
Vesalius, Andreas, lxxxviii, 239, 263, 273, 385, 386, 444
Vieussens, Raymond, 263, 444
View of the Creation, A, 313
View of the Soul, A (Saunders), 258
Villena, marquis of, 31
Viper, 180, 181, 424
Virtuoso, xliv, 11, 320, 399, 414, 426
Vis inertiae, 95

Vital flame, 76
Vitruvius Pollio, 238
Vives, Jean Luis, 297, 444
Voetius, Gisbertus, or Gijsbert Voet, 298, 444
Volcanoes, 106, 108, 345, 417
Volkamer, or Volckamer, Johann Georg, 65, 112, 444
Voltaire, François Marie Arouet de, lxxvii
Vossius, Gerardus, or Gerard Jan Voss, 268, 444-45
Vossius, or Voss, Isaac, 298, 445
Vox Corvi, 211n
Voyage Round the World, A. See *New Voyage*

Wainewright, Jeremiah, 75, 96, 277, 283, 445
Waldschmidt, or Waldschmiedt, Wilhelm, 146, 445
Walker, Joseph, lxiii, lxv, 32
Waller, Edmund, 76, 445
Waller, Richard, xli–xliv passim
Wallis, John, 58, 69, 71, 302, 406, 445
Wanley, Nathaniel, lxvi, 293, 445
Ward, Samuel, 120, 121, 445-46
Water, 102, 106; action of sun and moon on tides, 99-100; bed, 96; description of physical properties of, 96; distribution of, over the globe, 97; flux and reflux of, 98; force of, 96; fresh, 100; motion of, 97; and plants, 129-30
Wepfer, Johann Jakob, 222, 446
Wesley, John, c–ci, civ
Westminster Catechism, 7n, 319
Whales, lxvii, lxix, 188–89, 360
Wharton, Thomas, 270, 284, 446
Wheat, 133
Whiston, William, xxxvi, 339, 389
Whitaker, Robert, 150n
Whitaker, Tobias, 145, 446

Wier, or Weyer, Johann, 169, 446
Wilkins, John, li, 247, 446
Willis, Thomas: on birds, 198; circle of, 446, 447; on earthworms, 177; life and works of, 446-47; on man, 253, 264, 265, 270, 277, 278, 279, 283; on quadrupeds, 218, 224, 226
Willughby, Francis: on birds, lii, 193–94, 195, 196, 202; on fishes, lii–liii, 184; life and works of, 447; and Ray, l, li; on spiders, 157
Wind, 111; definition of, 77, 336; hurricanes, 79; monsoons, 79; origins of, 77–78; trade-, 78, 79; typhoons, 70; uses of, 79
Winthrop, John, IV, xcv
Wisdom, Apocryphal Book of, 41
Wisdom of God. See Ray, John
Witsius, or Wits, Herman, 298, 448
Wittie, or Witty, Robert, 39, 448
Wolff, Christian, ci–ciii
Wollaston, William, xcvi, xcvii, c
Woodward, John, xliv, lx–lxi; as a Boyle lecturer, 389; Cotton Mather's letters to, xli, lxxvii, 361–62; on earthquakes, 106, 345; on kinds of earths, 122, 249–50; life and works of, 448
Wormius, Olaus, or Ole Worm, 125, 180, 448
Worms, lxvii, 60, 154, 156, 168–70, 173, 181, 357–58, 380, 442

Zabians. See Sabeans
Zaphenath-paneah, 99
Zeiller, or Zeiler, Martin, 124, 448
Zenet. See Zeno
Zeno, Bartholomew, 108, 345–46
Zeno, Niccolò, 345–46, 448
Zoan, Field of, 284
Zoroaster, 119, 425, 448–49
Zucchi, Niccolò, 58, 449
Zwinger, Theodore, the Elder, 293, 298, 301, 449